Landslide Hazard and Environment Risk Assessment

Landslide Hazard and Environment Risk Assessment

Editors

Enrico Miccadei
Cristiano Carabella
Giorgio Paglia

MDPI • Basel • Beijing • Wuhan • Barcelona • Belgrade • Manchester • Tokyo • Cluj • Tianjin

Editors
Enrico Miccadei
"G. d'Annunzio" University of
Chieti-Pescara
Italy

Cristiano Carabella
"G. d'Annunzio" University of
Chieti-Pescara
Italy

Giorgio Paglia
"G. d'Annunzio" University of
Chieti-Pescara
Italy

Editorial Office
MDPI
St. Alban-Anlage 66
4052 Basel, Switzerland

This is a reprint of articles from the Special Issue published online in the open access journal *Land* (ISSN 2073-445X) (available at: https://www.mdpi.com/journal/land/special_issues/LandslideHazard_ERA).

For citation purposes, cite each article independently as indicated on the article page online and as indicated below:

LastName, A.A.; LastName, B.B.; LastName, C.C. Article Title. *Journal Name* **Year**, *Volume Number*, Page Range.

ISBN 978-3-0365-3693-4 (Hbk)
ISBN 978-3-0365-3694-1 (PDF)

Cover image courtesy of Gianluca Esposito and Enrico Miccadei.

© 2022 by the authors. Articles in this book are Open Access and distributed under the Creative Commons Attribution (CC BY) license, which allows users to download, copy and build upon published articles, as long as the author and publisher are properly credited, which ensures maximum dissemination and a wider impact of our publications.

The book as a whole is distributed by MDPI under the terms and conditions of the Creative Commons license CC BY-NC-ND.

Contents

About the Editors vii

Preface to "Landslide Hazard and Environment Risk Assessment" ix

Enrico Miccadei, Cristiano Carabella and Giorgio Paglia
Landslide Hazard and Environment Risk Assessment
Reprinted from: *Land* **2022**, *11*, 428, doi:10.3390/land11030428 1

Suhua Zhou, Shuaikang Zhou and Xin Tan
Nationwide Susceptibility Mapping of Landslides in Kenya Using the Fuzzy Analytic Hierarchy Process Model
Reprinted from: *Land* **2020**, *9*, 535, doi:10.3390/land9120535 7

Nikolaos Tavoularis, George Papathanassiou, Athanassios Ganas and Panagiotis Argyrakis
Development of the Landslide Susceptibility Map of Attica Region, Greece, Based on the Method of Rock Engineering System
Reprinted from: *Land* **2021**, *10*, 148, doi:10.3390/land10020148 29

Christos Polykretis, Manolis G. Grillakis, Athanasios V. Argyriou, Nikos Papadopoulos and Dimitrios D. Alexakis
Integrating Multivariate (GeoDetector) and Bivariate (IV) Statistics for Hybrid Landslide Susceptibility Modeling: A Case of the Vicinity of Pinios Artificial Lake, Ilia, Greece
Reprinted from: *Land* **2021**, *10*, 973, doi:10.3390/land10090973 61

Minu Treesa Abraham, Neelima Satyam, Revuri Lokesh, Biswajeet Pradhan and Abdullah Alamri
Factors Affecting Landslide Susceptibility Mapping: Assessing the Influence of Different Machine Learning Approaches, Sampling Strategies and Data Splitting
Reprinted from: *Land* **2021**, *10*, 989, doi:10.3390/land10090989 85

Samuele Segoni and Francesco Caleca
Definition of Environmental Indicators for a Fast Estimation of Landslide Risk at National Scale
Reprinted from: *Land* **2021**, *10*, 621, doi:10.3390/land10060621 109

Shirin Moradi, Thomas Heinze, Jasmin Budler, Thanushika Gunatilake, Andreas Kemna and Johan Alexander Huisman
Combining Site Characterization, Monitoring and Hydromechanical Modeling for Assessing Slope Stability
Reprinted from: *Land* **2021**, *10*, 423, doi:10.3390/land10040423 123

Okoli Jude Emeka, Haslinda Nahazanan, Bahareh Kalantar, Zailani Khuzaimah and Ojogbane Success Sani
Evaluation of the Effect of Hydroseeded Vegetation for Slope Reinforcement
Reprinted from: *Land* **2021**, *10*, 995, doi:10.3390/land10100995 147

Massimiliano Fazzini, Marco Cordeschi, Cristiano Carabella, Giorgio Paglia, Gianluca Esposito and Enrico Miccadei
Snow Avalanche Assessment in Mass Movement-Prone Areas: Results from Climate Extremization in Relationship with Environmental Risk Reduction in the Prati di Tivo Area (Gran Sasso Massif, Central Italy)
Reprinted from: *Land* **2021**, *10*, 1176, doi:10.3390/land10111176 171

Valentino Demurtas, Paolo Emanuele Orrù and Giacomo Deiana
Evolution of Deep-Seated Gravitational Slope Deformations in Relation with Uplift and Fluvial Capture Processes in Central Eastern Sardinia (Italy)
Reprinted from: *Land* **2021**, *10*, 1193, doi:10.3390/land10111193 . . . . **205**

Domenico Aringoli, Piero Farabollini, Gilberto Pambianchi, Marco Materazzi, Margherita Bufalini, Emy Fuffa, Matteo Gentilucci and Gianni Scalella
Geomorphological Hazard in Active Tectonics Area: Study Cases from Sibillini Mountains Thrust System (Central Apennines)
Reprinted from: *Land* **2021**, *10*, 510, doi:10.3390/land10050510 . . . . **235**

Marco Materazzi, Margherita Bufalini, Matteo Gentilucci, Gilberto Pambianchi, Domenico Aringoli and Piero Farabollini
Landslide Hazard Assessment in a Monoclinal Setting (Central Italy): Numerical vs. Geomorphological Approach
Reprinted from: *Land* **2021**, *10*, 624, doi:10.3390/land10060624 . . . . **257**

Gianluca Esposito, Cristiano Carabella, Giorgio Paglia and Enrico Miccadei
Relationships between Morphostructural/Geological Framework and Landslide Types: Historical Landslides in the Hilly Piedmont Area of Abruzzo Region (Central Italy)
Reprinted from: *Land* **2021**, *10*, 287, doi:10.3390/land10030287 . . . . **279**

About the Editors

Enrico Miccadei is Full Professor in Physical Geography and Geomorphology at Department of Engineering and Geology at the "G. d'Annunzio" University of Chieti-Pescara, Italy. He graduated in Geology at the University of Rome, La Sapienza, Italy, and began his career with Post Degree research grantships at the National Research Council, Consiglio Nazionale delle Ricerche CNR, Italy. Since 1995, he conducts teaching activity in the Physical Geography and Geomorphology scientific disciplinary sector at the University of Chieti-Pescara, Italy. His preferred research topics are the landscape and geomorphological evolution of Central Italy; geological–geomorphological mapping; applied geomorphology; educational activities and geological tourism; and sustainable territorial planning. He is the author of more than 250 scientific papers published in national and international academic journals and as several contributions to national and international conferences proceedings.

Cristiano Carabella is a fellow student at the Department of Engineering and Geology, "G. d'Annunzio" University of Chieti-Pescara, Italy. He holds a BSc in Geological Sciences and an MSc in Geological Sciences and Technologies. His main research topic is focused on the analysis and assessment of geomorphological hazard, with experiences in geomorphological mapping, landslide and flood hazard assessment, landslide susceptibility assessment, and GIS techniques.

Giorgio Paglia is a Ph.D. student in Earth Systems and Built Environments at the Department of Engineering and Geology, "G. d'Annunzio" University of Chieti-Pescara, Italy. He holds a BSc in Geological Sciences and an MSc in Geological Sciences and Technologies. His main research topic is focused on the analysis and assessment of climatic and geomorphological hazard over the Abruzzo Region, with experiences in geomorphological mapping, landslide hazard assessment, and GIS techniques.

Preface to "Landslide Hazard and Environment Risk Assessment"

This book presents a print version of the Special Issue of the journal *Land* dedicated to "Landslide Hazard and Environment Risk Assessment". The overall goal of this Special Issue was to present innovative approaches for the analysis and mapping of landslide phenomena. Methodologies for landslide susceptibility mapping, slope stability and environmental risk management in mass-movement-prone areas, and multidisciplinary approaches for landslide analysis in different geomorphological/morphostructural environments were the main research and targets that the papers published in this Special Issue aimed to address. In the twelve papers collected in this volume, interested readers will find a collection of scientific contributions providing a sample of the state-of-the-art and forefront research in these fields. Among the articles published in the Special Issue, the geographic distribution of the case studies is wide enough to attract the interest of an international audience of readers. The articles collected here will hopefully provide useful insights into advancements in scientific approaches for the landslide susceptibility mapping and slope stability at both the local and regional scales, highlighting new ideas and innovations in the analysis of various types of mass movements (e.g., DGSDs, snow avalanches, shallow landslides, and complex and historical landslides).

Enrico Miccadei, Cristiano Carabella, Giorgio Paglia
Editors

Editorial

Landslide Hazard and Environment Risk Assessment

Enrico Miccadei *, Cristiano Carabella and Giorgio Paglia

Department of Engineering and Geology, Università degli Studi "G. d'Annunzio" Chieti-Pescara, Via dei Vestini 31, 66100 Chieti Scalo, Italy; cristiano.carabella@unich.it (C.C.); giorgio.paglia@unich.it (G.P.)
* Correspondence: enrico.miccadei@unich.it

Citation: Miccadei, E.; Carabella, C.; Paglia, G. Landslide Hazard and Environment Risk Assessment. *Land* **2022**, *11*, 428. https://doi.org/10.3390/land11030428

Received: 8 March 2022
Accepted: 14 March 2022
Published: 16 March 2022

Publisher's Note: MDPI stays neutral with regard to jurisdictional claims in published maps and institutional affiliations.

Copyright: © 2022 by the authors. Licensee MDPI, Basel, Switzerland. This article is an open access article distributed under the terms and conditions of the Creative Commons Attribution (CC BY) license (https://creativecommons.org/licenses/by/4.0/).

1. Introduction

Landslides are among the most widespread and frequent natural hazards that lead to fatalities, socioeconomic losses, and property damage globally [1,2]. These phenomena also play essential roles in landscape evolution and occur in relation to peculiar predisposing factors (i.e., morphology, lithology, geological setting, land use, climate, etc.) and to triggering events (i.e., extreme rainfall events, earthquakes, wildfires, etc.) [3–6]. According to Varnes [7] and Cruden [8], a "landslide" can be defined as the movement of a mass of rock, debris, or earth downward and outward of a slope under the influence of gravity. The range of landslides is particularly wide, making them one of the most diversified and complex natural phenomena with impacts on territories in all geographic areas. The definition of their affectable areas and recurrence is complicated since they are linked to complex mass movements and to the difficulty in deriving historical data. Instead, discriminating the spatial distribution of potentially unstable areas can be easier by assessing the likelihood of landslides occurring in a region based on the local environmental conditions [9]. Spatial occurrence can be inferred from numerous approaches such as inventory-based mapping, deterministic and probabilistic techniques, heuristic approaches, statistical analysis, and multi-criteria decision-making analysis [10–12]. As a result, it is crucial to follow stepwise approaches mainly involving geomorphological field activities, remotely sensed analysis, numerical modelling, and innovative GIS techniques in order to provide correct landslide hazard assessments and zonation and to support best practices for long-term risk mitigation and reduction.

Given the above scenario, the overall goal of this Special Issue was to present innovative approaches for the analysis and mapping of landslide dynamics, mechanisms, and processes. In the collected papers, readers will find a compendium of scientific contributions providing a sample of the state-of-the-art and forefront research in landslide hazard assessment. Each article offers valuable advancements in scientific approaches for landslide susceptibility mapping and slope stability at the local and regional scales, highlighting new ideas and innovations in analysing various mass movements (e.g., DGSDs, snow avalanches, shallow landslides, and complex and historical landslides).

2. Overview of the Special Issue

This Special Issue reports scientific improvements in landslide hazard and environmental risk assessment with contributions written by authors from Italian regions and other countries, facilitating the interest of an international audience of readers. In detail, it contains 12 peer-reviewed papers focused on (i) methodologies for landslide susceptibility mapping, (ii) slope stability and environmental risk management in mass movement-prone areas, and (iii) multidisciplinary approaches for landslide analysis in different geomorphological/morphostructural environments.

2.1. Methodologies for Landslide Susceptibility Mapping

Zhou et al. [13] developed a landslide susceptibility map at the national level in Kenya. First, a hierarchical evaluation index system containing ten landslide contributing factors and their subclasses was established to produce a susceptibility map. Then, the weights of these indexes were determined through pairwise comparisons. The landslide inventory and landslide causative factors used in this study were collected from various sources. Triangular fuzzy numbers (TFNs) were widely employed to scale the relative importance based on experts' opinions. The entire Kenyan territory was divided into five susceptibility levels, highlighting regions in which further studies were conducted to improve planning and land resource management.

Tavoularis et al. [14] developed a landslide susceptibility map for the entire area of the Attica region (Greece) using the Rock Engineering System (RES). This semi-quantitative heuristic methodology was applied through an interaction matrix. Ten parameters, selected as controlling factors, were statistically correlated with the spatial distribution of slope failures. The generated model was validated using historical landslide data, field-verified slope failures, and a prototype technique developed by the Oregon Department of Geology and Mineral Industries. The resulting data allowed for the construction of an updated geodatabase and the definition of susceptibility levels, representing the basic steps in producing upcoming landslide hazard and risk maps.

Polykretis et al. [15] developed a hybrid landslide susceptibility model for the vicinity of the Pinios artificial lake (Ilia, Greece). In a GIS-based framework, the model was defined by integrating two different statistical analysis models: the multivariate Geographical Detector (GeoDetector) and the bivariate information value (IV). A landslide inventory of 60 past landslides and 14 conditioning factors was incorporated in the model and used to compose the spatial database. The resulting data confirmed the performance of GeoDIV in the definition of the spatial distribution of zones of potential landslides over the study area. The produced LS map represents a basis for regional or local authorities to develop both general (long-term) and emergency (short-term) strategies.

Abraham et al. [16] focused on the factors affecting landslide susceptibility mapping. Five different Machine Learning (ML) algorithms were used to investigate the Wayanad district in Kerala (India), involving different sampling strategies and training datasets. The results show that Naïve Bayes (NB) and Logistic Regression (LR) algorithms are less sensitive to the sampling strategy and data splitting. In contrast, the performance of the other three algorithms—Random Forest (RF), K Nearest Neighbors (KNN), and Support Vector Machine (SVM)—is considerably influenced by the sampling strategy. Hence, as shown in the final H-index plots, both the choice of algorithm and the sampling strategy are critical in obtaining the best-suited landslide susceptibility maps.

2.2. Slope Stability and Environmental Risk Management in Mass Movement-Prone Areas

Segoni and Caleca [17] proposed a new set of environmental indicators for the fast estimation of landslide risk at national scale. Italy was chosen as a test case. Landslide susceptibility maps and soil sealing/land consumption maps were combined to derive a spatially distributed indicator (LRI—Landslide Risk Index). Using GIS techniques, LRI was aggregated at the municipal scale to define the Average Landslide Risk index (ALR) and the Total Landslide Risk index (TLR). The proposed indexes cannot substitute a detailed quantitative risk assessment; nevertheless, they can provide a preliminary overview of the spatial distribution of landslide risk, offering valuable information to each Italian municipality in planning sustainable urban growth.

Moradi et al. [18] presented a multi-method approach of site characterization combined with field observation and a hydromechanical model. A failure-prone hillslope near Bonn (Germany) was chosen as a study site. The field investigation allowed for constructing a three-dimensional slope model with geological units derived from drilling and refraction seismic surveys. Mechanical and hydraulic soil parameters were derived from laboratory analysis; water dynamics were monitored through geoelectrical monitoring. The work presents a potential workflow to improve numerical slope stability analysis through multiple data sources and outlines the usage of such a system for a site-specific early warning system.

Emeka et al. [19] focused on the effect of hydroseeded vegetation for slope reinforcement. The article introduces vegetation establishment as a low-cost, practical measure for slope stability through the ground cover and the root of the vegetation. The study was conducted within the UPM, a government tertiary institution in Serdang, Selangor, Peninsular Malaysia. Twelve conditioning factors were used through the Analytic Hierarchy Process (AHP) model to produce a landslide susceptibility map. Four seed samples, namely ryegrass, rye corn, signal grass, and couch, were hydroseeded to determine the vegetation root and ground cover's effectiveness in stabilizing a high-risk susceptible slope, suggesting variable landslide control benefits.

Fazzini et al. [20] performed a multidisciplinary analysis of detailed climatic and geomorphological data integrated with GIS techniques to advance the snow avalanche hazard assessment. Mass movement phenomena widely affected the Prati di Tivo area (Abruzzo Region, Central Italy) with its well-developed tourist facilities. The resulting data properly defined the main steps for developing a risk mitigation protocol. It involved the provision of new data about the geomorphological setting of the study area and the definition of a technical-scientific basis for civil protection plans required to increase the knowledge of citizens and interested stakeholders about proper land management considering multi-hazard scenarios (i.e., snow avalanches and landslides).

2.3. Multidisciplinary Approaches for Landslide Analysis in Different Geomorphological/Morphostructural Environments

Demurtas et al. [21] focused on the evolution of deep-seated gravitational slope deformations in East Sardinia (Italy). Their article highlights the connections between Plio-Pleistocene tectonic activity and geomorphological changes in Pardu and Quirra River valleys. The use of LiDAR, high-resolution uncrewed aerial vehicle digital photogrammetry (UAV-DP) and geological–geomorphological surveys enabled a depth morphometric analysis and the creation of interpretative 3D models. Multi-source and multi-scale data showed that the state of activity of the DGSDs is closely related to uplift and geomorphological processes. It is recorded by geomorphological indicators, such as fluvial captures, engraved valleys, waterfalls, and heterogeneous water drainage.

Aringoli et al. [22] focused on a geomorphological hazard analysis in an active tectonic area. The Sibillini Mountains (Central Italy) sector was chosen as the study area for the complex tectonic-structural setting, recent seismic sequences, and evident traces of huge landslides. An aerophoto-geological analysis and geomorphological survey verified the link between gravitational occurrences and in-depth tectonic-structural elements. The resulting data show the relationship between tectonic structures, critical hydrogeological conditions, and high-relief energies versus huge gravitational movements. Moreover, these relationships could also play a significant role in differentiating the risk associated with seismic and hydrogeological events.

Materazzi et al. [23] contributed to evaluating the best procedure to be implemented for Landslide Hazard Assessment (LHA), comparing the results obtained using two different approaches (geomorphological and numerical). The area chosen for the analysis is located in a high hilly sector of the Adriatic side of the Central Apennines (Italy), characterized by monoclinal reliefs and cuesta morphologies formed by differential tectonic movements in a recent uplift area. Although preliminary, the results clarify the two different methods' role, usefulness, and limits. Moreover, they demonstrate how a combined approach can certainly provide mutual advantages by addressing the choice of the best numerical model through direct observations and surveys.

Esposito et al. [24] focused on the relationships between the morphostructural/geological framework and landslide types in the hilly piedmont area of the Abruzzo Region (Central Italy). A detailed analysis of three selected case studies was carried out to highlight the multitemporal geomorphological evolution of each landslide phenomenon. Historical landslides were analysed using an integrated approach combining literature data and landslide inventory analysis, relationships between landslide types and lithological units, detailed photogeological analysis, and geomorphological field mapping. The resulting data defined some advances in the understanding of the spatial interrelationship of landslide types, morphostructural setting, and climate regime in the study area.

3. Conclusions

Landslides are global geomorphological phenomena that occur in all geographic regions in response to many predisposing and triggering factors. Directly and indirectly, they impact territories, causing fatalities and huge socioeconomic losses due to environmental degradation and rapid population growth. Consequently, to support sustainable territorial plannings and operative activities, there is a clear need for valid land-use policies, and best long-term risk mitigation and reduction practices. The contributions to this Special Issue represent valuable scientific advances in geomorphological field activities, satellite remote sensing, landslide susceptibility mapping, and numerical modelling, offering practical support for mapping and monitoring of landslide dynamics at both the local and regional scales. All landslide types have been considered, from DGSDs to complex and historical landslides, from rockfalls to debris flows, and from slow-moving slides to shallow landslides. The results described in each article allow for the definition of mitigation activities needed to manage permanent settlements, recreation infrastructures, buildings, and ski facilities. Each paper provides a scientific and methodological basis used to support the idea that landslide hazard assessments must be accurately defined to help local administrations, decision-makers, and interested stakeholders in land planning, emergency planning, and protecting the environment and human life.

Author Contributions: Conceptualization and writing—original draft preparation, E.M.; writing—review and editing, E.M., C.C. and G.P. All authors have read and agreed to the published version of the manuscript.

Funding: This research received no external funding.

Acknowledgments: The Guest Editors express their gratitude to all of the authors who have kindly shared their scientific knowledge through their contributions and to all of the peer reviewers who have contributed to increasing the quality of the papers published in this Special Issue.

Conflicts of Interest: The authors declare no conflict of interest.

References

1. Petley, D. Global patterns of loss of life from landslides. *Geology* **2012**, *40*, 927–930. [CrossRef]
2. Turner, A.K. Social and environmental impacts of landslides. *Innov. Infrastruct. Solut.* **2018**, *3*, 70. [CrossRef]
3. Benz, S.A.; Blum, P. Global detection of rainfall-triggered landslide clusters. *Nat. Hazards Earth Syst. Sci.* **2019**, *19*, 1433–1444. [CrossRef]
4. Carabella, C.; Cinosi, J.; Piattelli, V.; Burrato, P.; Miccadei, E. Earthquake-induced landslides susceptibility evaluation: A case study from the Abruzzo region (Central Italy). *Catena* **2022**, *208*, 19. [CrossRef]
5. Carabella, C.; Miccadei, E.; Paglia, G.; Sciarra, N. Post-wildfire landslide hazard assessment: The case of the 2017 Montagna del Morrone fire (Central Apennines, Italy). *Geosciences* **2019**, *9*, 175. [CrossRef]
6. Tanyaş, H.; van Westen, C.J.; Allstadt, K.E.; Anna Nowicki Jessee, M.; Görüm, T.; Jibson, R.W.; Godt, J.W.; Sato, H.P.; Schmitt, R.G.; Marc, O.; et al. Presentation and Analysis of a Worldwide Database of Earthquake-Induced Landslide Inventories. *J. Geophys. Res. Earth Surf.* **2017**, *122*, 1991–2015. [CrossRef]
7. Varnes, D.J. Slope movement types and processes. In *Landslides, Analysis and Control*; Schuster, R.L., Krizek, R.J., Eds.; National Academy of Sciences: Washington, DC, USA, 1978; Volume Special Re, pp. 11–33, ISBN 0360859X.
8. Cruden, D.M. A simple definition of a landslide. *Bull. Int. Assoc. Eng. Geol.—Bull. l'Association Int. Géologie l'Ingénieur* **1991**, *43*, 27–29. [CrossRef]
9. Aleotti, P.; Chowdhury, R. Landslide hazard assessment: Summary review and new perspectives. *Bull. Eng. Geol. Environ.* **1999**, *58*, 21–44. [CrossRef]
10. Fell, R.; Corominas, J.; Bonnard, C.; Cascini, L.; Leroi, E.; Savage, W.Z. Guidelines for landslide susceptibility, hazard and risk zoning for land use planning. *Eng. Geol.* **2008**, *102*, 85–98. [CrossRef]
11. Marsala, V.; Galli, A.; Paglia, G.; Miccadei, E. Landslide susceptibility assessment of Mauritius Island (Indian Ocean). *Geosciences* **2019**, *9*, 493. [CrossRef]

12. Pardeshi, S.D.; Autade, S.E.; Pardeshi, S.S. Landslide hazard assessment: Recent trends and techniques. *Springerplus* **2013**, *2*, 11. [CrossRef] [PubMed]
13. Zhou, S.; Zhou, S.; Tan, X. Nationwide susceptibility mapping of landslides in Kenya using the fuzzy analytic hierarchy process model. *Land* **2020**, *9*, 535. [CrossRef]
14. Tavoularis, N.; Papathanassiou, G.; Ganas, A.; Argyrakis, P. Development of the landslide susceptibility map of Attica region, Greece, based on the method of rock engineering system. *Land* **2021**, *10*, 148. [CrossRef]
15. Polykretis, C.; Grillakis, M.G.; Argyriou, A.V.; Papadopoulos, N.; Alexakis, D.D. Integrating multivariate (Geodetector) and bivariate (iv) statistics for hybrid landslide susceptibility modeling: A case of the vicinity of Pinios artificial lake, Ilia, Greece. *Land* **2021**, *10*, 973. [CrossRef]
16. Abraham, M.T.; Satyam, N.; Lokesh, R.; Pradhan, B.; Alamri, A. Factors affecting landslide susceptibility mapping: Assessing the influence of different machine learning approaches, sampling strategies and data splitting. *Land* **2021**, *10*, 989. [CrossRef]
17. Segoni, S.; Caleca, F. Definition of environmental indicators for a fast estimation of landslide risk at national scale. *Land* **2021**, *10*, 621. [CrossRef]
18. Moradi, S.; Heinze, T.; Budler, J.; Gunatilake, T.; Kemna, A.; Huisman, J.A. Combining site characterization, monitoring and hydromechanical modeling for assessing slope stability. *Land* **2021**, *10*, 423. [CrossRef]
19. Emeka, O.J.; Nahazanan, H.; Kalantar, B.; Khuzaimah, Z.; Sani, O.S. Evaluation of the effect of hydroseeded vegetation for slope reinforcement. *Land* **2021**, *10*, 995. [CrossRef]
20. Fazzini, M.; Cordeschi, M.; Carabella, C.; Paglia, G.; Esposito, G.; Miccadei, E. Snow avalanche assessment in mass movement-prone areas: Results from climate extremization in relationship with environmental risk reduction in the Prati di Tivo area (Gran Sasso Massif, Central Italy). *Land* **2021**, *10*, 1176. [CrossRef]
21. Demurtas, V.; Orrù, P.E.; Deiana, G. Evolution of deep-seated gravitational slope deformations in relation with uplift and fluvial capture processes in Central Eastern Sardinia (Italy). *Land* **2021**, *10*, 1193. [CrossRef]
22. Aringoli, D.; Farabollini, P.; Pambianchi, G.; Materazzi, M.; Bufalini, M.; Fuffa, E.; Gentilucci, M.; Scalella, G. Geomorphological hazard in active tectonics area: Study cases from Sibillini Mountains thrust system (Central Apennines). *Land* **2021**, *10*, 510. [CrossRef]
23. Materazzi, M.; Bufalini, M.; Gentilucci, M.; Pambianchi, G.; Aringoli, D.; Farabollini, P. Landslide hazard assessment in a monoclinal setting (Central Italy): Numerical vs. geomorphological approach. *Land* **2021**, *10*, 624. [CrossRef]
24. Esposito, G.; Carabella, C.; Paglia, G.; Miccadei, E. Relationships between morphostructural/geological framework and landslide types: Historical landslides in the hilly piedmont area of Abruzzo Region (Central Italy). *Land* **2021**, *10*, 287. [CrossRef]

Article

Nationwide Susceptibility Mapping of Landslides in Kenya Using the Fuzzy Analytic Hierarchy Process Model

Suhua Zhou [1,2], Shuaikang Zhou [2] and Xin Tan [1,2,*]

1. National Center for International Research Collaboration in Building Safety and Environment, Hunan University, Changsha 410082, China; zhousuhua@hnu.edu.cn
2. College of Civil Engineering, Hunan University, Changsha 410082, China; Zhou0328@hnu.edu.cn
* Correspondence: xintan@hnu.edu.cn

Received: 18 November 2020; Accepted: 18 December 2020; Published: 21 December 2020

Abstract: Landslide susceptibility mapping (LSM) is a cost-effective tool for landslide hazard mitigation. To date, no nationwide landslide susceptibility maps have been produced for the entire Kenyan territory. Hence, this work aimed to develop a landslide susceptibility map at the national level in Kenya using the fuzzy analytic hierarchy process method. First, a hierarchical evaluation index system containing 10 landslide contributing factors and their subclasses was established to produce a susceptibility map. Then, the weights of these indexes were determined through pairwise comparisons, in which triangular fuzzy numbers (TFNs) were employed to scale the relative importance based on the opinions of experts. Ultimately, these weights were merged in a hierarchical order to obtain the final landslide susceptibility map. The entire Kenyan territory was divided into five susceptibility levels. Areas with very low susceptibility covered 5.53% of the Kenyan territory, areas with low susceptibility covered 20.58%, areas with the moderate susceptibility covered 29.29%, areas with high susceptibility covered 29.16%, and areas with extremely high susceptibility covered 15.44% of Kenya. The resulting map was validated using an inventory of 425 historical landslides in Kenya. The results indicated that the TFN-AHP model showed a significantly improved performance (AUC = 0.86) compared with the conventional AHP (AUC = 0.72) in LSM for the study area. In total, 31.53% and 29.88% of known landslides occurred within the "extremely high" and "high" susceptibility zones, respectively. Only 8.24% and 1.65% of known landslides fell within the "low" and "very low" susceptibility zones, respectively. The map obtained as a result of this study is beneficial to inform planning and land resource management in Kenya.

Keywords: Kenya; landslide susceptibility; fuzzy analytic hierarchy process; triangular fuzzy numbers; GIS

1. Introduction

Every year, landslides cause a large number of deaths and enormous property losses in mountainous areas [1]. Landslides are a prehistoric issue. It is currently receiving considerable attention since damage induced by landslides has risen in recent years. Estimated fatalities from landslides reached 32,322 between 2004 and 2010, though this value is likely underestimated [2]. The situation varies in different countries. Landslides are concentrated in developing countries or regions, such as the Himalayan region and its surrounding areas in China and African countries. Hence, more effort is required to reduce landslide risks within those countries. Within this topic, the preemptive identification of landslide-prone areas through landslide susceptibility mapping (LSM) is a very promising hazard mitigation approach.

The term landslide susceptibility is a quantitative measure of the likelihood of slope failures under a particular geological condition [3]. With the increasing availability of geospatial data and rapid developments in computational science, numerous LSM methods have been proposed in the last three decades. Most of these methods were built on geographic information systems (GISs). In a broad sense, these LSM models can be summarized as qualitative (knowledge-based or inventory-based) and quantitative (statistically or physically based). In qualitative LSM modeling, each landslide factor is weighted based on the knowledge of experts in geotechnical or geological fields. Afterward, the derived weights were combined to calculate the landslide susceptibility index (LSI). Typical qualitative LSM models include heuristic analysis, inventory analysis, and analytic hierarchy processing (AHP) [4]. As a comparison, statistical LSM models quantify the weights of each factor based on the spatial correlations of historical landslides and these factors. Building on the basic assumption that "the past predicts the future", the weights determined using historical landslides are used to predict the likelihood of future landslide occurring. Frequency ratios [5], logistic regressions [6], weights of evidence [7], artificial neural networks [8], and support vector machines [9] are frequently used statistically based LSMs in the literature. For physically based LSM models, slope stability models and groundwater flow models are integrated to calculate the safety factor for each slope unit. Several programs have been developed for LSM, such as SHALSTAB, SINMAP, and TRIGRS [10,11]. The advantages and disadvantages of different LSM models have been reviewed by Van Westren et al. [12] and Reichenbach et al. [13]. Comparative studies have shown that the optimized selection of LSM methods largely depends on the scale, nature, and data availability of the study area [14–16].

Similar to most African countries, landslide is ranked as the deadliest geohazard in Kenya [17,18]. Despite enormous damage induced by landslides, literature reviews indicate that very few attempts have been carried out to research landslides in Kenya. The studies of landslides performed in Kenya in the past few decades have concentrated on landslide inventory mapping [19], geological investigation of single landslide events and developing general overviews of landside phenomena [18]. It is noted in the literature that intensive precipitation is a dominant factor triggering landslides in Kenya [20]. Steep topography, weathered regolith, and human activities such as deforestation, overgrazing, and overfarming have been identified as causative factors of landslides in Kenya [18]. In recent years, the continuously growing population and expansive development of infrastructure have placed a heavy burden on the environment and land resources in Kenya. No systematic research on landslide susceptibility assessment in Kenya has been published yet. Filling this research gap is the reason why this study was performed.

Difficulties remain in developing LSM for the whole territory of a country because of inadequate availability of landslide inventories and related information. As illustrated in Table 1, a literature review of some such examples showed that qualitative methods, such as spatial multicriteria evaluation (SMCE) and heuristic weighting, are the most popular existing LSM on a national scale. Van Westren et al. [10] suggested that the most suitable methods for LSM at a medium scale are quantitative methods, while qualitative methods are more appropriate for LSM of large areas (small scale) [19]. The cell size of LSMs varies from coarse (1000 m) to medium (30 m) for qualitative and quantitative LSMs. The suitability of cell size is typically determined by data availability and the mapping scale [20].

The main objective of this work was to develop a landslide susceptibility map for Kenya. To conduct this, the fuzzy analytic hierarchy process (FAHP) method was adopted. This FAHP is a semiqualitative method suitable for LSM on a national scale. The landslide inventory and landslide causative factors used in this study were collected from a variety of sources. Regions highly susceptible to landslides in Kenya were highlighted as a basis for further studies of landslide hazards or risk assessments. Additionally, the output presented serves as an effective tool for the authorities involved in land planning and land resource management.

Table 1. Studies of landslide susceptibility mapping at nation scale.

No.	Country	LSM Method	Cell Size (m)	Landslide Inventory Availability	No. of Causative Factors	Source
1	Cuba	Integrating SMCE and AHP	90	No	5	Abella and Van Westen [21]
2	Romania	Heuristic Weighting	100	Yes	6	Bălteanu et al. [22]
3	Greece	Integrating Landslide Relative Frequency and R-mode	1000	Yes	10	Sabatakakis et al. [23]
4	France	Integrating SMCE and Expert Knowledge	90	Yes	3	Malet et al. [24]
5	Georgia	SMCE	100	Yes	9	Gaprindashvili and Van Westen [25]
6	Turkey	Heuristic Weighting	500	Yes	6	Okalp and Akgün [26]
7	Rwanda	SMCE	30	Yes	8	Nsengiyumva et al. [27]

2. Study Area

Kenya is an east African country (Figure 1). Kenya has a territorial area of 582,646 km² and a population of 41.8 million. The elevation of Kenya stretches from sea level in the coastal regions to over 5000 m above sea level (a.s.l.) at Mount Kenya. Geomorphologically, the landforms in Kenya are dominated by highlands in the central and the west regions, plains in the northeast and the coastal regions [20]. The Great Rift Valley (GRV) cuts through the western territory of Kenya from south to north, separating the highlands into two parts, the western highland and the eastern highland. The narrowest part of the rift valley basin is about 30 km near the Naivasha Lake, while the broadest of that is about 300 km in width near the Turkana Lake (Figure 1). The elevation difference in the GRV zone ranges from 500 to 1000 m.

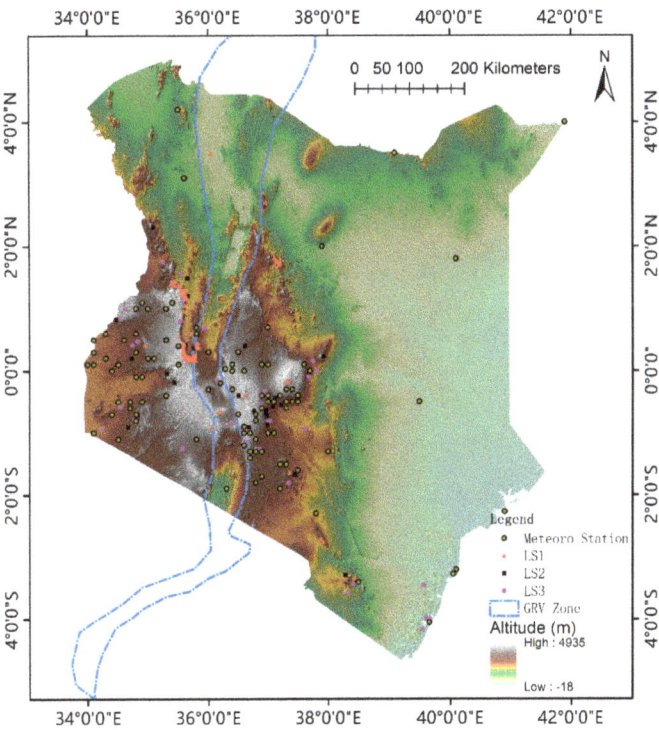

Figure 1. General conditions of Kenya and the spatial distribution of historical landslides in Kenya.

Geologically, Kenya is mainly constituted of five parts [20]: (1) the Archean rocks, of which the major rock types are: shales, mudstones, greywackes, phyllites, and conglomerates; (2) the Proterozoic rocks, of which the major rock types are: rhyolites, basalts, quartzites, and conglomerates; (3) the Paleozoic and Mesozoic sediments, which were dominant by rock types of granites, granodiorites, and leucogranites; (4) the Tertiary/Quaternary volcanic rocks and sediments, of which the major rock types are: sands, marls, clays, conglomerates, and limestones; (5) the Pleistocene to Recent deposits, in which clays, diatomite, shales, and silts are major rock types.

Kenya has a mild climate with annual temperatures ranging from 16 to 26 degrees Celsius. The mean annual precipitation (MAP) in Kenya ranges from <200 mm to 2500 mm. The precipitation of Kenya is characterized by its nonuniform distribution in both time and space. From the historical meteorological data of the Jomo Kenyatta weather station (Figure 2), two distinct rainy seasons can be observed. One is from March to May (the heavy rainy season), and the other is from October to December (the light rainy season). The rest is the dry season. The spatial distribution of MAP is illustrated in Figure 3a. Because of the sudden elevation changes in the GRV zone (from highland to valley then to highland again), there is a sharp transition between wet and dry regions across the GRV zone in southwestern Kenya. Both sides of the GRV in this region had the highest value of MAP, while the northeastern and northern parts of Kenya had the lowest MAP of less than 800 mm. Kenya has experienced a series of geohazards arising from floods, storms, landslides, and debris/mudflows. Most of these geohazards are related to climate extremes [18,20]. Rainfall and human activities (farming, devegetation, construction, etc.) have triggered the majority of the landslides that have occurred in Kenya.

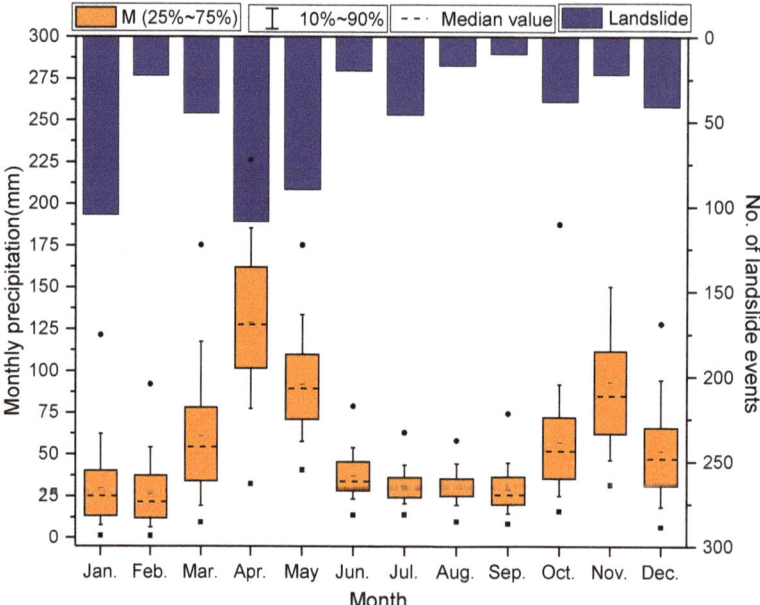

Figure 2. Statistics of monthly average precipitation and distribution of landslides in Kenya. Precipitation data was obtained from the Jomo Kenyatta weather station. Since no detailed information of the exact occurring moth of landslides in LS3, only LS1 and LS2 data were included in this figure.

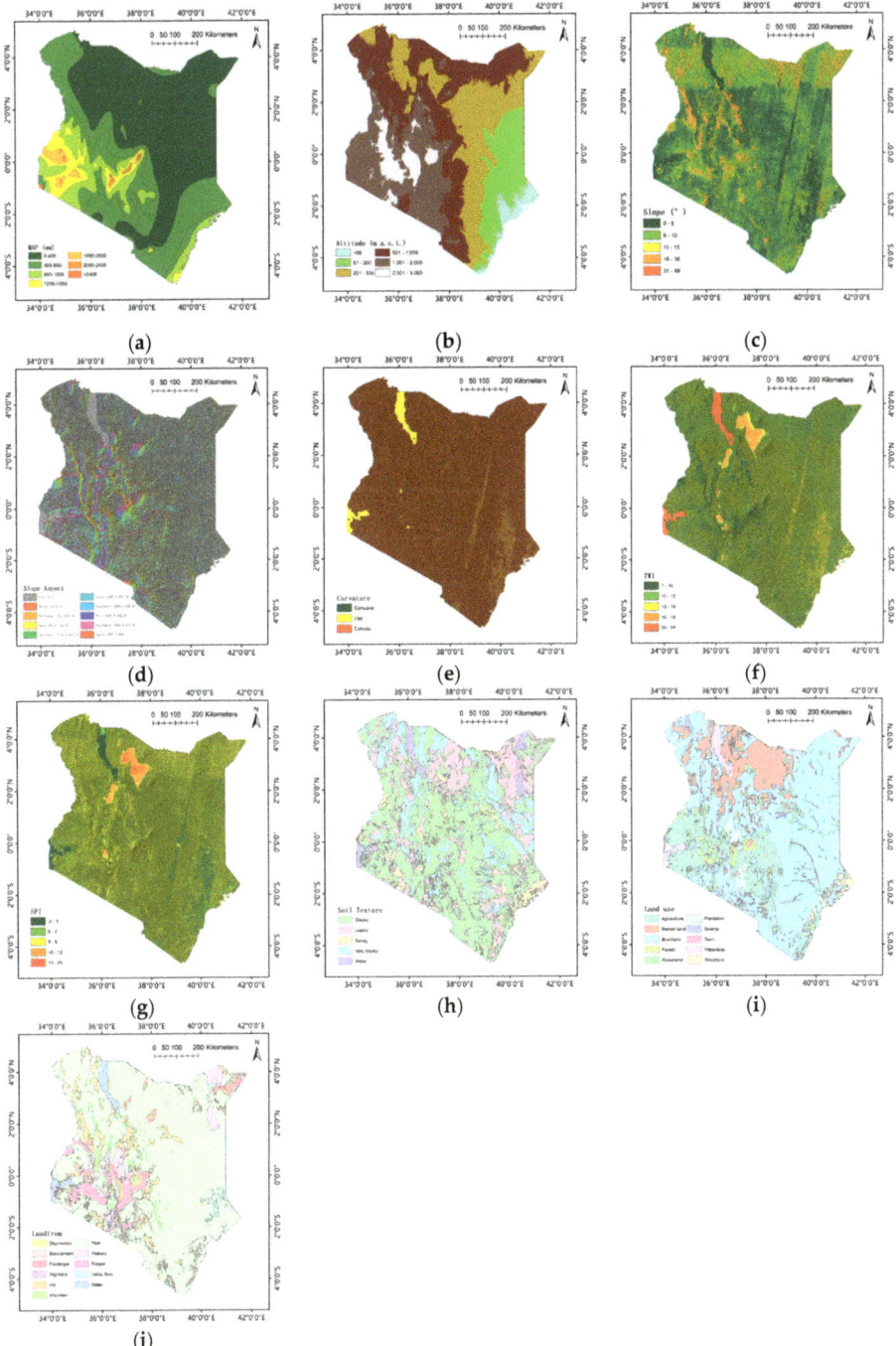

Figure 3. List of landslide contribution factors: (**a**) MAP; (**b**) altitude; (**c**) slope; (**d**) aspect; (**e**) curvature; (**f**) TWI; (**g**) SPI; (**h**) soil texture; (**i**) land use; (**j**) landform.

3. Materials

3.1. Historical Landslides in Kenya

A landslide inventory usually portrays the date, location, cause, type, and geometry of landslides. A detailed landslide inventory is a mandatory input for LSM using quantitative methods. Landslide inventories can be produced through field surveys, review of relevant documents, and remote sensing (satellite images, aerial photos, etc.). Since compiling a detailed landslide inventory remains a time- and labor-consuming task there has been no reliable landslide inventory with national coverage available in Kenya until now.

Considering the conditions mentioned above, the landslide inventory utilized in the present study was composed of three subsets from different sources, as illustrated in Figure 1. The first subinventory (LS1) was obtained from the Kenya Open Data of the Ministry of Information, Communications, and Technology of Kenya. LS1 contained 39 historical landslides that occurred in Kenya during the 1999–2013 period. Landslide information regarding the specific longitude/latitude, dates of occurrence, number of people affected, and estimated economic losses were provided in the LS1. The second subinventory (LS2) was derived from the global fatal landslide database (GFLD) (version 2) extracted from Froude and Petley [28]. Similar to LS1, LS2 also provided the location, date of landslide occurrence, induced fatalities, general description, and related reports of landslides. In total, 63 landslides were recorded in LS2. The third subinventory (LS3) was extracted from a landslide inventory of Africa, which was compiled by Broeckx et al. [29]. Within Kenya's territory, 323 landslides were detected in LS3. The majority of landslides in LS3 were mapped through visual interpretation of Google Earth imagery. This was time-consuming work and involved subjective interpretation. In contrast to LS1 and LS2, only the locations of landslides were provided in LS3. For the present study, a total of 425 landslides were stored in the inventory.

As indicated in Figure 2, the landslide occurrences were mostly concentrated in the periods from April to May and from November to December. It should also be noted that the monthly distribution of landslides was consistent with the monthly distribution of precipitation.

3.2. Landslide Contributing Factors (LCFs)

From the perspective of geotechnical engineering, landslides are a comprehensive consequence of several contributing factors. Nevertheless, there are no global rules for selecting these factors. In a typical LSM, such LCFs are chosen on the basis of data availability, characteristics and scale of the study area, as well as the expert knowledge or experience. In this study, ten LCFs were utilized in LSM over a nationwide area of Kenya, including four topographic factors (namely, the altitude, aspect, slope, and curvature), two hydrological factors (namely, the topographic wetness index (TWI), stream power index (SPI)), soil texture, precipitation, land use, and landform (as shown in Figure 3). To perform the LSM, all factors were rasterized into 1×1 km grids and classified into several classes using ArcGIS software (Version 10.2). In what follows, a brief description of each landslide contributing factor is given.

3.2.1. Mean Annual Precipitation (MAP)

Rainfall is among the dominant inducing factors for landslides not only in Kenya but also in many countries because it increases the soil mass and decreases the soil shear strength. As shown in Figure 2, there are strong correlations between landslide occurrences and precipitation in Kenya. For this study, the MAP data were adopted and categorized into nine levels using a 400 mm interval, as shown in Figure 3a. To initially obtain the MAP factor, monthly total rainfall data from 87 meteorological stations in Kenya were collected and filtered. Only data from stations operated in all types of weather were kept. Then, the filtered monthly rainfall data were summed and averaged annually for each station. Finally, through the inverse distance weighting (IDW) of data for each station, the MAP map of Kenya was derived.

3.2.2. Topographic Factors

Topographic factors are most frequently used in LSM. In this study, several topographic factors were created using a 12.5-m digital elevation model (DEM) obtained from the Alaska Satellite Facility (ASF) [30], which included altitude, slope gradient, slope aspect, and curvature information. Because of variations in temperature, humidity, and vegetation, the degree and type of weathering also varied with altitude. Therefore, altitude has been employed as an LCF in previous studies [8,13,15]. Considering the setting of geomorphology and landforms present in Kenya, the altitude factor was categorized as (1) <50 m, (2) 50–200 m, (3) 200–500 m, (4) 500–1000 m, (5) 1000–2000 m, and (6) >2000 m (Figure 3b). Steepness directly affects slope stability because slopes become more susceptible to landslides as the slope gradient increases. Reviews of LSM studies have suggested that the slope gradient factors is usually categorized using a 5° interval. Thus, the slope LCF was categorized as follows: <5°, 5–10°, 10–15°, 15–20°, 20–25°, 25–30°, 30–35°, 35–40°, and >40° (Figure 3c). Since parameters such as sunlight, precipitation, and vegetation cover vary, slope aspects may also have an effect on landslide occurrences. Consistent with previous studies [15,16,31], the aspect factor was classified into 9 subclasses, as shown in Figure 3d. By controlling the water flow and erosion type in curved terrains, curvature is also a commonly used topographic factor that is associated with landslides. A positive curvature value indicates an upwardly convex terrain, while a negative value indicates an upwardly concave terrain. Terrains with values of zero for the curvature factor were classified as flat (Figure 3e).

3.2.3. Hydrological Factors

Hydrological factors played a determinant role in affecting slope stability. For this study, the TWI and the SPI were utilized as predictors of landslides. Both the TWI and the SPI are secondary attributes of the DEM and can be calculated using Equations (1) and (2), respectively. The TWI measures the topographic control on groundwater flow and accumulation. Terrains with a higher TWI values are more likely to become saturated during rainfall events. The SPI quantifies the erosion power of flowing water. Gullies are more likely to form at locations with high SPI values. As indicated in Figure 3f,g, the TWI of the study area ranged from 6.80 to 34.72, and the SPI ranged from −2.41 to 25.13. The factors of TWI and SPI were finally divided into five categories through the "natural breaks" function in GIS.

$$\text{TWI} = \ln\left(\frac{A}{\tan \beta}\right) \tag{1}$$

$$\text{SPI} = \ln(A \cdot \tan \beta) \tag{2}$$

where A is the unit upstream accumulation area, β is the slope gradient, and ln is the natural logarithm.

3.2.4. Environmental Factors

Given the poor data availability in Kenya, three environmental LCFs were considered in this study to produce LS maps. These two factors were soil texture and land use. Soil texture indicates the proportional composition of sand, silt, and clay content in the soil. Because high clay soils usually contain high organic matter content, which is favorable for soil resistance against detachment, soils characterized by high sand or loam content are more susceptible to land sliding than clayey soils [32]. A nationwide soil property GIS database developed by the International Soil Reference and Information Centre (ISRIC) [33] was utilized in this study. Types of soil texture were classified as follows: (1) very clayed, (2) clayed, (3) loamy, and (4) sandy (Figure 3h). Land use and landform type contribute significantly to slope stability. Specifically, vegetation roots may enhance slope stability by altering the cohesive forces and hydrologic properties. The degradation of forests and vegetation increases the degree of susceptibility of the area to landslides [34]. As for the factor of landform, steep and hill/mountain terrains are prone to landslides compared with flat terrains such as plains, valley floor and foot slope. In addition, from the perspective of geomorphology, landslide itself also plays as a driving role in landform evolution [35,36]. The Kenya National Land Use Dataset (KNLD) [37] was utilized in this study to obtain the LCFs of land use and landform type. As indicated in Figure 3i, the KNLD contains ten land use types (Figure 3i) and eleven landform types (Figure 3j).

4. Methodology

A hybrid model of the conventional AHP and fuzzy theory was utilized to conduct the LSM in this study. The AHP has shown good capacity in solving a multicriteria decision-making problem by incorporating expert knowledge into quantitative analysis. To reduce subjectivities involved in conventional AHP analysis, the fuzzy set theory was adopted to handle blurry sets or categories. Hence, the hybrid use of fuzzy sets and conventional AHPs effectively addresses the decision-making issues under multiple criteria. The theoretical background of the conventional AHP and fuzzy theory were briefly introduced in Sections 4.1 and 4.2, respectively. After that, the process of incorporating fuzzy theory into AHP was given in Section 4.3.

4.1. The Theoretical Background of the Conventional AHP

The AHP, originally developed by Saaty [38], has shown great potential for handling multicriteria decision-making (MCDM) issues. Implemented in GIS, the AHP has been successfully employed in LSM in many previous studies [13,29,31]. A detailed description of the AHP application steps in LSM was introduced by Van et al. [31] and can be summarized as follows:

Step 1: Dividing the decision problem into a hierarchical structure

In this step, a complex decision problem was decomposed into a hierarchical structure, including an "objective" level on the top, one or more "criterion" level(s) in the middle, and several decision alternatives at the bottom level. Although there are no universal rules to be followed in constructing such a hierarchy, it was suggested by Saaty [39] that the hierarchy be built based on the decision maker's knowledge and experience with the problem.

Step 2: Constructing the pairwise comparison matrix

In this step, a comparison matrix was constructed with each element indicating the pairwise comparison between all the decision elements. By asking the decision maker how important alternative A is compared to alternative B, the pairwise comparison results (relative importance) are usually rated using a linguistic variable, such as "Slightly Important", "Moderately Important", or "Extremely Important" (Table 2).

Table 2. Triangular fuzzy scale used in this study.

Linguistic Variables	Intensity of Conventional AHP	Reciprocal of Intensity	Triangular Fuzzy Number (TFN)	Reciprocal of TFN
Equally Important (EQI)	1	1	(1,1,2)	(1/2,1,1)
Slightly Important (SLI)	3	1/3	(2,3,4)	(1/4,1/3,1/2)
Moderately Important (MOI)	5	1/5	(4,5,6)	(1/6,1/5,1/4)
Very Important (VEI)	7	1/7	(6,7,8)	(1/8,1/7,1/6)
Extremely Important (EXI)	9	1/9	(8,9,9)	(1/9,1/9,1/8)
Intermediate value	2	1/2	(1,2,3)	(1/3,1/2,1)
	4	1/4	(3,4,5)	(1/5,1/4,1/3)
	6	1/6	(5,6,7)	(1/7,1/6,1/5)
	8	1/8	(7,8,9)	(1/9,1/8,1/7)

Step 3: Calculate the weights of each decision element and check its consistency

For each comparison matrix, the relative weights of each decision element were calculated using the eigenvalue method (or some other methods). Weights could be used only if consistency had been satisfied.

Step 4: Hierarchically aggregate weights from all "criterion" levels

In this step, the score of each alternative with respect to the final goal was calculated by aggregating the weights of decision elements' weights from all "criterion" levels.

The numerical intensity scale for the relative importance between two decision elements, proposed by Saaty [40], has been broadly used in the AHP. Table 2 shows that in this study, the importance of "Equally Important" to "Extremely Important" was scaled from 1 to 9. The numbers 2, 4, 6, and 8 were used to describe intermediate importance. Inverse importance was scaled using the reciprocals of the numbers from 1 to 1/9. The eigenvalue method was adopted to calculate the weights. In this regard, the consistency index (CI) was calculated as follows:

$$CI = \frac{\lambda_{max} - n}{n - 1} \qquad (3)$$

where λ_{max} represents the largest eigenvalue of a matrix.

For evaluation of the CI, the term consistency ratio (CR) was introduced. The CR was defined as the ratio of a given CI and that of a randomly generated reciprocal matrix (RI). Consistency is satisfied if CR < 0.1.

$$CR = \frac{CI}{RI} \qquad (4)$$

4.2. Triangular Fuzzy Number (TFN)

In the practical application of the conventional AHP for LSM, the determination of the exact relative importance of two factors (A and B) is more difficult than to identify one factor as being more important to another. Given this, fuzzy theory was employed to extend the conventional AHP by scaling the experts' decisions as fuzzy numbers. Thus, assigning exact ratio values to pairwise comparison results was avoided. There are many types of fuzzy numbers. For this study, triangular fuzzy numbers (TFNs) were used. Concepts for the TFN-AHP are briefly introduced in the following.

Let $\widetilde{M}$ be a TFN on R; then, its member function $x \in \widetilde{M}$, $\mu_M(x) : R \rightarrow [0, 1]$ can be defined as follows:

$$\mu_M(x) \begin{cases} 0 & x < a \text{ or } x > c \\ (x-a)/(b-a), & a \leq x \leq b \\ (c-x)/(c-b), & b \leq x \leq c \end{cases} \qquad (5)$$

where a, b, and c represent the left, modal, and right values of $\widetilde{M}$, respectively (see Figure 4).

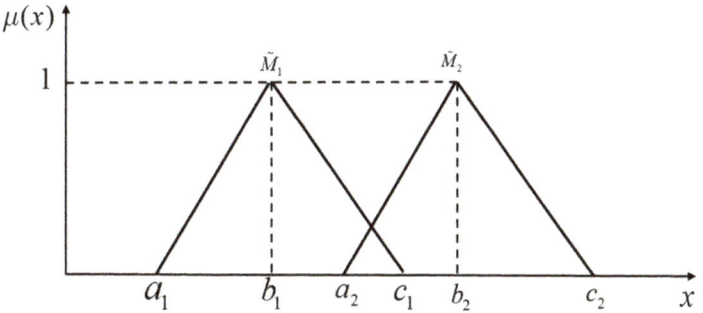

Figure 4. Illustration of the membership function of TFNs.

A TFN can be denoted by $\widetilde{M} = (a, b, c)$. Let $\widetilde{M}_1 = (a_1, b_1, c_1)$ and $\widetilde{M}_2 = (a_2, b_2, c_2)$ be two TFNs, where $a_1, a_2 > 0$, $b_1, b_2 > 0$ and $c_1, c_1 > 0$. The laws of the operations can be defined as follows:

- Summation of two TFNs:
$$\widetilde{M}_1 \oplus \widetilde{M}_2 = (a_1 + a_2, b_1 + b_2, c_1 + c_2) \quad (6)$$

- Subtraction of a TFN from another TFN:
$$\widetilde{M}_1 \ominus \widetilde{M}_2 = (a_1 - a_2, b_1 - b_2, c_1 - c_2) \quad (7)$$

- Multiplication of two TFNs:
$$\widetilde{M}_1 \otimes \widetilde{M}_2 = (a_1 \times a_2, b_1 \times b_2, c_1 \times c_2) \quad (8)$$

- Multiplication of a number and a TFN:
$$\lambda \otimes \widetilde{M}_1 = (\lambda \times a_1, \lambda \times b_1, \lambda \times c_1) \quad (9)$$

- Division of a TFN by another TFN:
$$\widetilde{M}_1 O \widetilde{M}_2 = (a_1, b_1, c_1) O (a_2, b_2, c_2) = \left(\frac{a_1}{c_2}, \frac{b_1}{b_2}, \frac{c_1}{a_2}\right) \quad (10)$$

- Reciprocal of a TFN:
$$\widetilde{M}_1^{-1} = (a_1, b_1, c_1)^{-1} = \left(\frac{1}{c_1}, \frac{1}{b_1}, \frac{1}{a_1}\right) \quad (11)$$

4.3. Integration of the AHP and TFN

The integration of fuzzy sets with the AHP has shown great potential not only for use in LSM but also in many other multicriteria decision making processes, such as hospital location selection and tourist risk evaluation. Very reliable results have been obtained in these applications. The following sections will describe the TFN-AHP theory.

In TFN-AHP theory, experts' judgments are scaled using a TFN rather than a definite number. Then, the TFN comparison matrix is defined as follows:

$$\widetilde{K} = (\widetilde{k}_{ij})_{n \times n} = \begin{bmatrix} \widetilde{1} & \widetilde{k}_{12} & \cdots & \widetilde{k}_{1n} \\ \widetilde{k}_{21} & \widetilde{1} & \cdots & \widetilde{k}_{2n} \\ \vdots & \vdots & \ddots & \vdots \\ \widetilde{k}_{n1} & \widetilde{k}_{n2} & \cdots & \widetilde{1} \end{bmatrix} = \begin{bmatrix} \widetilde{1} & \widetilde{k}_{12} & \cdots & \widetilde{k}_{1n} \\ \widetilde{k}_{12}^{-1} & \widetilde{1} & \cdots & \widetilde{k}_{2n} \\ \vdots & \vdots & \ddots & \vdots \\ \widetilde{k}_{1n}^{-1} & \widetilde{k}_{2n}^{-1} & \cdots & \widetilde{1} \end{bmatrix} \quad (12)$$

where $\widetilde{k}_{ij} = (a_{ij}, b_{ij}, c_{ij})$ denotes the fuzzy comparison value of criterion i to criterion j and $\widetilde{k}_{ij}^{-1} = \left(\frac{1}{c_{ij}}, \frac{1}{b_{ij}}, \frac{1}{a_{ij}}\right)$ denotes the reciprocal value for $i, j = 1, 2, \cdots, n$ and $i \neq j$.

If such judgments are made by consulting more than one expert, the element value is calculated by the average of their decisions as follows:

$$k_{ij} = \frac{1}{h} \otimes \left(k_{ij}^1 + k_{ij}^1 + \ldots + k_{ij}^1\right) \quad (13)$$

where $k_{ij}^p = [l_{ij}^p, m_{ij}^p, u_{ij}^p]$ $p \in [1, 2, \cdots, h]$ and h is the number of experts.

Consistent with Chang (1996) [41], the required steps to compute the weight vector for a TFN comparison matrix can be expressed using the following procedure:

First, calculate the fuzzy synthetic extent with respect to the *i*th alternative by normalization of the row sums of the fuzzy comparison matrix as follows:

$$\widetilde{S_i} = \sum_{j=1}^{n} \widetilde{k}_{ij} \otimes \left[\sum_{l=1}^{n} \sum_{j=1}^{n} \widetilde{k}_{lj} \right]^{-1}, i = 1, 2, \cdots, n \quad (14)$$

Then, calculate weight vectors concerning each decision element under a certain criterion using the degree of possibility of $\widetilde{M}_i \geq \widetilde{M}_j$, which is defined as follows:

$$V(\widetilde{M}_i \geq \widetilde{M}_j) = \sup_{y \geq x} \left\{ \min\left[\widetilde{M}_j(x), \widetilde{M}_i(y) \right] \right\} \quad (15)$$

This equation can be equivalently expressed as follows:

$$V(\widetilde{M}_i \geq \widetilde{M}_j) = \begin{cases} 1 & b_i \geq b_j \\ \frac{c_i - a_j}{(c_i - b_i) + (b_j - a_j)} & a_j \geq c_i \\ 0 & \text{otherwise} \end{cases} \quad (16)$$

Finally, calculate the normalized vector of weights $\widetilde{W} = (\widetilde{w}_1, \widetilde{w}_2, \cdots, \widetilde{w}_n)$ of the TFN comparison matrix $\widetilde{K}$ as follows:

$$w_i = \frac{V(\widetilde{M}_i \geq \widetilde{M}_j | j = 1, 2, \cdots, n; i \neq j)}{\sum_{k=1}^{n} V(\widetilde{M}_i \geq \widetilde{M}_j | j = 1, 2, \cdots, n; i \neq j)} \quad (17)$$

The typical LSM is performed based on raster cells. For the TFN-AHP application in LSM, the criteria refer to a series of LCFs, whereas the alternatives refer to the raster cells within the study area. To perform the LSM using the TFH-AHP, a weighted linear combination (WLC) is conducted to calculate the LSI for each raster pixel as follows:

$$LSI = \sum_{i=1}^{n} w_i \cdot s_i^{k(x,y)} \quad (18)$$

where w_i is the weight of *i*th criterion (LCF) *i*, $s_i^{k(x,y)}$ is the weight of the *k*th subcriteria (subclass for the LCF) in the *i*th criterion, and *k* is determined by the spatial location (*x*, *y*) of the raster cell.

4.4. Accuracy Validation

A review of the literature has shown that the receiver operating characteristic (ROC) curve is a popular method to evaluate the goodness of fit for classification [1,4–7]. The area under this curve (AUC) is adopted to measure the generalization performance of the LSM model. The value of AUC usually ranges between 0.5 to 1.0. A higher AUC value, closer to 1.0, indicates a better performance of the classification model.

$$TPR = \frac{TP}{TP + FN} \quad (19)$$

$$FPR = \frac{FP}{FP + TN} \quad (20)$$

The first step in performing the ROC analysis was to construct the validation dataset, which contained both landslide and non-landslide events. For this study, 425 known landslides were used for validation. Additionally, 425 non-landslides were randomly chosen for validation within the study area. Then, by setting different threshold LSI values, the dataset was separated into four groups according to the actual label and precited label. As shown in Table 3, the four groups were true positive (TP), true negative (TN), false positive (FP), and false-negative (FN) events. After that, two indexes were employed; one was the true positive rate (TPR) computed using Equation (19), and the other was the false positive rate (FPR) computed using Equation (20). Eventually, the ROC curve was drawn by plotting the FPR and the TPR on the horizontal and vertical axes, respectively.

Table 3. Labeling of data according to its predicted label and actual label.

Actual Label	Predicted Lable	
	Positive	Negative
Positive	True Positive(TP)	False Negative(FN)
Negative	False Positive(FP)	True Negative(TN)

4.5. Flowchart of Conducting the LSM

In general, the process to conduct the LSM using the proposed TFN-AHP method can be summarized as following 5 steps. Firstly, the 10 LCFs (criteria) that were required to perform the LSM were chosen (as described in Section 3). Then, a two-level hierarchical model was developed with 10 criteria and 41 subcriteria. Next, 11 comparison matrices were established to calculate the criteria weights. After that, using a WLC of weights of all levels, an LSI map was created and reclassified. Finally, the accuracy of the obtained map was validated using ROC curve and the known historical landslides.

5. Results

5.1. Weights of LCFs and Their Subclasses

The weighting vector derivation plays a central role in multicriteria decision making. For the present study, weights were assigned to each LCF. For this purpose, the geotechnical experts were called upon to make a pairwise comparison of the LCFs based on their experiences and knowledge. As illustrated in Table 4, the pairwise comparison matrix for the ten LCFs was constructed by considering expert opinion and similar previous studies [21–23,26,27]. From the matrix, the weight vector for the criteria was computed using Equation (14) to Equation (17) and is presented in Table 5. After normalization, the weights for each criterion were derived using Equation (18) and are shown in Table 6. The CR was calculated using Equations (3) and (4). When the CR = 0.086 < 0.1, the judgment was deemed to be consistent.

For the subcriteria (subclasses) under a certain uplevel criterion (an LCF), the weights were derived using the same procedures. The sum of subcriteria weights under each corresponding uplevel criterion should be 1.0. Hence, 10 comparison matrices were created. Additionally, the final weights for the subclasses within each LCF were calculated and are shown in Table 6. Before using these calculated weights, a consistency check was conducted for each comparison matrix. Only if CR < 0.1 was the derived weight accepted. Since all CR values were less than 0.1 (Table 6), a consistency check of the 10 matrices indicated that all the judgments were consistent.

From the TFN-AHP analysis, slope gradient (0.1923), MAP (0.1884), and curvature (0.1651) were considered to be the three most important factors contributing to landslide occurrence, whereas the least important factors were land use (0.0220) and aspect (0.0315). For the factor of slope gradient, the terrain steeper than 30° was most susceptible to landslides (weight for this subclass is 0.364313), while the category of 5–10° obtained the lowest weights (0.071825) in determining the landslide occurrences. It also can be seen from results that barren land (0.160743), bush land (0.136464), and grassland (0.121685) were most susceptible to landslides compared with other land use types. In case of curvature, both concave and convex terrain were more prone to landsliding than flat area. Convex terrain (subclass weight is 0.570014) was more favorable for landsliding than concave terrains (0.356956).

Table 4. Pairwise comparison matrix of the LCFs.

LCF	Land Use	AMP	Aspect	Soil Texture	TWI	SPI	Curvature	Altitude	Landform	Slope
Land use	(1,1,1)	(1/3,1/2,1/1)	(1,2,3)	(1/3,1/2,1)	(1/3,1/2,1)	(1/3,1/2,1)	(1/5,1/4,1/3)	(1/3,1/2,1)	(1/3,1/2,1)	(1/5,1/4,1/3)
AMP	(1,2,3)	(1,1,1)	(1,2,3)	(2,3,4)	(2,3,4)	(2,3,4)	(1,2,3)	(2,3,4)	(3,4,5)	(2,3,4)
Aspect	(1/3,1/2,1)	(1/3,1/2,1)	(1,1,1)	(1/3,1/2,1)	(1/3,1/2,1)	(1/3,1/2,1)	(1/3,1/2,1)	(1,2,3)	(1/3,1/2,1)	(1/4,1/3,1/2)
Soil Texture	(1,2,3)	(1/4,1/3,1/2)	(1,2,3)	(1,1,1)	(1,2,3)	(1,2,3)	(1/4,1/3,1/2)	(1/5,1/4,1/3)	(1,2,3)	(1/5,1/4,1/3)
TWI	(1,2,3)	(1/4,1/3,1/2)	(1,2,3)	(1/3,1/2,1)	(1,1,1)	(1,1,2)	(1/3,1/2,1)	(1,2,3)	(1,2,3)	(1/5,1/4,1/3)
SPI	(1,2,3)	(1/4,1/3,1/2)	(1,2,3)	(1/3,1/2,1)	(1/2,1,1)	(1,1,1)	(1/3,1/2,1)	(1/3,1/2,1)	(1,2,3)	(1/5,1/4,1/3)
Curvature	(3,4,5)	(1/3,1/2,1)	(1,2,3)	(2,3,4)	(1,2,3)	(1,2,3)	(1,1,1)	(1,2,3)	(2,3,4)	(1,2,3)
Altitude	(1,2,3)	(1/4,1/3,1/2)	(1/3,1/2,1)	(3,4,5)	(1/3,1/2,1)	(1,2,3)	(1/3,1/2,1)	(1,1,1)	(2,3,4)	(1/5,1/4,1/3)
Landform	(1,2,3)	(1/5,1/4,1/3)	(1,2,3)	(1/3,1/2,1)	(1,2,3)	(1/3,1/2,1)	(1/4,1/3,1/2)	(1,2,3)	(1,1,1)	(1/3,1/2,1)
Slope	(3,4,5)	(1/4,1/3,1/2)	(2,3,4)	(3,4,5)	(3,4,5)	(3,4,5)	(1/3,1/2,1)	(1/4,1/3,1/2)	(1,2,3)	(1,1,1)

Table 5. The calculation of degree possibility for $S_i \geq S_j$ and weight of LCFs.

LCFs		$S_{Land\ Use}$	S_{MAP}	S_{Aspect}	$S_{Soil\ Texture}$	S_{TWI}	S_{SPI}	$S_{Curvature}$	$S_{Altitude}$	$S_{Landform}$	S_{Slope}
	$S_{land\ use}$	–	1.0000	1.0000	1.0000	1.0000	1.0000	1.0000	1.0000	1.0000	1.0000
	S_{MAP}	0.1858	–	0.2301	0.5237	0.5175	0.4054	0.8836	0.6087	0.2794	1.0000
	S_{Aspect}	0.9753	1.0000	–	1.2867	1.0000	1.0000	1.0000	1.0000	1.0000	1.0000
	$S_{Soil\ Texture}$	0.6708	1.0000	0.7063	–	0.9748	0.8960	1.0000	1.0000	0.7685	1.0000
S_j	S_{TWI}	0.6915	1.0000	0.7274	1.0270	–	0.9223	1.0000	1.0000	0.7919	1.0000
	S_{SPI}	0.7732	1.0000	0.8059	1.0997	1.0000	–	1.0000	1.0000	0.8721	1.0000
	$S_{Curvature}$	0.3190	1.0000	0.3616	0.6558	0.6453	0.5405	–	0.7401	0.4138	1.0000
	$S_{Altitude}$	0.5626	1.0000	0.6035	0.9146	0.8926	0.8016	1.0000	–	0.6664	1.0000
	$S_{Landform}$	0.9008	1.0000	0.9288	1.2194	1.0000	1.0000	1.0000	1.0000	–	1.0000
	S_{Slope}	0.1146	0.9796	0.1637	0.4773	0.4723	0.3505	0.8585	0.5674	0.2155	–
min{$V(S_i \geq S_j)$}		0.1146	0.9796	0.1637	0.4773	0.4723	0.3505	0.8585	0.5674	0.2155	1.0000

Table 6. Calculated weights for the LCFs and their subclasses.

LCF	Weight				Weights for Subclasses						
Altitude	0.1091	Subclass	<50	50-200	200-500	500-1000	1000-2000	>2000			
		Sub-weight	0.06005	0.116325	0.132193	0.332756	0.248017	0.110659			
Slope	0.1923	Subclass	0-5	5-10	10-15	15-30	>30				
		Sub-weight	0.272576	0.071825	0.103312	0.187974	0.364313				
Aspect	0.0315	Subclass	East	North	South	Flat	Southeast	Northeast	Northwest	Southwest	West
		Sub-weight	0.051693	0.148905	0.104221	0.035564	0.064601	0.102414	0.153685	0.130982	0.197935
Curvature	0.1651	Subclass	Concave	Flat	Convex						
		Sub-weight	0.356956	0.07303	0.570014						
TWI	0.0908	Subclass	6.8-9.87	9.88-12.06	12.07-14.68	14.69-18.73	18.74-34.72				
		Sub-weight	0.193617	0.145043	0.130591	0.251617	0.279132				
SPI	0.0674	Subclass	-2.41-4.61	-2.41-4.61	-2.41-4.61	-2.41-4.61	-2.41-4.61				
		Sub-weight	0.154267	0.072107	0.236641	0.21716	0.319825				
Soil Texture	0.0918	Subclass	Very clayed	Clayed	Loamy	Sandy	Water				
		Sub-weight	0.347343	0.240127	0.064201	0.19557	0.152758				
Land use	0.0220	Subclass	Grassland	Barren land	Bushland	Waterbody	Plantation	Agriculture	Town	Forrest	Woodland
		Sub-weight	0.121685	0.160743	0.136464	0.059885	0.105435	0.098511	0.069957	0.093253	0.081596
Landform	0.0414	Subclass	Depression	Escarpment	Water	Highland	Hill	Mountain	Plain	Ridges	Valley floor	Foot slope
		Sub-weight	0.149353	0.175107	0.022159	0.099969	0.069948	0.162601	0.010686	0.210827	0.038302	0.061047
AMP	0.1884	Subclass	0-400	400-800	800-1200	1200-1600	1600-2000	2000-2400	>2400		
		Sub-weight	0.153978	0.148163	0.038692	0.069628	0.166877	0.198069	0.224593		

5.2. Landslide Susceptibility Maps

Using Equation (18), the LSI value for each raster cell within Kenya was calculated. As shown in Figure 5, the resultant LSI map was reclassified into five susceptibility levels using the "natural break" function ArcGIS. In total, 15.44% and 29.16% of the Kenyan territory were mapped as extremely high susceptibility zones. A total of 29.16% of the total area was predicted as a high susceptibility zone. Low and very low susceptibility classes covered 20.58% and 5.53% of the study area, respectively. The remaining 29.29% of the study area was determined to be moderately susceptible to landslides (Figure 6). The distribution of susceptibility classes differed in each province. As illustrated in Figure 7, the Rift Valley Province and Eastern Province had the highest percentages of EH landslide susceptibility coverage (21% and 19%, respectively), while the Central Province and Nyanza Province had the lowest percentages of EH landslide susceptibility (5% and 6%, respectively).

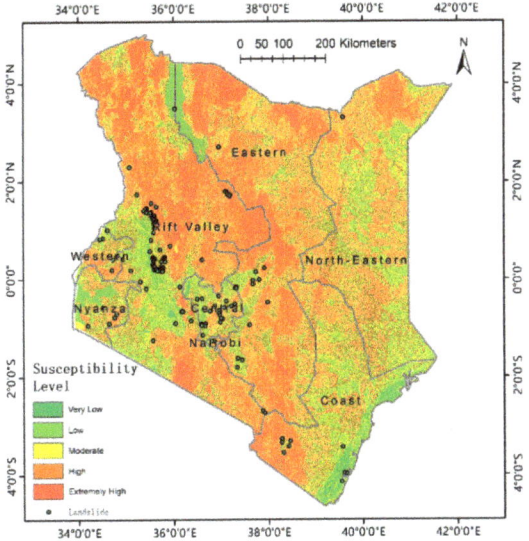

Figure 5. The landslide susceptibility map produced using the TFN-AHP.

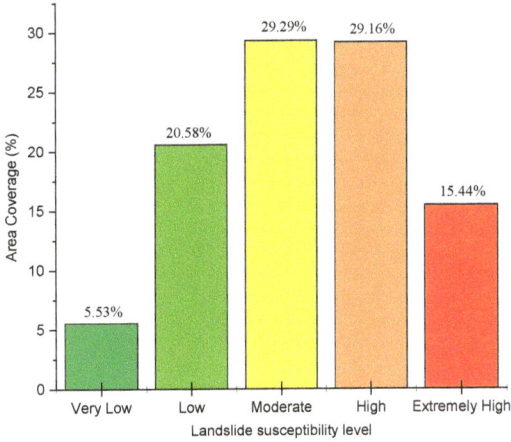

Figure 6. Area coverage of the five landslide susceptibility levels in Kenya.

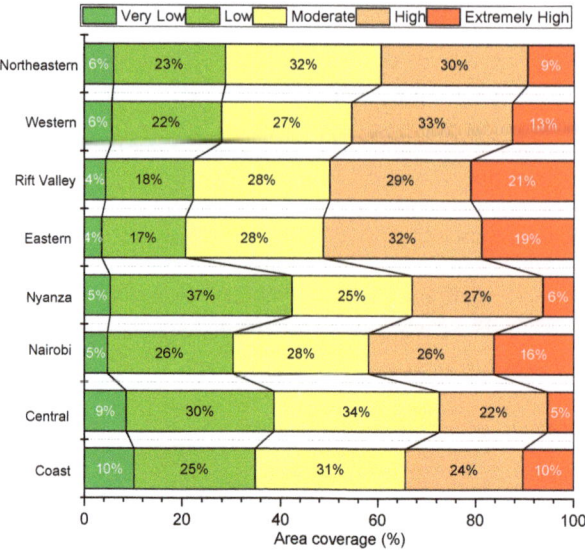

Figure 7. Area coverage of the five landslide susceptibility levels in each province.

5.3. Accuracy Validation

Figure 8 shows that over 60% of the landslides had occurred in the extremely high (31.53%, 134 of 425) and high (29.88%, 127 of 425) landslide susceptibility areas, respectively. Less than 10% of the total landslides occurred in the area mapped as low (8.24%, 35 of 425) and very low (1.65%, 7 of 425) susceptibility levels. In line with the procedures described in Section 4.5, the ROC curve was drawn as shown in Figure 9, and the AUC value was 0.86.

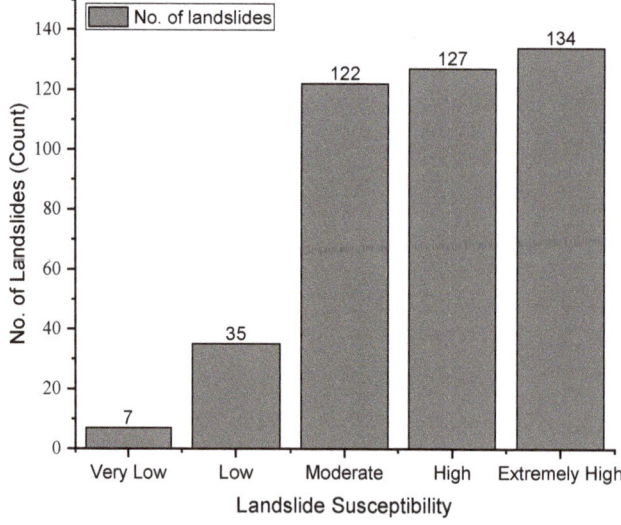

Figure 8. Distribution of known landslides in each landslide susceptibility level.

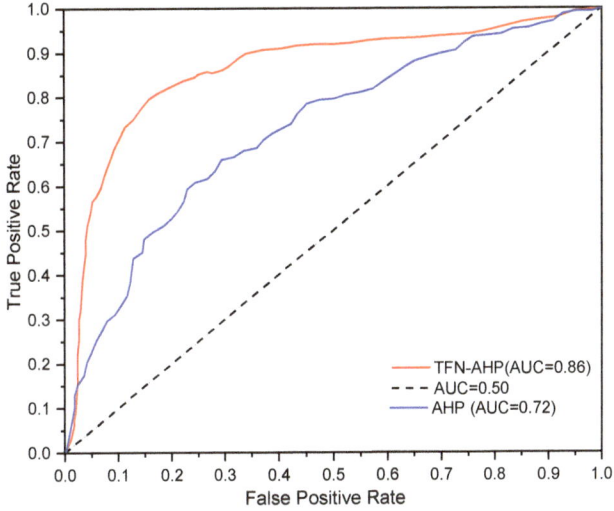

Figure 9. ROC curve validation for the obtained landslide susceptibility maps.

6. Discussion

This research was proposed to apply the TFN-AHP method to map landslide susceptibility in Kenya. Figure 5 directly displays the visible landslide susceptibility information for the entire Kenyan territory, indicating the likelihood of potential landslides. As a developing country, such information would greatly benefit Kenyan efforts to minimize landslide-induced losses and develop optimized land management policies. From Figure 6, it was observed that 44.6% of Kenya is classified as high- and extremely high-susceptibility zones, whereas 26.11% of Kenya was mapped as having low and very low susceptibility. High and extremely high landslide susceptibility zones predominantly cover the rift valley region and its surrounding areas. This finding can be attributed to plentiful rainfall, steep terrains, and fractured ground. Low and very low landslide susceptibility areas are primarily distributed in the southwestern and coastal regions. The distribution of susceptibility classifications also varies in different provinces (Figure 7). The Rift Valley Province had a majority of the historical landslides. This province has the largest area coverage of extremely high- and high-susceptibility zones.

Dozens of methods have been used in LSM at different scales. For large areas with poor availability of historical landslide inventories, the spatial multicriteria evaluation (SMCE) method has exhibited overwhelming advantages over statistical and the physically based methods [15,42]. As a representative SMCE method, a review of previous studies (as displayed in Table 1) has suggested that the AHP and its fuzzy extensions are one of the favorable methods in LSM for large areas (e.g., for a whole country). One limitation of the application of the statistical method in this study is the incompleteness of the historical landslide inventory, which reduces the reliability of the results. Despite this, the historical landslide inventory can still be used for validation in a better than no sense.

The validation results demonstrate that the adopted TFN-AHP resulted in promising accuracy with an AUC value of 0.86 (Figure 9). Despite no strict rules for the evaluation of this accuracy, the resultant accuracy seems to be good compared to similar studies in different areas [1,3,8]. For the LSM, an ideal result map should include as many historical landslides as possible in "high" or "extremely high" susceptibility regions. Additionally, few historical landslides should occur in the "low" or "very low" susceptibility region. Figure 8 shows that the concentration of known landslides decreases from the extremely high category to the very low category. For decision making under multiple criteria, it is difficult for humans to quantify criteria weights using extract numbers. However, rational decisions can be made by skilled experts through a certain value with some uncertainties to capture human subjectivity. Given this, the TFN-AHP makes the comparison process more flexible to minimize the objectivities and uncertainties involved in the conventional AHP process. For the purpose of comparison, the map produced using the conventional AHP is illustrated in Figure 10. The ROC analysis, with an AUC value of 0.72, is also plotted in Figure 9. To perform the conventional AHP, the element value of the TFN-AHP comparison matrix is replaced by a single number according to Table 2. The comparison matrix for the conventional AHP analysis is shown in Appendix A. It should be noted that another source of the subjectivities involved in this study and other similar studies using the SMCE methods or statistical methods may originate from the selection of the LCFs. As shown in Table 1, the number of factors used for LSM ranges from 3 to 10. As discussed in many case studies [1,5,9,16,25,27] and more recently reviewed in [13], the selection of LCFs largely depends on conditions such as data availability, scale, and nature of the study area.

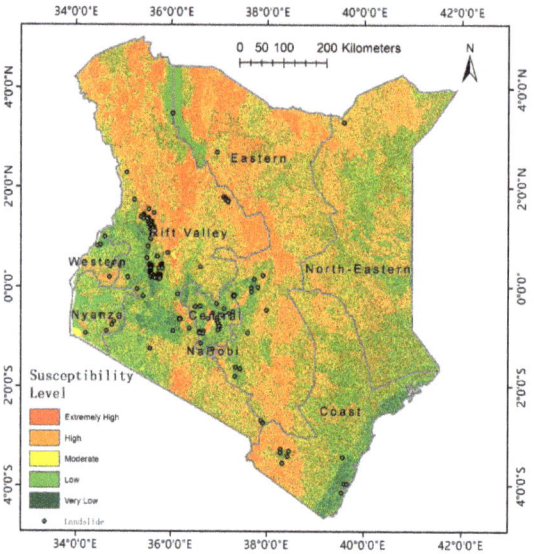

Figure 10. The landslide susceptibility map produced using the conventional AHP.

Even for skilled departments from China and many other developing countries, no universal rules have been proposed. Hence, for a given study area, comparative studies have always been conducted to select the best maps.

7. Conclusions

In this research, an integrated method of fuzzy theory and conventional AHP analysis was employed for the LSM of Kenya. A two-level hierarchical index system was established to predict landslide susceptibility with a GIS platform. Ten factors contributing to landslide occurrence were included in the first level of the evaluation system. These contributing factors included slope, altitude, aspect, SPI, TWI, curvature, land use, MAP, landform, and soil texture. For the second level, each of these factors was divided into several subclasses. The weights of these factors and their subclasses were determined using the adopted TFN-AHP theory. A nationwide landslide susceptibility map for the entire Kenyan territory was produced with five different levels ranging from extremely high susceptibility to very low susceptibility. Extremely high and high landslide susceptibility zones primarily covered the rift valley and its nearby regions. Validation results using ROC curves indicated that the TFN-AHP method performed well for developing LS maps of the study area. This method resulted in a higher AUC accuracy than the conventional AHP using the same datasets.

This study was the first attempt to identify landslide susceptibility zones in Kenya on a national scale. The produced map can be used as a general indicator of the relative landslide susceptibility for larger areas rather than an accurate susceptibility measure for each specific site. The results would be helpful in various land resources-related fields to inform decision making, such as regional landslide hazard mitigation, land use management, and infrastructure planning.

Author Contributions: X.T. contributed to the conception of the study, S.Z. (Suhua Zhou) performed the data analyses. S.Z. (Shuaikang Zhou) wrote the manuscript. All authors have read and agreed to the published version of the manuscript.

Funding: This work was supported by the National Natural Science Foundation of China (grant number 51708199); the Science and Technology Infrastructure Program of Guizhou Province (grant number 2020-4Y047: 2018-133-042); the Fundamental Research Funds for the Central Universities (grant number 531107050969); the Science and Technology Program of Beijing (grant number Z181100003918005). All these financial supports were acknowledged.

Acknowledgments: The authors would like to express their gratitude to Fang and Li Hongquan in Central South University for their kind help in evaluating the LCFs.

Conflicts of Interest: The authors declare no conflict of interest.

Appendix A

Table A1. Pairwise comparison matrix and normalized weight of ten LCFs using the conventional AHP.

LCF	Landuse	AMP	Aspect	Soil Texture	TWI	SPI	Curvature	Altitude	Landform	Slope	w
Landuse	1	1/2	2	1/2	1/2	1/2	1/4	1/2	1/2	1/4	0.0455
AMP	2	1	2	3	3	3	2	3	4	3	0.2073
Aspect	1/2	1/2	1	1/2	1/2	1/2	1/2	2	1/2	1/3	0.0552
Soil Texture	2	1/3	2	1	2	2	1/3	1/5	3	1/4	0.0776
TWI	2	1/3	2	1/2	1	1	1/2	2	3	1/4	0.0786
SPI	2	1/3	2	1/2	1	1	1/2	1/2	2	1/5	0.0615
Curvature	4	1/2	2	3	2	2	1	2	3	2	0.1527
Altitude	2	1/3	1/2	5	1/2	2	1/2	1	2	1/4	0.0914
Landform	2	1/4	2	1/3	1/3	1/2	1/3	1/2	1	1/2	0.0498
Slope	4	1/3	3	4	4	5	1/2	4	2	1	0.1803

Table A2. Weights for subclasses within each LCF calculated using the conventional AHP.

LCF	Subclass	Subclass Indicator	C1	C2	C3	C4	C5	C6	C7	C8	C9	C10	w
Altitude (CR = 0.0561)	<50	C1	1	1/2	1	1/5	1/4	1					0.0835
	50–200	C2	2	1	1/2	1/3	1/2	1					0.1094
	200–500	C3	1	2	1	1/2	1	1/2					0.1385
	500–1000	C4	5	3	2	1	2	2					0.3209
	1000–2000	C5	4	2	1	1/2	1	2					0.2104
	>2000	C6	1	1	2	1/2	1/2	1					0.1374
Slope (CR = 0.0665)	0–5	C1	1	2	1	3	1/2						0.2357
	5–10	C2	1/2	1	1	1/3	1/3						0.1010
	10–15	C3	1	1	1	1	1/3						0.1438
	15–30	C4	1/3	3	1	1	1/2						0.1634
	>30	C5	2	3	3	2	1						0.3561
Aspect (CR = 0.0983)	East	C1	1	1/3	1/2	2	2	1/3	1/2	1/2	1/4		0.0642
	North	C2	3	1	3	2	2	2	1	1/3	1/2		0.1316
	South	C3	2	1/3	1	2	1	3	1/2	1/3	1/3		0.0894
	Flat	C4	1/2	1/2	1/2	1	1	1/3	1/3	1	1/4		0.0544
	Southeast	C5	1/2	1/2	1	1	1	1	1/3	2	1/2		0.0838
	Northeast	C6	3	1/2	1/3	3	1	1	1/2	1	1/3		0.0892
	Northwest	C7	2	1	2	3	3	2	1	1	1/2		0.1396
	Southwest	C8	2	3	3	1	1/2	1	1	1	1		0.1442
	West	C9	4	2	3	4	2	3	2	1	1		0.2036
Curvature (CR = 0.0028)	Concave	C1	1	2	1/2								0.2970
	Flat	C2	1/2	1	1/3								0.1634
	Convex	C3	2	3	1								0.5396

Table A2. Cont.

LCF	Subclass	Subclass Indicator	C1	C2	C3	C4	C5	C6	C7	C8	C9	C10	w
TWI (CR = 0.0261)	6.8–9.87	C1	1	2	1	1/2	1/2						0.1635
	9.88–12.06	C2	1/2	1	1	1/2	1/2						0.1234
	12.07–14.68	C3	1	1	1	1/2	1/2						0.1394
	14.69–18.73	C4	2	2	2	1	1/2						0.2468
	18.74–34.72	C5	2	2	2	2	1						0.3270
SPI (CR = 0.0738)	−2.41–4.61	C1	1	1	1/2	1	1/2						0.1370
	4.62–6.66	C2	1	1	1/2	1/2	1/3						0.1069
	6.67–9.04	C3	2	2	1	1/2	1						0.2152
	9.05–12.28	C4	1	2	2	1	1/3						0.2076
	12.29–25.13	C5	2	3	1	3	1						0.3333
Soil Texture (CR = 0.0366)	Veryclayed	C1	1	2	3	2	3						0.3692
	Clayed	C2	1/2	1	2	1	2						0.2085
	Loamy	C3	1/3	1/2	1	1/2	1/2						0.0958
	Sandy	C4	1/2	1	2	1	1						0.1796
	Water	C5	1/3	1/2	2	1	1						0.1469
Landuse (CR = 0.0929)	Grassland	C1	1	1/3	1/2	1	2	1	1	3	3	2	0.1146
	Barrenland	C2	3	1	2	3	2	3	2	2	3	2	0.1905
	Bushland	C3	2	1/2	1	2	1	2	3	2	2	2	0.1379
	Waterbody	C4	1	1/3	1/2	1	1/2	1/2	2	1/2	1/3	1	0.0618
	Plantation	C5	1/2	1/2	1	2	1	1	2	2	1/2	2	0.0990
	Agriculture	C6	1	1/3	1/2	2	1	1	1	1	3	1	0.0891
	Town	C7	1	1/2	1/3	1/2	1/2	1	1	2	2	1/3	0.0741
	Forrest	C8	1/3	1/2	1/2	2	1/2	1	1/2	1	2	3	0.0854
	Swamp	C9	1/3	1/3	1/2	3	2	1/3	1/2	1/2	1	1/2	0.0686
	Woodland	C10	1/2	1/2	1/2	1	1/2	1	3	1/3	2	1	0.0790
Landform (CR = 0.0999)	Depression	C1	1	1/3	3	2	2	1/2	4	1/3	3	3	0.1192
	Escarpment	C2	3	1	3	2	2	1	4	1	3	2	0.1609
	Water	C3	1/3	1/3	1	1/2	1/2	1/4	2	1/5	1	1/2	0.0433
	Highland	C4	1/2	1/2	2	1	1	1/3	3	1/4	2	2	0.0794
	Hill	C5	1/2	1/2	2	1	1	1/3	3	1/4	2	1/2	0.0693
	Mountain	C6	2	1	4	3	3	1	3	1/2	3	1/2	0.1434
	Plain	C7	1/4	1/4	1/2	1/3	1/3	1/3	1	1	1	2	0.0573
	Ridges	C8	3	1	5	4	4	2	1	1	3	3	0.1989
	Valleyfloor	C9	1/3	1/3	1	1/2	1/2	1/3	1	1/3	1	2	0.0518
	Footslope	C10	1/3	1/2	2	1/2	2	2	1/2	1/3	1/2	1	0.0766
AMP (CR = 0.0931)	0–400	C1	1	1	2	2	1/3	2	1/2				0.1425
	400–800	C2	1	1	1	3	2	1/3	1/2				0.1350
	800–1200	C3	1/2	1	1	1/2	1/2	1/3	1/3				0.0683
	1200–1600	C4	1/2	1/3	2	1	1/2	1/3	1/3				0.0700
	1600–2000	C5	3	1/2	2	2	1	1	1/3				0.1541
	2000–2400	C6	1/2	3	3	3	1	1	1				0.1901
	>2400	C7	2	2	3	3	3	1	1				0.2400

References

1. Pham, B.T.; Shirzadi, A.; Bui, D.T.; Prakash, I.; Dholakia, M. A hybrid machine learning ensemble approach based on a radial basis function neural network and rotation forest for landslide susceptibility modeling: A case study in the Himalayan area. *Int. J. Sediment Res.* **2018**, *33*, 157–170. [CrossRef]
2. Petley, D. Global patterns of loss of life from landslides. *Geology* **2012**, *40*, 927–930. [CrossRef]
3. Brabb, E.E. *Innovative Approaches to Landslide Hazard Mapping*; Landslides: Toronto, ON, Canada, 1984; pp. 307–324.
4. Razandi, Y.; Pourghasemi, H.R.; Neisani, N.S.; Rahmati, O. Application of analytical hierarchy process, frequency ratio, and certainty factor models for groundwater potential mapping using GIS. *Earth Sci. Inform.* **2015**, *8*, 867–883. [CrossRef]
5. Son, J.; Suh, J.; Park, H.D. GIS-based landslide susceptibility assessment in Seoul, South Korea, applying the radius of influence to frequency ratio analysis. *Environ. Earth Sci.* **2016**, *75*, 310. [CrossRef]
6. Dailey, L.A.; Fuhrmann, S. GIS-Based Logistic Regression for Landslide Susceptibility Analysis in Western Washington State. *Int. J. Appl. Geospat. Res.* **2017**, *8*, 1–19. [CrossRef]
7. Hong, H.; Tsangaratos, P.; Ilia, I.; Liu, J.; Zhu, A.-X.; Chen, W. Application of fuzzy weight of evidence and data mining techniques in construction of flood susceptibility map of Poyang County, China. *Sci. Total. Environ.* **2018**, *625*, 575–588. [CrossRef] [PubMed]

8. Shahri, A.A.; Spross, J.; Johansson, F.; Larsson, S. Landslide susceptibility hazard map in southwest Sweden using artificial neural network. *Catena* **2019**, *183*, 104225. [CrossRef]
9. Zhou, S.; Fang, L. Support vector machine modeling of earthquake-induced landslides susceptibility in central part of Sichuan province. *Geoenvironmental Disasters* **2015**, *2*, 117. [CrossRef]
10. Godt, J.; Baum, R.; Savage, W.; Salciarini, D.; Schulz, W.; Harp, E. Transient deterministic shallow landslide modeling: Requirements for susceptibility and hazard assessments in a GIS framework. *Eng. Geol.* **2008**, *102*, 214–226. [CrossRef]
11. Oliveira, S.C.; Zêzere, J.L.; Lajas, S.; Melo, R. Combination of statistical and physically based methods to assess shallow slide susceptibility at the basin scale. *Nat. Hazards Earth Syst. Sci.* **2017**, *17*, 1091–1109. [CrossRef]
12. Van Westen, C.J.; Castellanos, E.; Kuriakose, S.L. Spatial data for landslide susceptibility, hazard, and vulnerability assessment: An overview. *Eng. Geol.* **2008**, *102*, 112–131. [CrossRef]
13. Reichenbach, P.; Rossi, M.; Malamud, B.D.; Mihir, M.; Guzzetti, F. A review of statistically-based landslide susceptibility models. *Earth Sci. Rev.* **2018**, *180*, 60–91. [CrossRef]
14. Cabral, G.G.; Oliveira, A.L.I. *Artificial Intelligence in Theory and Practice II*; Springer Science & Business Media: Santiago, Chile, 2008; pp. 245–254.
15. Barella, C.F.; Sobreira, F.G.; Zêzere, J.L. A comparative analysis of statistical landslide susceptibility mapping in the southeast region of Minas Gerais state. *Bull. Int. Assoc. Eng. Geol.* **2018**, *78*, 3205–3221. [CrossRef]
16. Zhu, A.-X.; Miao, Y.; Wang, R.; Zhu, T.; Deng, Y.; Liu, J.; Yang, L.; Qin, C.-Z.; Hong, H. A comparative study of an expert knowledge-based model and two data-driven models for landslide susceptibility mapping. *Catena* **2018**, *166*, 317–327. [CrossRef]
17. Davies, T.C. Landslide research in Kenya. *J. Afr. Earth Sci.* **1996**, *23*, 541–545. [CrossRef]
18. Paron, P.; Olago, D.O.; Omuto, C.T. *Kenya: A Natural Outlook: Geo-Environmental Resources and Hazards*; Newnes: Boston, MA, USA, 2013; pp. 293–314.
19. Waithaka, E.H.; Agutu, N.; Mbaka, J.G.; Ngigi, T.G.; Waithaka, E.H. Landslide scar/soil erodibility mapping using Landsat TM/ETM+ bands 7 and 3 normalised difference index: A case study of central region of Kenya. *Appl. Geogr.* **2015**, *64*, 108–120. [CrossRef]
20. Ngecu, W.M.; Mathu, E.M. The El-Nino-triggered landslides and their socioeconomic impact on Kenya. *Environ. Earth Sci.* **1999**, *38*, 277–284. [CrossRef]
21. Abella, E.A.C.; Van Westen, C.J. Generation of a landslide risk index map for Cuba using spatial multi-criteria evaluation. *Landslides* **2007**, *4*, 311–325. [CrossRef]
22. Bălteanu, D.; Chendeş, V.; Sima, M.; Enciu, P. A country-wide spatial assessment of landslide susceptibility in Romania. *Geomorphology* **2010**, *124*, 102–112. [CrossRef]
23. Sabatakakis, N.; Koukis, G.; Vassiliades, E.; Lainas, S. Landslide susceptibility zonation in Greece. *Nat. Hazards* **2012**, *65*, 523–543. [CrossRef]
24. Margottini, C.; Canuti, P.; Sassa, K. *Landslide Science and Practice: Volume 1: Landslide Inventory and Susceptibility and Hazard Zoning*; Springer: Berlin/Heidelberg, Germany, 2013; pp. 303–311.
25. Gaprindashvili, G.; Van Westen, C.J. Generation of a national landslide hazard and risk map for the country of Georgia. *Nat. Hazards* **2015**, *80*, 69–101. [CrossRef]
26. Okalp, K.; Akgün, H. National level landslide susceptibility assessment of Turkey utilizing public domain dataset. *Environ. Earth Sci.* **2016**, *75*, 847. [CrossRef]
27. Nsengiyumva, J.B.; Luo, G.; Nahayo, L.; Huang, X.; Cai, P. Landslide susceptibility assessment using spatial multi-criteria evaluation model in Rwanda. *Int. J. Environ. Res. Public Health* **2018**, *15*, 243. [CrossRef] [PubMed]
28. Froude, M.J.; Petley, D.N. Global fatal landslide occurrence from 2004 to 2016. *Nat. Hazards Earth Syst. Sci.* **2018**, *18*, 2161–2181. [CrossRef]
29. Broeckx, J.; Vanmaercke, M.; Duchateau, R.; Poesen, J. A data-based landslide susceptibility map of Africa. *Earth Sci. Rev.* **2018**, *185*, 102–121. [CrossRef]
30. Alaska Satellite Facility. Available online: www.asf.alaska.edu/ (accessed on 13 December 2020).
31. Van, N.T.H.; Van Son, P.; Khanh, N.H.; Binh, L.T. Landslide susceptibility mapping by combining the analytical hierarchy process and weighted linear combination methods: A case study in the upper Lo River catchment (Vietnam). *Landslides* **2016**, *13*, 1285–1301. [CrossRef]

32. Sharma, L.P.; Patel, N.; Debnath, P.; Ghose, M. Assessing landslide vulnerability from soil characteristics—A GIS-based analysis. *Arab. J. Geosci.* **2011**, *5*, 789–796. [CrossRef]
33. ISRIC-World Soil Information. Available online: www.data.isric.org (accessed on 13 December 2020).
34. Reichenbach, P.; Mondini, A.C.; Rossi, M. The influence of land use change on landslide susceptibility zonation: The Briga catchment test site (Messina, Italy). *Environ. Manag.* **2014**, *54*, 1372–1384. [CrossRef]
35. Roering, J. Tectonic geomorphology: Landslides limit mountain relief. *Nat. Geosci.* **2012**, *5*, 446–447. [CrossRef]
36. Larsen, I.J.; Montgomery, D.R. Landslide erosion coupled to tectonics and river incision. *Nat. Geosci.* **2012**, *5*, 468–473. [CrossRef]
37. ICPAC-Geoportal. Available online: www.geoportal.icpac.net/ (accessed on 13 December 2020).
38. Saaty, T.L. A scaling method for priorities in hierarchical structures. *J. Math. Psychol.* **1977**, *15*, 234–281. [CrossRef]
39. Saaty, T.L. *Fundamentals of Decision Making and Priority Theory with AHP*; RWS Publications: Pittsburg, CA, USA, 2000.
40. Saaty, T.L. *The Analytic Hierarchy Process: Planning, Priority Setting*; Resource Allocation (Decision Making Series); McGraw-Hill: New York, NY, USA, 1980.
41. Chang, D.Y. Applications of the extent analysis method on fuzzy AHP. *Eur. J. Oper. Res.* **1996**, *95*, 649–655. [CrossRef]
42. Glade, T.; Crozier, M.J. A review of scale dependency in landslide hazard and risk analysis. *Landslide Hazard Risk* **2012**, *75*, 75–138. [CrossRef]

Publisher's Note: MDPI stays neutral with regard to jurisdictional claims in published maps and institutional affiliations.

© 2020 by the authors. Licensee MDPI, Basel, Switzerland. This article is an open access article distributed under the terms and conditions of the Creative Commons Attribution (CC BY) license (http://creativecommons.org/licenses/by/4.0/).

Article

Development of the Landslide Susceptibility Map of Attica Region, Greece, Based on the Method of Rock Engineering System

Nikolaos Tavoularis [1,*], George Papathanassiou [2], Athanassios Ganas [3] and Panagiotis Argyrakis [4]

1. Regional Administration of Attica, Directorate of Technical Works, L. Syggrou St., 80-88, 117 41 Athens, Greece
2. Department of Civil Engineering, Polytechnic School, Democritus University of Thrace, 671 00 Xanthi, Greece; gpapatha@civil.duth.gr
3. Institute of Geodynamics, National Observatory of Athens, 118 10 Athens, Greece; aganas@gein.noa.gr
4. Department of Informatics and Telecommunications, Faculty of Economics and Technology, University of Peloponnese, 221 31 Tripolis, Greece; pargyrak@noa.gr
* Correspondence: ntavoularis@metal.ntua.gr; Tel.: +30-21-3206-5894

Citation: Tavoularis, N.; Papathanassiou, G.; Ganas, A.; Argyrakis, P. Development of the Landslide Susceptibility Map of Attica Region, Greece, Based on the Method of Rock Engineering System. *Land* **2021**, *10*, 148. https://doi.org/10.3390/land10020148

Academic Editors: Enrico Miccadei, Giorgio Paglia and Cristiano Carabella

Received: 29 December 2020
Accepted: 29 January 2021
Published: 3 February 2021

Publisher's Note: MDPI stays neutral with regard to jurisdictional claims in published maps and institutional affiliations.

Copyright: © 2021 by the authors. Licensee MDPI, Basel, Switzerland. This article is an open access article distributed under the terms and conditions of the Creative Commons Attribution (CC BY) license (https://creativecommons.org/licenses/by/4.0/).

Abstract: The triggering of slope failures can cause a significant impact on human settlements and infrastructure in cities, coasts, islands and mountains. Therefore, a reliable evaluation of the landslide hazard would help mitigate the effects of such landslides and decrease the relevant risk. The goal of this paper is to develop, for the first time on a regional scale (1:100,000), a landslide susceptibility map for the entire area of the Attica region in Greece. In order to achieve this, a database of slope failures triggered in the Attica Region from 1961 to 2020 was developed and a semi-quantitative heuristic methodology called Rock Engineering System (RES) was applied through an interaction matrix, where ten parameters, selected as controlling factors for the landslide occurrence, were statistically correlated with the spatial distribution of slope failures. The generated model was validated by using historical landslide data, field-verified slope failures and a methodology developed by the Oregon Department of Geology and Mineral Industries, showing a satisfactory correlation between the expected and existing landslide susceptibility level. Having compiled the landslide susceptibility map, studies focusing on landslide risk assessment can be realized in the Attica Region.

Keywords: interaction matrix; heuristic; susceptibility; inventory; Greece

1. Introduction

Landslide hazard assessment requires a multi-hazard approach, since the types of landslides that will occur usually have different characteristics with different spatial, temporal, and causal factors [1]. The first step towards the evaluation of landslide hazards on a regional scale (e.g., 1:25,000–1:250,000) is the assessment of the relevant susceptibility, which is defined as the likelihood of a landslide occurring in an area in relation to the local geomorphological conditions [2]. In addition, the landslide susceptibility map can be used as an end product in itself [1]. In order to develop a susceptibility map, it is mandatory to first compile an inventory map where the spatial distribution of existing slope failures is shown. It should additionally be pointed out that on a regional scale map is not feasible to discriminate in detail the type of landslide and delineate the runout per failure.

Having developed the landslide inventory map, the likelihood of slope failures i.e., susceptibility, can be assessed by both qualitative and quantitative methods. The former group of methods includes the knowledge-driven methods (direct and indirect mapping), and the latter group includes the data-driven and the physically-based ones [1]. Considering the regional and local scale maps, the knowledge and data-driven approaches are suggested to be applied; for the former approach a geoscientist i.e., geomorphologist, can directly determine the level of susceptibility based on his/her experience and information related to terrain conditions, while the data-driven mapping statistical models are used in order to forecast likely to landslide areas, based on information obtained from the interrelation between the spatial distribution of landslide conditioning factors and the landslide zones [3]. The most widely applied data-driven approaches are [1]: bivariate statistical analysis, multivariate statistical models and data integration methods like Artificial Neural Network analysis. Bivariate statistical methods (e.g., fuzzy logic, Bayesian combination rules, weights of evidence modelling) are considered as an important tool that can be used in order to analyze which factors play a significant role in slope failure, without taking into account the interdependence of parameters. Multivariate statistical models evaluate the combined relationship between the slope failure and a series of landslide controlling factors. In this type of analysis, all relevant landslide parameters are sampled either on a grid basis or in a slope unit and the presence or absence of landslides is evaluated. These techniques have become standard in regional-scale landslide susceptibility assessment.

Nowadays, the majority of the studies considering landslide susceptibility mapping makes use of digital tools for handling spatial data such as Geographical Information Systems (GIS). Specifically, the GIS-based techniques are considered very suitable for the landslide susceptibility mapping, in which the predisposing factors (e.g., geology, topography) are entered into the GIS environment and combined with the spatial distribution of slope failures i.e., landslide inventory map [3–6]. For the purposes of this study, the semi-quantitative methodology of Rock Engineering System (RES) originally introduced by Hudson [7] was implemented in Greece, particularly in the Attica region for the assessment of landslide susceptibility. This region, which is a county with a size of approximately 3800 km^2, was selected due to the following reasons:

(i) in this region, many cases of slope failures have been reported (Figure 1); the well-known historical landslide of Malakasa (1995) [8] caused serious economic consequences due to the cut-off connection between Athens (the capital city of Greece) and the northern part of Greece; the dangerous, due to rockfalls, segment (located in Kakia Skala) of the National motorway connecting Athens to Patras, some other characteristic rockfall sites such as Alepochori–Psatha, and Alepochori–Schino in Western Attica. Furthermore, rockfalls at particular segments of main streams due to erosion and flash floods, landslides and rockfalls at Attica islands (e.g., Kithira, Salamina, Aegina, Spetses, Hydra, Poros), are some of the most characteristic slope failures that already took place in the administrative region of Attica. Thus, adopting the principle that *"slope failures in the future will be more likely to occur under the conditions which led to past and present instability"* [9], and inventorying and mapping the susceptible to failure slopes provides crucial information for evaluating the future occurrence of landslides in this region.

(ii) the existing information considering the landslide occurrences in Attica Region was dispersed in more than one public agency, and was mainly focused on landslides documented along the road network and residential areas, while only a few cases were georeferenced. The slope failures induced at the mountainous areas and at sites that are not directly affecting the manmade environment were either not recorded or probably under-reported. Thus, there is a need for gathering every slope failure that happened till nowadays, for generating reliable hazard maps in order to use them for civil protection actions.

(iii) the Attica region concentrates almost half of the Greek population, more than 60% of the industrial production in Greece and high-value properties and infrastructure. For this reason, mapping areas prone to slope failure helps public authorities associated with public works in taking mitigation measures against the increase of risk in potentially dangerous areas, leading to losses of life and investments in such a densely populated county.

(iv) the completeness and quality of the available slope failures and thematic geodata.

(v) to the author's knowledge, this is the first time that a landslide susceptibility analysis has been conducted on a regional scale (1:100,000), for the whole territory of the Attica

Region. Furthermore, the generated landslide susceptibility map will serve for many authorities related to public works, as a dynamic map for the planning, design, and implementation of a long-term landslide reduction strategy as well as identifying the areas where more detailed investigations will be required for the planning of critical infrastructure.

(vi) taking into account that the next five to ten years, very important civil engineering projects are about to be constructed in Attica county (such as transports network elements: highways, railroads, metro-tunnels, hospitals, administrative buildings, security/emergency structures, residential buildings) the existence of a regional-scale landslide susceptibility map could be a very useful tool for supporting decisions in order to prevent the location of high-value constructions in unsuitable locations.

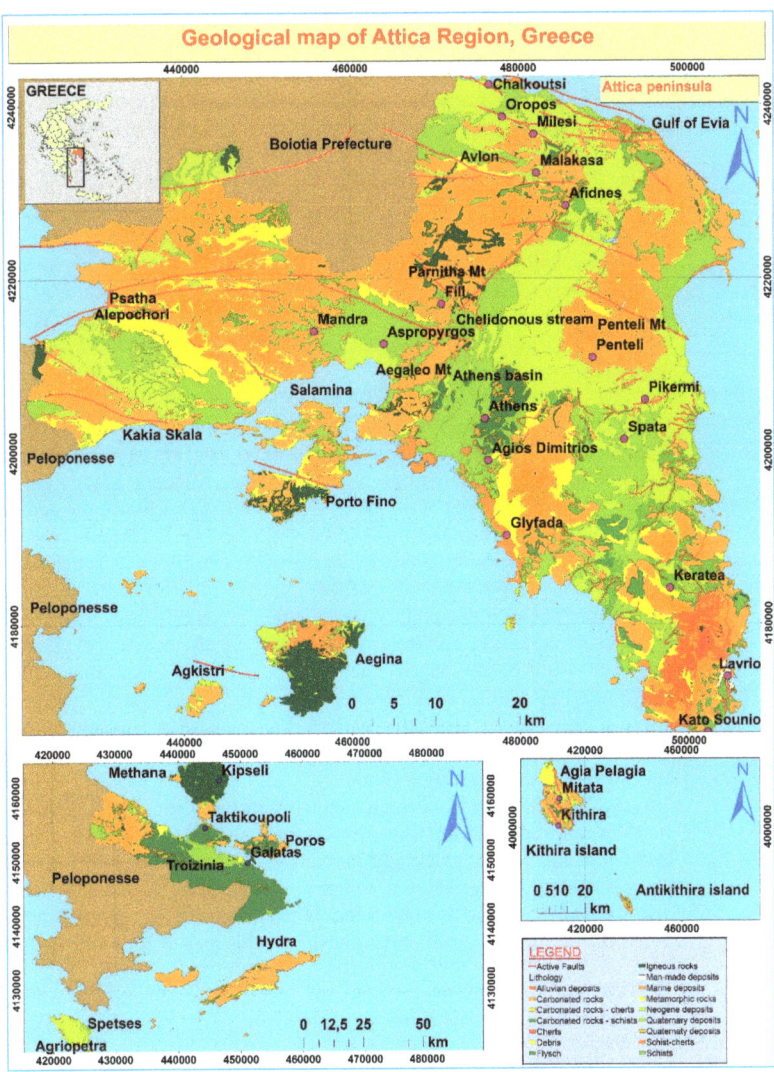

Figure 1. Simplified geological map of Attica region, based on the official Greek projection system (EGSA 87). Active faults were inserted in this map from the National Observatory of Athens (NOAFAULTs, https://zenodo.org/record/4304613#.YAmJbugza1Z).

Regarding the above-mentioned, the scope of this study that is part of the project "Landslide Risk Assessment of Attica Region (DIAS)", is twofold: (i) construct a uniform and updated geodatabase of slope failures induced the last sixty years in the whole territory of Attica Region, and (ii) compile a landslide susceptibility map, being the basic step to produce the upcoming landslide hazard and risk maps.

2. Geology and Tectonic Setting of Attica

Attica is located in the back-arc area of the Hellenic Arc. The geology of Attica comprises Alpine basement rocks, both metamorphic and non-metamorphic, and post-Alpine sediments (Figure 1). The Alpine rocks belong to the high-pressure metamorphic units of the Cyclades and Almyropotamos that extend from Penteli Mt, east Attica [10] to the southern Gulf of Evia and to the non-metamorphic units of Eastern Greece/Sub-Pelagonian units that outcrop in Parnitha Mt and in west Attica. The southern parts of Attica are also underlain by schists and marbles of the Cycladic Metamorphic Belt. An 8.2 Ma granodiorite outcrops in the Lavrion area of SE Attica. The post-alpine (syn-rift) formations consist of alternating beds of marls, lacustrine limestone marls and sandstones. Quaternary deposits are talus cones, sandy–clayey soils, scree, and unconsolidated clays. [11].

Rifting started in Middle-Upper Miocene and continues until the present day resulting in the formation of several basins. According to Freyberg [12], in the western part of the Athens Basin, the Pliocene formation (with a considerable thickness reaching locally more than 300–400 m) can be found, such as clays, sands and sandstones, and gravels in alternation with white limestone. The dating of the synrift ranges from Upper Miocene to Holocene times. There are also Quaternary volcanic formations consisting of loose volcanic extrusive rocks with tuff blocks, dacitic and andesite domes as well as alluvial fan deposits and steep talus cones covering parts of Aegina island, Poros island and almost the entire Methana peninsula.

The Athens basin is the main neotectonic feature in Attica, elongated in a NE-SW direction. An important tectonic structure is the NNE-SSW, west-dipping detachment fault that separates the metamorphic units to the east from the un-metamorphic units to the west [13,14]. The fault was active in Late Miocene-Early Pliocene and produced several hundred meters of debris-flow deposits. In addition, the active normal faults of Avlon-Malakasa, Afidnes, Milesi, Pendeli, Kakia Skala, Thriassion and Fili dominate the area [15,16]. These faults present characteristic features such as prominent scarp linearity, considerable scarp height, unweathered scarp appearance and fault-slip kinematics that are compatible with the regional stress–strain fields (N-S to NNE-SSW) [17,18].

Based on their morphotectonic features [16], all normal NW-SE trending major faults of Attica could be considered "active structures". Overall, the northern part of Attica is bounded by a series of north-dipping active fault segments, while the central part by south-dipping active faults, respectively [16,19–22]. The slip rates of active faults are less than 1 mm/year [15,21,22] and average earthquake recurrence intervals are expected in the order of a few thousands years.

An interesting part of the geological setting of the Attica region is Kithira and Antikithira islands, which are the southeastern islands of the Ionian Sea between Peloponnese and Crete and belong to the administration of Attica Regional Authority. The geological formations that are found there, comprise metamorphic rocks as well as carbonate rocks of Tripolis and Pindos geotectonic zone. Both islands are surrounded by N-S oriented active faults due to ongoing east-west extension in this area of the Hellenic Arc.

3. Materials and Methods

3.1. Landslide Inventory of Attica

The first step towards the compilation of a landslide susceptibility map is the development of a landslide inventory [23]. In this study, the generated inventory map, and the landslide geodatabase, cover a chronological period from 1961 up to the present.

The methods that were used for the generation of the inventory are classified into the following approaches:

- An in-depth collection and review of technical reports (analog and electronic copies) from public authorities, research institutes and newspaper articles
- Field surveys and validation of previously mapped landslides by the authors of past reports
- Airborne and satellite image analysis and interpretation using (a) multi-temporal optical images from Google Earth Pro, (b) processed hillshade imagery extracted from a high-resolution Digital Elevation Model (pixel size of 5 m). we used the 5-m Digital Elevation Model for mapping older landslide features and identify new potential ones. Those landslide areas were delineated based on the guidelines recommended by the protocol of Special Paper 42 from the Oregon Department of Geology and Mineral Industries [24]. The identified slope failures were imported in the ArcGIS database, georeferenced, based on the official Greek projection system (EGSA 87), as: (1) spatial data (mapped as points, lines and polygons) and (2) tabular (descriptive) data in text or numeric form, stored in rows and columns in a database and linked to spatial data [24]. Characteristic examples of the slope failures that were reported in the Attica Region and employed in the DIAS geodatabase are shown in the following Figure 2.

Figure 2. *Cont.*

Figure 2. (a) Rockfalls at the coastal areas of Alepochori–Psatha (North-Western Attica), (b) earth fall on bank slopes subjected to undercutting by Chelidonous stream (North of Athens, Kifisia municipality). (c) Rockfalls occurred in Attica islands such as Spetses (e.g., Agriopetra) and (d) Kithira (e.g., Galani spring-Agia Pelagia). In Figure 2c, the blue circles around rocks emphasize the great possibility for rockfalls. (e) Complex slope failure in Salamina island (Porto Fino site), (f) A rock topple failure in Hydra (adjacent to Miaoulis statue), (g,h) An earth slide from Penteli area (Ntrafi site) at northeastern of Athens. The toponyms of each characteristic site are depicted in Figure 1.

Following the terminology defined by the Working Party on World Landslide Inventory (1990) [25], the majority of the depicted slope failure sites hold information on location, dimensions-geometry, landslide-movement type, trigger mechanism, damage caused, slope and aspect, lithological composition, movement date, older activation, seismic risk zone, meteorological data, hydrogeological behavior, consequences, proposed remedial measures, the confidence of landslide identification, mass movement date–field survey date, bibliographic reference and characteristic photos for each slope failure. The developed landslide inventory map is shown in Figure 3, where slope failures are interpreted as points (220 sites), polygons (98 areas delineated based on the Oregon Protocol) and erosion lines based on data provided by the Hellenic Survey of Geology and Mineral Exploration (H.A.G M.E.), assigning a unique identifier and a number of attributes to each landslide. Taking into account Varnes classification (1978) [26], the movement type of the 220 slope failures, shown as points, can be characterized as follows (Table 1):

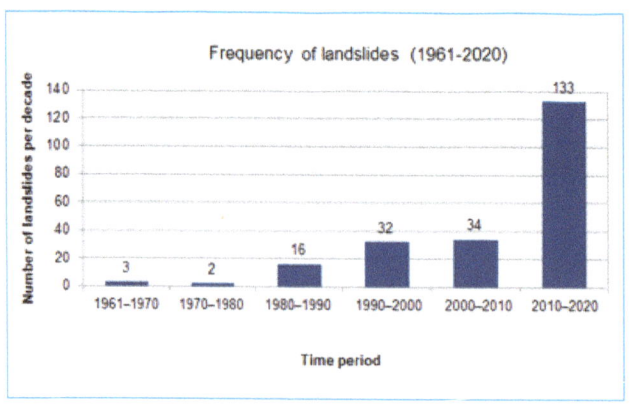

Figure 3. Frequency of landslides of Attica for the period 1961–2020. Bin-size ten years.

Table 1. Movement type of the 220 inventoried slope failures in Attica Region (based on Varnes nomenclature). Each type is associated with a specific recorded number of failures. Each slope failure is depicted in Figure 4.

Movement Type	Rock: 52	Debris: 58	Earth: 110
Fall	1. Rock fall: 40	2. Debris fall: 41	3. Earth fall: 67
Topple	4. Rock topple: 3	5. Debris topple: 3	6. Earth topple: 8
Rotational sliding	7. Rock slump: -	8. Debris slump: -	9. Earth slump: 27
Translational sliding	10. Block slide: 3	11. Debris slide: -	12. Earth slide: 6
Lateral spreading	13. Rock spread: -	-	14. Earth spread: -
Flow	15. Rock creep: -	16. Talus flow: -	21. Dry sand flow: -
		17. Debris flow: 1	22. Wet sand flow: -
		18. Debris avalanche: -	23. Quick clay flow: -
		19. Solifluction: -	24. Earth flow: 1
		20. Soil creep: 13	25. Rapid earth flow: -
			26. Loess flow: 1
Complex	27. Rock slide-debris avalanche: 6	28. Cambering, valley bulging: -	29. Earth slump-earth flow: -

The geodata within the DIAS database followed the EU Inspire Directive and is maintained in a digital format that can be adapted and updated for future use. Furthermore, from the DIAS geodatabase, some more extra remarks can be deduced about the frequency of slope failures per decade from 1961–2020 (Figure 3). It is noted that the number of recorded slope failures increased in the 2000–2010 and 2010–2020 decades in comparison to the pre-2000 data, and this can be explained due to intensive climate change and due to the execution of more detailed field and remote sensing surveys from public authorities, research institutes and consulting agencies.

In the following Figure 4, the developed landslide inventory map is shown. In the legend of the map, the slope failures depicted with green circles correspond to landslides that have already manifested at Attica Region in the past. Slope failures in red polygon shapes are those that are delineated through the methodology described by the protocol of Special Paper 42, developed by the Oregon Department of Geology and Mineral Industries. Finally, erosion lines were provided by the Hellenic Survey of Geology and Mineral Exploration (H.S.G.M.E.) through a research project which proposed flooding mitigation measures in the Mandra area, west of Athens.

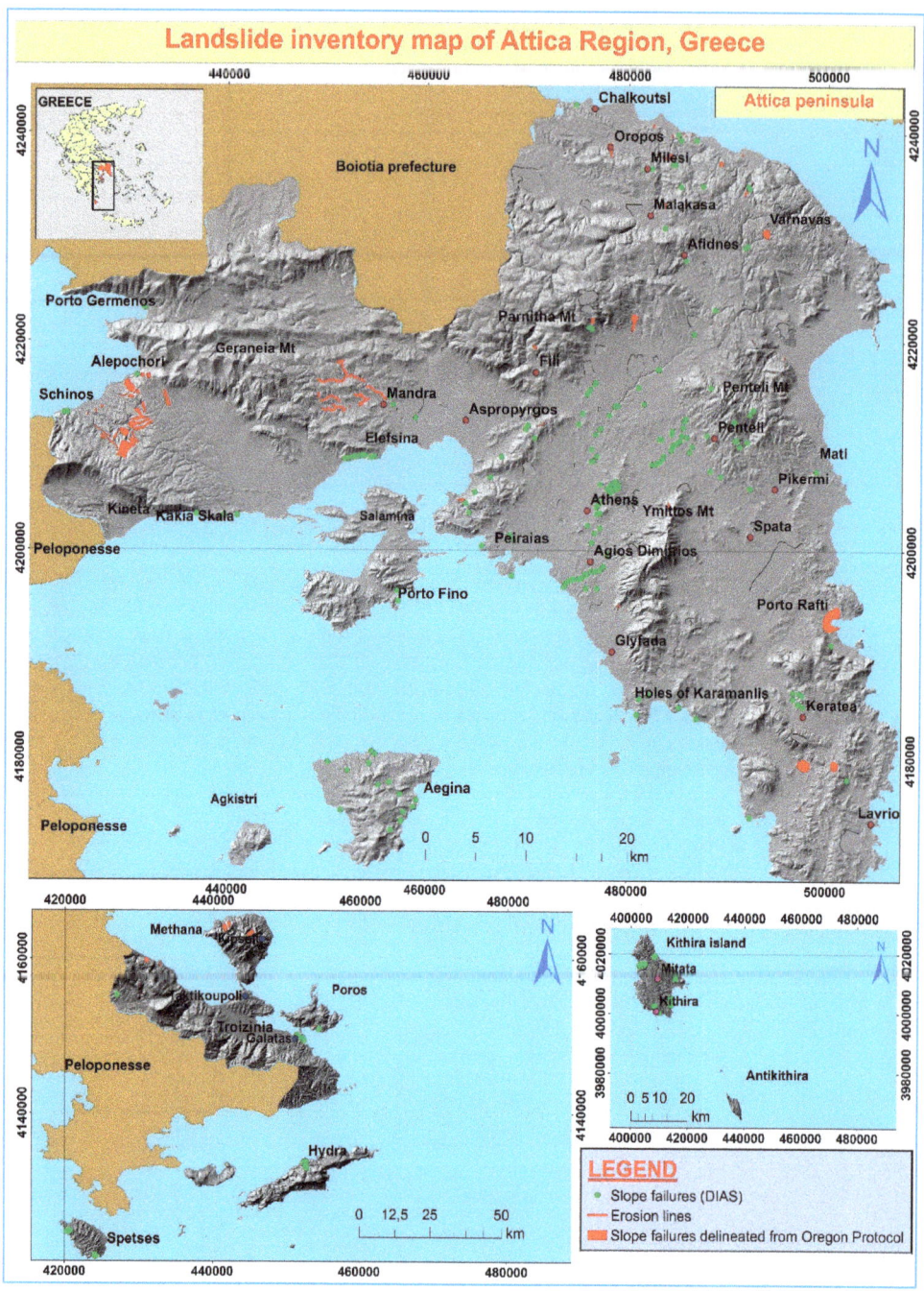

Figure 4. The developed by this study landslide inventory map of Attica Region for the period 1961–2020. Background hillshade image is derived from a high-resolution Digital Elevation Model.

3.2. Assessment of Landslide Susceptibility at the Attica Region

For the purposes of this study, a semi-quantitative heuristic methodology called Rock Engineering Systems (RES), originally introduced by J. Hudson (1992) [7], was applied to assess the landslide susceptibility. The Rock Engineering System approach has been used for a wide variety of rock engineering and other topics, such as surface blasting, natural slope instability, earthquake and rainfall-induced natural slope instability, road-cut induced slope instability, rockfall assessment, engineering geology zonation, coastal landslides, TBM performance, Metro tunnel stability, and many more applications in engineering modelling and design [27]. Furthermore, regarding recent findings of the implementation of RES generally in geotechnical engineering applications, it can be mentioned that:

(i) R. Rafiee et al. (2018) [28], have used fuzzy RES in order to apply system thinking-based techniques for assessment of the rock mass cavability in block caving mines.
(ii) J. Wang et al. (2018) [29], have implemented RES to evaluate sandy soil liquefaction.
(iii) M. Ferentinou and M. Fakir (2018) [30], used RES in accordance with self-organising maps (e.g., artificial neural networks), so as to assess the stability performance of newly open pit slopes.
(iv) Finally, M. Elmouttie and P. Dean (2020) [31], used RES and a system theoretic process analysis in order to design the control system for the slope stability monitoring in an open cut mining.

In Greece, the RES methodology has been applied in different geological settings and scales. For example, Rozos et al. (2006) [32] have used RES for a study in Karditsa prefecture, Greece (scaled in 1:50,000), Rozos et al. (2011) [33] have compared RES and Analytical Hierarchy Process (AHP), Tavoularis et al. (2017) [34] tested RES on Malakasa (1995) and Tsakona (2003), Greece in site-specific scale (1:1,000 to 1:5,000), Tavoularis (2017) [35] implemented RES in a regional scale area (Geological Sheet of Megalopolis, Greece scaled in 1:50,000) in complex geological setting and tectonic regime environment.

In this study, an attempt is made to implement RES in a larger coverage area (scaled in 1:100,000) than those previously mentioned with many different geological settings (active faults, places adjacent to dormant volcanic eruptions, streams banks eroded by flash floods), densely populated and surrounded by many important infrastructure facilities.

3.2.1. The RES Approach

A crucial problem of any engineering design is ensuring that all the necessary parameters are included and that the interactions among them are understood. John Hudson was the researcher that originally introduced the Rock Engineering Systems (RES) approach in 1992. The RES methodology is a synthetic approach which studies the problem (e.g., landslide), breaks it down into its constituent variables (e.g., predisposing parameters, estimation of landslide instability index), and assesses their significance (e.g., calculation of susceptibility analysis). In most slopes, that kind of analysis is complicated due to different interacting factors, complexity of geological formations, different scale of the instability events as well as a scarcity of detailed geodata. These problems can be solved through the use of RES, where its use can take into account the particular problems at any investigated site so as to identify critical sites in order to support decisions on land use and planning development [27].

For consideration of a specific engineering project–system (in our research the landslide susceptibility of the Attica region), some parameters are expected to show a greater effect on the project–system than others and some parameters will in their turn be significantly affected by the system. The RES methodology uses a table (i.e., interaction matrix) with x_i rows and y_j columns, in which the selected n parameters are selected as leading diagonal terms and the interactions between them are considered as off-diagonal terms. In Figure 5, the row passing through the parameter Pi represents the influence of Pi on all the other parameters in the system, whereas the column through Pi represents the influence of the other parameters on Pi. Afterward, we study this so-called influence by coding the off-diagonal components in order to express their importance. A semi-quantitative coding method was used with values ranging from 0 to 4 corresponding to: 0-No interaction (most stable conditions); 1—Weak interaction; 2—Medium interaction; 3—Strong interaction and 4—Critical interaction (most favorable condition for slope failure), respectively. For eliminating the subjectivity, this coding method can be used by one or more experts familiar with the project being considered [7]. Next, the sum of each row (named as "cause-C") and each column (named as "effect-E") can be determined and designated as co-ordinates (C, E) in the diagram of Figure 5. The meaning behind this diagram is that C represents the way in which Pi affects the system; and E represents the effect that the system has on Pi, by indicating a parameter's interaction intensity (as the distance along the diagonal) and dominance (the perpendicular distance from this diagonal to the parameter point). By these two words, we quantify parameter significance inside the matrix system (i.e., landslide).

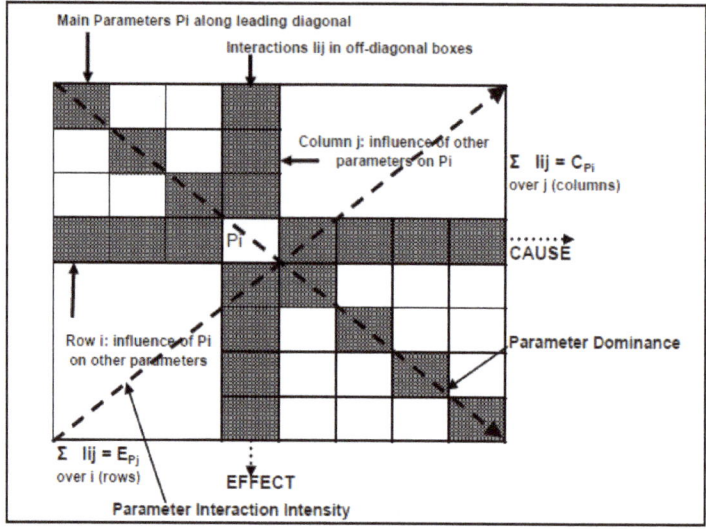

Figure 5. Interaction matrix. The dashed lines correspond to the terms "interaction intensity" and "dominance" respectively [7].

According to Hudson (1992) [7], there are many "constellations" that could occur, the two main ones being mainly along the C = E line or mainly along a line perpendicular to it. If the parameter points are scattered along the C = E line but close to it, then they can be ranked according to their parameter interaction intensity; in other words, they can be listed in order of interactive importance (Figure 6a). If, on the other hand, the parameter points are scattered on a line perpendicular to the C = E line, they will have similar interaction intensities but widely differing dominance values (Figure 6b). In the former case, it might be possible to use five or six parameters in such a scheme; in the latter case, all the parameters must be used.

The cause versus effect diagram reveals the influential role of each parameter on slope failure which is expressed by the term "weighted of coefficient influence". Respectively, the role of the system's interactivity is expressed from the histogram of the interactive intensity [cause (C) + effect (E)] against the parameters. This intensity is transformed into weighting coefficients, which express the proportional share of each factor in slope failure and normalized by dividing with the maximum rating (4), giving the a_i%, as it is explained in the next paragraph.

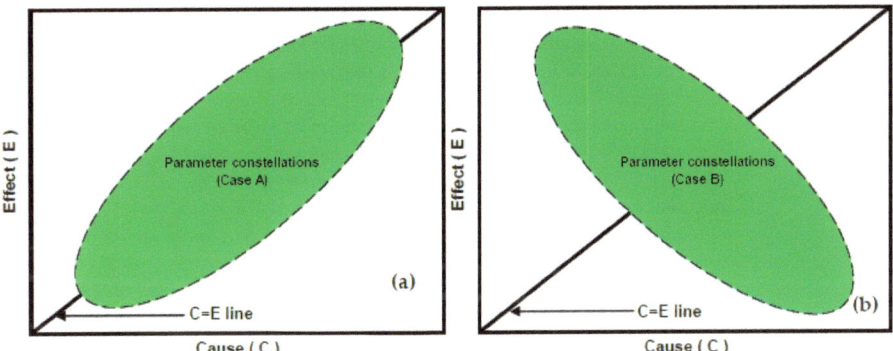

Figure 6. Interaction intensity–dominance diagram, with different forms (**a**,**b**) of parameters constellations [34].

The next step is to compute the instability index (Ii) for each examined slope, by using the following equation:

$$Ii = \Sigma ai \times Pij \qquad (1)$$

where Pij is the rating value assigned to the different category of each parameter's separation, i refers to parameters (from 1 to 10 corresponding to this research, and generally from 1 to n, in other case studies where a different number of landslide parameters are selected), j refers to the examined slope and ai is the weighting coefficient of each parameter provided by the formula:

$$ai = 1/4 * [(C + E)/(\Sigma iC + \Sigma iE)]\% \qquad (2)$$

normalized to the maximum rating of 4. It should be noted that the instability index is an expression of the potential instability of the slope, with values ranging between 0 (no slope failure at all) and 100 which refers to the most unfavorable conditions (i.e., landslide).

3.2.2. Selection of the Parameters Controlling the Slope Failures

Ayalew and Yamagishi (2005) have reported that there are five basic concepts for the chosen parameters regarding the assessment of landslide susceptibility [36]. Parameters should: (i) vary spatially, (ii) be measurable, (iii) be related to the presence or absence of landslides, (iv) be representative of the entire study area, and (v) not account for double consequences in the final outcome. Ten parameters were selected as independent controlling factors for the landslide manifestation of the Attica Region, and classified into five classes. These factors which were utilized for the RES methodology are the (i) distance from roads, (ii) slope inclination, (iii) slope orientation (aspect), (iv) lithology, (v) hydrogeological conditions, (vi) rainfall, (vii) land use, (viii) distance from streams, (ix) distance from tectonic elements and (x) elevation.

In order to decide and consequently select the above-mentioned parameters, (a) we studied a huge amount of published and unpublished engineering geology reports, (b) we applied very interesting landslide research based on statistical analysis gained in Greek territory [37], and (c) we took into consideration the field observations conducted in Attica region in the frame of this study [38]. In the following paragraphs, the importance of each selected parameter for the initiation of a landslide and an analysis of what do we mean by the terms "dependence" and "independent" are provided.

The meanings of "dependance" variable and "independent" parameter are related to the role each one has inside the whole system we study. The system can be a slope failure or underground stability and support or the selection of the right type of tunnel boring machine or any other geotechnical engineering problem that can be addressed by using this semi-quantitative heuristic methodology of RES.

To be more specific, referring to "dependance" variable is meant the occurrence or not of a slope failure. For example, we study the interaction of ten landslide parameters and according to RES methodology, we calculate the weighted coefficient of each landslide parameter, estimating the instability index for each examined slope. If the calculated index is over a critical accepted threshold (as the one that we will present in the following section of this paper), this means that the selected parameters are crucial for the slope failure occurrence, and subsequently, measures must be taken in order to minimize their effect on slope instability. Otherwise, if the estimated index is under this critical threshold, then, no landslide is about to happen and immediately we conclude that those parameters that we selected are not crucial for landslide initiation.

On the other side, as "independent variables" are characterized the landslide controlling factors (such as geology, distance from roads, hydrogeological conditions, distance from tectonic elements), and each other is tested on how dominant or how interactive can be with the other selected landslide parameters.

RES studies the interaction of each parameter to the other and vice versa, by quantifying the different importance of these interactions. This is justified because some parameters will have a greater effect on the system (e.g., in our case the landslide susceptibility in Attica county,) than others and some parameters will in their turn be largely affected by the system. Thus, talking for example about the interaction of hydrogeological condition on lithology, it is meant how lithology can be affected by the permeability status that dictates the geological formation that constitutes the examined slope and vice versa how a specific type of rock or soil of the examined slope will affect the hydrogeological equilibrium of the slope. In another case, we examine how the distance from a road affects the amount of vegetation that exists around this. To be more specific, if a public authority plans to construct a new highway in a place where forest or a grassland area already exists in that particular zone, then it is proved that buffer zones of highway that are in a distance 50 or 200 m from the surrounded slopes affect the existence of vegetation dramatically [33]. Vice versa, the influence of vegetation on slopes that are in an x distance from roads is less important.

In the following paragraphs, a brief comment on the importance of each predisposing landslide parameter is presented.

(i) Distance from roads

During the construction of the road network, vegetation removal, and the application of external loads as well as extensive excavation are some of the most common human intervention actions which are taking place, and result in landslide triggering [39]. It should be mentioned that the digital record of Attica county roads for the generation of DIAS geodatabase was provided by the General Secretary of Civil Protection Agency of Greece. Buffer zones were created around the roads. According to many studies but mostly based on Rozos et al. [33], the slopes that are at a distance of 50 m from a road are more prone to failure.

(ii) Slope inclination

Slope gradient influences on a high grade the slope proneness to failure due to a combination of reasons such as the weathering processes, the internal geometry of geological formations as well as the intensity of meteorological conditions [34]. Through the use of digital elevation model and geographical information systems processing, the slope layer was derived and classified into five classes, as follows: (1) 0°–5°, (2) 5°–15°, (3) 15°–30°, (4) 30°–45°, and (5) >45°, with the higher rating (4) to be given to the slopes with the higher inclination (>45°) [33].

(iii) Slope orientation (aspect)

Another morphological characteristic that influences landslide initiation is the slope orientation (i.e., aspect). Since vegetation and moisture retention depends on aspect, in their turn may affect soil strength and as a result the proneness to landslides. Furthermore, since specific orientations are associated with increased snow concentrations and consequently longer periods for freeze and thaw processes, (not to mention that significant amount of rainfall falling on a slope may vary depending on its orientation [40]), make everybody accept that this is a very crucial parameter for the estimation of landslide susceptibility. The classification of the slope aspect is shown in Table 2 and its rating is based on Koukis and Ziourkas [37]. According to them, in statistical analysis for landslides in Greece took place in the period 1949–1991, the classes 0°–45°, 45°–90°, are associated more frequently with slope failures. Thus, in this study, the highest rating corresponds to rating 4.

Table 2. Parameters and their rating selected to be employed in the model.

Parameters.	Grade	Parameters	Grade	Parameters	Grade
1. Distance from roads		5. Hydrogeological conditions		9. Distance from tectonic elements	
Distant (>200 m)	0	Impermeable formations (Marl, siltstone)	0	Distant (>200 m)	0
Moderately distant (151–200 m)	1	Fractured formations characterized as having low to negligible permeability (Flysch, schists)	1	Moderately distant (151–200 m)	1
Immediate (101–150 m)	2	Volcanic rocks, conglomerate	2	Immediate (101–150 m)	2
Less immediate (51–100 m)	3	Carbonate formations with medium to high permeability	3	Less immediate (51–100 m)	3
Close (0–50 m)	4	Debris, alluvial–marine deposits	4	Close (0–50 m)	4
2. Slope's inclination		6. Rainfall		10. Elevation	
0°–5°	0	<400 mm	1	>1000 m	1
6°–15°	1	400–800 mm	4	0–200 m	2
15°–30°	2	800–1000 mm	3	600–1000 m	3
30°–45°	3	1000–1400 mm	2	200–600 m	4
>45°	4	7. Land Use			
3. Slope's orientation		Barren areas	0		
270°–315°	1	Urban areas	1		
90°–135°, 135°–180°, 225°–270°	2	Forest areas	2		
180°–225°, 315°–0°	3	Shrubby areas-Natural grassland	3		
0°–45°, 45°–90°	4	Cultivated areas	4		
		8. Distance from streams			
4. Lithology		Distant (>200 m)	0		
Carbonate rocks (e.g., limestones, marbles), schist, cherts	1	Moderately distant (151–200 m)	1		
Metamorphic rocks exhibiting schistocity	2	Immediate (101–150 m)	2		
Loose soil formations (alluvial, etc.)	3	Less immediate (51–100 m)	3		
Flysch, marine deposits	4	Close (0–50 m)	4		

(iv) Lithology

According to Koukis and Ziourkas [37], lithology in Greek territory is classified into six classes as follows: (a) igneous rocks, (b) cherts, schists, (c) carbonate rocks (e.g., limestones, marbles), (d) metamorphic formations exhibiting schistosity, (e) loose soil formations (alluvial, etc.) and (f) flysch. They concluded that flysch is the geological formation that is associated with the most frequent landslide incidents in Greek territory (36% frequency of landslides), and accordingly it was decided to correspond this complex formation (intercalations mostly of sandstone, siltstone and limestone) to rating 4. In this research, the geologic map of the entire county of Attica region, provided by the Hellenic Survey of Geology and Mineral Exploration (H.S.G.M.E.) was taken into consideration. This map, comprises a digital mosaic of twenty-one (21) geological sheets scaled in 1:50,000.

(v) Hydrogeological conditions

In this research, the classification is based on River basin management plans from the Greek Ministry of Environment, Energy and Climate Change/Special Secretariat for Water (2012) [41], where the highest rating (4) was given to debris, alluvial–marine deposits whose permeability is crucial for slope failure.

(vi) Rainfall

It is well known that high precipitation can increase both the groundwater level and the pore pressure in a soil mass/weathered mantle or aquifer, and accordingly it constitutes the main triggering causal factor of landslides [39]. The data that we used were provided by Attica meteorological stations of the National Observatory of Athens (NOA). NOA has published reports presenting the locally encountered conditions [42]. Those data were analyzed using kriging interpolation in order to acquire a rainfall layer of information for the upcoming GIS geoprocessing. In addition, the rating was based on the statistical analysis made by Lalioti and Spanou (2001) for Greece during the period 1991–1998 [43]. In this research, the class 400–800 mm is the one with the greater amount of rain (mean annual) in the Greek territory, so the highest rating for this study corresponds to 4.

(vii) Land Use

Land use is a crucial parameter in controlling soil erosion as it is related to the vegetation covering which in its turn provides a protective layer on the earth and regulates the transfer of water from the atmosphere to the surface, soil and underlying rocks [44]. The vegetation data used in this study was extracted from the EU Corine Land Cover 2018 database and its rating is based on Rozos et al. [33]. According to them, the higher rating was given to the cultivated areas, due to the maximum percentage of landslide density that is observed.

(viii) Distance from streams

The closer a slope is to a stream, the less stable it is. This happens, due to the fact that streams may adversely affect stability by eroding and saturating the bottom zones of the slopes [45]. The hydrographic network for DIAS geodatabase was generated using the digital elevation model of 5 m pixel size resolution as well as ArcGIS algorithms referring to hydrology processing (Fill, Accumulate, Flow direction based on Strahler classification).

For the examination of this parameter, buffer zones were created around the streams at distances of 50, 100, 150 and 200 m. The classes of the buffer zones are shown in Table 2 and its ranking was based on Rozos et al. [33], suggesting that the most prone class to landslide is that of 0–50 m. This implies that as the distance from the hydrographic axes decreases, the highest percentage of landslide density increases.

(ix) Distance from tectonic elements (e.g., faults)

There is an increase in the occurrence of slope failures at areas close to fault zones, because as the distance from a tectonic element decreases, the fracture of the rock and the degree of weathering increases [46], while the structure of the surficial material is affected causing selective erosion and forcing the movement of water along fault planes to decrease slope stability [47,48]. In Attica Region, many active faults were mapped particularly in west and northeastern part of its peninsula as well as in some islands (such as those of Salamina, Kithira). The digital fault database was provided by the Hellenic Survey of Geology and Mineral Exploration and from the National Observatory of Athens [49]. The classes of the buffer zones are shown in Table 2, with the most prone class to landslide to be that of 0–50 m (rating: 4) [33].

(x) Elevation

The combination of elevation, precipitation and erosion-weathering process contribute to landslide manifestation. The elevation data used in the model were derived from high-resolution DEM (5 m pixel analysis) provided by the Greek Cadastre S.A. The classes of the buffer zones are shown in Table 2 and its ranking was based on the landslide statistical analysis made by Koukis and Ziourkas (1991) for Greece during the period 1949–1991 [37]. In this research, the category 200–600 m is related to the highest number of slope failures that happened in Greece, so this class is associated with a rating of 4.

The above data were rated so as to be used in the development of the interaction matrix (Table 2).

4. Results

4.1. Implementation of RES for the Estimation of Weighted Coefficients

In this section, the results of the application of the RES method in the Attica Region are presented, such as the interactions among the selected parameters, the calculation of their weighting coefficients and finally the instability index accompanying with charts and tables which they decode and translate the geodata. As it was previously presented, the interaction matrix shown in Table 3 was coded using the Expert Semi-Quantitative method. For example, regarding the effect of lithology (P4) on rainfall (P6), it can be stated that there is no influence at all (coding: 0), whereas rainfall does affect lithology through the infiltrating and weathering-erosion process that may alter not only the mineralogical composition of a specific rock or soil of the slope but also influence their hydrogeological behavior too (coding: 2).

Note that, in Table 3, the sum of cause-and-effect (C + E) value for each parameter represents the "interaction intensity" term, which means how active that parameter is within the matrix system (i.e., the slope stability). On the contrary, the (C − E) value represents how dominant the variable is within the system: positive values of (C − E) represent a dominant variable, whereas negative values of (C − E) represent that the system is affecting the variable more than the variable is affecting the system [7]. More specifically, from Table 3 and Figure 7, it can be seen that the hydrogeological conditions are the most interactive parameter (C + E = 39) [e.g., has the greatest value (concerning C + E)], meaning those conditions play the most decisive role for landslide activation, whereas elevation is the least interactive (C + E = 18). This suggests that elevation does not depend on the influence of the other parameters, but it is an independent agent.

Table 3. Coding values for the interaction matrix of Attica Region.

Interaction Matrix of Attica Region											
P1	3	1	0	1	0	2	0	0	0	7	
2	P2	1	0	1	0	2	2	1	0	9	
1	2	P3	1	2	2	2	2	0	0	12	
1	3	2	P4	4	0	2	3	2	2	19	
2	2	2	2	P5	0	3	3	1	0	15	Cause (C)
4	3	0	2	4	P6	4	3	0	0	20	
0	1	0	1	2	0	P7	1	0	0	5	
2	1	1	1	4	0	2	P8	1	0	12	
4	3	1	2	4	0	0	2	P9	0	16	
2	2	0	1	2	4	3	2	0	P10	16	
18	20	8	10	24	6	20	18	5	2		
Effect (E)											

P1 = Distance from roads	P2 = Slope	P3 = Aspect	P4 = Lithology	P5 = Hydrogeological conditions
P6 = Rainfall	P7 = Land Use	P8 = Distance from streams	P9 = Distance from tectonic elements	P10 = Elevation

	Parameters	C + E	C−E	[(C + E)/Σ(C + E)]*100%	Maximum rating	Weighted coefficient $a_i = [(C + E)/\Sigma(C + E)]*100\%/4$
1	Distance from roads	25	−11	9.54	4	2.39
2	Slope	29	−11	11.07	4	2.77
3	Aspect	20	4	7.63	4	1.91
4	Lithology	29	9	11.07	4	2.77
5	Hydrogeological conditions	39	−9	14.89	4	3.72
6	Rainfall	26	14	9.92	4	2.48
7	Land Use	25	−15	9.54	4	2.39
8	Distance from streams	30	−6	11.45	4	2.86
9	Distance from tectonic elements	21	11	8.02	4	2.00
10	Elevation	18	14	6.87	4	1.72
Total	Σ(C + E)	262		100.00		25.00

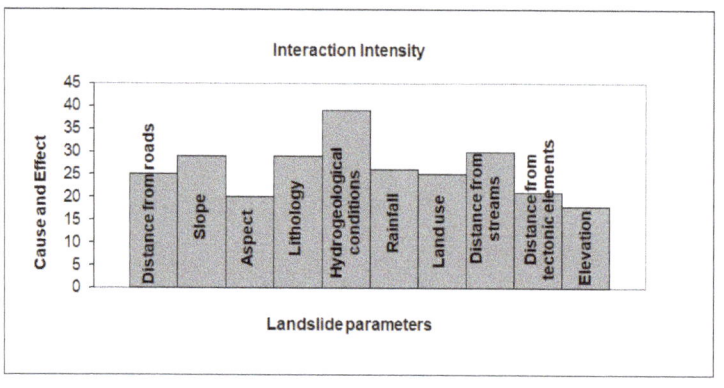

Figure 7. Histogram of interaction intensity.

From the RES model and by focusing on the weights assigned to each parameter, it can be clearly reported that hydrogeological conditions contribute the most to landslide occurrence out of all the factors, followed by distance from streams, lithology, slope angle, rainfall, distance from roads, land use, distance to fault lines, aspect and elevation.

In Figure 8, the form of C vs. E constellation in relation to C = E line, defines the number of parameters that will be needed for calculating the instability index. So, according to the interaction intensity–dominance diagram (Figure 6b), the form of the C vs. E constellation is (almost) perpendicular to the C = E line, which means that (based on the aforementioned RES analysis) there is little range in parameter interaction intensity. On the contrary, there is a wide range in dominance (C − E values), so all the selected parameters will be required for the calculation of the instability index for each examined slope.

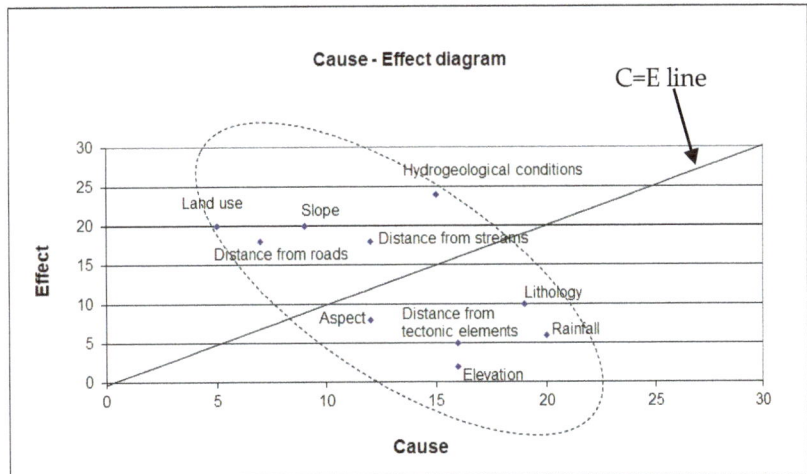

Figure 8. Cause–Effect diagram.

Supplementary, the following Table 4, decode and "translate" simultaneously the geodata acquired from our research and contribute in giving the necessary objective answer to the prognosis of the potential instability of the examined slopes of Attica Region. This can be accomplished by the estimation of the instability index, as clearly explained in Section 3.2.1.

A characteristic sample, 10 out of 220 cases of the computation results regarding the instability index, is given in Table 4. In this table, each examined slope (is depicted in the column "Slopes") is ranked according to Table 1 rating, taking into account in parallel the specific geological conditions that characterize it according to either the ad-hoc technical report we collected or field study we carried out. Afterward, for each slope site, every ranking of each parameter (each parameter is depicted in the second line under the title "Parameters", named as 1, 2, 3, . . . , 10) is multiplied by its weighted coefficient (last line of the Table) respectively and each outcome, based on Equation (1) is added in order to yield the instability index for each slope. For example, the instability index of Slope (1) is estimated as follows:

$$\Sigma \, [\text{Parameter (1): } 4 * 2.39 + \text{Parameter (2): } 1 * 2.77 + \ldots + \text{Parameter (10): } 2 * 1.72] = 71 \qquad (3)$$

In Table 5, the classification for relative landslide susceptibility is listed as proposed by Brabb et al. (1972) [50].

Table 4. Calculation of Instability Index based on Rock Engineering System methodology for a characteristic sample of 10 slope failures out of 220 ones in Attica Region.

Slopes/Coordinates (Greek Projection EGSA 87)	Parameters										Instability Index
	1	2	3	4	5	6	7	8	9	10	
1 (476,117–4,215,245)	4	1	3	4	4	4	1	4	0	2	71
2 (476,790–4,216,087)	4	0	3	4	4	4	1	4	0	2	68
3 (483,219–4,208,555)	4	1	4	4	1	4	1	4	0	4	65
4 (482,341–4,208,253)	4	1	3	4	1	4	1	4	0	4	63
5 (483,441–4,208,871)	4	1	2	4	1	4	1	4	0	4	62
6 (458,846–4,212,690)	4	1	4	3	1	1	4	4	0	2	59
7 (477,287–4,211,687)	4	2	3	4	4	4	1	4	0	2	74
8 (476,938–476,938)	4	1	3	4	4	4	1	4	0	2	71
9 (475,095–4,212,107)	4	1	3	4	4	4	1	4	0	2	71
10 (457,187–4,195,149)	4	3	2	1	3	1	1	4	4	2	63
Maximum Pij rating	4	4	4	4	4	4	4	4	4	4	
[(C + E)/Σ(C + E)] * 100%	9.54	11.07	7.63	11.07	14.89	9.92	9.54	11.45	8.02	6.87	100
Weigh. Coeff. (ai) = C + E)/Σ(C + E)] * 100%/4	2.39	2.77	1.91	2.77	3.72	2.48	2.39	2.86	2.00	1.72	

Table 5. Classification for relative landslide susceptibility proposed by Brabb et al. (1972) [50].

% Failed Area	0–1	2–8	9–25	25–42	42–53	53–70	70–100
Relat. Susceptib.	I	II	III	IV	V	VI	L
	Negligible	Low	Middle	High	Very high	Extremely high	Landslide

As it is shown in this table, the generated instability index that is greater than 53%, corresponds to extremely high relative susceptibility up to slope failure and that this is the crucial point for a planner or a researcher for producing a landslide susceptibility map for a particular examined area. This remark is going to be used extensively in the following sessions of this study.

4.2. Correlation of Spatial Distribution of Slope Failures with the Predisposing Factors Using Statistical Analysis

Based on the information of Table 4, and according to the ranking of parameters of Table 2, the following useful findings come out during the generation of the susceptibility map of the Attica region. Based on this analysis, it can be concluded that 211 out of 220 (96%) slope failures are in a distance from roads up to 50 m.

Concerning the aspect parameter, 37% of the examined slopes are primarily more abundant on Southeast-facing and secondly on Northwest-facing (34%). Based on the rating assigned to each geological formation (e.g., lithology), the highest (40%) one is observed at flysch (and debris) and secondly to carbonate rocks (37%). This remark was expected since the former ones are the most statistically frequent formations prone to landslides in Greek territory, whereas the latter ones are associated mainly with rockfall incidents in many parts of Greece.

Regarding hydrogeological conditions, carbonate rocks with medium to high permeability due to karstification and secondary fragmentation correspond to the highest (35%) category of permeable rocks in this study. Based on the comparison among rainfall data and landslide occurrences, it was established that landslides are more likely to take place when the mean annual rainfall is between 400–800 mm. As far as land use parameter is concerned, landslides reported mostly in urban areas (62%) while based on the results given for the elevation, it was found that the landslides develop preferentially on 0–200 m of altitude (63%).

Furthermore, a large portion of landslides (58%) are located near to the hydrographic network in relation especially to the undermining of the banks between 0 m and 50 m. Such places were recorded in many streams (mostly) in the Athens basin (such as those of Kifisos river, Chelidonous, Sapfous, Penteli, Eschatia stream).

Summarizing, the percentages of landslides per each class of predisposing factor are illustrated in the following Figure 9.

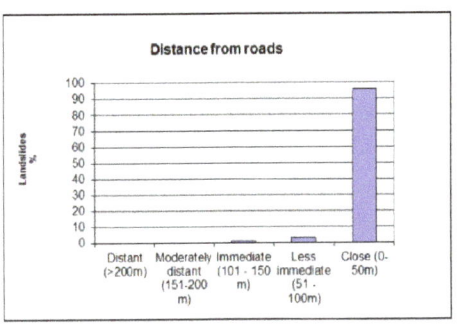
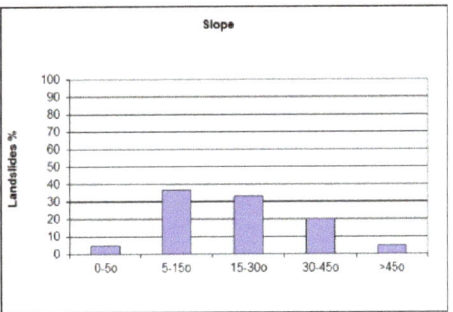
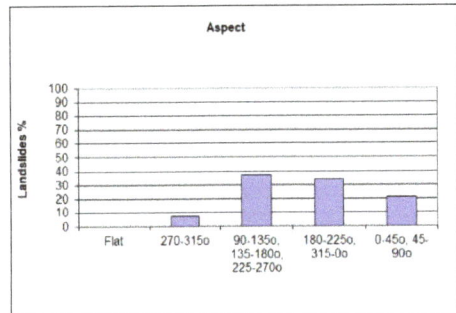
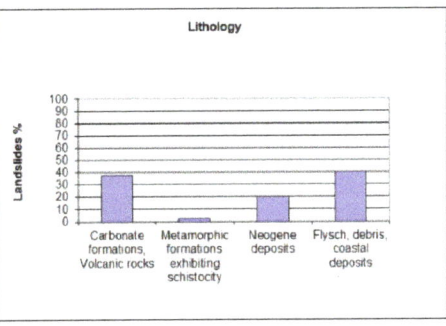
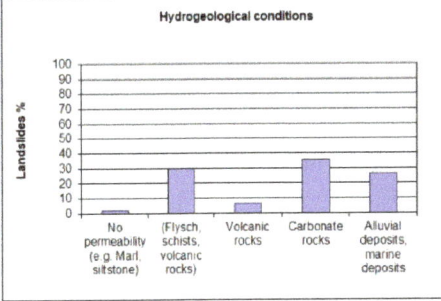
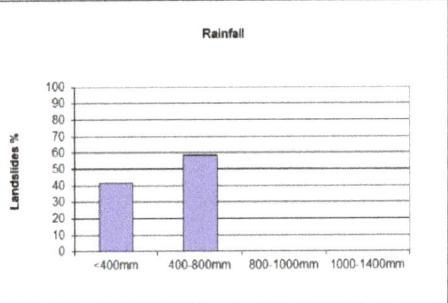

Figure 9. *Cont.*

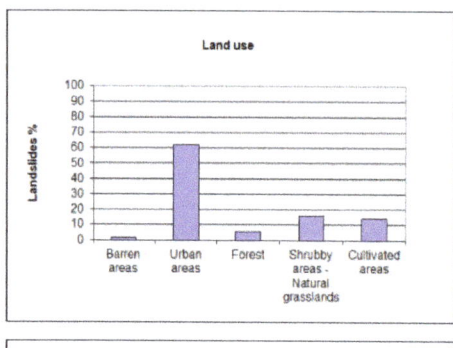
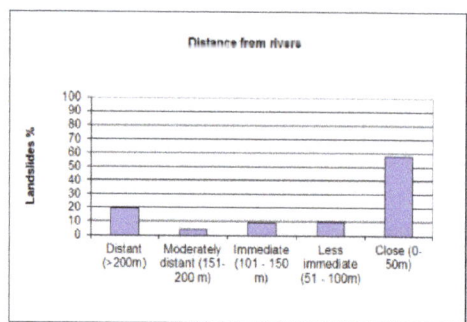
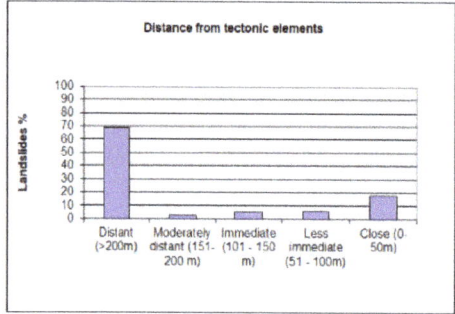
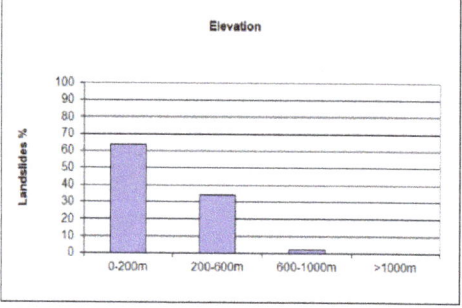

Figure 9. Percentage of landslides in each class of the causal and triggering factor of landslide occurrence.

4.3. Landslide Susceptibility Map

The subdivision of the predisposing parameters into subclasses (from Table 2) was used for the evaluation of the final slope failure susceptibility map. This map was generated in a GIS environment, through the use of different layers-thematic maps (Figure 10a–j). The data used for the preparation of these layers were obtained from different geodata sources among which are the Digital Elevation Model from Hellenic Cadastre S.A. and a mosaic geological map from the Hellenic Survey of Geological and Mineral Exploration. All data layers were digitized either from the original thematic maps or derived from spatial GIS calculations and finally were converted into grids with a cell size of 20 × 20 m. Afterward, weights and rank values to the reclassified raster layers (representing predisposing factors) and to the classes of each layer were assigned, respectively. This was realized with the use of the previously extended analyzed methodology of RES. Finally, the weighted raster thematic maps with the assigned ranking values for their classes were multiplied by the corresponding weights and added up (through the ArcGIS tool of the weighted sum) to yield the slope failure map where each cell has a certain landslide susceptibility index value. The reclassification of this map represents the final susceptibility map of the study area, divided into susceptibility zones according to Brabb et al. (1972) [50] classification (Figure 11). The landslide susceptibility index (LSI) values in the final susceptibility map were classified into five categories, namely "Low-Middle", with Instability index (Ii) < 25, "High" with 25 < Ii < 42, "Very High" with 42 < Ii < 53, "Extremely high" with 53 < Ii < 70", and "Landslide" with Ii > 70%. From this classification, it can be clearly notified that the higher the LSI, the more susceptible the area is to landslides (instability index higher than 70%).

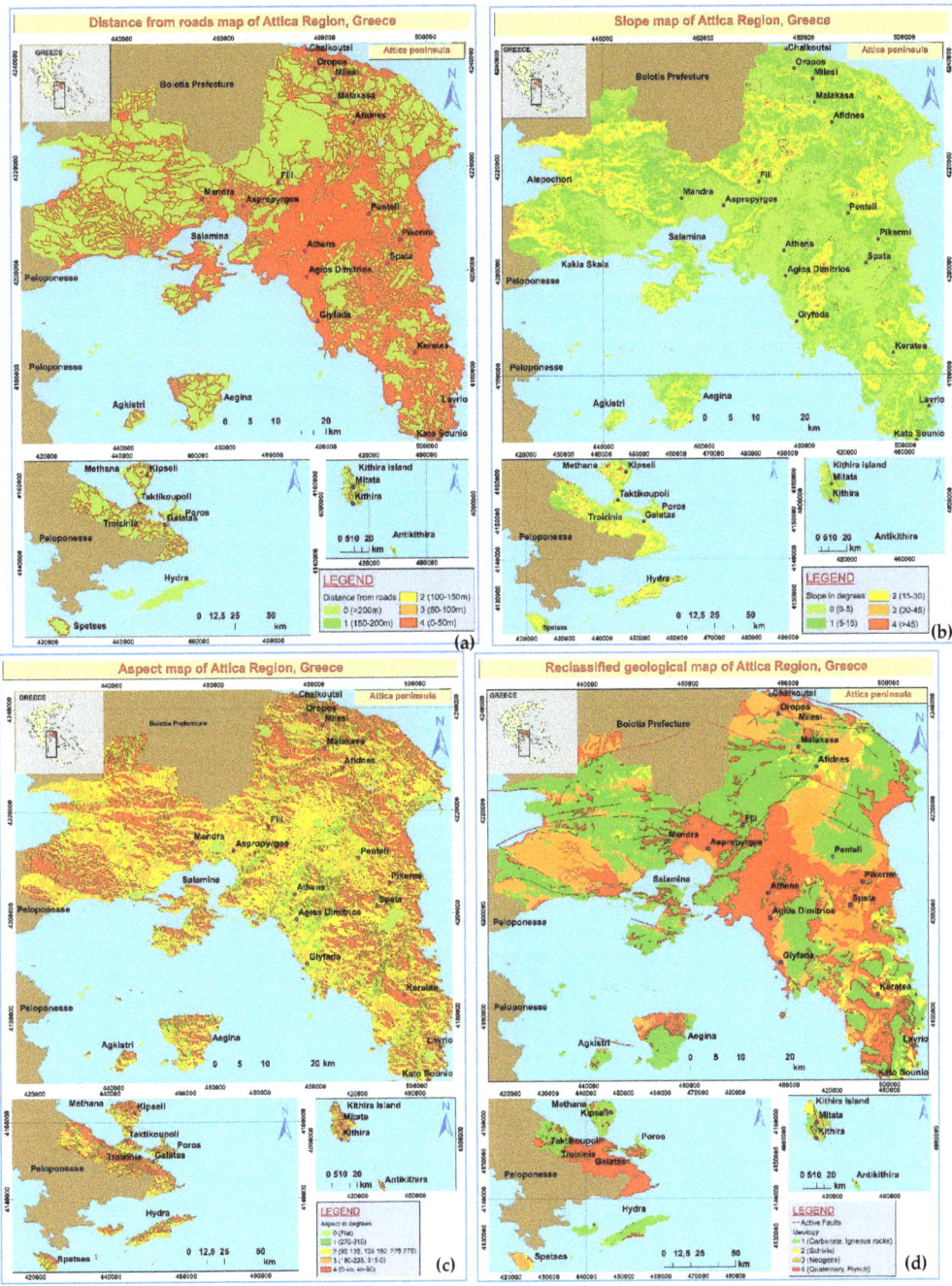

Figure 10. *Cont.*

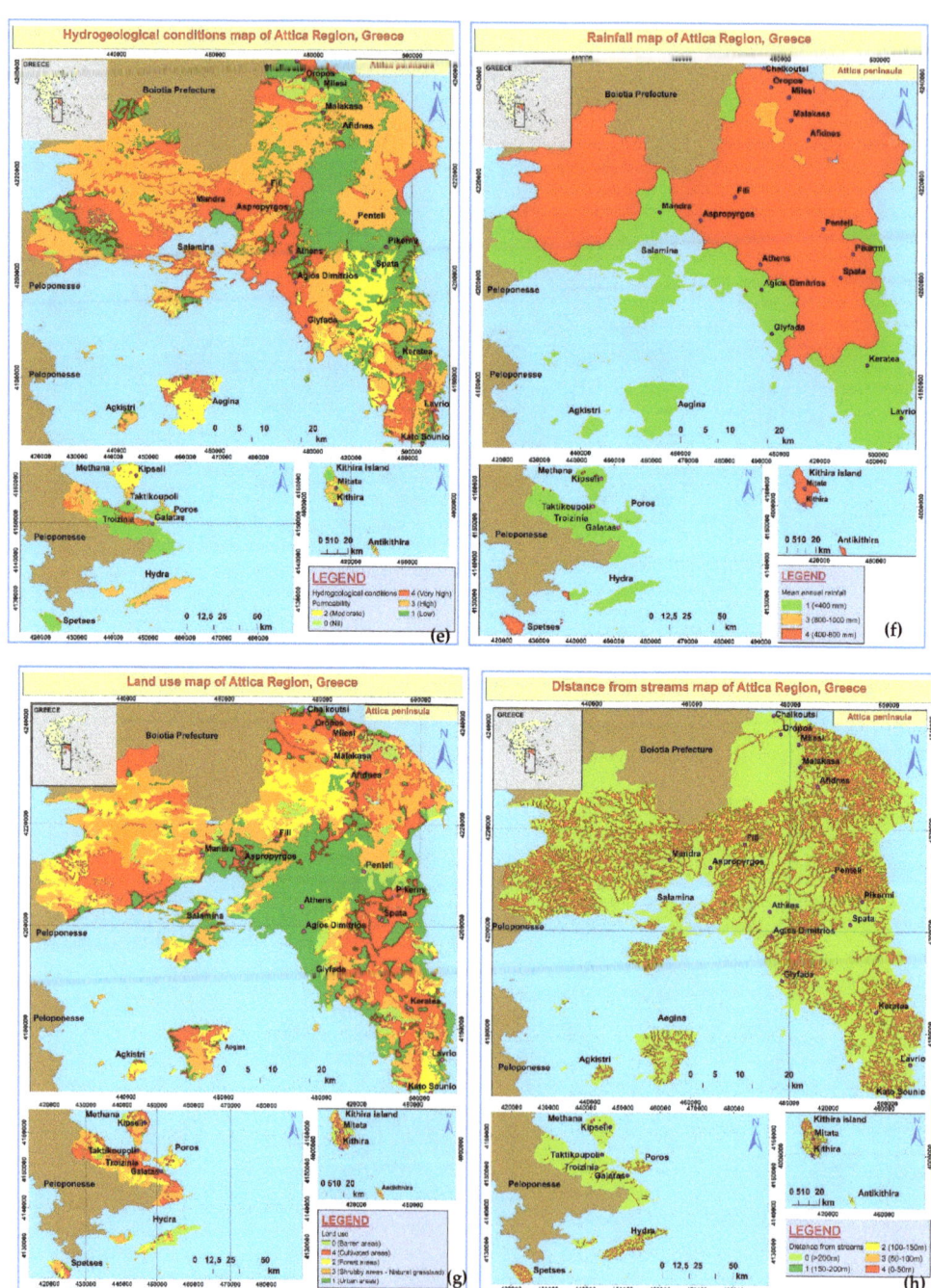

Figure 10. Cont.

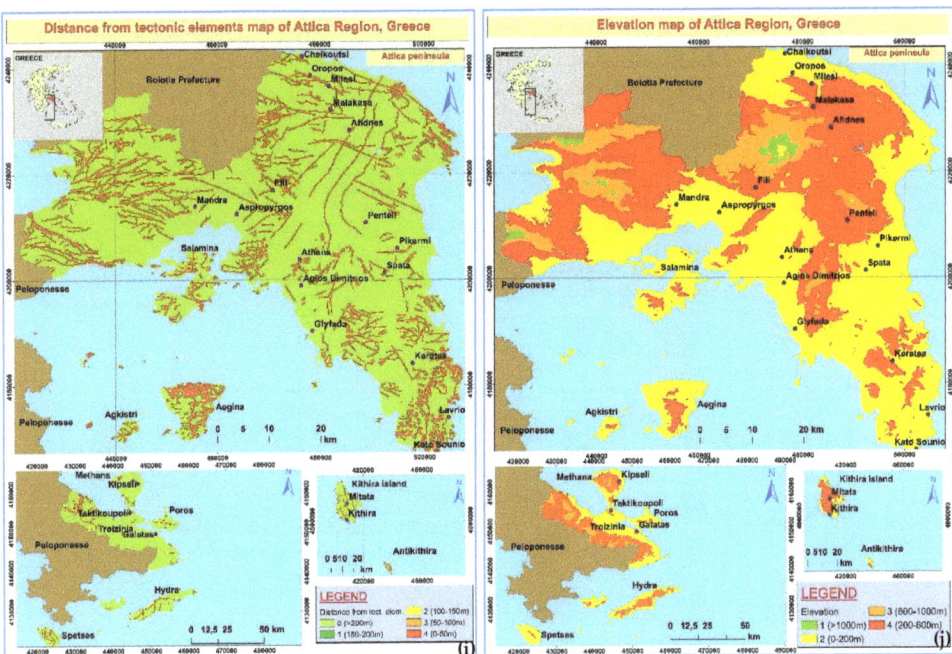

Figure 10. Thematic raster maps of the ten (10) landslide parameters used for the estimation of Attica region susceptibility: (**a**) Distance from roads, (**b**) Slope, (**c**) Aspect, (**d**) Reclassified geological map, (**e**) Hydrogeological conditions, (**f**) Rainfall, (**g**) Land use, (**h**) Distance from streams, (**i**) Distance from tectonic elements, (**j**) Elevation.

From Figure 11, some further findings that come out are as follows (Table 6, Figure 12):

Table 6. Correlation between instability index and susceptibility coverage class in km^2.

Instability Index Category	Susceptibility Coverage Class in km^2
<25%	5 (0.13%)
25.01–42%	585 (15.54%)
42.01–53%	1552 (41.23%)
53.01–70%	1500 (39.85%)
70.01–100%	122 (3.24%)
	Total examined area: 3.764 km^2

From the above pie diagram, it is clear that 43.09% (39.85% + 3.24%) of the examined area is associated with an instability index greater than 53%. Furthermore, it can be added that 122 km^2 (3.24%) of the total examined area are correlated to potential landslide occurrence. Public authorities responsible for auditing and supervising technical works should be aware of these findings, so as to take the appropriate advance, mitigation measures against the possible initiation of potential disastrous landslide phenomena taken place in these proposed, for slope failures, areas.

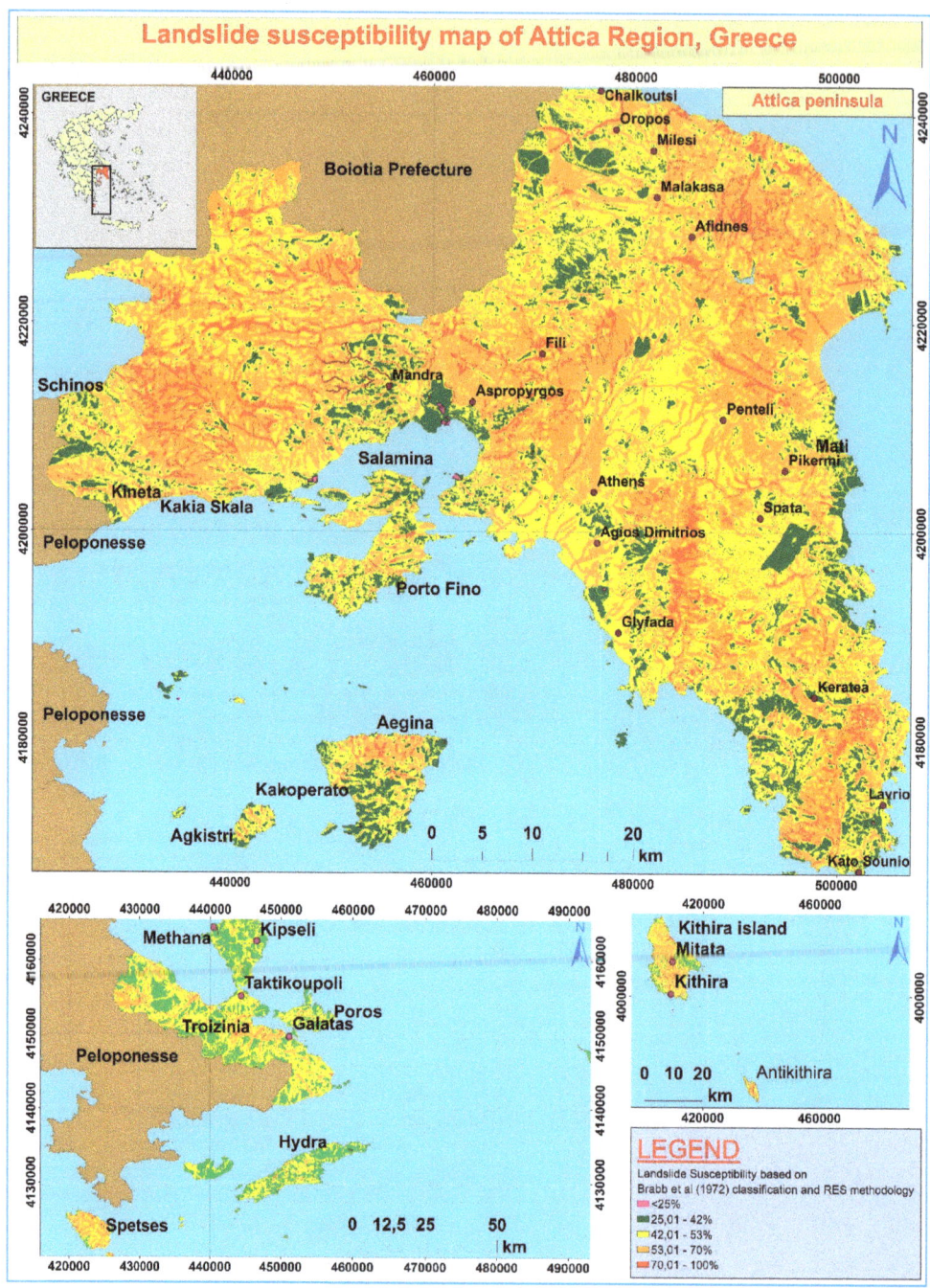

Figure 11. The Susceptibility map of Attica Region.

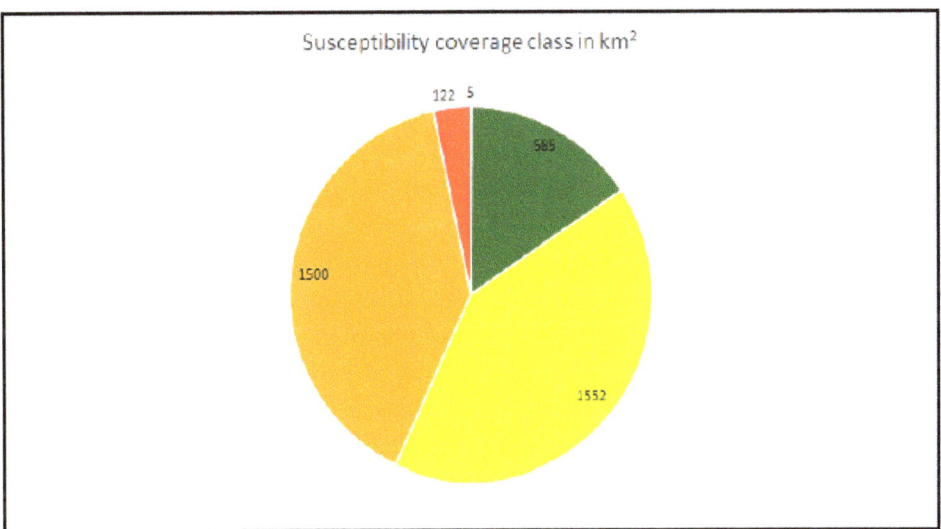

Figure 12. Pie diagram depicting the landslide susceptibility coverage class in km² for Attica region.

4.4. Validation of the Landslide Susceptibility Map

For having scientific significance in any generated model, the most important component in prediction modelling, is to implement a validation of the prediction results [51]. Thus, in the final landslide susceptibility map, we compared the results with the distribution of the 220 slope failure events that had occurred in the examined area. The predicted map showed very satisfactory results and particularly, at the susceptibility map of the Attica region, 68% of the locations of actual and potential landslides correspond to the "Extremely high" and 21% are associated with a landslide (Figure 13, Table 7).

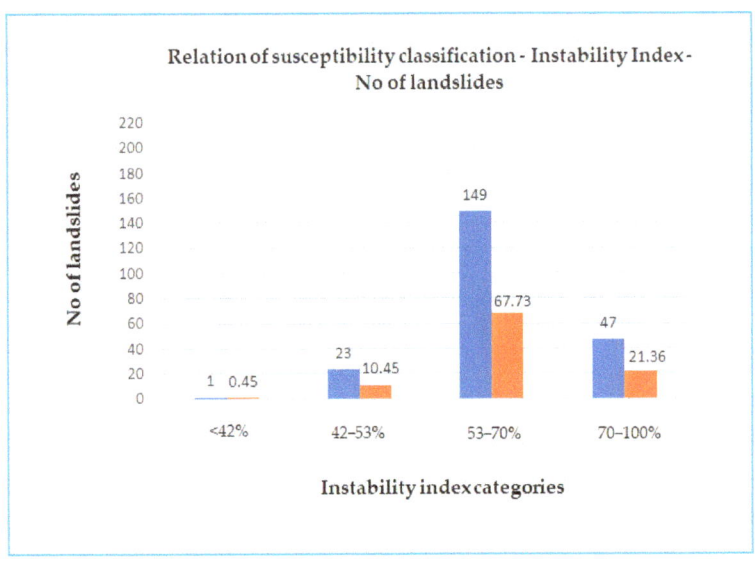

Figure 13. Correlation among number of examined slope failures, instability index, and susceptibility classification. Blue color corresponds to the number of slope failures, orange color is linked with instability index percentage associated with susceptibility categories.

Table 7. Correlation among number of examined slope failures, instability index, and susceptibility classification.

Relative Susceptibility Classification	Number (No) of Slope Failures	Instability Index (%)
<42% (Low to moderate)	1	0.45
42.01–53% (High)	23	10.45
53.01–70% (Extrem. high)	149	67.73
70.01–100% (Landslide)	47	21.36
	Total: 220	100

Moreover, another method for validating the above mentioned, was the implementation of the confusion matrix. It is a table that is often used to describe the performance of a classification model on a set of test data for which the true values are known [52]. We used the confusion matrix for a binary classifier e.g., (α) the existence of landslides with instability index greater than 53% and (b) the no existence of landslides (with instability index less than 53%). Each row of the matrix represents the instances in an actual class while each column represents the instances in a predicted class (or vice versa). In our case, In Table 8, four different combinations of predicted and actual values were used.

Table 8. Confusion matrix of the landslide susceptibility map validation.

Total Population: n = 220 Slope Failures		Predicted Conditions	
		Predicted NO	Predicted YES
True conditions (Observed)	Actual NO	3 (TN)	21 (FP)
	Actual YES	22 (FN)	174 (TP)

Where: TN means when the examined slope does not correspond to landslide, how often does it predict no, FP means when the examined slope does not correspond to landslide, how often does it predict yes, FN means the falsely predicted landslide, TP means when the examined slope correspond to landslide, how often does it predict yes.

The following is a list of rates that were computed from the confusion matrix for a binary classifier:

- Accuracy: Overall, how often is the classifier correct?

$$(TP + TN)/total = (174 + 3)/220 = 0.80 \quad (4)$$

- Precision: Out of all the positive classes we predicted correctly, how many are actually positive.

$$TP/predicted\ yes = 174/195 = 0.89 \quad (5)$$

- Prevalence: How often does the yes condition actually occur in our sample?

$$actual\ yes/total = 196/220 = 0.89 \quad (6)$$

From the above, analytically presented, it is clear that the described RES methodology has 89% precision.

In addition, the validation of the generated susceptibility map was tested with two additional landslide databases. These are (a) the 98 polygons derived from Oregon methodology as previously mentioned, and (b) erosion lines derived from a project delivered by the Hellenic Survey of Geological and Mineral Exploration concerning the Mandra area flooding susceptibility [53]. Particularly, it was found that regarding the Oregon protocol, in the generated landslide susceptibility map of the Attica region, 49% of the defined polygons correspond to the "Extremely high" category and 33% are associated with landslides. Concerning the rest of the delineated areas, it is proposed to conduct geological–geotechnical investigations to define the potential of the slopes to failure.

Finally, the erosion lines which were defined by the aforementioned research institute were in accordance with the instability index greater than 70%.

Practical use of the final susceptibility map is the implementation it may have during the planning, design and construction of various important infrastructure projects. Even though it is not advisable to be used for local or site-specific planning, C.J. Van Westen (2016) [54], recommends the following use of the above-mentioned susceptibility classes.

Low susceptibility zones

In those areas, with respect to planning and constructing civil engineering projects, no special care should be taken by planners and engineers.

Moderate susceptibility zone

This zone is the most problematic for spatial planning and construction infrastructure and it is encouraged to implement geotechnical/geophysical investigation for critical civil engineering projects (e.g., highways, important public buildings such as hospitals).

High and very high susceptibility zones

Slope failures are expected to occur within these zones. The best is to avoid these areas regarding the development of future residential areas or crucial infrastructure projects. However, if this is not possible, a detailed level of geotechnical investigation of landslide hazard is required for these areas before allowing new constructions. In the present study, areas of this category can be found in the northeastern, southeastern and western part of the Attica peninsula as well as in the northern part of Aegina island, and the central part of Salamina and Kithira islands.

5. Discussion

Using RES and GIS techniques, the landslide susceptibility of the Attica Region was assessed by correlating ten parameters and producing the final susceptibility map for the whole Attica peninsula (with its islands included) in Greece. The validity of this approach was tested using the slope failures that were recorded during the last sixty years in this region. In particular, 68% of the recorded 220 slope failures were found to be in the "Extremely high susceptibility" and 21% in the "Landslide" zone respectively of the developed map. Studying more carefully this map, some more remarks can be extracted.

Initially, it is shown in the susceptibility map that slope failure incidents are located mostly in areas where Neogene and Quaternary sediments outcrop. Secondarily, slope failures are associated with carbonate rocks basically due to rockfalls. In order to preliminary assess the potential landslide risk in respect to settlements, the villages and cities at the study area were plotted on the susceptibility map (Figure 11).

This correlation suggests that 16 settlements are entirely located within "Landslide" and 201 urban areas are in the "Extremely high landslide susceptibility" zone. To be more specific, in the "Landslide" zone, places such as Chalkoutsi, Grammatiko, Kato Alepochori, Schinos can be found. In the "Extremely high landslide susceptibility" zone, characteristic sites are Mesagros (Aegina Island), Varnavas, Galaniana (Antikithira island).

In addition, many defined slope failure areas are associated with the existence of faults. This result should be taken into consideration by public authorities responsible for the construction of public technical works, regarding urban planning and design of new infrastructure projects (e.g., highways, tunnels, major buildings).

According to the generated susceptibility map, areas associated with an instability index greater than 70% are located in many sites around the Attica region (islands included). For that reason, public authorities responsible for civil protection need to get advice from such maps to make emergency plans at different administrative levels, useful for the pre-event of the landslide risk management cycle.

Moreover, the landslide susceptibility map can be used with the already produced potential highly flood hazard zoning maps of Attica Region authorized by the Greek Ministry of Environment and Energy, and with the produced flooded area maps, delivered by the Copernicus Emergency Management Service-Mapping.

Concerning localities that were affected by catastrophic forest fires in previous years such as those of Kineta (2018) and Mati (2018) (Figure 11) [38], and studying the generated susceptibility map, it is realized that such fires can cause in the immediate future "secondary" hazards like earth slides, debris flows and flash floods. Those two areas are associated with an instability index greater than 53% and this means that the drastically changed environmental conditions due to the fires may increase the landslide activity in the area in the near future.

Finally, it should be pointed out that the developed susceptibility map is at a regional scale (1:100,000) and its practical use is to be applied in conjunction with site-specific work, from experts such as experienced geologists, geotechnical engineers before development takes place. Additionally, it should be mentioned that even though susceptibility analysis does not define either the time and the type of the failure, or the volume of the mass involved, it is necessary for the estimation of hazard and risk index and zoning, respectively.

For all these reasons, the applied methodology (RES and Oregon Protocol methodology) should be accompanied each time by the appropriate fieldwork as well as the necessary geotechnical desk study, so as to acquire the most accurate geological model of the ad hoc examined area susceptible to slope failure [55].

6. Conclusions

This study presents the landslide susceptibility analysis for Attica Region, which is the most densely populated area in Greece. The produced susceptibility map is a cartographic product in a regional scale (1:100,000) generated for the Attica county via a semi-quantitative heuristic methodology named Rock Engineering System and a prototype technique originally developed by the Oregon Department of Geology (USA). To the author's knowledge, this is the first time that such an in-depth analysis has been conducted for the whole of Attica county. Furthermore, for the compilation of this map, RES methodology was applied as a simple and fast tool for the calculation of the instability index of each examined slope failure recorded in a well-organized geodatabase according to the EU Inspire Directive.

Considering the mentioned previously, it should be noted that 68% of the locations of actual and potential landslides correspond to the "Extremely high" and 21% are associated with a landslide. Responding to the previous remark, particular sites in Northeastern Attica (e.g., Kapandriti, Varnavas, Oropos, Kalamos), historical slopes in Western Attica (such as those of Alepochori-Psatha, Alepochori-Schinos, Kakia Skala), the most well-known historical landslide of Malakasa, characteristic places in Attica islands (e.g., Kithira-Kapsali, Aegina-Kakoperato, Salamina-Porto Fino) were validated through the above–mentioned methodology and it was found that all of them were confirmed as landslides (Figure 11). Furthermore, this correlation suggests that 16 settlements are entirely located within "Landslide" and 201 urban areas are in the "Extremely high landslide susceptibility" zone.

As in Section 3.2 is mentioned, RES methodology was applied in different physiographic environments with a variety of geological and tectonic settings and scales. In the present study, the previous statement was confirmed by implementing RES in an area with complex geological settings (e.g., active faults, many different streams based on Strahler classification as well as a variation of geological formations). Thus, it is suggested that this procedure (i.e., RES, GIS techniques, Oregon protocol-Special Paper 42) could be used in other regions with different geological environments and tectonic characteristics.

Summarizing, the DIAS geodatabase represents the spatial distribution of over 300 landslides (rockfalls, falls, erosion lines included) based on published and unpublished information, field observations and remote sensing techniques. The intention is that the database should be updated constantly. The outcome of the DIAS project will be accessible to the public, through a web-based platform using an open-source G.I.S. software so as to aid awareness of landslides among different stakeholders (e.g., landslide experts, government agencies, planners, citizens). Moreover, the DIAS project can facilitate the role of Civil Protection Authorities, by providing inputs for prevention and preparedness.

Taking into consideration the previous outcomes, the upcoming steps of this research (DIAS project) will be the generation of hazard and risk maps using triggering dynamic factors like earthquake and rainfall data, as well as different elements of risk, respectively, in specific areas.

Author Contributions: Conceptualization, N.T. and G.P.; methodology, N.T. and G.P.; software, N.T. and P.A.; validation, N.T., G.P. and A.G.; formal analysis, N.T., G.P. and A.G.; investigation, N.T.; resources, N.T., G.P. and A.G.; data curation, N.T., G.P., A.G. and P.A.; writing—original draft preparation, N.T.; writing—review and editing, N.T., G.P., A.G. and P.A.; visualization, N.T.; supervision, G.P., A.G.; project administration, N.T., G.P.; funding acquisition, N.T., G.P., A.G. and P.A. All authors have read and agreed to the published version of the manuscript.

Funding: This research was co-funded by Greece and the European Union (European Social Fund-ESF) through the Operational Program "Human Resources Development, Education and Lifelong Learning 2014–2020" in the context of the project "Landslide Risk Assessment of Attica Region" MIS (5050327).

Institutional Review Board Statement: Not applicable.

Informed Consent Statement: Not applicable.

Acknowledgments: The authors are grateful to the Greek Ministry of Environment, Region of Attica, Hellenic Survey for Geology and Mineral Exploration (H.S.G.M.E.), General Secretary of Civil Protection of Greece and Greek Cadastre S.A. for providing valuable technical landslide reports as well as crucial digital geodata records. Furthermore, rainfall data were provided by the Institute for Environmental and Sustainable Development Research (IEPBA) of the National Observatory of Athens.

Conflicts of Interest: The authors declare no conflict of interest. The funders had no role in the design of the study; in the collection, analyses, or interpretation of data; in the writing of the manuscript, or in the decision to publish the results.

References

1. Corominas, J.; Van Western, C.; Frattini, P.; Cascini, L.; Malet, J.; Fotopoulou, S.; Catani, F.; Van Den Eeckhaut, M.; Mavrouli, O.; Agliardi, F.; et al. Recommendations for the quan- titative analysis of landslide risk. *Bull. Eng. Geol. Environ.* **2014**, *73*, 209–263.
2. Brabb, E. Innovative approaches to landslide hazard and risk mapping. In Proceedings of the IVth ISL, Toronto, ON, Canada, 23–31 August 1984; Volume 1, pp. 307–324.
3. Papathanassiou, G.; Valkaniotis, S.; Ganas, A.; Pavlides, S. GIS-based statistical analysis of the spatial distribution of earthquake-induced landslides in the island of Lefkada, Ionian Islands, Greece. *Landslides* **2013**, *10*, 771–783. [CrossRef]
4. Bonham-Carter, G.F. Geographic Information System for Geoscientists: Modelling with GIS. In *Computer Methods in the Geosciences*; Pergamon Press: Oxford, UK, 1994; Volume 3.
5. Chung, C.J.; Fabbri, A.; Van Westen, C.J. Multivariate regression analysis for landslide hazard zonation. In *Geographical Information Systems in Assessing Natural Hazards*; Carrara, A., Guzetti, F., Eds.; Kluwer: Alphen aan den Rijn, The Netherlands, 1995; pp. 107–133.
6. Van Westen, C.J.; Rengers, N.; Soeters, R. Use of geomorphological information in indirect landslide susceptibility assessment. *Nat. Hazards* **2003**, *30*, 399–419. [CrossRef]
7. Hudson, J. *Rock Engineering Systems: Theory and Practice*, 1st ed.; Ellis Horwood Limited: Chichester, West Sussex, UK, 1992; pp. 1–185.
8. Marinos, P.; Yannatos, M.; Sotiropoulos, E.; Cavounidis, S. Increasing the stability of a failed slope by pumping, Malakasa Landslide, Athens, Greece. In *Engineering Geology and the Environment*; Marinos, P., Koukis, G., Tsiambaos, G., Stoumaras, G., Eds.; Balkema: Rotterdam, The Netherlands, 1997; pp. 853–857.
9. Varnes, D.; IAEG Commission on Landslides and Other Mass-Movements. *Landslide Hazard Zonation: A Review of Principles and Practice*; UNESCO Press: Paris, France, 1984; 63p.

10. Lozios, S. Tectonic Analysis of the Metamorphic Rocks in NE Attica. Ph.D. Thesis, Department of Geology, University of Athens, Athens, Greece, 1993; 299p. (In Greek).
11. Mettos, A.; Ioakim, C.; Rondoyanni, T. *Palaeoclimatic and Palaeogeographic Evolution of Attica Beotia (Central Greece)*; Special Publication 9; Geological Society of Greece: Athens, Greece, 2000; pp. 187–196.
12. von Freyberg, B. *Das Neogen Gebiet Nordwestlich Athen*; Publication of Subsurface Research Department, Ministry of Coordination: Athens, Greece, 1951.
13. Papanikolaou, D.; Papanikolaou, I. Geological, geomorphological and tectonic structure of NE Attica and seismic hazard implications for the northern edge of the Athens plain. *Bull. Geol. Soc. Greece* **2007**, *40*, 425–438. [CrossRef]
14. Krohe, A.; Mposkos, E.; Diamantopoulos, A.; Kaouras, G. Formation of basins and mountain ranges in Attica (Greece): The role of Miocene to Recent low-angle normal detachment faults. *Earth-Sci. Rev.* **2010**, *98*, 81–104. [CrossRef]
15. Ganas, A.; Pavlides, S.B.; Sboras, S.; Valkaniotis, S.; Papaioannou, S.; Alexandris, G.A.; Plessa, A.; Papadopoulos, G.A. Active Fault Geometry and Kinematics in Parnitha Mountain, Attica, Greece 2004. *J. Struct. Geol.* **2004**, *26*, 2103–2118. [CrossRef]
16. Ganas, A.; Pavlides, S.; Karastathis, V. DEM-based morphometry of range-front escarpments in Attica, central Greece, and its relation to fault slip rates. *Geomorphology* **2005**, *65*, 301–319. [CrossRef]
17. Ambraseys, N.N.; Jackson, J.A. Seismicity and associated strain of central Greece between 1890 and 1988. *Geophys. J. Int.* **1990**, *101*, 663–708. [CrossRef]
18. Chousianitis, K.; Ganas, A.; Gianniou, M. Kinematic interpretation of present-day crustal deformation in central Greece from continuous GPS measurements. *J. Geodyn.* **2013**, *71*, 1–13. [CrossRef]
19. Papanikolaou, D.; Mariolakos, I.; Lekkas, E.; Lozios, S. Morphotectonic observations on the Asopos Basin and the coastal zone of Oropos. Contribution on the neotectonics of Northern Attica. *Bull. Geol. Soc. Greece* **1988**, *20*, 252–267.
20. Goldsworthy, M.; Jackson, J.; Haines, J. The continuity of active fault systems in Greece. *Geophys. J. Int.* **2002**, *148*, 596–618. [CrossRef]
21. Grützner, C.; Schneiderwind, S.; Papanikolaou, I.; Deligiannakis, G.; Pallikarakis, A.; Reicherter, K. New constraints on extensional tectonics and seismic hazard in northern Attica, Greece: The case of the Milesi Fault. *Geophys. J. Int.* **2016**, *204*, 180–199. [CrossRef]
22. Deligiannakis, G.; Papanikolaou, I.D.; Roberts, G. Fault specific GIS based seismic hazard maps for the Attica region, Greece. *Geomorphology* **2018**, *306*, 264–282. [CrossRef]
23. Casagli, N.; Catani, F.; Puglisi, C.; Delmonaco, G.; Ermini, L.; Margottini, C. An inventory-based approach to landslide susceptibility assessment and its application to the Virginio River Basin, Italy. *Environ. Eng. Geosci.* **2004**, *10*, 203–216. [CrossRef]
24. Burns, W.; Madin, I. *Protocol for Inventory Mapping of Landslide Deposits from Light Detection and Ranging (LiDAR) Imagery*; Special paper 42; Oregon Department of Geology and Mineral Industries: Portland, OR, USA, 2009.
25. Fell, R.; Lacerda, W.; Cruden, D.M.; Evans, S.G.; LaRochelle, P.; Martinez, F.; Watanabe, M. A suggested method for reporting a landslide. *Bull. Int. Assoc. Eng. Geol.* **1990**, *41*, 5–12.
26. Varnes, D.J. Slope movement types and processes. In *Landslides, Analysis and Control, Special Report 176: Transportation Research Board*; Schuster, R.L., Krizek, R.J., Eds.; National Academy of Sciences: Washington, DC, USA, 1978; pp. 11–33.
27. Hudson, J. *A Review of Rock Engineering Systems (RES) Applications over the Last 20 Years*; Department of Earth Science and Engineering, Imperial College: London, UK, 2013.
28. Rafiee, R.; Mohammadi, A.; Ataei, M.; Khalookakaie, R. Application of fuzzy RES and fuzzy DEMATEL in the rock behavioral systems under uncertainty. *Geosyst. Eng.* **2018**, *22*, 1–12. [CrossRef]
29. Wang, J.; Deng, Y.; Wu, L.; Liu, X.; Yin, Y.; Xu, N. Estimation model o sandy soil liquefaction based on RES model. *Arab. J. Geosci.* **2018**, *11*, 565. [CrossRef]
30. Ferentinou, M.; Fakir, M. Integrating Rock Engineering Systems device and Artificial Neural Networks to predict stability conditions in an open pit. *Eng. Geol.* **2018**, *246*, 293–309. [CrossRef]
31. Elmouttie, M.; Dean, P. Systems Engineering Approach to Slope Stability Monitoring in the Digital Mine. *Resources* **2020**, *9*, 42. [CrossRef]
32. Rozos, D.; Tsagaratos, P.; Markantonis, C.; Skias, S. An application of Rock Engineering System (R.E.S.) method for ranking the instability potential of natural slopes in Achaia County, Greece. In Proceedings of the XIth International Congress for Mathematical Geology, Liege, Belgium, 3–8 September 2006.
33. Rozos, D.; Bathrellos, G.; Skillodimou, H. Comparison of the implementation of rock engineering system and analytic hierarchy process methods, upon landslide susceptibility mapping, using GIS: A case study from the Eastern Achaia County of Peloponnesus, Greece. *Environ. Earth Sci.* **2011**, *63*, 49–63. [CrossRef]
34. Tavoularis, N.; Koumantakis, I.; Rozos, D.; Koukis, G. The contribution of landslide susceptibility factors through the use of Rock Engineering System (RES) to the prognosis of slope failures. An application in Panagopoula and Malakasa landslide areas in Greece. *Geotech. Geol. Eng. Int. J.* **2017**, *36*, 1491–1508. [CrossRef]
35. Tavoularis, N.; Koumantakis, I.; Rozos, D.; Koukis, G. Landslide susceptibility mapping using Rock Engineering System approach and GIS technique: An example from southwest Arcadia, Greece. *Eur. Geol. J.* **2017**, *44*, 19–27. Available online: https://eurogeologists.eu/tavoularis-landslide-susceptibility-mapping-using-rock-engineering-system-approach-gis-technique-example-southwest-arcadia-greece/ (accessed on 7 December 2017).
36. Ayalew, L.; Yamagishi, H. The application of GIS-based logistic regression for landslide susceptibility mapping in the Kakuda-Yahiko Mountains, Central Japan. *Geomorphology* **2005**, *65*, 15–31. [CrossRef]

37. Koukis, G.; Ziourkas, C. Slope instability phenomena in Greece: A statistical analysis. *Bull. IAEG* **1991**, *43*, 47–60. [CrossRef]
38. Tavoularis, N.; Argyrakis, P.; Papathanassiou, G.; Ganas, A. Towards the development of an updated GIS-based landslide hazard map in Attica region, Greece; DIAS project. In Proceedings of the SafeGreece 2020 7th International Conference on Civil Protection & New Technologies, 14–16 October 2020. Available online: https://safegreece2020_proceedings.pdf (accessed on 13 November 2017).
39. Popescu, M.E. A suggested method for reporting landslide causes. *Bull. Int. Assoc. Eng. Geol.* **1994**, *50*, 71–74. [CrossRef]
40. Dai, F.C.; Lee, C.F.; Ngai, Y.Y. Landslide risk assessment and management: An overview. *Eng. Geol.* **2002**, *64*, 65–87. [CrossRef]
41. Greek Ministry of Environment, Energy and Climate Change/Special Secretariat. *Water River Basin Management Plans*; Greek Ministry of Environment, Energy and Climate Change/Special Secretariat: Athens, Greece, 2012.
42. Founda, D.; Giannakopoulos, C.; Pierros, F.; Kalimeris, A.; Petrakis, M. Observed and projected precipitation variability in Athens over a 2.5 century period. *Atmos. Sci. Lett.* **2013**, *14*, 72–78. [CrossRef]
43. Lalioti, V.; Spanou, N. Recording, analysis and evaluation of landslides phenomena in Greek territory during the period 1991–1998. Master's Thesis, Sector of Engineering Geology, Geology Department, University of Patras, Patras, Greece, 2001.
44. Greenway, D.R. Vegetation and slope stability. In *Slope Stability*; Anderson, M.G., Richards, K.S., Eds.; Wiley: New York, NY, USA, 1987; pp. 187–230.
45. Gokceoglou, C.; Aksoy, H. Slide susceptibility mapping of the slopes in the residual soils of the Mengen Region (Turkey) by deterministic stability analyses and image-processing techniques. *Eng. Geol.* **1996**, *44*, 147–161. [CrossRef]
46. Kanungo, D.P.; Arora, M.K.; Sarkar, S.; Gupta, R.P. A comparative study of conventional, ANN black box, fuzzy and combined neural and fuzzy weighting procedures for landslide susceptibility zonation in Darjeeling Himalayas. *Eng. Geol.* **2006**, *85*, 347–366. [CrossRef]
47. Gemitzi, A.; Eskioglou, P.; Falalakis, G.; Petalas, C. Evaluating landslide susceptibility using environmental factors, fuzzy membership functions and GIS. *Glob. Nest J.* **2011**, *13*, 28–40.
48. Foumelis, M.; Lekkas, E.; Parcharidis, I. Landslide susceptibility mapping by GIS-based qualitative weighting procedure in Corinth area. *Bull. Geol. Soc. Greece* **2004**, *36*, 904–912. [CrossRef]
49. Ganas, A.; Tsironi, V.; Tsimi, C.; Delagas, M.; Konstantakopoulou, E.; Kollia, E.; Efstathiou, E.; Oikonomou, A. *NOAFaults v3.0: New Upgrades of the NOA Geospatial Database of Active Faults in the Broader Aegean Area*; Institute of Geodynamics, National Observatory of Athens: Athens, Greece, 2020.
50. Brabb, E.; Bonilla, M.G.; Pampeyan, E. *Landslide Susceptibility in San Mateo County, California*. US Geological Survey Miscellaneous Field Studies, Map MF-360, Scale 1:62,500; US Geological Survey: Reston, VA, USA, 1972; reprinted in 1978.
51. Chung, C.F.; Fabbri, A.G. Validation of Spatial Prediction Models for Landslide Hazard Mapping. *Nat. Hazards* **2003**, *30*, 451–472. [CrossRef]
52. Data School. Simple Guide to Confusion Matrix Terminology, 25 March 2014. Available online: https://dataschool.io (accessed on November 2017).
53. Hellenic Survey of Geology and Mineral Exploration. *Geological Investigation of 2017 Mandra Flooding and Proposed Mitigation Measures*; Unpublished Technical Report; 2018.
54. Van Westen, C.J. *National Scale Landslide Susceptibility Assessment for SVG. CHARIM Caribbean Handbook on Risk Information Management*; World Bank Group GFDRR, ACP-EU Natural Risk Reduction Program; 2016; Available online: www.charim.net (accessed on November 2017).
55. Tavoularis, N.; Koumantakis, I.; Rozos, D.; Koukis, G. An implementation of rock engineering system (RES) for ranking the instability potential of slopes in Greek territory. An application in Tsakona area (Peloponnese—Prefecture of Arcadia). *Bull. Geol. Soc. Greece* **2015**, *XLIX*, 38–58. [CrossRef]

Article

Integrating Multivariate (GeoDetector) and Bivariate (IV) Statistics for Hybrid Landslide Susceptibility Modeling: A Case of the Vicinity of Pinios Artificial Lake, Ilia, Greece

Christos Polykretis *, Manolis G. Grillakis, Athanasios V. Argyriou, Nikos Papadopoulos and Dimitrios D. Alexakis

Laboratory of Geophysical—Satellite Remote Sensing and Archaeo-Environment (GeoSat ReSeArch Lab), Institute for Mediterranean Studies (IMS), Foundation for Research and Technology—Hellas (FORTH), 74100 Rethymno, Greece; grillakis@hydrogaia.gr (M.G.G.); nasos@ims.forth.gr (A.V.A.); nikos@ims.forth.gr (N.P.); dalexakis@ims.forth.gr (D.D.A.)
* Correspondence: polykretis@ims.forth.gr; Tel.: +30-283-110-6251

Citation: Polykretis, C.; Grillakis, M.G.; Argyriou, A.V.; Papadopoulos, N.; Alexakis, D.D. Integrating Multivariate (GeoDetector) and Bivariate (IV) Statistics for Hybrid Landslide Susceptibility Modeling: A Case of the Vicinity of Pinios Artificial Lake, Ilia, Greece. *Land* **2021**, *10*, 973. https://doi.org/10.3390/land10090973

Academic Editors: Enrico Miccadei, Cristiano Carabella and Giorgio Paglia

Received: 20 August 2021
Accepted: 13 September 2021
Published: 15 September 2021

Publisher's Note: MDPI stays neutral with regard to jurisdictional claims in published maps and institutional affiliations.

Copyright: © 2021 by the authors. Licensee MDPI, Basel, Switzerland. This article is an open access article distributed under the terms and conditions of the Creative Commons Attribution (CC BY) license (https://creativecommons.org/licenses/by/4.0/).

Abstract: Over the last few years, landslides have occurred more and more frequently worldwide, causing severe effects on both natural and human environments. Given that landslide susceptibility (LS) assessments and mapping can spatially determine the potential for landslides in a region, it constitutes a basic step in effective risk management and disaster response. Nowadays, several LS models are available, with each one having its advantages and disadvantages. In order to enhance the benefits and overcome the weaknesses of individual modeling, the present study proposes a hybrid LS model based on the integration of two different statistical analysis models, the multivariate Geographical Detector (GeoDetector) and the bivariate information value (IV). In a GIS-based framework, the hybrid model named GeoDIV was tested to generate a reliable LS map for the vicinity of the Pinios artificial lake (Ilia, Greece), a Greek wetland. A landslide inventory of 60 past landslides and 14 conditioning (morphological, hydro-lithological and anthropogenic) factors was prepared to compose the spatial database. An LS map was derived from the GeoDIV model, presenting the different zones of potential landslides (probability) for the study area. This map was then validated by success and prediction rates—which translate to the accuracy and prediction ability of the model, respectively. The findings confirmed that hybrid modeling can outperform individual modeling, as the proposed GeoDIV model presented better validation results than the IV model.

Keywords: landslides; susceptibility; hybrid modeling; Geographical Detector; information value; Greece

1. Introduction

A landslide is a gravity-driven environmental process which involves the movement of rocks, debris, earth, or a combination of them down a slope [1]. According to official data, landslides constituted the third (after floods and storms, and before earthquakes) most frequent natural disaster worldwide in 2020 [2]. Generally, the extreme weather events due to climate change, and the high seismic activity in combination with the poorly planned expansion of human activities (deforestation of slopes, uncontrolled irrigation, etc.), have contributed to a global upward tendency in landslide occurrence in the recent years [3].

Due to their occurring without warning and seriously threatening both natural and human environments, landslides are a major problem. Due to severe damage, or even destruction, of infrastructure and properties, they generate larger annual economic losses (billions euro) than any other natural disaster in many countries. In addition, a considerable number of people each year are injured and, in some cases, killed by them. It is indicative that during 1998–2017, totally 4.8 million people were affected by landslides worldwide, with 18,414 of them being killed [4]. In addition, the environmental effects of landslides are mainly changes in terrain morphology, and increased sediment loads in rivers and subsequent transport to dams.

The increased frequency of landslides and the severity of their effects have led to growing interest from international scientific community. Since predictions of occurrence and intensity remain challenging, most of the attention has been given to the determination of potential spatial locations. The acquisition of this spatial information can be achieved through landslide susceptibility (LS) assessments and mapping. LS refers to the potential landslide activity as a result of terrain conditions [5]. An assessment depends on the spatial distribution of past landslides in an area and their relation to its terrain conditions, in order to generate spatial predictions for areas that are not landslide-affected but have similar conditions. The output is a map presenting the region of interest divided into homogeneous zones of susceptibility [6]. LS maps with high levels of accuracy and reliability are considered crucial tools that can then be used as inputs for disaster management plans.

The advancements in the geospatial tools of geographic information systems (GIS) and remote sensing (RS), assisted by improvements in computer processing power, have improved LS modeling over the last few decades. Based on the literature, a considerable number of models are currently available for assessing LS at different spatial scales. In terms of degree of objectivity and necessity for landslide occurrence data, all these models can be separated into two different groups, the qualitative and quantitative models. The qualitative (or semi-quantitative) models estimate a susceptibility score on the basis of weights assigned to landslide conditioning factors from one or more expert(s). They suffer from low objectivity associated with the experts' subjective judgements [7]. On the other hand, the quantitative models decrease bias in the weight assignments, since they depend on fixed mathematical rules, regardless of any expert judgement [8]. Particularly, the impacts of different conditioning factors on past occurrences are quantitatively determined, resulting in high objectivity.

The current capability for acquiring multi-temporal landslide occurrence data through RS-based approaches has led to wide use of the data-driven quantitative models. These models range from complicated geotechnical and advanced machine learning models to more conventional statistical analysis models. Based on mechanical laws for the calculation of a safety factor, the geotechnical models [9,10] examine the slope stability from the perspective of the mechanical properties of the slope. Being based on human learning procedures, machine learning models are used to solve problems characterized by nonlinear functions and data. Commonly applied machine learning models are artificial neural networks (ANN), support vector machines (SVM), random forests (RF) and decision trees (DT) [11–13].

Regarding statistical analysis models, their fundamental principle is to estimate the probability of a landslide under the existence of spatial associations between the conditioning factors and past landslides [14]. Depending on the examination of factors individually or cumulatively, they can be either bivariate or multivariate. In bivariate modeling, weights are calculated for the classes of each individual factor by their levels of association with landslides in a historic dataset. Frequency ratio (FR), information value (IV) and weights of evidence (WoE) constitute the main representatives of bivariate models [15,16]. Conversely, in multivariate modeling, all the factors are sampled, and the presence or absence of landslide is determined for each of the sampling units [17]. Then, weights are calculated for the factors via statistical means. Among the multivariate models, logistic regression (LR) is doubtless the most used [18,19]. However, models such as LR consider the factors as explanatory variables without taking into account the spatial information contained in them and exploring their impacts on landslide occurrence (dependent variable) from a spatial perspective. In order to overcome this limitation, new spatially-based multivariate models have been put forward recently. These models can address the specificities of each space and consider that spatial variations in landslides may cause different responses to variations in the factor variables. Such a model is the Geographical Detector (GeoDetector). Although GeoDetector has been tested in various studies of health, social and environmental sciences [20–22] over the last few years, its use in landslide-related research has been quite limited. Since it provides an effective way to identify and eliminate redundant variables, GeoDetector has been used in a few relevant studies [14,23,24] for factor selection purposes.

In general, all the quantitative models have been proven beneficial for identifying locations that are prone to landslides; however, some shortcomings still characterize them. The geotechnical models require detailed mechanical data of soil or rock, and as a result they are only suitable for studying small regions or single slopes. Although the statistical models are easy to understand and perform well in most cases, they find it difficult to solve situations with large amounts of data. Moreover, despite their ability to handle large amounts of nonlinear data, machine learning models are not significantly better than the statistical ones, and cannot perform well under different conditions and in different areas [25]. In order to produce the most reliable LS map for a region of interest, one possible solution is to compare different models and select the optimum in terms of accuracy and prediction ability. Several studies have compared two or more different models to recognize the most suitable for a specific region [26–28].

The aforementioned shortcomings tend to increase the uncertainty and reduce the efficiency of models when applied individually. Thus, another solution has gained popularity recently, which is the development of hybrid models. Hybrid modeling can resolve the shortcomings of individual models and improve performance. This type of modeling has been gradually applied in LS assessment studies over the last decade. For instance, in the work of Arabameri et al. [8], the efficiency of the integration of statistical (FR) and machine learning (RF) models was explored for LS mapping in northern Iran. For assessing the LS in a region of India, Saha et al. [29] integrated a statistical and a machine learning model to improve on their individual accuracies. Chen et al. [30] applied a combination of bivariate (WoE) and multivariate (LR) statistical models with a machine learning model (RF) for LS mapping of a mountainous region of China. Roy et al. [31] delineated LS zones in districts of India by integrating bivariate statistical (WoE) and machine learning (SVM) models. Chowdhuri et al. [32] introduced hybrid models from statistical and machine learning model integrations for predicting spatially the landslide occurrence in a basin of India. In addition, some studies have improved the performances of machine learning models by combining them with optimization or meta-heuristic algorithms [33,34].

In Greece, landslide activity has been highly facilitated by the frequent occurrence of intense rainfall and seismic events. Along with them, its complex geo-morphological settings (strained geological formations and steep slopes) and the uncontrolled land-use in landslide-prone areas have contributed. As a result, the interest in and awareness of the importance of LS assessments for regions of Greece have increased, particularly over the last decade. However, the majority of relevant studies has focused on the implementation of individual statistical and machine learning models [35–37], rather than integrated approaches. It could be mentioned that the work of Chalkias et al. [38] constitutes an exception.

The region of Peloponnese has experienced severe natural disasters, including floods, earthquakes, landslides and wildfires. Specifically, landslides have highly damaged settlements within its boundaries (mainly in its northern and western parts), resulting in partial destruction and necessary re-locations to nearby geologically stable lands. Considering all the above, the present study aimed to assess the LS and create a reliable map of a wetland in northwestern Peloponnese. Therefore, a hybrid LS modeling is proposed based on the integration of two different statistical models, the multivariate GeoDetector and bivariate IV. Past landslide occurrence and conditioning factor datasets were incorporated into the hybrid model, named GeoDIV, and analyzed in a GIS environment to determine the spatial distribution of susceptibility. In order to confirm the targeted reliability of LS map, the performance of proposed GeoDIV model was compared with that of the individual IV model in a validation procedure.

2. Study Area

The surrounding area of the Pinios artificial lake was selected for investigation in this study. It is located in the western part of Greece and the northwestern part of geographical area of Peloponnese (Figure 1), covering a total extent of approximately 239 km². It belongs administratively to the Prefecture of Ilia and hydrologically to the drainage basin of Pinios River. The boundaries of the study area are defined in the north and south by the basin's boundaries, and in the west and east by altitude contours of 100 and 200 m, respectively. The Pinios artificial lake was created in 1960, after the construction of a dam on the homonym river, and is the largest in Peloponnese (with a total extent of approximately 20 km²). Its water is used for the irrigation of the plain of Ilia, and hence it is considered one of the most important land improvement projects in the entire prefecture. The total quantity of water withdrawn from the lake annual for irrigation and water supply purposes amounts to 126 million m³.

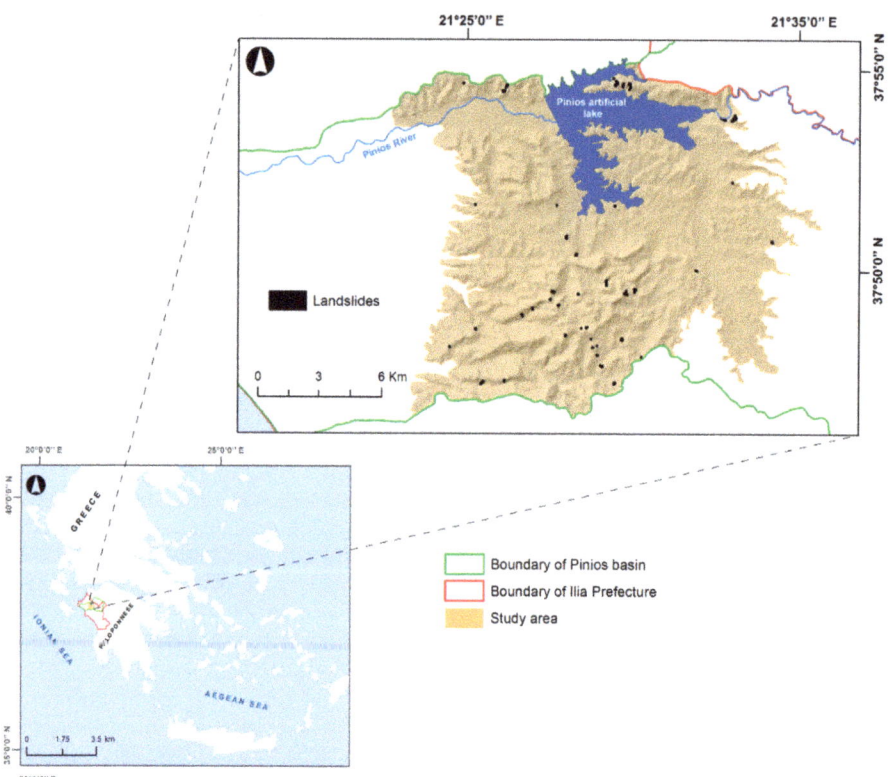

Figure 1. The study area and the locations of events from the landslide inventory.

Following the typical landscape of Ilia Prefecture, the study area can be characterized as an agricultural region at a low altitude (mean altitude at 154 m above sea level). Heterogeneous croplands or fields mixed with natural vegetation represent the predominant agricultural lands. More than 30 settlements are situated within its boundaries, containing 5400 inhabitants according to the official 2011 census [39].

The climate is Mediterranean mild with a mean temperature ranging from 20 to 25 °C in the summer months, and from 4 to 10 °C in the winter months [40]. Long-term rainfall records including the period of the last two decades show a mean annual value reaching approximately 500 mm. From a geological perspective, the study area is mainly covered by

Neogene and Quaternary loose deposits varying in thickness and consistency. Confined granular aquifer systems have been formed inside alluvial deposits, and unconfined aquifers have been developed in Quaternary deposits where groundwater flows to the direction of the sea [41].

3. Data and Methods

In this study, a hybrid model was developed for LS assessment based on the integration of two different individual models, the GeoDetector and IV. A spatial database was created in GIS to be used in hybrid modeling, including: (a) the landslide inventory dataset and (b) the conditioning factor datasets.

3.1. Landslide Inventory

Such a dataset provides information about the landslide events that occurred in the past in a given region. Hence, this information is crucial for any quantitative LS modeling effort. A database maintained by the Laboratory of Engineering Geology at the Department of Geology at University of Patras referring to landslides that occurred between 2000 and 2015 [42], and field surveys, were initially exploited for the spatial locations of past landslides in the study area. Then, multi-temporal Google Earth satellite imagery (Figure 2) was used for their delimitation. Based on the classification proposed by Varnes [1], for this study, the term landslide included shallow debris flows and earth rotational slides, varying in extent from some hundreds to several thousands of meters squared (Table 1). Since it is not always possible to differentiate the depletion and accumulation zones of these landslide types in an inventory map [18], these zones were mapped together in an entire area forming a single polygon feature for each landslide. Therefore, 60 landslide polygons were eventually represented in the relevant inventory map (Figure 1).

Table 1. Types and basic morphometrical parameters of the landslides in the study area.

Landslide Type	Amount of Events	% of Total Landslide Events	Area (sq. m)			Altitude (m)			Slope Angle (Degrees)		
			Max	Min	Mean	Max	Min	Mean	Max	Min	Mean
Debris flows	40	67	4187	103	1268	348	110	222	60	15	26
Earth rotational slides	20	33	18,000	240	4188	332	89	199	45	15	30

3.2. Conditioning Factors

Landslide occurrence is considered to be affected by a variety of natural and anthropogenic factors representing the conditions of a given region. These conditioning factors can be separated into two main categories: (a) the preparatory factors which create suitable conditions for a landslide by changing the state of a slope from stable to marginally stable, and (b) the triggering factors which initiate a landslide by changing the state of a slope from marginally stable to unstable [43]. Morphological and hydro-lithological conditions of the region of interest are represented by natural preparatory factors, whereas the human interventions on it are represented by anthropogenic preparatory factors. The triggering factors mainly represent climatic and seismic conditions related to rainstorms and earthquakes, respectively.

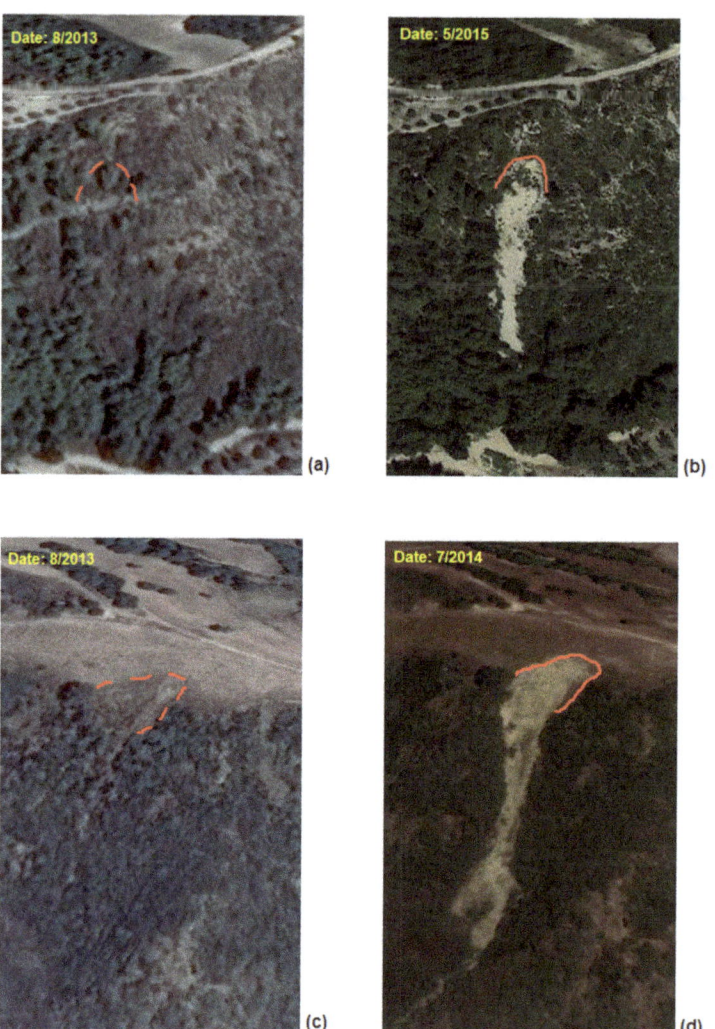

Figure 2. Multi-temporal Google Earth images: (**a,c**) before the landslides; (**b,d**) after the landslides. The red dashed lines indicate the location of the landslide before it happened, and the red solid line shows the scar of the landslide after it occurred.

Since no official guidelines are used by the scientific community for the selection of factors, the characteristics of the study area, data availability and a literature review [29,30] were taken into account for this study. In total, fourteen conditioning factors were selected, including both preparatory and triggering factors. In particular, the altitude, slope angle, slope aspect, profile curvature, plan curvature, stream density, stream power index (SPI), topographical wetness index (TWI), lithology, proximity to faults and soil type were used as natural preparatory factors; the land use/cover and proximity to roads were used as anthropogenic preparatory factors, and the mean annual rainfall was used as a triggering factor.

Defined as the height above a reference point (typically above the mean sea level), altitude is an important conditioning factor due to its gravitational potential energy. In general, the higher the slope angle is, the higher the likelihood of failure. Therefore, steep slopes are more prone to failures. The slope aspect is defined as the azimuth-based orientation of terrain and is highly related to exposure to sunlight; evapotranspiration; and rainfall's effects on weathering, soil, vegetation cover and root development [44]. Expressed by different types, such as plan and profile, the curvature indicates the runoff and erosion factors of water. The plan curvature is perpendicular to the maximum slope direction, whereas the profile curvature is parallel to the same direction [45]. By retaining more rainfall water and erosion-induced sediment than convex slopes, concave slopes are correlated with higher likelihoods of failure.

Considering its effects on groundwater recharge, stream density constitutes another important factor for landslide activity. This factor determines the ratio of the total length of streams to the extent of the study area. A high stream density is linked to low surface water infiltration and thus mass movements with high velocity [46]. SPI is another hydrological factor that measures the erosive power of the streams. On the other hand, TWI quantifies the moisture content of the surface [32].

Lithology is one of the most crucial factors for LS assessments, since different lithological formations have different slope instability performances in terms of strength and permeability. In a tectonically active country such as Greece, the faults seem to be associated with extensive fractured zones and steep relief anomalies presenting favorable conditions for landslides [35]. Hence, landslides are usually found in proximity to faults. Additionally, different soil types can have different impacts on surface infiltration and groundwater flow, depending on their particular physical and mechanical properties [47].

Changes in land use/cover as a result of human activities such as cultivation, deforestation and forest logging can significantly affect the occurrence of landslides. Proximity to roads can also reflect the human impact on landslides, as road construction at the base of a slope tends to degrade its stability.

Rainfall—causing an increase in the pore water pressure and a reduction in the shear strength of the soil [48]—is a basic triggering mechanism for not only the development of new landslides but also the re-activation of old ones. Particularly in Greece, rainfall-triggered landslides are among the most frequent and devastating disasters [38]. It is worth mentioning that since the majority of earthquakes that occurred in the study area during the last two decades were characterized by relatively low magnitudes (with M_w between 3.0 and 3.5) and great depth (greater than 15 km) [49], seismic factor was not included in the analysis.

As is shown in Table 2, all the above conditioning factors were represented by GIS-supported data formats. Most of them were in raster format (grids), but others were converted from vector (point, line, or polygon features) to a raster format with 25 m spatial resolution.

3.3. Geographical Detector (GeoDetector)

GeoDetector is a spatially-based multivariate statistical model which was developed in 2010 by Wang et al. [50]. It can detect the spatially stratified heterogeneity of a given phenomenon according to the basic principle that if a determinant is associated with the phenomenon, then there may be some similarities between their spatial distributions. Furthermore, it can reveal the driving forces behind the phenomenon by quantifying the impacts of individual determinants and of their pairwise interactions. The phenomenon under investigation as a dependent variable can be represented by either numerical continuous or discrete classified (stratified) data, and the determinants as explanatory variables exclusively by classified data.

Table 2. Summary of the datasets representing the conditioning factors.

Factor	Dataset	Data Source	Spatial/Temporal Scale/Resolution	Primary Format
Altitude	EU-DEM (v1.1)	"Copernicus" Land Monitoring Service	25 m/2011	Raster (grid)
Slope angle	DEM derivative		25 m/2011	Raster (grid)
Slope aspect	DEM derivative		25 m/2011	Raster (grid)
Plan & Profile curvatures	DEM derivative		25 m/2011	Raster (grid)
Stream density	Rivers and streams	General Use Map of Greece (Hellenic Military Geographical Service)	1:50,000/1989	Vector (line)
SPI	DEM-based hydrological analysis		25 m/2011	Raster (grid)
TWI				
Lithology	Lithological formations	Geological Map of Greece (Institute of Geology and Mineral Exploration)	1:50,000/1993	Vector (polygon)
Proximity to faults	Faults			
Soil type	Soil types	Hellenic Ministry of Environment and Energy	1:50,000/1997	Vector (polygon)
		Soil Map of Greece (Aristotle University of Thessaloniki)	1:500,000/2015	
Land use/cover	"CORINE" features	"Copernicus" Land Monitoring Service	1:100,000/2018	Vector (polygon)
Proximity to roads	Main roads	"OpenStreetMap"	–/2020	Vector (line)
Mean annual rainfall	"E-OBS" daily precipitation	"Copernicus" Climate Change Service	0.1 degrees/2000–2015	Raster (grid)

In the case of LS, GeoDetector can detect whether a conditioning factor (explanatory variable) causes the spatial stratified heterogeneity of landslide occurrence (presence or absence of a landslide, dependent variable) or not. In particular, it can quantify the degree of impact of each factor on the landslide occurrence using a q-statistic calculated as follows [51]:

$$q = 1 - \frac{\sum_{h=1}^{L} N_h \sigma_h^2}{N\sigma^2} \quad (1)$$

where $h = 1, 2, \ldots, L$ is a given class (stratum) of an explanatory variable; L is the number of classes; N_h and N are the numbers of samples in class h and entire study area, respectively; and σ_h and σ are the variance of dependent variable in class h and entire study area, respectively. Ranging from 0 to 1, the higher the q value is, the more this explanatory variable contributes to the dependent variable. A p-statistic, an indicator of statistical significance for each explanatory variable, is also calculated by a non-central F-distribution:

$$p(q < x) = p\left(F < \frac{N-L}{L-1} \frac{x}{1-x}\right) = 1 - a \quad (2)$$

where a is the probability of q being higher than or equal to x. In a 95% confidence interval, an explanatory variables with a p value greater than 0.05 is considered to have a statistically insignificant relationship with the dependent variable and could be eliminated from the model.

By estimating the value of q-statistic corresponding to the interaction of two explanatory variables, GeoDetector can also quantify the degree of the interactive impact of each pair of conditioning factors on landslide occurrence. As is shown in Table 3, based on the comparison of this value with the individually estimated values, the type of interaction can be then determined.

Table 3. Types of interaction between two explanatory variables (X1 and X2).

Interaction Type	Description
Nonlinear-weaken	q(X1∩X2) < Min(q(X1), q(X2))
Univariate-weaken	Min(q(X1), q(X2)) < q(X1∩X2) < Max(q(X1), q(X2))
Bivariate-enhanced	q(X1∩X2) > Max(q(X1), q(X2))
Independent	q(X1∩X2) = q(X1) + q(X2)
Nonlinear-enhanced	q(X1∩X2) > q(X1) + q(X2)

3.4. Information Value (IV)

IV is a bivariate statistical model which was initially proposed by Yin and Yan [52] and later modified by van Westen [53]. It includes class-level estimations of weight values based on the spatial associations between the landslide occurrence and each class of each conditioning factor. The IV for a given factor class is derived from a mathematical formula of the ratio of landslide density in this class to the landslide density in entire study area (or factor):

$$IV = ln\left(\frac{Npix(Si)/Npix(Ni)}{\sum Npix(Si)/\sum Npix(Ni)}\right) \quad (3)$$

where $Npix(Si)$ is the number of landslide pixels within the factor class i, and $Npix(Ni)$ is the number of all pixels in the same class. The calculated value can be either positive or negative, and the higher (or lower) it is, the more (or less) significant the contribution of the relevant factor class to landslide occurrence.

4. LS Assessment by Hybrid Modeling

Considering the functionalities and data requirements of the two models composing the GeoDIV hybrid model, two GIS-based data processing procedures initially took place under the general methodological framework (Figure 3). These procedures were the (non)landslide sampling and the factor preparation. For sampling, the landslide inventory dataset was divided into two subsets used as inputs in the model's training (training dataset) and validation (validation dataset), respectively. Among the amount of 60 landslides contained in the inventory, 80% of them (48 in amount) were randomly selected for the training dataset in this study. The remaining 20% (12 in number) constituted the validation dataset. Based on the sizes of mapped landslides and the spatial resolution of obtained factor data, the entire study area was then tiled into grid pixels of 25 × 25 m as the basic analysis unit, resulting in 188 training and 41 validation landslide pixels. The IV model required only a landslide dataset, whereas the GeoDetector model required both landslide and non-landslide datasets. Hence, in order to construct the dependent variable for GeoDetector, an equal number of pixels from the not landslide-affected part of study area were also selected in a random way for the training dataset (totally 376 pixels). The target values of 0 and 1 were assigned to the non-landslide and landslide pixels, respectively, making the dependent variable a binary classified dataset.

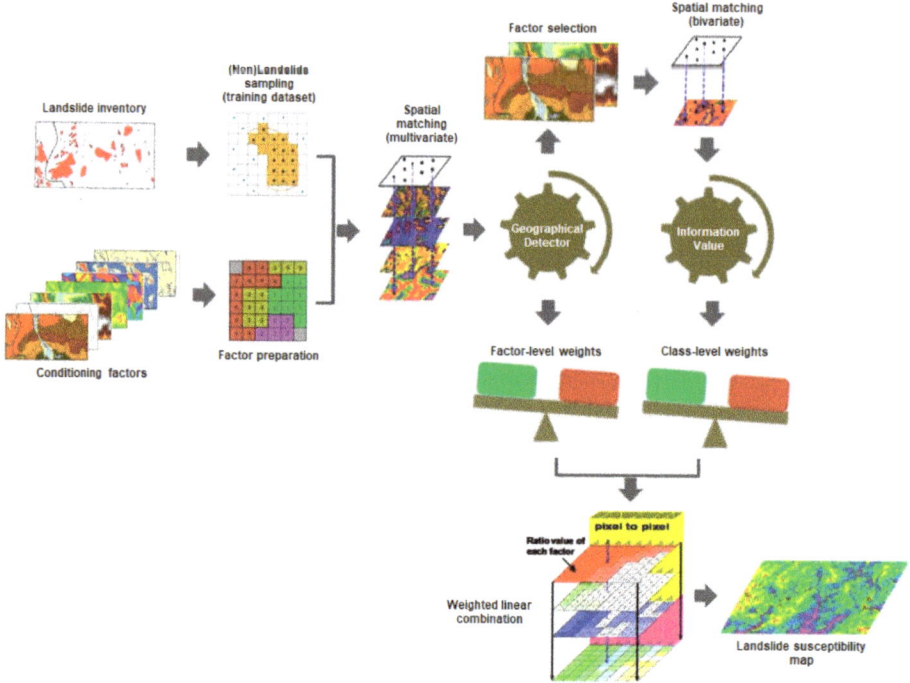

Figure 3. Methodological framework for the development of the hybrid GeoDIV model.

In regard to factor preparation, the raster layers of conditioning factors on a continuous numerical scale (altitude, slope angle, profile curvature, plan curvature, stream density, SPI, TWI, proximity to faults, proximity to roads and mean annual rainfall) were divided into a number of discrete classes (Figure 4). In this study, the number of categories and their relative break values were mainly determined by the "natural breaks (Jenks)" classification method [54]. In this method, class breaks identify the most similar within-group values and maximize the differences between classes according to the deviations about the median [55]. Additionally, the raster layers of factors originally on a discrete classified scale (slope aspect, lithology, soil type, and land use/cover) were prepared by grouping them into more or less common initial classes (Figure 4).

After the data processing procedures, the GeoDIV model was implemented. A database was firstly created as the result of the matching of the sample of 376 training data with each factor layer. Including the fourteen classified factors as independent variables and the landslide presence or absence (binary target value of 0 and 1) as the dependent variable were determined in the GeoDetector software, developed by Xu and Wang [56], to determine the impacts of the factors and their pairwise interactions on the spatial stratified heterogeneity of landslide occurrence represented by the training sample. This determination included the calculation of q values for the factors and their pairwise interactions (Tables 4 and 5). To incorporate in the model only the factors with statistically significant relationships with landslide occurrence, the estimated p values (Table 4) of the factors were also exploited for factor selection. Despite the requirement for p values less than 0.05 in the 95% confidence interval, factors such as altitude, slope angle, plan curvature, stream density, TWI, proximity to faults, proximity to roads, lithology, soil type and land use/cover remained in the model. Conversely, slope aspect, profile curvature, SPI and mean annual rainfall were not qualified to be further analyzed by the model, indicating that there were statistically insignificant relationships (i.e., p values greater than 0.05) between them and landslide occurrence in the same confidence interval.

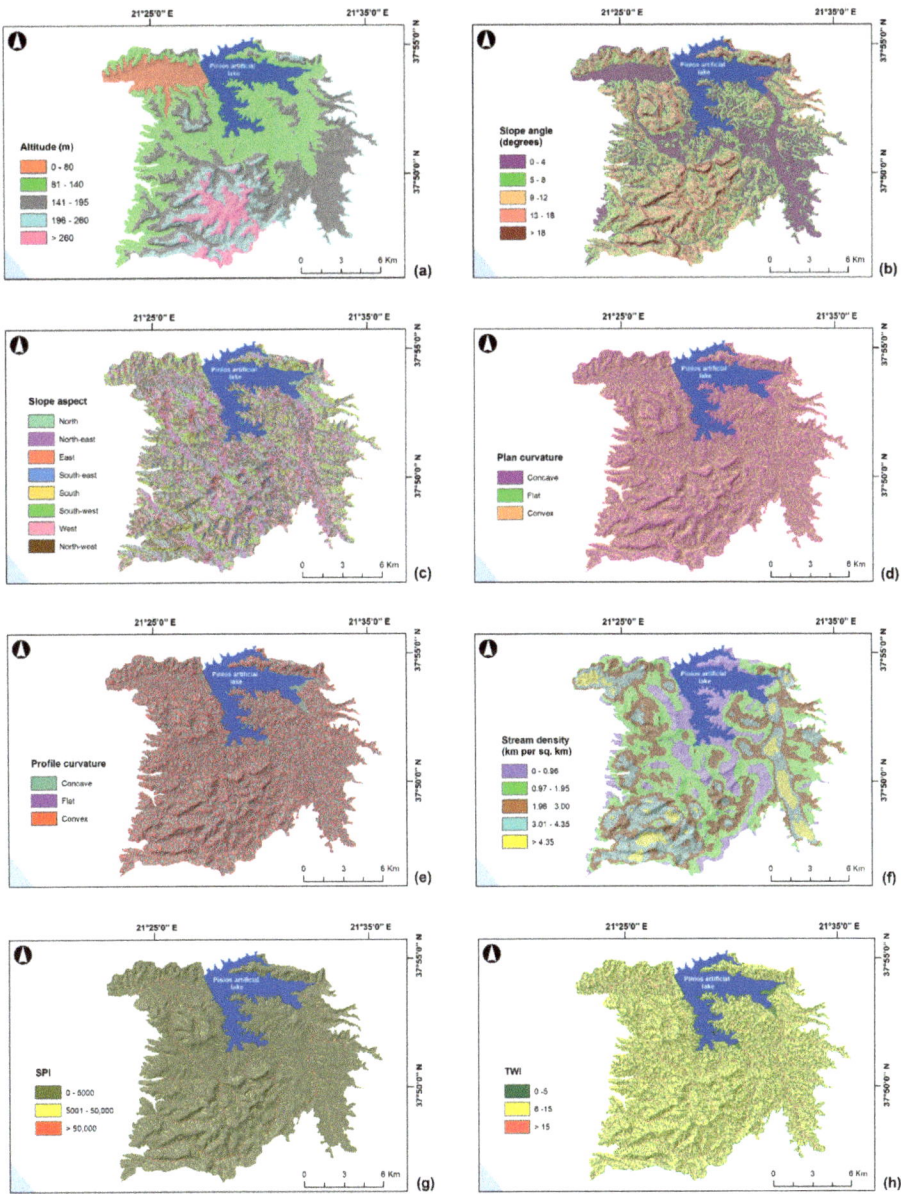

Figure 4. *Cont.*

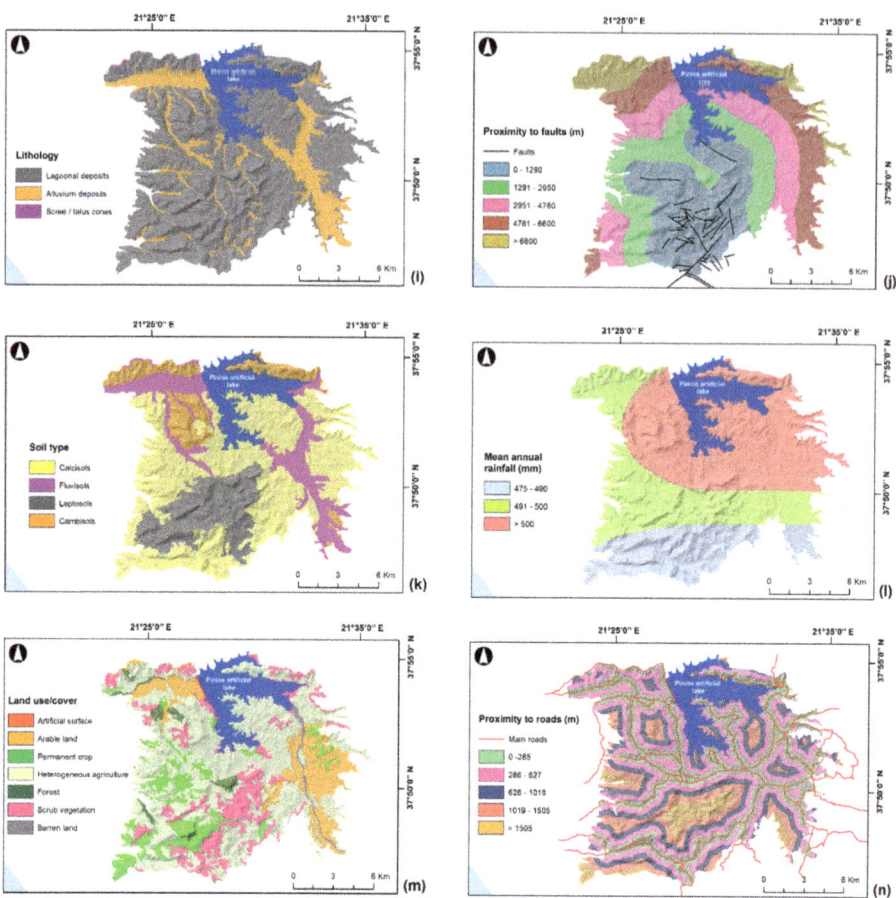

Figure 4. Conditioning factors: (**a**) altitude; (**b**) slope angle; (**c**) slope aspect; (**d**) plan curvature; (**e**) profile curvature; (**f**) stream density; (**g**) SPI; (**h**) TWI; (**i**) lithology; (**j**) proximity to faults; (**k**) soil type; (**l**) mean annual rainfall; (**m**) land use/cover; (**n**) proximity to roads.

Table 4. The q and p-statistic values for the conditioning factors, calculated using GeoDetector.

Factor	q Value	p Value
Altitude	0.078	0.00
Slope angle	0.264	0.00
Slope aspect	0.038	0.06 *
Plan curvature	0.016	0.01
Profile curvature	0.021	0.06 *
Stream density	0.065	0.00
SPI	0.003	0.30 *
TWI	0.019	0.04
Lithology	0.053	0.00
Proximity to faults	0.147	0.00
Soil type	0.072	0.00
Land use/cover	0.151	0.00
Proximity to roads	0.174	0.00
Mean annual rainfall	0.001	0.85 *

* indicate the factors eliminated from GeoDetector according to the p values.

Table 5. The q-statistic values for the pairwise interactions between the conditioning factors, calculated using GeoDetector.

Factor	Altitude	Slope Angle	Slope Aspect	Plan Curvature	Profile Curvature	SPI	TWI	Proximity to Roads	Proximity to Faults	Stream Density	Mean Annual Rainfall	Lithology	Soil Type	Land Use/Cover
Altitude	0.370													
Slope angle	0.268	0.423												
Slope aspect	0.106	0.294	0.098											
Plan curvature	0.101	0.300	0.074	0.034										
Profile curvature	0.093	0.274	0.061	0.022	0.027									
SPI	0.104	0.298	0.110	0.032	0.033	0.021								
TWI	0.290	0.488	0.347	0.196	0.211	0.190	0.204							
Proximity to roads	0.299	0.398	0.388	0.184	0.188	0.153	0.197	0.319						
Proximity to faults	0.289	0.317	0.237	0.092	0.121	0.083	0.097	0.310	0.377					
Stream density	0.155	0.301	0.137	0.030	0.041	0.012	0.027	0.224	0.243	0.227				
Mean annual rainfall	0.123	0.278	0.133	0.067	0.078	0.058	0.070	0.242	0.224	0.134	0.072			
Lithology	0.195	0.341	0.275	0.090	0.101	0.085	0.099	0.299	0.319	0.274	0.163	0.107		
Soil type	0.215	0.350	0.285	0.175	0.177	0.168	0.180	0.326	0.273	0.278	0.170	0.177	0.286	
Land use/cover														

Subsequently, by matching only the 188 landslide training data with each layer of statistically significant factors, the landslide density for each of their classes was estimated. The IVs were then calculated by Equation (2) to determine the impact of each class on landslide occurrence (Figure 5).

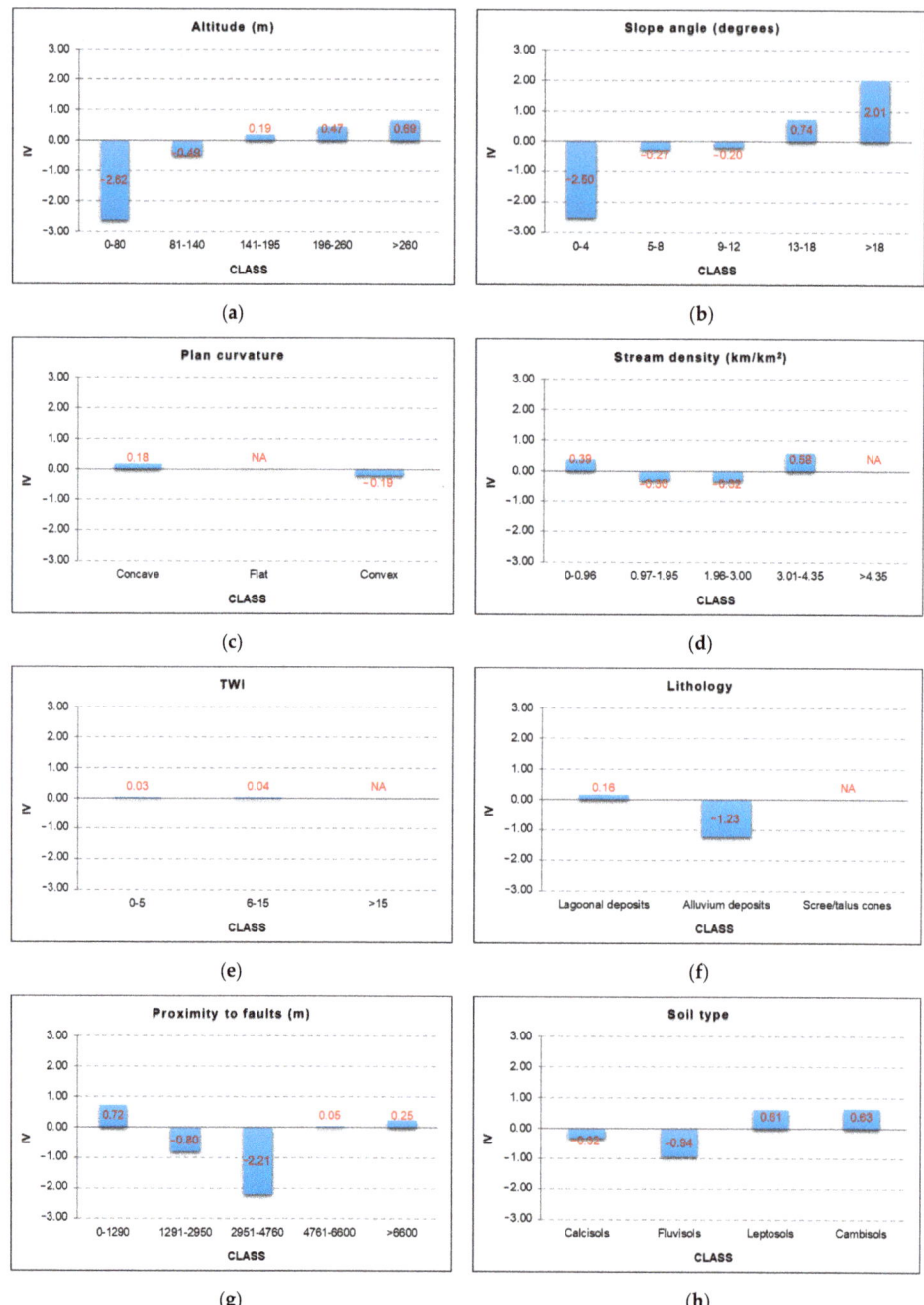

Figure 5. *Cont.*

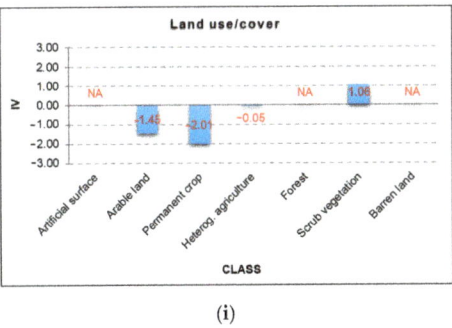 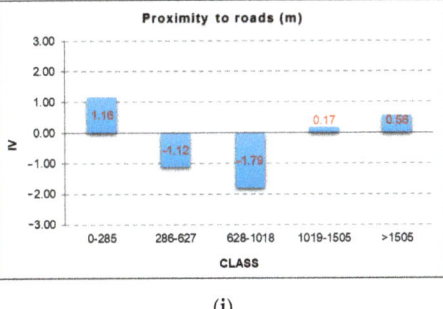

(i)   (j)

Figure 5. The estimated IVs for the classes of conditioning factors qualified from the factor selection: (**a**) altitude; (**b**) slope angle; (**c**) plan curvature; (**d**) stream density; (**e**) TWI; (**f**) lithology; (**g**) proximity to faults; (**h**) soil type; (**i**) land use/cover; (**j**) proximity to roads. NA values (or no bars) indicate "not applicable" for these classes.

By using the q values from GeoDetector as factor-level weights and IVs as class-level weights, the overall landslide susceptibility (LS) score was estimated through a GIS-based weighted linear combination of statistically significant factors:

$$LS = \sum_{j=1}^{n} W_J \times s_{i,j} \qquad (4)$$

where W_j is the weight of a given factor j, $s_{i,j}$ is the weight for a given class i of factor j and n is the number of factors. The spatial distribution of the estimated overall score was visualized by a LS map divided into five classes ("very low", "low", "moderate", "high" and "very high" susceptibility) according to the "natural breaks (Jenks)" method (Figure 6).

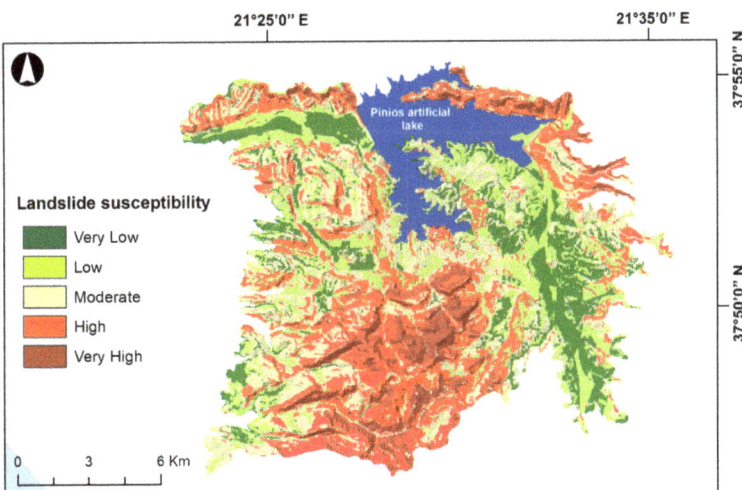

Figure 6. The landslide susceptibility map produced by the hybrid GeoDIV model.

5. Results

The weights from the GeoDIV model are summarized in Table 4 and Figure 5. Among the conditioning factors that eventually remained in the model, the highest factor-level weight was obtained from slope angle (q value of 0.264). It was followed by proximity to roads, land use/cover and proximity to faults (q values of 0.174, 0.151 and 0.147, respectively). For these factors, the classes with the highest class-level weights were the "greater than 18 degrees" (IV = 2.01) for slope angle, "0 to 285 m" (IV = 1.16) for proximity to roads, "scrub vegetation" (IV = 1.06) for land use/cover and "0–1290 m" (IV = 0.72) for proximity to faults. The rest of conditioning factors were found to have much lower factor-level weights (q values below 0.10). Plan curvature was the factor with the lowest weight (q value of 0.016).

According to the correlations between Tables 3 and 5, the impact degree and types of the different pairwise interactions of factors were determined. The interaction between slope angle and proximity to roads presented the highest weight value (q value of 0.488). This value was greater than the sum of their individual values, indicating that their interaction type was nonlinearly enhanced. Generally, the weights of all the factors (even the lowest of plan curvature) were significantly increased by slope angle, achieving either nonlinear enhancement or bivariate enhancement.

The LS map from GeoDIV model is illustrated in Figure 6. It shows that the "high" and "very high" susceptibility zones are mainly in the southern and northern parts of the vicinity of Pinios artificial lake, with some large pockets of "high" susceptibility in the western part. These two zones cover 25% and 12% of the lake's vicinity, respectively.

Validation and Comparison

In order to evaluate the performance of a model applied for LS assessment and mapping, a validation step is required. Since it can provide information about the accuracy and prediction ability of the model, and thus the reliability of its LS output, this step is crucial for any relevant research effort. A standard validation procedure is one based on success and prediction rates [28,45,48]. This specific procedure depends on the creation of two rate curves explaining the percentages of landslides that fall into defined LS ranks. These curves are graphically presented in cumulative frequency diagrams, with respect to the two different datasets of landslide inventory. For the success rate curve, the landslide training dataset was used to indicate how well the model fits to the training data. On the contrary, for the prediction rate curve, the "independent" landslide validation dataset was used to show how well the model can predict the distribution of future landslides [57].

To obtain the success and prediction rate curves in this study, the overall LS score (Equation (3)) was initially sorted in descending order (from high to low). Then, the ordered LS score was divided into 100 classes with 1% cumulative intervals. The resultant LS ranks (0–100%, where a higher rank means a lower LS score) were plotted on the x-axis, whereas their cumulative percentages of training and validation landslide data are on the y-axis. An area under curve (AUC) value was eventually calculated for each of the two rate curves indicating the accuracy and prediction ability of GeoDIV model, respectively. With a range of 0.5–1.0, this value reflects the model's performance.

Aiming to confirm the potential "superiority" of the hybrid modeling against the individual modeling and explore the impact of GeoDetector-based factor selection on LS assessment, the individual IV model was also applied, and its validation results were compared with those of GeoDIV model. In this context, IVs were additionally calculated for the classes of statistically insignificant factors (not included in GeoDIV model). The overall LS score (presented also by classes, in Figures 7 and 8) was then obtained by the summation of all the fourteen IV-weighted factors as follows:

$$LS = \sum_{j=1}^{n} IV_j \qquad (5)$$

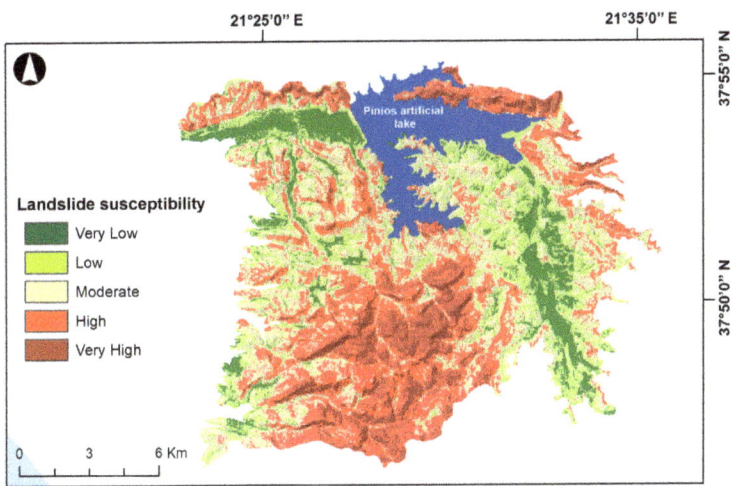

Figure 7. The landslide susceptibility map produced by the individual IV model.

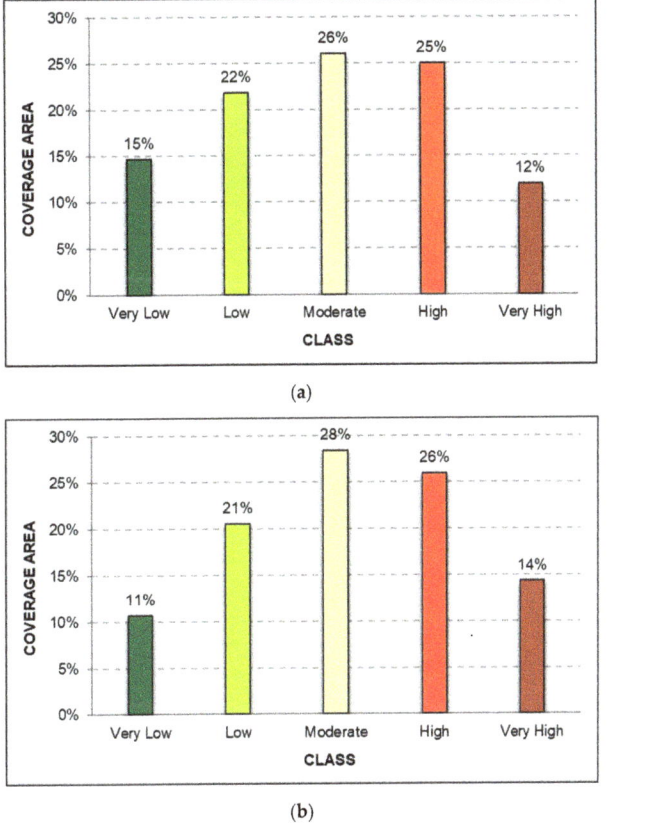

Figure 8. Diagrams with the coverage area percentages of the landslide susceptibility classes for the models: (**a**) hybrid GeoDIV model; (**b**) individual IV model.

Based on the *LS* score rank, the success and prediction rate curves were created, and the relative AUC values were calculated for individual *IV* model as well.

The results of validation procedure for both models are presented in Figure 9. The success and prediction rate curves indicate that the first 30% of the LS ranks derived from GeoDIV model can explain about 70% of landslide training data and 60% of landslide validation data, respectively. Moreover, the relevant AUC values of 0.78 and 0.76 revealed remarkable accuracy (data fitting) and prediction ability of the model. All these results were found to be worse for individual IV model, with explanation percentages of 60% and 50%, respectively, and AUC values of 0.72 and 0.71, respectively.

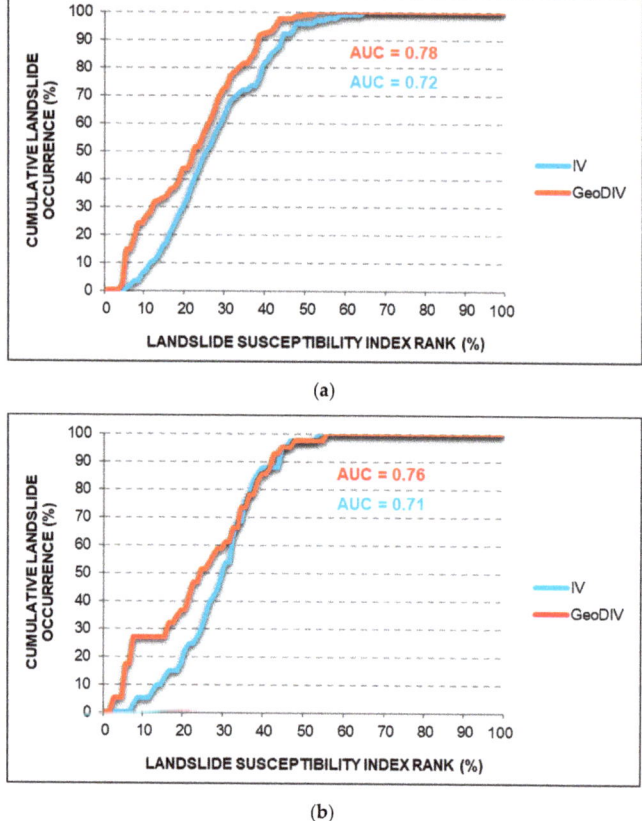

Figure 9. The results of the validation procedure: (**a**) success rate curves; (**b**) prediction rate curves.

6. Discussion

Due to the observed upward tendency in landslide occurrence, authorities of all the administration levels (national, regional and local) are called on to collaborate with the scientific community to spatially determine potential landslide instances and mitigate, or even prevent, the damage and losses that they may cause. LS assessment and mapping is the first and most basic step for effective risk management and disaster response [58]. Several LS assessment models have been developed and applied, with their own advantages and disadvantages [59]. A current tendency is the integration of these individual models to enhance their benefits and overcome their weaknesses. The consequent hybrid models are expected to reduce the uncertainty and improve the reliability of the output LS maps [60]. In order to address this statement, in the present study, a hybrid model based on the integration of two different statistical analysis models, multivariate GeoDetector and bivariate IV, was proposed for LS assessment and mapping. In general, GeoDetector, as a new spatial model, has been rarely used in LS studies compared with other models. Hence, its integrated applications are even more limited. To the best of our knowledge and without ignoring the research works of Luo and Liu [7] and Yang et al. [61], the proposed integration had not been tested hitherto for the development of hybrid LS modeling.

A variety of natural and anthropogenic conditioning factors and a landslide inventory for a Greek wetland around the Pinios artificial lake, were analyzed as inputs in the hybrid model named GeoDIV. It can be stated that the advantages (or disadvantages) of GeoDIV model are "inherited" from the two individual models which it was based on. Under strict, prior defined data assumptions, the IV model is capable of evaluating the impact of each class of many conditioning factors due to the occurrence of past landslides; however, the mutual relationship between the factors is mostly neglected [62]. Without any assumptions on the distribution of data, the GeoDetector model is capable of exploring this relationship but not evaluating individually the impact of each factor class.

In addition to the models, factor selection also plays a major role in the LS results [14]. Too many redundant factors may lead to less realistic and reliable results. Therefore, the capability of significance statistics-based factor selection provided by GeoDetector makes it an ideal option for selecting the most proper factors and then assigning objective weights to them with regard to their different contributions to past landslide occurrence. By incorporating this property of GeoDetector in hybrid GeoDIV model, among the fourteen conditioning factors initially collected, four of them (slope aspect, profile curvature, SPI, and mean annual rainfall) were identified as statistically insignificant and were not finally included in LS assessment. Similar factors were also eliminated as redundant in [14,23,24].

Focusing on the factors qualified from factor selection, slope angle was highlighted by the factor-level weights (q values) of the GeoDIV model. In GeoDetector's terminology, slope angle can be characterized as the factor which most explains the spatial stratified heterogeneity of landslide occurrence in the study area. In simple words, its weight was found to be much higher than the rest of factors, revealing that slope angle has the greatest impact on landslide activity. This is in line with findings from other studies in Greece which, on the basis of using either qualitative or quantitative models at different scales (national and regional), also indicated slope angle as one of the most important factors [38,63,64].

When slope angle interacted to some degree with proximity to roads, an even greater impact was detected, according to the interaction weights. Generally, the single impacts of all other factors were shown to be significantly improved from their interactions with the slope angle. Except for the particularly influential role of the specific factor, this finding also confirms the "nature" of landslides as a phenomenon that, to a great extent, constitutes the result of interactions between multiple conditioning factors. From a sub-factor perspective, as it was derived from the class-level weights (IVs), the steep parts of study area being very close to roads and covered by scrub vegetation seem to be more prone to landslides.

The output map of the GeoDIV model illustrated the spatial distribution of the estimated LS. It shows that extensive parts, mainly located in south and north, are most likely to have landslides in the future. In comparison with the relevant map from the individual IV model, it can be mentioned that despite the preservation of the general spatial pattern, there was displacement of the pockets from low susceptibility in GeoDIV's map to higher susceptibility in IV's map. This "overestimation" from the IV model may have been due to the inclusion of the additional four conditioning factors, confirming the above statement about the negative impacts of redundant factors on the reliability of LS results.

Regarding the performance of proposed hybrid model, it has to be firstly noted that GeoDIV provided far more than satisfactory validation results in terms of accuracy (success rate) and prediction ability (prediction rate), considering the scale of analysis. Compared to the IV model, although both models seemed to converge to approximate results, the convergence of GeoDIV was found to be faster. This finding proves the expected "superiority" of hybrid against the individual modeling and is in agreement with previous studies, concluding that the integration of bivariate with multivariate statistical models improved the performance of former ones [30,47].

Some assumptions and limitations of the present study have to be pointed out. The quality of LS assessment and mapping is highly related to both the landslide inventory and conditioning factors. By using Google Earth satellite imagery, the landslides with identifiable signs in the images were mainly mapped in landslide inventory. Hence, the inventory cannot totally represent the landslide-contributing factors of the study area. Moreover, although the preparation of different susceptibility maps for the various types of landslides can provide more realistic predictions [65], the different types of mapped landslides were not considered in this study. On the other hand, the differentiation between landslide source and deposition zones enabled the models to accurately identify source areas and hence to precisely define the factors that contribute to the initiation of a landslide. The lack of this differentiation resulted to a study's assumption concerning the existence of similar terrain conditions within these zones and thus the representation of each landslide by a single polygon feature. Considerable simplification of these polygons had to be then undertaken by converting them to grid pixels. In this way, an underestimation of landslide data may have taken place in some cases. Additionally, the sampling procedure for the creation of the landslide training and validation datasets can affect the model's efficiency. On the basis of appropriate sizes, a sufficient amount of data should be included in the training dataset, and a remaining "independent" amount of data in the validation dataset. From the perspective of conditioning factors, their spatial resolution and classification can affect the precision of the spatial matching between the landslide and factor data. Therefore, the examination of alternatives for these parameters could lead to different results.

7. Conclusions

A hybrid model named GeoDIV was applied to produce a reliable LS map for the vicinity of Pinios artificial lake (Ilia, Greece). Based on the analysis of landslide and factor conditioning data, the GeoDIV framework exploited the multivariate GeoDetector to eliminate redundant factors and objectively quantify the individual and interactive impacts of the remaining ones (factor-level weights) on landslide occurrence. The bivariate IV was used for objectively quantifying the impacts of their classes (class-level weights). In practice, the integration of these two models increased their efficiency. The findings confirmed that hybrid modeling outperform modeling: the GeoDIV model yielded better results than the individual IV model in terms of accuracy and prediction ability. Thus, GeoDIV can be considered as a promising and robust model which can be beneficial not only to the current study area, but also to other regions with similar or even different conditions and settings.

In general, it was revealed that hybrid LS modeling assisted by multiple geospatial tools (RS and GIS) can contribute well to the production of reliable maps. The LS map produced by the GeoDIV model could be important basis for the regional or local authorities in order to develop both general (long-term) and emergency (short-term) strategies centered on "space design" disaster management. Knowledge about the potential for landslides in a region is valuable for policy makers, as it can allow them to select safe locations while planning land use and approving construction projects. Policy makers could also identify threatened settlements and roads, and in response take drastic disaster management measures (including building engineered structures, planning evacuation routes and issuing early warnings).

Future research work will focus on testing the proposed hybrid modeling for LS assessments of other regions characterized by different environmental and/or human settings, with various landslide densities. Comparisons with other advanced models, such as machine learning models, will be also performed.

Author Contributions: Conceptualization, C.P.; methodology, C.P.; software, C.P.; validation, C.P.; formal analysis, C.P., M.G.G. and A.V.A.; data curation, C.P. and M.G.G.; visualization, C.P. and A.V.A.; writing—original draft preparation, C.P., M.G.G., A.V.A., N.P. and D.D.A.; writing—review and editing, C.P., M.G.G., A.V.A., N.P. and D.D.A.; supervision, N.P. and D.D.A. All authors have read and agreed to the published version of the manuscript.

Funding: This research received no external funding.

Acknowledgments: The authors extremely appreciate the significant contributions of the journal's editor and reviewers to the handling and revision of this article.

Conflicts of Interest: The authors declare no conflict of interest.

References

1. Varnes, D.J. Slope movement types and processes. In *Landslides: Analysis and Control*; Schuster, R.L., Krizek, R.J., Eds.; Transportation Research Board Special Report 176: Washington, DC, USA, 1978; pp. 11–33.
2. Centre for Research on the Epidemiology of Disasters—CRED. *Disaster Year in Review 2020: Global Trends and Perspectives*; Cred Crunch; Université Catholique de Louvain: Brussels, Belgium, 2021; Volume 62, p. 2.
3. Psomiadis, E.; Papazachariou, A.; Soulis, K.X.; Alexiou, D.-S.; Charalampopoulos, I. Landslide mapping and susceptibility assessment using geospatial analysis and earth observation data. *Land* **2020**, *9*, 133. [CrossRef]
4. Centre for Research on the Epidemiology of Disasters—CRED; United Nations International Strategy for Disaster Reduction—UNISDR. *Economic Losses, Poverty & Disasters (1998–2017)*; Université Catholique de Louvain: Brussels, Belgium, 2018; p. 33.
5. Brabb, E.E. Innovative approaches to landslide hazard mapping. In *Proceedings of the 4th International Symposium on Landslides*; Canadian Geotechnical Society: Torornto, ON, Canada, 1984; pp. 307–324.
6. Varnes, D.J. *Landslide Hazard Zonation: A Review of Principles and Practice*; UNESCO: Paris, France, 1984; Volume 3.
7. Luo, W.; Liu, C.-C. Innovative landslide susceptibility mapping supported by geomorphon and geographical detector methods. *Landslides* **2018**, *15*, 465–474. [CrossRef]
8. Arabameri, A.; Pradhan, B.; Rezaei, K.; Lee, C.-W. Assessment of landslide susceptibility using statistical- and artificial intelligence-based FR-RF integrated model and multiresolution DEMs. *Remote Sens.* **2019**, *11*, 999. [CrossRef]
9. Ciurleo, M.; Mandaglio, M.C.; Moraci, N. A quantitative approach for debris flow inception and propagation analysis in the lead up to risk management. *Landslides* **2021**, *18*, 2073–2093. [CrossRef]
10. Vieira, B.C.; Fernandes, N.F.; Filho, O.A.; Martins, T.D.; Montgomery, D.R. Assessing shallow landslide hazards using the TRIGRS and SHALSTAB models, Serra do Mar, Brazil. *Environ. Earth Sci.* **2018**, *77*, 260. [CrossRef]
11. Su, C.; Wang, L.; Wang, X.; Huang, Z.; Zhang, X. Mapping of rainfall-induced landslide susceptibility in Wencheng, China, using support vector machine. *Nat. Hazards* **2015**, *76*, 1759–1779. [CrossRef]
12. Taaleb, K.; Cheng, T.; Zhang, Y. Mapping landslide susceptibility and types using Random Forest. *Big Earth Data* **2018**, *2*, 159–178. [CrossRef]
13. Gorsevski, P.V.; Brown, M.K.; Panter, K.; Onasch, C.M.; Simic, A.; Snyder, J. Landslide detection and susceptibility mapping using LiDAR and an artificial neural network approach: A case study in the Cuyahoga Valley National Park, Ohio. *Landslides* **2016**, *13*, 467–484. [CrossRef]
14. Xie, W.; Li, X.; Jian, W.; Yang, Y.; Liu, H.; Robledo, L.F.; Nie, W. A novel hybrid method for landslide susceptibility mapping-based geodetector and machine learning cluster: A case of Xiaojin county, China. *ISPRS Int. J. Geo-Inf.* **2021**, *10*, 93. [CrossRef]
15. Kayastha, P. Landslide susceptibility mapping and factor effect analysis using frequency ratio in a catchment scale: A case study from Garuwa sub-basin, East Nepal. *Arab. J. Geosci.* **2015**, *8*, 8601–8613. [CrossRef]
16. Borrelli, L.; Ciurleo, M.; Gullà, G. Shallow landslide susceptibility assessment in granitic rocks using GIS-based statistical methods: The contribution of the weathering grade map. *Landslides* **2018**, *15*, 1127–1142. [CrossRef]
17. Baeza, C.; Corominas, J. Assessment of shallow landslide susceptibility by means of multivariate statistical techniques. *Earth Surf. Process. Landf.* **2001**, *26*, 1251–1263. [CrossRef]
18. Dagdelenler, G.; Nefeslioglu, H.A.; Gokceoglu, C. Modification of seed cell sampling strategy for landslide susceptibility mapping: An application from the Eastern part of the Gallipoli Peninsula (Canakkale, Turkey). *Bull. Eng. Geol. Environ.* **2016**, *75*, 575–590. [CrossRef]
19. Mondal, S.; Mandal, S. RS & GIS-based landslide susceptibility mapping of the Balason River basin, Darjeeling Himalaya, using logistic regression (LR) model. *Georisk* **2018**, *12*, 29–44.
20. Polykretis, C.; Alexakis, D.D. Spatial stratified heterogeneity of fertility and its association with socio-economic determinants using Geographical Detector: The case study of Crete Island, Greece. *Appl. Geogr.* **2021**, *127*, 102384. [CrossRef]
21. Xiong, J.; Pang, Q.; Fan, C.; Cheng, W.; Ye, C.; Zhao, Y.; He, Y.; Cao, Y. Spatiotemporal characteristics and driving force analysis of flash floods in Fujian province. *ISPRS Int. J. Geo-Inf.* **2020**, *9*, 133. [CrossRef]
22. Zhou, C.; Zhu, N.; Xu, J.; Yang, D. The contribution rate of driving factors and their interactions to temperature in the yangtze river delta region. *Atmosphere* **2020**, *11*, 32. [CrossRef]
23. Yang, J.; Song, C.; Yang, Y.; Xu, C.; Guo, F.; Xie, L. New method for landslide susceptibility mapping supported by spatial logistic regression and GeoDetector: A case study of Duwen Highway Basin, Sichuan Province, China. *Geomorphology* **2019**, *324*, 62–71. [CrossRef]

24. Rong, G.; Li, K.; Han, L.; Alu, S.; Zhang, J.; Zhang, Y. Hazard mapping of the rainfall–landslides disaster Chain based on GeoDetector and Bayesian Network Models in Shuicheng County, China. *Water* **2020**, *12*, 2572. [CrossRef]
25. Pourghasemi, H.R.; Teimoori Yansari, Z.; Panagos, P.; Pradhan, B. Analysis and evaluation of landslide susceptibility. A review on articles published during 2005–2016 (periods of 2005–2012 and 2013–2016). *Arab. J. Geosci.* **2018**, *11*, 193. [CrossRef]
26. Ciurleo, M.; Cascini, L.; Calvello, M. A comparison of statistical and deterministic methods for shallow landslide susceptibility zoning in clayey soils. *Eng. Geol.* **2017**, *223*, 71–81. [CrossRef]
27. Aditian, A.; Kubota, T.; Shinohara, Y. Comparison of GIS-based landslide susceptibility models using frequency ratio, logistic regression, and artificial neural network in a tertiary region of Ambon, Indonesia. *Geomorphology* **2018**, *318*, 101–111. [CrossRef]
28. Barella, C.F.; Sobreira, F.G.; Zêzere, J.L. A comparative analysis of statistical landslide susceptibility mapping in the southeast region of Minas Gerais state, Brazil. *Bull. Eng. Geol. Environ.* **2019**, *78*, 3205–3221. [CrossRef]
29. Saha, S.; Arabameri, A.; Saha, A.; Blaschke, T.; Ngo, P.T.T.; Nhu, V.H.; Band, S.S. Prediction of landslide susceptibility in Rudraprayag, India using novel ensemble of conditional probability and boosted regression tree-based on cross-validation method. *Sci. Total Environ.* **2021**, *764*, 142928. [CrossRef]
30. Chen, W.; Sun, Z.; Han, J. Landslide susceptibility modeling using integrated ensemble weights of evidence with logistic regression and random forest models. *Appl. Sci.* **2019**, *9*, 171. [CrossRef]
31. Roy, J.; Saha, S.; Arabameri, A.; Blaschke, T.; Bui, D.T. A novel ensemble approach for landslide susceptibility mapping (LSM) in Darjeeling and Kalimpong districts, West Bengal, India. *Remote Sens.* **2019**, *11*, 2866. [CrossRef]
32. Chowdhuri, I.; Pal, S.C.; Arabameri, A.; Ngo, P.T.T.; Chakrabortty, R.; Malik, S.; Das, B.; Roy, P. Ensemble approach to develop landslide susceptibility map in landslide dominated Sikkim Himalayan region, India. *Environ. Earth Sci.* **2020**, *79*, 476. [CrossRef]
33. Paryani, S.; Neshat, A.; Javadi, S.; Pradhan, B. Comparative performance of new hybrid ANFIS models in landslide susceptibility mapping. *Nat. Hazards* **2020**, *103*, 1961–1988. [CrossRef]
34. Balogun, A.-L.; Rezaie, F.; Pham, Q.B.; Gigović, L.; Drobnjak, S.; Aina, Y.A.; Panahi, M.; Yekeen, S.T.; Lee, S. Spatial prediction of landslide susceptibility in western Serbia using hybrid support vector regression (SVR) with GWO, BAT and COA algorithms. *Geosci. Front.* **2021**, *12*, 101104. [CrossRef]
35. Sabatakakis, N.; Koukis, G.; Vassiliades, E.; Lainas, S. Landslide susceptibility zonation in Greece. *Nat. Hazards* **2013**, *65*, 523–543. [CrossRef]
36. Tsangaratos, P.; Ilia, I. Landslide susceptibility mapping using a modified decision tree classifier in the Xanthi Perfection, Greece. *Landslides* **2016**, *13*, 305–320. [CrossRef]
37. Kavoura, K.; Sabatakakis, N. Investigating landslide susceptibility procedures in Greece. *Landslides* **2020**, *17*, 127–145. [CrossRef]
38. Chalkias, C.; Polykretis, C.; Ferentinou, M.; Karymbalis, E. Integrating expert knowledge with statistical analysis for landslide susceptibility assessment at regional scale. *Geosciences* **2016**, *6*, 14. [CrossRef]
39. Hellenic Statistical Authority (ELSTAT). Population and Housing Census: Resident Population. Available online: https://www.statistics.gr/el/statistics/pop (accessed on 1 June 2021).
40. Lainas, S.; Sabatakakis, N.; Koukis, G. Rainfall thresholds for possible landslide initiation in wildfire-affected areas of western Greece. *Bull. Eng. Geol. Environ.* **2016**, *75*, 883–896. [CrossRef]
41. Dimitriadou, S.; Katsanou, K.; Charalabopoulos, S.; Lambrakis, N. Interpretation of the factors defining groundwater quality of the site subjected to the wildfire of 2007 in Ilia prefecture, South-Western Greece. *Geosciences* **2018**, *8*, 108. [CrossRef]
42. Laboratory of Engineering Geology, Department of Geology, University of Patras. Landslide Management System of Western Greece. Available online: http://landslide.engeolab.gr/ (accessed on 9 March 2021).
43. Popescu, M.E. Landslide causal factors and landslide remediatial options. In Proceedings of the 3rd International Conference on Landslides, Slope Stability and Safety of Infrastructures, Singapore, 11–12 July 2002; pp. 61–81.
44. Shirzadi, A.; Bui, D.T.; Pham, B.T.; Solaimani, K.; Chapi, K.; Kavian, A.; Shahabi, H.; Revhaug, I. Shallow landslide susceptibility assessment using a novel hybrid intelligence approach. *Environ. Earth Sci.* **2017**, *76*, 60. [CrossRef]
45. Rabby, Y.W.; Li, Y. Landslide susceptibility mapping using integrated methods: A case study in the Chittagong hilly areas, Bangladesh. *Geosciences* **2020**, *10*, 483. [CrossRef]
46. Pradhan, A.M.S.; Kim, Y.T. Relative effect method of landslide susceptibility zonation in weathered granite soil: A case study in Deokjeok-ri Creek, South Korea. *Nat. Hazards* **2014**, *72*, 1189–1217. [CrossRef]
47. Li, R.; Wang, N. Landslide susceptibility mapping for the Muchuan county (China): A comparison between bivariate statistical models (WoE, EBF, and IoE) and their ensembles with logistic regression. *Symmetry* **2019**, *11*, 762. [CrossRef]
48. Yan, F.; Zhang, Q.; Ye, S.; Ren, B. A novel hybrid approach for landslide susceptibility mapping integrating analytical hierarchy process and normalized frequency ratio methods with the cloud model. *Geomorphology* **2019**, *327*, 170–187. [CrossRef]
49. Institute of Geodynamics—National Observatory of Athens (IG-NOA). Earthquake Inventories and Maps. Available online: http://www.gein.noa.gr/el/seismikotita/ (accessed on 16 April 2021).
50. Wang, J.-F.; Li, X.-H.; Christakos, G.; Liao, Y.-L.; Zhang, T.; Gu, X.; Zheng, X.-Y. Geographical detectors-based health risk assessment and its application in the neural tube defects study of the Heshun region, China. *Int. J. Geogr. Inf. Sci.* **2010**, *24*, 107–127. [CrossRef]
51. Wang, J.-F.; Zhang, T.-L.; Fu, B.-J. A measure of spatial stratified heterogeneity. *Ecol. Indic.* **2016**, *67*, 250–256. [CrossRef]
52. Yin, K.J.; Yan, T.Z. Statistical prediction model for slope instability of metamorphosed rocks. In Proceedings of the 5th international Symposium on Landslides, Lausanne, Switzerland, 13 September 1988; Volume 2, pp. 1269–1272.

53. Van Westen, C.J. *Application of Geographical Information System to Landslide Hazard Zonation*; ITC-Publication No. 15; ITC-Publication; International Institute for Geo-Information Science and Earth Observation: Enschede, The Netherlands, 1993; p. 245.
54. Calvello, M.; Ciurleo, M. Optimal use of thematic maps for landslide susceptibility assessment by means of statistical analyses: Case study of shallow landslides in fine grained soils. In *Landslides and Engineered Slopes. Experience, Theory and Practice, Proceedings of the 12th International Symposium on Landslides, Napoli, Italy, 12–19 June 2016*; Aversa, S., Cascini, L., Picarelli, L., Scavia, C., Eds.; CRC Press: London, UK, 2016; Volume 2, pp. 537–544.
55. Jenks, G.F. *Optimal Data Classification for Choropleth Maps*; University of Kansas: Lawrence, KS, USA, 1977.
56. Xu, C.-D.; Wang, J.-F. Geodetector: Software for Measure and Attribution of Stratified Heterogeneity (SH). Available online: http://www.geodetector.cn/ (accessed on 19 March 2020).
57. Chung, C.F.; Fabbri, A.G. Probabilistic prediction models for landslide hazard mapping. *Photogramm. Eng. Remote Sens.* **1999**, *65*, 1389–1399.
58. Xing, Y.; Yue, J.; Guo, Z.; Chen, Y.; Hu, J.; Travé, A. Large-scale landslide susceptibility mapping using an integrated machine learning model: A case study in the Lvliang Mountains of China. *Front. Earth Sci.* **2021**, *9*, 722491. [CrossRef]
59. Du, G.-L.; Zhang, Y.-S.; Iqbal, J.; Yang, Z.-H.; Yao, X. Landslide susceptibility mapping using an integrated model of information value method and logistic regression in the Bailongjiang watershed, Gansu Province, China. *J. Mt. Sci.* **2017**, *14*, 249–268. [CrossRef]
60. Luo, X.; Lin, F.; Chen, Y.; Zhu, S.; Xu, Z.; Huo, Z.; Yu, M.; Peng, J. Coupling logistic model tree and random subspace to predict the landslide susceptibility areas with considering the uncertainty of environmental features. *Sci. Rep.* **2019**, *9*, 15369. [CrossRef]
61. Yang, Y.; Yang, J.; Xu, C.; Xu, C.; Song, C. Local-scale landslide susceptibility mapping using the B-GeoSVC model. *Landslides* **2019**, *16*, 1301–1312. [CrossRef]
62. Zhang, G.; Cai, Y.; Zheng, Z.; Zhen, J.; Liu, Y.; Huang, K. Integration of the statistical index method and the analytic hierarchy process technique for the assessment of landslide susceptibility in Huizhou, China. *Catena* **2016**, *142*, 233–244. [CrossRef]
63. Sakkas, G.; Misailidis, I.; Sakellariou, N.; Kouskouna, V.; Kaviri, G. Modeling landslide susceptibility in Greece: A weighted linear combination approach using analytic hierarchical process, validated with spatial and statistical analysis. *Nat. Hazards* **2016**, *84*, 1873–1904. [CrossRef]
64. Polykretis, C.; Chalkias, C. Comparison and evaluation of landslide susceptibility maps obtained from weight of evidence, logistic regression, and artificial neural network models. *Nat. Hazards* **2018**, *93*, 249–274. [CrossRef]
65. Regmi, A.D.; Devkota, K.C.; Yoshida, K.; Pradhan, B.; Pourghasemi, H.R.; Kumamoto, T.; Akgun, A. Application of frequency ratio, statistical index, and weights-of-evidence models and their comparison in landslide susceptibility mapping in Central Nepal Himalaya. *Arab. J. Geosci.* **2014**, *7*, 725–742. [CrossRef]

Article

Factors Affecting Landslide Susceptibility Mapping: Assessing the Influence of Different Machine Learning Approaches, Sampling Strategies and Data Splitting

Minu Treesa Abraham [1], Neelima Satyam [1], Revuri Lokesh [1], Biswajeet Pradhan [2,3,*] and Abdullah Alamri [4]

[1] Department of Civil Engineering, Indian Institute of Technology Indore, Indore 453552, India; phd1901204011@iiti.ac.in (M.T.A.); neelima.satyam@iiti.ac.in (N.S.); ce170004028@iiti.ac.in (R.L.)
[2] Centre for Advanced Modelling and Geospatial Information Systems (CAMGIS), Faculty of Engineering and Information Technology, University of Technology Sydney, Sydney P.O. Box 123, Australia
[3] Earth Observation Center, Institute of Climate Change, Universiti Kebangsaan Malaysia, Bangi 43600, Malaysia
[4] Department of Geology & Geophysics, College of Science, King Saud University, P.O. Box 2455, Riyadh 11451, Saudi Arabia; amsamri@ksu.edu.sa
* Correspondence: Biswajeet.Pradhan@uts.edu.au or Biswajeet24@gmail.com

Citation: Abraham, M.T.; Satyam, N.; Lokesh, R.; Pradhan, B.; Alamri, A. Factors Affecting Landslide Susceptibility Mapping: Assessing the Influence of Different Machine Learning Approaches, Sampling Strategies and Data Splitting. *Land* **2021**, *10*, 989. https://doi.org/10.3390/land10090989

Academic Editors: Enrico Miccadei, Cristiano Carabella and Giorgio Paglia

Received: 26 August 2021
Accepted: 17 September 2021
Published: 19 September 2021

Publisher's Note: MDPI stays neutral with regard to jurisdictional claims in published maps and institutional affiliations.

Copyright: © 2021 by the authors. Licensee MDPI, Basel, Switzerland. This article is an open access article distributed under the terms and conditions of the Creative Commons Attribution (CC BY) license (https://creativecommons.org/licenses/by/4.0/).

Abstract: Data driven methods are widely used for the development of Landslide Susceptibility Mapping (LSM). The results of these methods are sensitive to different factors, such as the quality of input data, choice of algorithm, sampling strategies, and data splitting ratios. In this study, five different Machine Learning (ML) algorithms are used for LSM for the Wayanad district in Kerala, India, using two different sampling strategies and nine different train to test ratios in cross validation. The results show that Random Forest (RF), K Nearest Neighbors (KNN), and Support Vector Machine (SVM) algorithms provide better results than Naïve Bayes (NB) and Logistic Regression (LR) for the study area. NB and LR algorithms are less sensitive to the sampling strategy and data splitting, while the performance of the other three algorithms is considerably influenced by the sampling strategy. From the results, both the choice of algorithm and sampling strategy are critical in obtaining the best suited landslide susceptibility map for a region. The accuracies of KNN, RF, and SVM algorithms have increased by 10.51%, 10.02%, and 4.98% with the use of polygon landslide inventory data, while for NB and LR algorithms, the performance was slightly reduced with the use of polygon data. Thus, the sampling strategy and data splitting ratio are less consequential with NB and algorithms, while more data points provide better results for KNN, RF, and SVM algorithms.

Keywords: landslide; susceptibility; machine learning; GIS; Kerala

1. Introduction

Catastrophic landslides in mountainous terrains interact with human environment and cause adverse impacts on lives and properties [1]. Aids for managing the risk due to landslides is a topic of which several decades of research has been devoted [2,3]. Mapping the spatial distribution of landslide hazard is one of the most-adopted strategies for risk management, as the landslide susceptibility maps can be used by the government for strategic planning and development [4]. With the recent advancements in Machine Learning (ML) techniques and computational facilities, Landslide Susceptibility Mapping (LSM) have become much easier.

Data driven methods are extensively used for LSM, and the earlier statistical methods using Geographical Information System (GIS)-based approaches are now being replaced by advanced ML algorithms. Different ML algorithms are being widely used for this purpose [5], and the literature shows that no single ML algorithm can be said to be the best for LSM. The choice of an ML algorithm for a particular region is subjected to the scientific goals and objectives of the LSM [5]. Five different algorithms are considered in this study, viz., Naïve Bayes (NB), Logistic Regression (LR), K Nearest Neighbors (KNN), Random Forest (RF), and Support Vector Machines (SVM). All the algorithms are popular in LSM, but the best suited model for each scenario has to be decided by a quantitative comparison of the model performances. The data used for training and testing of the ML algorithm should be prepared with utmost care, as the quality of data is the key parameter which decides the performance of any ML model. The data includes the landslide inventory and the Landslide Conditioning Factors (LCF). The LCFs are selected considering the topographical and meteo-geological conditions of the study area, and most conditioning factors are often derived from Digital Elevation Models (DEM), satellite data, and existing regional maps. In most cases, landslide inventories are obtained based on satellite images and field investigations [6].

Even though the quality of LCFs are found to be satisfactory and with good resolution DEMs available from satellite-based missions such as TanDEM-X and ALOS, the landslide inventories are often incomplete [7]. The quality of the landslide inventory is subjected to the positional accuracy and sampling strategy. In many studies, the inventories are prepared using points representing landslide crowns. The training and testing data for LSM are prepared using the data from all LCFs extracted using the landslide points. Hence, the positional accuracy of the inventory significantly affects the dataset used for testing and training. When the region is affected by shallow landslides only, the Crown Point provides a satisfactory representation of the landslide-affected area. However, when a region is affected by long runout landslide events, such as debris flows and avalanches, the runout zones cannot be represented using single point information [7], and the events can cause adverse effects in downslope areas [1,8]. The LCFs of the initiation zones and runout zones are entirely different, and a model which is trained using only the initiation zones will ignore the runout zones that may be affected by landslides [9,10]; however, in most studies, landslides are represented using point data, due to the limitations in data availability [11]. Hence, in this study, both point data (single point at the crown of landslide) and polygon data (cluster of points covering the area affected by landslides) are used for LSM. Each point in the cluster represents a cell in the landslide body and is used for LSM. The difference between both the approaches is that the point data considers only the crown area, while the polygon data considers the whole area affected by landslides, including the crown and the runout zone.

The resampling technique of cross validation is a recent advancement in ML, applied to test cases with limited data samples [5]. k-fold cross validation techniques are being widely used for LSM applications, in which the data is split into k parts and are internally resampled such that k−1 parts are used for training and 1 part of testing at each stage of sampling. Even though the method is being widely used for the purpose of validation, there are no guidelines for the number of k to be chosen for an analysis, and, in most studies, the value is chosen as 5 or 10 arbitrarily [12]. The number of k decides the ratio of train to test data, which can affect the performance of the ML model. Hence, in this study, the value of k is also varied from 2 to 10 in order to find the optimum value of k for each algorithm.

To test the objectives, the Wayanad district in Kerala, India, was selected as the test site. The district has suffered from a number of landslides after the incessant rains that occurred during monsoon seasons of 2018, and the landslide inventory data of 2018 was used for LSM.

2. Study Area

The Wayanad district is in the southern part of India (Figure 1), which belongs to Western Ghats, the most prominent orographic feature of the peninsular India. This district is highly prone to landslides [13,14] and has a total area of 2130 km^2, of which 40% is covered by forests. The topography falls mostly in plateau region sloping towards east, for this hilly district is located at the southern tip of Deccan plateau. A major share of the district contributes to the east-flowing river Kabani and its tributaries (Figure 1). The natural drainage system is constituted by a number of streams, rivulets, and small springs, and the district landscape with flood plains and ridges is formed by this drainage system. Many debris flows that have occurred in the district have runout distances of a few hundred meters, and the longest one ranges up to 3 km. All these slides have contributed to the process of landscape evolution in the district, and minor order streams are originated along the debris flow paths. Thus, the development of drainage paths and watersheds are highly related to the occurrence of landslides, especially debris flows in the region. The flood plains are formed by alluvial deposits with a thickness of more than 10 m. The northwest, southwest, and western parts of the region are formed by higher elevation hill ranges, with steep slopes and a rugged topography. Most of the forest areas are also along these hilly regions. The continuous erosion, transportation, and deposition of the rocks have resulted in the formation of valleys in between the hill ranges. The long runout debris flows that are common in the region also contribute to this process of landscape evolution. Geologically, the district is composed of a peninsular gneissic complex, charnockite group, Wayanad group, and the migmatite complex [15]. Bands of the Wayanad group are found in the northern part of the district, while the rocks of south and southeast are formed by the charnockite group [15]. The northcentral part is composed of a peninsular gneissic complex and the southcentral part is of the migmatite complex.

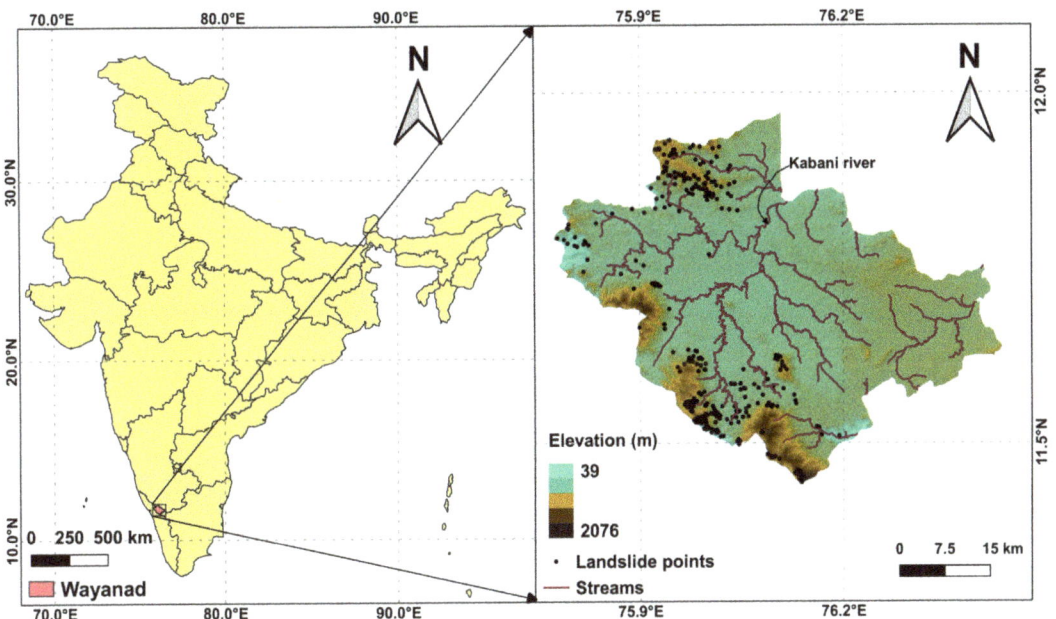

Figure 1. Location map of Wayanad.

A major share of the district is covered by reddish-brown lateritic soil with higher fine content. The forest zones are covered by forest soil with rich organic content, and the riverbanks are formed by thick alluvial deposits. The larger regolith thickness often leads to the bed erosion and bulking of landslides, which increases the landslide volume and destruction potential [16].

The district is highly affected by geohazards such as landslides and floods due to its topographic and geomorphological conditions. The highly dissected hills and valleys along the west, northwest, and southwest parts of the district are highly prone to landslides. During August of 2018, the district was affected by a number of landslides due to torrential rains [17]. A total of 388 landslides (Figure 1) were mapped within the district using government reports and pre- and post-event satellite images from Google Earth, and have been verified using a recently published dataset [18]. The inventory data were prepared separately for LSM, and are different from the dataset used for previous studies conducted by the authors [13] in which they derived the rainfall thresholds for the region. For deriving rainfall thresholds, multiple landslides occurring on the same day were considered to be a single landslide event, and approximate locations were used, as the focus was on the day of occurrence of the landslide event. However, the inventory data of LSM needs to be accurate, and the spatial distribution of landslides is more important than the time of occurrence of landslides. Hence, the high resolution satellite images available from Google Earth were utilized to prepare a separate landslide inventory database of 388 landslides which occurred in 2018 alone. The district faced major setbacks during the disaster and the catastrophic landslides repeated in the years 2019 and 2020 as well. The increasing frequency of landslides in the districts calls for an updated landslide susceptibility map using data-driven approaches.

3. Methodology

This study aims at evaluating the uncertainties in LSM using ML by adopting different ML algorithms, sampling strategies, and train to test ratios. The first step was the preparation of the dataset, starting from the landslide inventory. The data has to be preprocessed before using it for training and testing. Five different ML approaches were used in this study for comparison.

3.1. Machine Learning Algorithms

Data-based methods are often used to solve real-world problems when the knowledge of the theoretical part is limited and the data is of a large size [19]. Being a non-linear problem, ML models are highly suitable for LSM. The algorithms can learn the association between the occurrence or non-occurrence of landslides and the LCFs using the landslide and no landslide points used for training. Five different ML algorithms are considered in this study, which are explained as follows:

3.1.1. Naïve Bayes

The name of the NB algorithm is formed by two words, 'Naïve' and 'Bayes'. While the latter word stands for the Bayes (named after Thomas Bayes) theorem, which is used for calculating the conditional probability of the occurrence of landslides, in NB, the first term stands for the assumption that the algorithm naively considers all parameters to be independent of each other. The use of simple Bayes' theorem helps the model to have good mathematical control and the results can be achieved fast by using an NB algorithm [20]. The equation for calculating conditional probability of occurrence of landslide (L), subject to the occurrence of conditioning factors C (C_1 to C_n) is given in the following equation:

$$P(L|\ C_1, C_2, \ldots\ldots C_n) = \frac{P(L) \times P(C_1, C_2, \ldots C_n|L)}{P(C_1, C_2, \ldots C_n)} \tag{1}$$

The advantage of an NB algorithm is its simplicity and lower calculation time. The model does not require any hyper parameter tuning and can be easily implemented on any dataset. The major limitation is its assumption of independent parameters. The assumption does not hold true for most of the real-world problems and hence the algorithm may not provide reliable results when the parameters are highly dependent on each other. The algorithm has been used in LSM for more than a decade [21].

3.1.2. Logistic Regression

An LR algorithm is formed from regression analyses, deriving a linear relationship amongst the LCFs by using coefficients [22]. This algorithm, which is derived from statistics, produces a regression output in the form of a mathematical function, and can calculate the probability of the occurrence of landslides. The sigmoid function or logistic function, which is used in this algorithm, is where the name of LR originates. The sigmoid function in 'S' shape is a core part of LR, which sets an asymptote, based on the positive or negative values of x. For positive values of x, an asymptote is set to $y = 1$, and for negative values of x, asymptote is set for $y = 0$.

The algorithm is easy to implement and does not require any hyper parameter tuning. The model finds its application in LSM due to this simplicity and its usage of probability to predict the solution. A non-linear relationship is established with the landslide and non-landslide points and LCFs and finds a fitting function. The probability of the occurrence of landslides $P(L)$ is calculated by LR as follows:

$$P(L) = \frac{e^x}{1+e^x} \qquad (2)$$

where, x is a linear fitting function, using the LCFs, given by:

$$z = a_0 + a_1C_1 + a_2C_2 + \cdots + a_nC_n \qquad (3)$$

where, a_0 is the intercept, $a_1, a_2 \ldots a_n$ are the regression coefficients, and $C_1, C_2, \ldots C_n$ are the LCFs. For dependent variables in binary form and large input data with minimum duplicates and minimum multi collinearity, the algorithm can produce satisfactory results in LSM [23].

3.1.3. K-Nearest Neighbors

The classification of a data point using a KNN algorithm is carried out by using the properties of the neighboring data points [20]. It is a more efficient form of the ball tree concept [24], which can be applied to larger dimensions. The algorithm is widely used in LSM applications [25] and the probability of a data point to be allocated in any class is determined by the classification of its nearest neighbors [26]. The data point takes the classification in which the maximum number of its neighbors is classified. The number of K shall be decided by tuning process for better results.

KNN is classified as a non-parametric model, as the computation process does not depend upon the distributions of the dataset. This is another advantage while using KNN for LSM applications where the number of parameters is more and the data seldom fits to standard distributions. For a set of unclassified points, the algorithm calculates the distance from each point to find K closest neighbors. The classification of these neighbors are then used for voting, and the classification with the maximum votes is assigned to the unclassified data point.

3.1.4. Random Forest

As the name indicates, RF is a combination of many Decision Trees (DT) and the concept was developed in 1995 [27]. Each DT has nodes and branches. The decisions are made at nodes and the classification continues on a particular branch based on the decision. The decisions are continued by considering all LCFs, and each DT assigns a class for the object. RF then considers the class predicted by all DTs and assigns a class for the object based on voting. Each DT is a subset of the whole dataset, and is independently sampled by bootstrapping. The randomness of selection at each node is the major advantage of RF model, which often results in highly accurate predictions, making it suitable for LSM [21,28–30].

The use of splitting at nodes, bootstrapping, and several number trees reduces overfitting in RF by increasing randomness. The model can be fine-tuned by varying the depth of trees, number of trees to be combined, and the number of features considered at each node.

3.1.5. Support Vector Machines

The SVM algorithm classifies a data point using a hyperplane in a multidimensional space, first proposed by Vapnik and Lerner [31–33]. The hyperplanes are boundaries that decide the classification of an object. The number of LCFs used for the analysis determines the dimensions of the hyperplane. For each dataset, multiple hyperplanes are possible, which can classify the points into different classes. Hence, the SVM algorithm should choose a hyperplane which can maximize the distance between the data points of both classes using statistical learning theory [31,34]. The distance is maximized in order to accommodate the future data points. The data points which are located near to the hyperplane determine the orientation and position of the hyperplane, and these data points are called support vectors.

The SVM algorithm classifies the objects by using different kernel functions, and the choice of kernel function is critical in the results produced by the algorithm. The algorithm is widely used for LSM applications [29,35] and has been in practice since the 2000s [34].

3.2. Data Collection and Sampling Strategies

The landslide inventory map for the study was prepared manually after interpreting satellite images before and after the event. A total of 388 landslides which occurred in 2018 were identified within the boundary of Wayanad. The 2018 disaster was chosen for the study as the district was widely affected by this particular event. The locations where historical landslides were reported were affected, and many new landslides were also reported. Two datasets were prepared from the landslide data collected (Figure 2). In the first approach, the landslide was represented by a point in the crown area and, in the second method, the shape of landslide was demarcated using pre- and post-satellite images; the polygon was marked as inventory data. The district was highly affected by long runout debris flows, as 309 events out of the total 388 were classified as debris flow events. Among the remaining events, 68 were shallow landslides and 11 were rock falls or rockslides. The 388 landslides were represented by 388 cells in the first sampling strategy (Figure 2a) and 9431 cells using the second strategy (Figure 2b). The developed landslide susceptibility map thus provides the probability of occurrence of any of these landslide typologies in the region, and it is not specific for any single landslide typology. The debris flows have very long runout distances [16], and even the locations which are a few kilometers away from the crown points, with entirely different LCFs, were also affected. Hence, using point data for the training and testing of the model might ignore the probability of the occurrence of hazards in the runout zones. To avoid this issue, polygon inventory data was also used in the analysis. The polygon data represents all the cells affected by landslides, unlike the single point used in the first approach. However, the polygon does not differentiate between the crown area and the runout zone. The objective is to train the ML model to predict the probability of the occurrence of a landslide in each cell, and the focus of this manuscript is to compare the probabilities predicted by different approaches. The methodology does not differentiate between crown and landslide body, and checks only if the cell is affected by landslide or not.

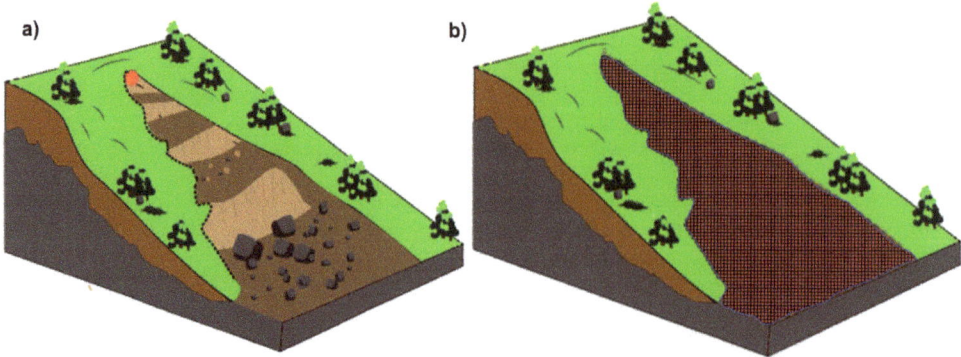

Figure 2. Different sampling strategies adopted in this study: (**a**) point data, and (**b**) polygon data.

The DEM for the study was collected from an Advanced Land Observing Satellite–Phased Array type L-band Synthetic Aperture Radar (ALOS PALSAR) [36], with a resolution of 12.5 m. All the other layers were also prepared in the same resolution as the DEM, and all GIS operations were carried out using QGIS version 3.10. The LCFs, such as slope, aspect, Stream Power Index (SPI), and Topographic Wetness Index (TWI) were derived from the DEM. The first LCF used in this study was elevation, which was directly obtained from the DEM. Slope angle is another significant factor which is critical in triggering the landslides. Slope is defined as the ratio of the vertical to horizontal distance between two points, expressed in terms of the tangent angle in degrees. The value of slope may vary between 0 and 90 degrees. The orientation of the sloping face is expressed using the direction and is termed as aspect. From previous studies, it was found that the value of aspect is critical when landslides occur after the formation of tension cracks in clay [37] and hence it is considered as an LCF. The value of aspect ranges from 0 to 360 degrees and it is classified into 9 categories based on the orientation.

The drainage map for the district was also prepared using the DEM. The locations of the streams were then verified using satellite images and were used for calculating the distance from the streams' layer, which is considered as an LCF. The observation from the inventory data was that many of the long runout debris flows occurred near the streams in the locality. The DEM was also used to create the flow accumulation map and the SPI and TWI layers were developed using the values of flow accumulation. Both SPI and TWI are significant in the process of the initiation of landslides, as SPI represents the power of a flowing water source to erode the material. As the values of the SPI ranges over multiple orders, the natural logarithm of SPI was used for calculation. TWI indicates the wetness of the location, which quantifies the topographic control on different hydrological processes.

The Normalized Difference Vegetation Index (NDVI) is considered to be an important LCF, as it indicates the amount of greenness of a location [38]. When the NDVI values are higher, it represents the presence of vegetation [39,40] and can be correlated with the canopy cover [23]. Thus, the NDVI values are maximum for forest regions and minimum for water bodies and non-vegetated surfaces. Most landslides have occurred within the forest region itself, and the long runout debris flows have originated in the forest area. The net cropped area is 1129.76 km^2, and a major share of cropped area is being used for perennial crops such as coffee, arecanut, and coconut [41]. The cash crops such as coffee and tea and spices such as cardamom are widely cultivated along the hill slopes, while the other crops are cultivated in flatter areas. The NDVI value was calculated using two bands of the electromagnetic spectrum, the Near Infra-Red (NIR) and Red (R) bands [42]. For Landsat 8 images, Band 5 represents NIR and Band 4 represents R. Hence, for this study, the NDVI values were calculated from Landsat 8 images captured in December 2017 and January 2018. As a major share of the cultivated areas is dedicated to perennial crops, the collected images can also satisfactorily represent the conditions at the time of landslides. From the collected images, NDVI is derived using the following formula:

$$NDVI = \frac{(Band\ 5 - Band\ 4)}{Band\ 5 + Band\ 4} \quad (4)$$

The rainfall data for the Wayanad district was collected from the Indian Meteorological Department (IMD) [43]. The data from four different rain gauge stations from 2010 to 2018 were interpolated using inverse distance weighted method of interpolation to get the average annual rainfall values across the district.

The geology, geomorphology, road network, and lineaments of the district were collected from maps published by the Geological Survey of India (GSI). The lineaments and roads were first rasterized and then used to develop the distance rasters, which were used as LCFs. The geology and geomorphology layers were classified and rasterized. The geology was classified into 7 groups, such as migmatite complex, charnockite, younger intrusive, basic intrusive, wayanad group, acid intrusive and peninsular gneissic complex (Figure 3). Geomorphologically, the region was classified into four categories: the highly dissected hills and valleys, moderately dissected hills and valleys, low dissected hills and valleys, and pediment complex. The collected layers are shown in Figure 3. The layers were then further processed to prepare the database for LSM.

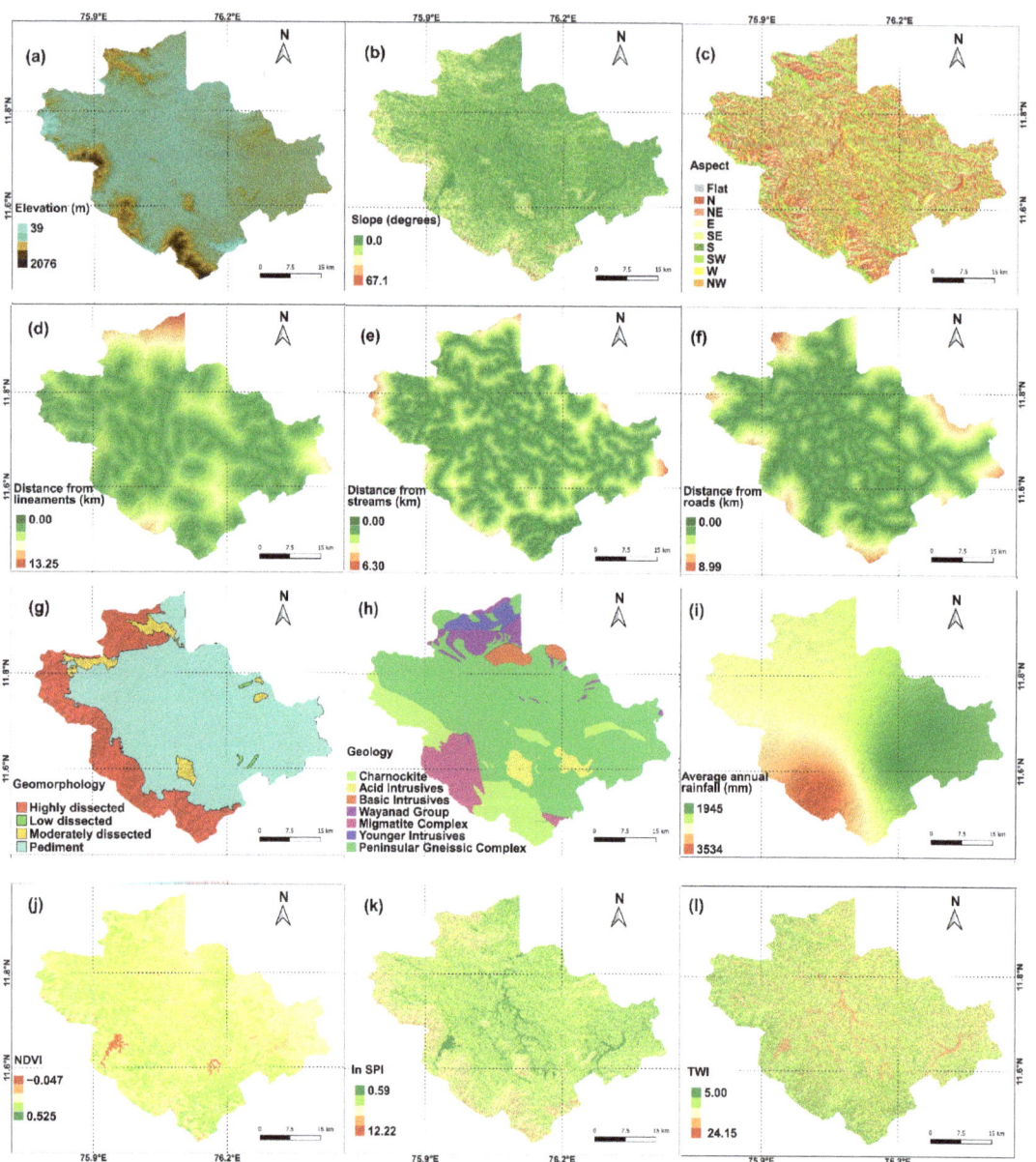

Figure 3. Different LCFs used for LSM: (**a**) elevation, (**b**) slope, (**c**) aspect, (**d**) distance from lineaments, (**e**) distance from streams, (**f**) distance from roads, (**g**) geomorphology, (**h**) geology, (**i**) rainfall, (**j**) NDVI, (**k**) ln SPII, and (**l**) TWI.

The processing of different LCFs is depicted in detail in Figure 4. The processing is different for raster and vector layers. The vector layers are first rasterized and then converted to XYZ format. For roads, streams, and lineaments, the distance from each feature is first calculated, and the distance rasters were used as LCF (Figure 3).

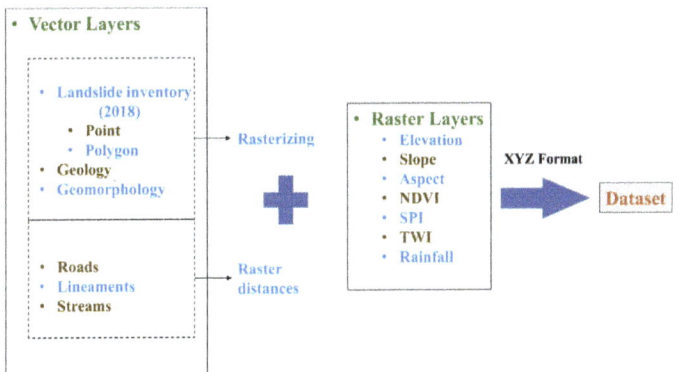

Figure 4. Schematic representation of dataset preparation from different spatial layers collected.

After preparing the landslide inventory data, an equal number of no landslide points were also prepared for the purpose of training and testing for both the sampling methods (point and polygon data). The landslide cells are represented using 1 and no landslide cells using 0 in the dataset. The data from all LCFs were then extracted for the landslide and no landslide points to develop the training and testing dataset. The derived model was later applied to the whole dataset to develop the landslide susceptibility map for the study area.

3.3. K-Fold Cross Validation and Data Splitting

Validation techniques are used to evaluate the performance of ML models. When the dataset is limited, cross validation techniques are often adopted to overcome the limitations associated with the size of the dataset. For k-fold cross validation, the value of k is the only input required, and the dataset is then divided into k different subsets or folds (Figure 5). Among the k-folds, k−1 folds are used for training the model and the last fold is used for testing. The process is repeated k−1 times so that each subset in the dataset is considered for testing.

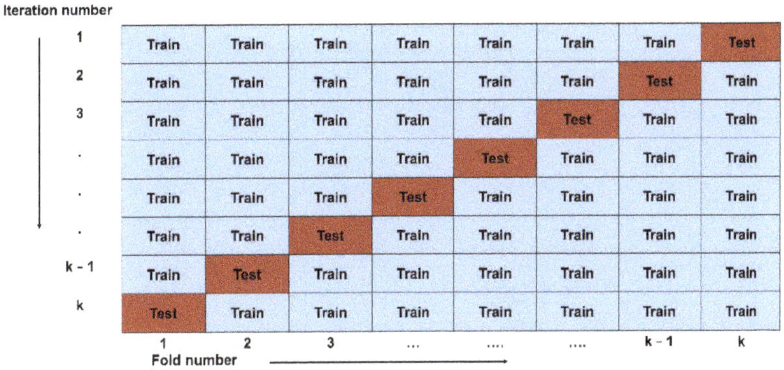

Figure 5. k-fold cross validation represented graphically.

The number of k decides the ratio of train to test ratio of validation and, in most studies, the value of k is randomly is chosen as 5 (train to test ratio 80:20) or 10 (train to test ratio 90:10) [44]. However, detailed studies on performance of cross validation suggest that repeated cross validation should be carried out to determine the optimum value of k [12].

3.4. Quantitative Comparison

The Receiver Operating Characteristic (ROC) curve approach is used for the quantitative comparison of different models. The curve is a plot between the False Positive Rate (FPR) on the x axis and the True Positive Rate (TPR) on the y axis. These parameters are calculated using a conventional confusion matrix where true positives are correctly predicted landslide points, true negatives are correctly predicted, no landslide points, false positives are incorrectly predicted, no landslide points and false negatives are landslide points missed by the model. From these four values, TPR and FPR are calculated as follows:

$$TPR = \frac{True\ Positives}{True\ Positives + False\ Negatives} \quad (5)$$

$$FPR = \frac{False\ Positives}{False\ Positives + True\ Negatives} \quad (6)$$

The plot with maximum Area Under the Curve (AUC) had the best performance. The landslide susceptibility maps were then prepared using the probabilities predicted by the derived ML models. The model predicts the probability of the occurrence of landslides in each cell, varying from 0 to 1. Based on the probability, the district is categorized into five [45–47] (0.0 to 0.2, 0.2 to 0.4, 0.4 to 0.6, 0.6 to 0.8, and 0.8 to 1.0) and the corresponding susceptibility classes are defined as very low, low, medium, high, and very high. The classification based on equal interval was chosen over the other approaches such as natural break and quantiles, as this study focuses on the comparison of probabilities predicted by different approaches. By using equal interval, the susceptibility classes predicted by each approach can be compared directly to evaluate the agreement or disagreement between the predicted probability values. In other approaches, relative values predicted by each model are used separately for defining the classes and hence the comparison of predicted probabilities is difficult. The statistical attributes such as accuracy and AUC do not provide insights into the agreement and disagreement between the different landslide susceptibility maps prepared. Hence, another parameter, called the Empirical Information Entropy (EIE), or H index, is used to evaluate the agreement between different maps. H index can be calculated as:

$$H = -\sum_{I=1}^{n} P(i) \log(P(i)) \quad (7)$$

where, $P(i)$ is the likelihood of the susceptibility class (very low, low, etc.) i, which is numbered from 1 to 5 in this study (1 is very low and 5 is very high), and n is the number of classes (5 in this case). When all the maps agree with each other, the value of H is zero and as the value increases; the disagreement also increases.

The value of the H-index can be used as an indication to quantify the mutual agreement between the landslide susceptibility maps considered [4]. When two landslide susceptibility maps are compared, there are two outcomes. When both the outcomes are same, the probability of occurrence of one susceptibility class becomes 1 and that of all the other classes are zero. Hence, the H-index becomes zero. In cases where both the outcomes are different, the probability of occurrence of two susceptibility classes is 0.5 and that of remaining classes are zero. The H-index value is the absolute value of twice the product of 0.5 and log(0.5); i.e., 0.30. When five landslide susceptibility maps are compared, the possible combinations of outcomes and H index values are given in Table 1 below. The number of landslide susceptibility maps predicting each class is interchangeable along the row, and all combinations result in the same value of H index.

From Table 1, it is clear that, as the value of H-index increases, the entropy increases [48], i.e., the disagreement between landslide susceptibility maps increases [4]. Hence, the value can be used to quantify the agreement amongst the results. If more landslide susceptibility maps predict the same class for a cell, the predicted results can be considered to be highly reliable.

Table 1. Possible H index values while comparing the landslide susceptibility maps produced using five algorithms.

Number of Landslide Susceptibility Maps Predicting Each Class					H-Index
Class 1	Class 2	Class 3	Class 4	Class 5	
5	0	0	0	0	0.00
4	1	0	0	0	0.22
3	2	0	0	0	0.29
3	1	1	0	0	0.41
2	2	1	0	0	0.46
2	1	1	1	0	0.58
1	1	1	1	1	0.70

The numbers in rows three to nine can be interchanged among the first five columns. The resulting H-index will remain the same.

4. Results

The performance of the test dataset was first evaluated using the ROC approach to find out the model with best performance. The analysis was carried out with the values of k ranging from 2 to 10 for algorithms, using both a point and polygon dataset, and the ROC curves are plotted in Figure 6.

The minimum and maximum accuracy of the model with NB algorithm and point data are 82.70% and 83.30%, respectively, and the corresponding AUC values are nearly the same, i.e., 0.903 and 0.904. The accuracy values remained the same, while the AUC values reduced when the polygon data is used with the NB algorithm. The trend is nearly the same for the LR algorithm as well. The AUC values are slightly better than NB, with the maximum value of 0.920 with point data. The pattern is different for the other three algorithms, and the performance is significantly improved with polygon data in all the three cases. With the point data, the maximum accuracy values are 84.71%, 88.12%, and 86.63% for KNN, RF, and SVM, respectively, while the maximum AUC values are 0.911, 0.954, and 0.930. With the use of polygon data, the maximum accuracy of KNN increased up to 95.22%, while that of RF became 98.14% and the same for SVM became 91.61%. The AUC values also increased up to 0.981, 0.993, and 0.963 for KNN, RF, and SVM, respectively. Another important observation is that the performance of SVM is better than KNN while using point data, with a difference of 1.92% in accuracy, albeit when polygon data is used. KNN performed better than SVM, with a difference of 3.61% accuracy (Table 2). In both the cases, the RF model outperforms the other models with the highest values of accuracy and AUC.

From Figure 6, it can be observed that the AUC values of KNN, RF and SVM have improved significantly by using polygon inventory data, while the variation is minimum in the case of NB and LR. Moreover, the effect of varying the value of k in k-fold cross validation is insignificant while using polygon data for NB, LR, and SVM algorithms, while, in the case of KNN and RF, variation in the number of folds can result in a variation of approximately 2% accuracy with polygon data. Even though the variation is not significant, the best performance of all models was obtained at k = 8, using point data. A summary of quantitative comparison is provided in Table 2, with the k values corresponding to minimum and maximum performances in the brackets.

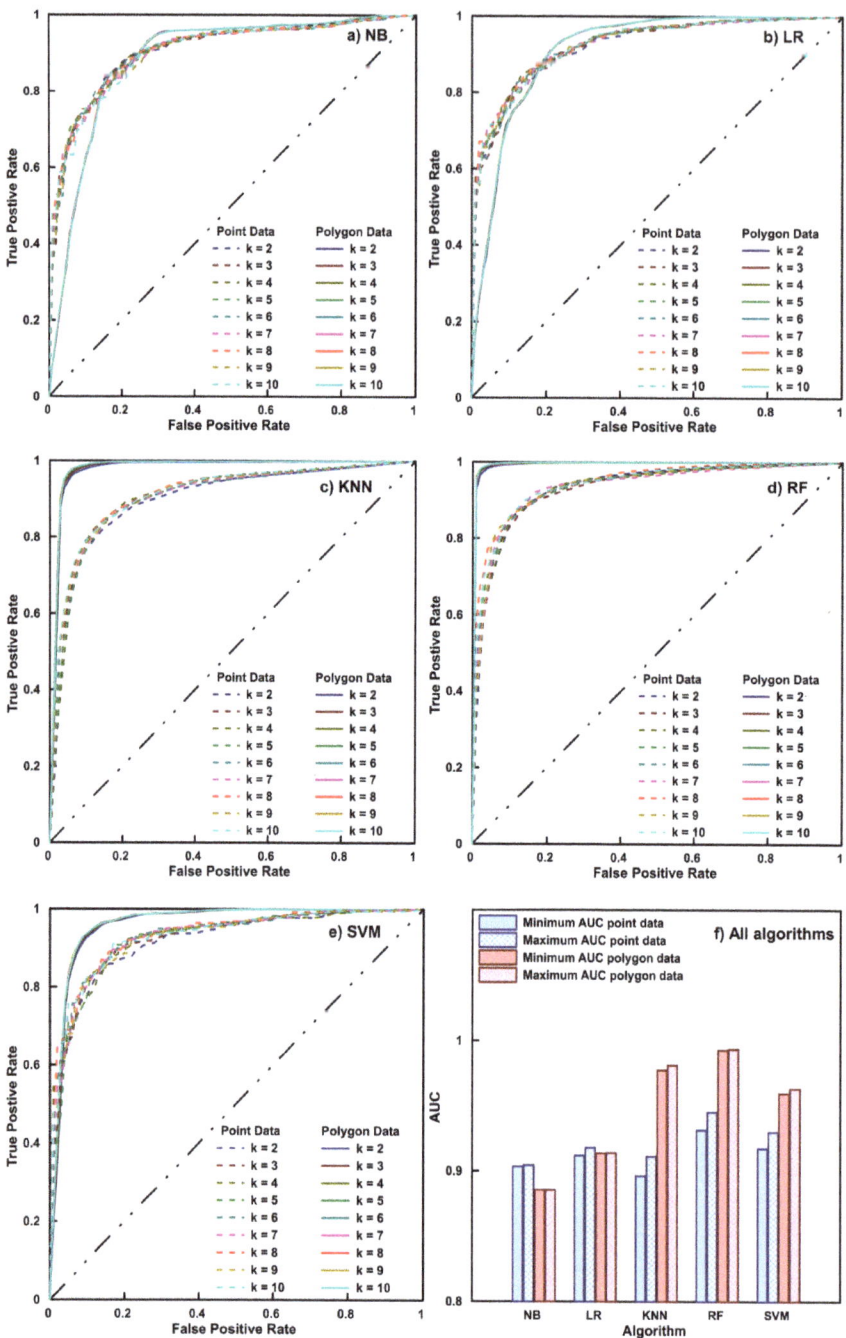

Figure 6. ROC curves, AUC, and accuracy of different models: (**a**) Naïve Bayes, (**b**) Logistic Regression, (**c**) K Nearest Neighbors, (**d**) Random Forest, (**e**) SVM, and (**f**) comparison of AUC of all five algorithms.

Table 2. Quantitative comparison of different algorithms, sampling strategies and data splitting using accuracy and AUC values.

Algorithm	NB	LR	KNN	RF	SVM
Point Data					
Min Accuracy (%) (k)	82.70 (3)	86.67 (3)	83.00 (2)	86.20 (3)	84.80 (2)
Max Accuracy (%) (k)	83.30 (8)	87.41 (8)	84.71 (8)	88.12 (8)	86.63 (8)
Min AUC (k)	0.903 (3)	0.912 (3)	0.896 (2)	0.932 (3)	0.917 (2)
Max AUC (k)	0.904 (8)	0.920 (8)	0.911 (8)	0.954 (8)	0.930 (8)
Polygon Data					
Min Accuracy (%) (k)	83.32 (2)	83.44 (2)	93.23 (2)	96.13 (2)	91.00 (2)
Max Accuracy (%) (k)	83.34 (6)	83.45 (5)	95.22 (8)	98.14 (9)	91.61 (9)
Min AUC (k)	0.885 (2)	0.914 (2)	0.977 (2)	0.992 (2)	0.959 (2)
Max AUC (k)	0.885 (6)	0.914 (5)	0.981 (8)	0.993 (9)	0.963 (9)

From the comparison of statistical performance obtained as per Figure 6 and Table 2, it can be observed that the RF algorithm with polygon inventory data is performing better than all other models. The performance of KNN and RF are comparable while using polygon data and the scores of RF and SVM are comparable while using point data. Still, the best suited model cannot be selected on the basis of statistical scores only. The choice needs a detailed understanding of the distribution of susceptibility classes and a detailed evaluation based on practical perspectives. The purpose of landslide susceptibility maps is to help the planners and authorities in making strategic decisions for future development. Hence, it is important to provide clear information about the susceptibility classes. Based on the value of probability of the occurrence of landslides, the district is divided into five susceptibility classes: very low, low, medium, high, and very high. The statistical attributes provide the prediction performance on the test data only [49]. From a practical perspective, a landslide susceptibility map with an acceptable performance should classify all the landslides correctly within the very high, high, or medium classes. At the same time, the model cannot be too conservative, which may restrict the developmental activities within a larger area. The landslide susceptibility maps prepared using both point and polygon data using each algorithm with the best performing model are evaluated in detail along with the H-index map for a better understanding of spatial agreement.

The number of pixels in each category and the number of landslides that occurred in each class are also important concerns. By using a reliable landslide susceptibility map, the landslides should occur within medium, high, and very high susceptible zones. The landslides which occur outside these zones are missed events, which should be considered with utmost care. Any model with an increased number of missed alerts fails to predict the possible occurrence of landslides.

The landslide susceptibility maps prepared using NB algorithm classifies 15.07% of the total area in the very high category with point data and 18.29% with polygon data (Figure 7). It can also be understood from Figure 7 that, among the 388 landslides considered, 72.64% occurred in very high classified areas, itself with point data, and the percentage increased to 80.64% using polygon data. Exactly 74.49% of the total area is classified as very low using point data and 73.27% using polygon data. The performance of the model is slightly reduced while using polygon data due to the increased number of false alarms within the increased percentage area covered by very high and high category. Considering the mutual agreement between the predictions made by both sampling strategies, 86.42% of the total predictions are in perfect agreement with each other (Figure 7c), while the classification of susceptibility predicted by both methods are different in the remaining area.

The LR algorithm classifies 6.90% of the total area as very high, 9.04% as high, 10.55% as medium, 22.21% as low, and 51.30% as very low susceptible classes using point data

(Figure 8). The number of landslides that occurred in the very high classified locations are reduced to 58.60% when compared with NB, but, at the same time, the number of landslides that occurred in the very low category was also reduced to 6.78%, which in turn slightly improved the performance of LR. While using polygon data, LR algorithm classifies 8.63% of the total area as very high, 7.82% as high, 9.03% as medium, 14.58% as low, and 59.95% as very low. Even though the missed alarms are reduced by this case, the increased number of false alarms resulted in a marginal decrease in accuracy and the AUC values. For 72% of the total area, the susceptibility class predicted using both point data and polygon data perfectly agreed with each other, with an H-index of 0.

From the AUC values (Figure 6), it is evident that the performance of KNN is comparable with NB and LR algorithms while using point data, but it has increased significantly while using polygon data. The reason for this is the drop in the areas classified into very high, high, and medium classes to 3.48%, 3.27%, and 3.44% while using polygon data when compared to 7.15%, 7.60%, and 7.82% while using point data (Figure 9). This reduction has resulted in a considerable reduction of false alarms and in the improvement of accuracy and AUC values. The variation is also reflected in the H-index map, as only 68.70% of the total area agrees with the prediction made using different sampling methods.

Similar to KNN, RF also shows a significant improvement in performance while using polygon data when compared to the point data. The reason is also very similar, as the percentage of very high, high, and medium classified points are reduced while using the polygon data. With the use of point data, 7.86% of the total area was classified under the very high category, which comprises 61.26% of the total landslide occurrences (Figure 10). However, with polygon data, 97.90% of the total landslides are happening within the 1.06% of the total area, which are classified into the very high category. The number of missed events is also reduced by using polygon data as only 0.13% and 0.06% of landslides occurring in the low and very low classified areas, respectively. The mutual agreement between the landslide susceptibility maps produced by point and polygon data is also the least in case of RF algorithm, as 71.20% of the total area has been classified into different categories by using different sampling strategies.

Similar to NB and LR, SVM also shows an increase in percentage of area classified into the very high category with the use of polygon data when compared with the landslide susceptibility map prepared using point data (Figure 11). However, the percentage increase in this category does not result in false alarms, as in the case of NB and LR, as most pixels classified as high and medium categories using point data were classified as in the very high category while using polygon data. Thus, the true positives have increased, and false negatives have been reduced by using polygon data, which in turn resulted in an increase in performance using polygon data. For 75.28% of the area, the categorization is same when using both point and polygon data, as depicted by the H-index plot.

While comparing the performance of different models, RF provides better performance by using both point and polygon data. Moreover, while using polygon data, the performance of KNN and RF are comparable and, while using point data, the performance of SVM and RF are comparable. Apart from statistical comparison, a better understanding of the pixel-wise distribution of susceptibility classes and mutual agreement between the landslide susceptibility maps can help in deciding the best suited landslide susceptibility map for a region.

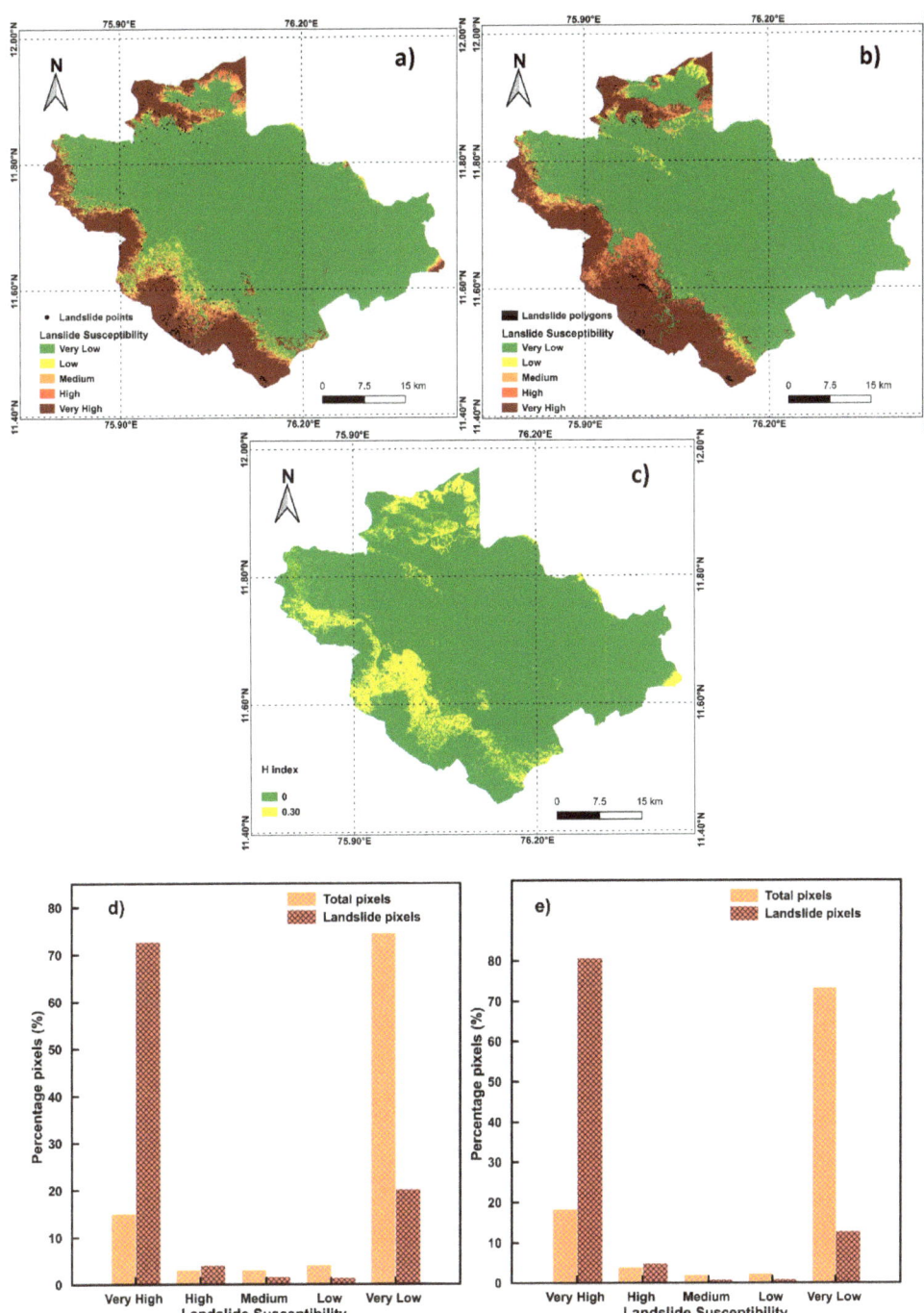

Figure 7. Details of landslide susceptibility maps prepared using NB algorithm: (**a**) using point data, (**b**) using polygon data, (**c**) H-index plot, (**d**) percentage distribution of using point data, and (**e**) percentage distribution of pixels using polygon data.

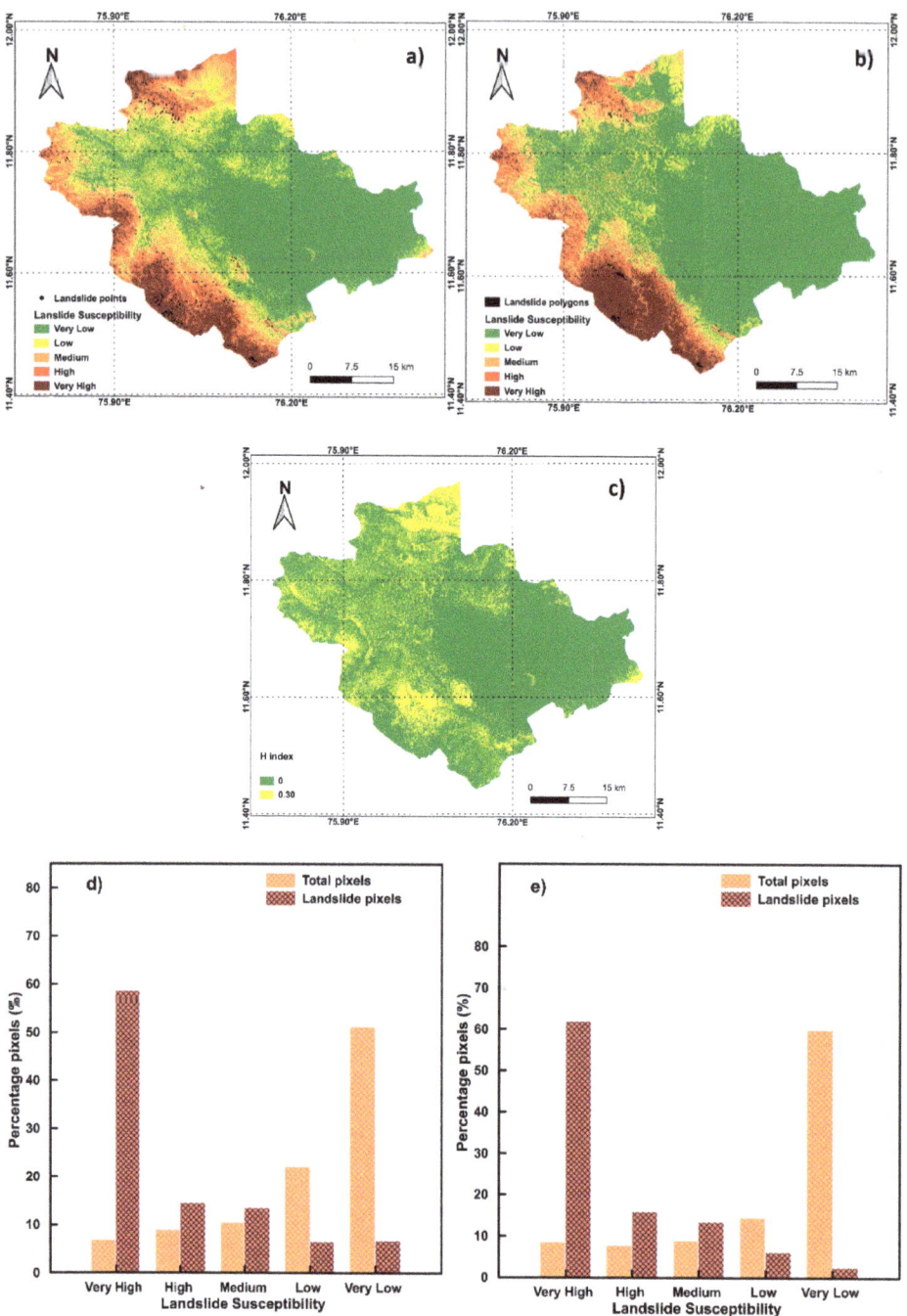

Figure 8. Details of landslide susceptibility maps prepared using LR algorithm: (**a**) using point data, (**b**) using polygon data, (**c**) H-index plot, (**d**) percentage distribution of pixels using point data, and (**e**) percentage distribution of pixels using polygon data.

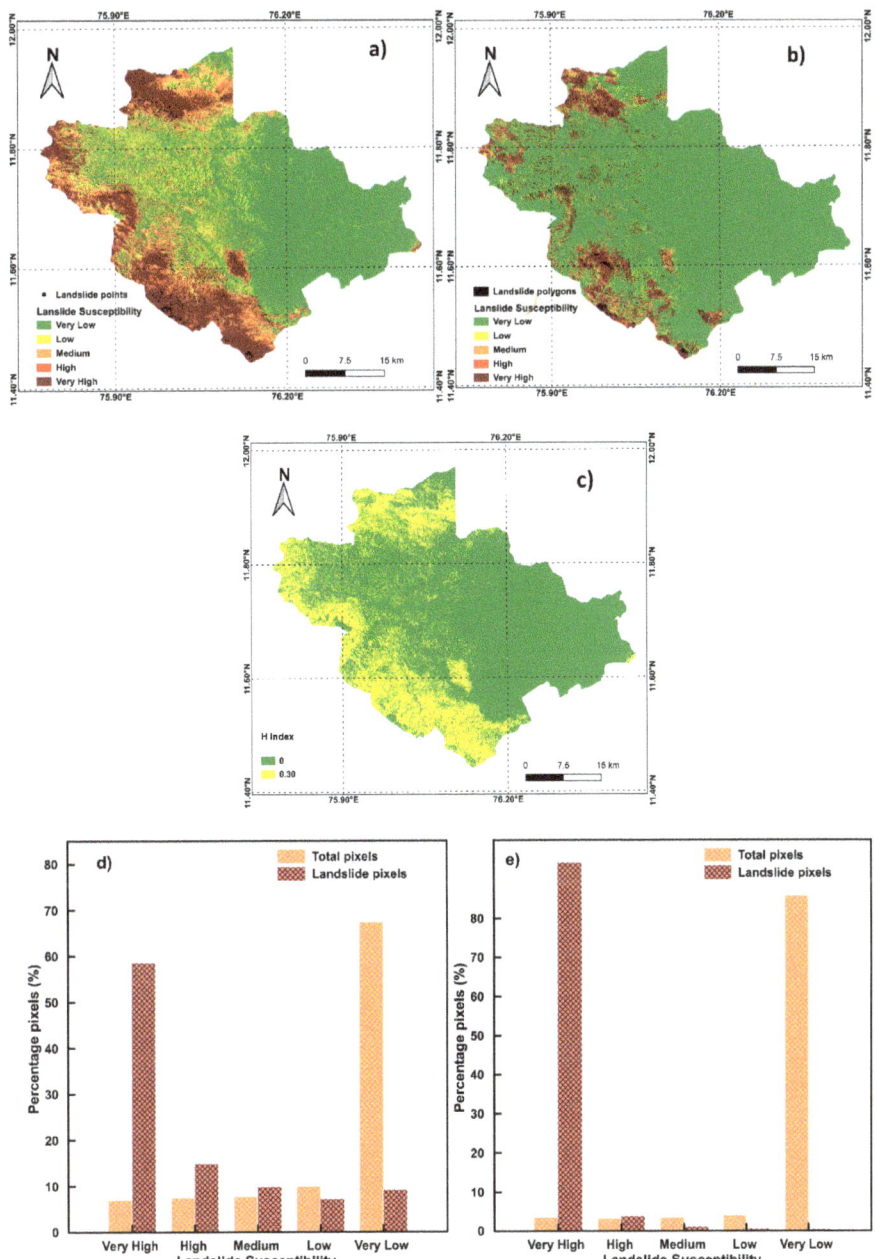

Figure 9. Details of landslide susceptibility maps prepared using KNN algorithm: (**a**) using point data, (**b**) using polygon data, (**c**) H-index plot, (**d**) percentage distribution of pixels using point data, and (**e**) percentage distribution of pixels using polygon data.

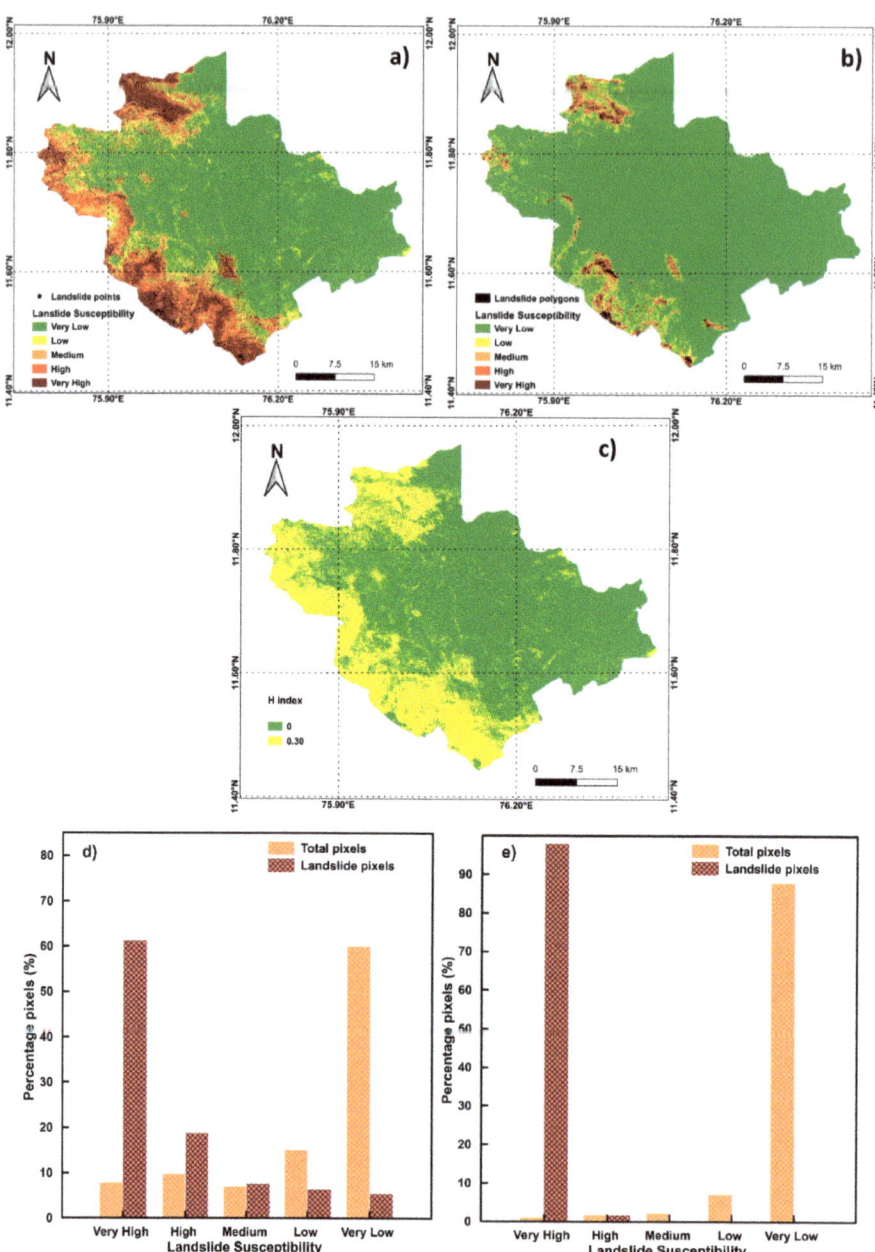

Figure 10. Details of landslide susceptibility map prepared using RF algorithm: (**a**) using point data, (**b**) using polygon data, (**c**) H-index plot, (**d**) percentage distribution of pixels using point data, and (**e**) percentage distribution of pixels using polygon data.

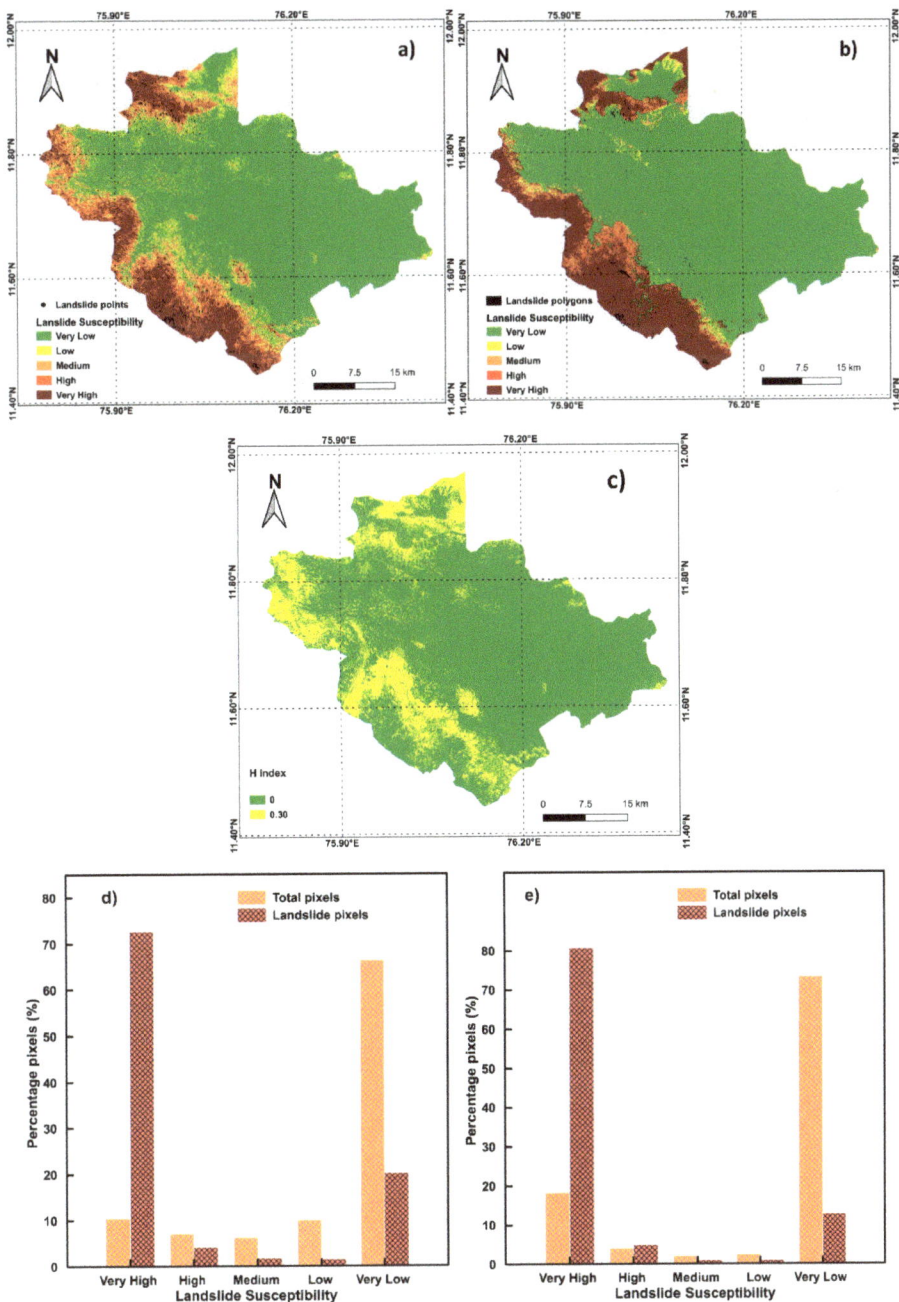

Figure 11. Details of landslide susceptibility maps prepared using SVM algorithm: (**a**) using point data, (**b**) using polygon data, (**c**) H-index plot, (**d**) percentage distribution of pixels using point data, and (**e**) percentage distribution of pixels using polygon data.

5. Discussion

From the obtained results (Table 2), it is evident that the choice of algorithm and sampling strategies can affect the prediction performance of a landslide susceptibility map significantly. The effect of data splitting is crucial for only RF, KNN, and SVM algorithms while using the point data for sampling. The landslide susceptibility maps and H-index plots provide more insights into the effects of different sampling strategies in the performance of different algorithms. From the H-index maps and AUC values, it is evident that the sampling strategy is least effective in the case of NB and most effective in the case of RF.

Figure 12 shows the H-index plots prepared to understand the mutual agreement between different algorithms using the same sampling strategy. It can be observed that, in the case of low susceptible area, the different algorithms are in good agreement with each other, and the LR algorithm classifies the least area in the very low category, which is 51.30% of the total area. While using point data, all algorithms agree in the classification of 47.56% of the total area and all algorithms differ in the case of 0.17% of the total area.

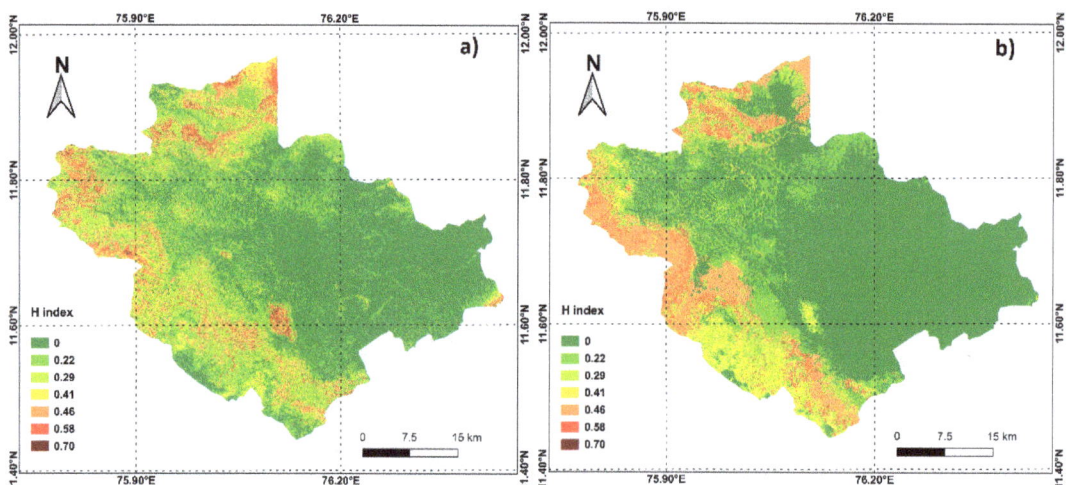

Figure 12. H-index maps plotted using all five algorithms with: (**a**) point data, and (**b**) polygon data.

The percentage distribution of each value of H-index is provided in Table 3 below. While using polygon data, the mutual agreement between algorithms is improved, with perfect agreement in 58.06% of the total area. In no pixels, the classification of all algorithms is entirely different and at least two algorithms agree with the predicted classification. As can be observed from Figure 12b and Table 3, there are no pixels with a H-index value of 0.70 when polygon data is used.

For NB and LR algorithms, the performance is reduced when a greater number of data points in the polygon dataset is used. This is a result of increased correlation between the LCFs with more data points, which violates the basic assumption of independent variables in both the cases. The use of linear fitting function in the case of LR also results in a slight decrease in the accuracy and AUC values with the increased number of data points. However, the advantage of using these algorithms is the reduced computational time involved, as they do not require any hyper parameter tuning.

Table 3. Percentage distribution of H-index values in the total area, using different sampling strategies: comparison between all algorithms.

H Index	Point Data	Polygon Data
	Percentage Pixels (%)	
0.00	47.56	58.06
0.22	19.31	13.25
0.29	13.04	7.47
0.41	8.93	6.96
0.46	7.52	11.48
0.58	3.47	2.77
0.70	0.17	0.00

In the case of KNN, SVM, and RF, the ratio of the train to test dataset can also result in a performance variation while using point data. The performance of these algorithms is significantly increased with the use of polygon data. The improvement in performance can be attributed to the improved size of data used for training the model. All three models demand a long time for the fine-tuning process. The models are highly sensitive to the parameters, train to test ratio, and the size of the dataset [5]. All the three models are widely used for LSM and, hence, if computational facilities are available, the train to test ratio should also be varied to produce the best results from these algorithms.

Even though the performance is comparable with KNN and RF, a higher number of landslides in the very low category make the landslide susceptibility maps made using SVM unsuitable for practical applications. This is an important aspect to be considered. From Figure 11, it is evident that the model using polygon data with an AUC of 0.963 is classifying 13% of the landslides in the very low susceptible zone. This is visible in the landslide susceptibility maps in Figure 11b. The performance can be further improved by using different data sampling approaches and ensemble algorithms and neural networks. In the case of RF, even though the results are statistically better from a practical perspective, the very high, high, and medium classes are bounded by the polygon data used for training and the model is too optimistic, which does not leave room for possible landslides in the surrounding areas in the future. The same issue is observed with the landslide susceptibility map prepared with the KNN algorithm using polygon data. Even though these three algorithms (KNN, SVM, and RF) are having the highest statistical attributes, they cannot be considered to be the best suited for the landslide susceptibility map, due to the limited part of the study area classified into very high, high, and medium classes. The landslide susceptibility map must be conservative, which considers the possible occurrence of landslides in areas other than the ones used for training and testing, and, at the same time, should not classify the safe zones as landslide-susceptible regions. The landslide susceptibility map produced using the RF algorithm with point data is an optimum solution with good statistical performance (AUC = 0.952 and accuracy = 88.12%) and practical applications. It classifies 7.87% of the total area into the very high category and 9.79%, 7.09%, 15.17%, and 60.08% into the high, medium, low, and very low categories, while the best performing model is developed using RF with polygon dataset, with an accuracy of 97.30% and an AUC of 0.993.

From the results, it can be inferred that both the choice of algorithm and sampling strategy can influence the prediction performance of LSM, but the choice of the landslide susceptibility map should not be based on the statistical performance only.

6. Conclusions

The influence of the choice of the ML algorithm, sampling strategies, and data splitting for LSM is evaluated in detail using a case study from the Wayanad district in Kerala.

12 LCFs were used to develop different models using five different ML algorithms (NB, LR, KNN, RF, and SVM), two sampling strategies (point data and polygon data), and different values of k in k-fold cross validation. The results show that data splitting is least effective among the considered parameters. The performance of NB and LR are unaffected by the variation of k values, but the performance of KNN, RF, and SVM are slightly varied by k values, with the best performance at k = 8 in all cases using point data.

The performance of NB and LR did not improve with the use of a large dataset with polygon inventory. The inter dependency of parameters is a critical factor affecting the performance of these algorithms while, in the case of KNN, RF, and SVM, the performance is significantly improved with the use of polygon data. By comparing the H index values, it was observed that the landslide susceptibility maps perfectly agreed with each other in the case of 47.56% of the total area while using point data and 58.06% while using polygon data.

The results produced by KNN and RF using the polygon dataset have a very good statistical performance with very high values for accuracy and AUC. The best performing model developed using an RF algorithm and polygon dataset has an accuracy of 97.30% and an AUC of 0.99.

Author Contributions: Conceptualization, M.T.A., N.S. and B.P.; methodology, M.T.A. and N.S.; data curation, M.T.A. and R.L.; writing—original draft preparation, M.T.A.; writing—review and editing, B.P. and A.A.; supervision, N.S. and B.P. All authors have read and agreed to the published version of the manuscript.

Funding: The study is supported by the Centre for Advanced Modelling and Geospatial Information Systems (CAMGIS), University of Technology Sydney. This research was also supported by Researchers Supporting Project number RSP-2021/14, King Saud University, Riyadh, Saudi Arabia.

Institutional Review Board Statement: Not applicable.

Informed Consent Statement: Not applicable.

Data Availability Statement: The publicly archived datasets used for the analysis are cited in the manuscript. The analysis has been carried out at Indian Institute of Technology Indore, and the derived data can be provided upon request to the corresponding author (Biswajeet.Pradhan@uts.edu.au).

Acknowledgments: The authors express their sincere gratitude to Geological Survey of India, Kerala SU, District Soil Conservation Office Wayanad and Kerala State Disaster Management Authority (KSDMA) for their support throughout the study. Authors would like to thank three anonymous reviewers for their critical reviews which helped to improve the quality of the manuscript.

Conflicts of Interest: Authors declare no conflict of interest.

References

1. Froude, M.J.; Petley, D.N. Global fatal landslide occurrence from 2004 to 2016. *Nat. Hazards Earth Syst. Sci.* **2018**, *18*, 2161–2181. [CrossRef]
2. Dou, J.; Yunus, A.P.; Merghadi, A.; Shirzadi, A.; Nguyen, H.; Hussain, Y.; Avtar, R.; Chen, Y.; Pham, B.T.; Yamagishi, H. Different sampling strategies for predicting landslide susceptibilities are deemed less consequential with deep learning. *Sci. Total Environ.* **2020**, *720*, 137320. [CrossRef]
3. Abraham, M.T.; Satyam, N.; Rosi, A.; Pradhan, B.; Segoni, S. Usage of antecedent soil moisture for improving the performance of rainfall thresholds for landslide early warning. *Catena* **2021**, *200*, 105147. [CrossRef]
4. Pradhan, B.; Sameen, M.I. *Effects of the Spatial Resolution of Digital Elevation Models and Their Products on Landslide Susceptibility Mapping*, 1st ed.; Pradhan, B., Ed.; Springer: Cham, Switzerland, 2017; ISBN 978-3-319-55341-2.
5. Merghadi, A.; Yunus, A.P.; Dou, J.; Whiteley, J.; Pham, B.T.; Bui, D.T.; Avtar, R.; Abderrahmane, B. Machine learning methods for landslide susceptibility studies: A comparative overview of algorithm performance. *Earth Sci. Rev.* **2020**, *207*, 103225. [CrossRef]
6. Reichenbach, P.; Rossi, M.; Malamud, B.D.; Mihir, M.; Guzzetti, F. A review of statistically-based landslide susceptibility models. *Earth Sci. Rev.* **2018**, *180*, 60–91. [CrossRef]
7. Korup, O.; Stolle, A. Landslide prediction from machine learning. *Geol. Today* **2014**, *30*, 26–33. [CrossRef]
8. Li, Y.; Liu, X.; Han, Z.; Dou, J. Spatial proximity-based geographically weighted regression model for landslide susceptibility assessment: A case study of Qingchuan area, China. *Appl. Sci.* **2020**, *10*, 1107. [CrossRef]

9. Simon, N.; Crozier, M.; de Roiste, M.; Rafek, A.G. Point based assessment: Selecting the best way to represent landslide polygon as point frequency in landslide investigation. *Electron J. Geotech. Eng.* **2013**, *18*, 775–784.
10. Süzen, M.L.; Doyuran, V. Data driven bivariate landslide susceptibility assessment using geographical information systems: A method and application to Asarsuyu catchment, Turkey. *Eng. Geol.* **2004**, *71*, 303–321. [CrossRef]
11. Tien Bui, D.; Shirzadi, A.; Shahabi, H.; Geertsema, M.; Omidvar, E.; Clague, J.; Thai Pham, B.; Dou, J.; Talebpour Asl, D.; Bin Ahmad, B.; et al. New ensemble models for shallow landslide susceptibility modeling in a semi-arid watershed. *Forests* **2019**, *10*, 743. [CrossRef]
12. Rodríguez, J.D.; Pérez, A.; Lozano, J.A. Sensitivity analysis of k-fold cross validation in prediction error estimation. *IEEE Trans. Pattern Anal. Mach. Intell.* **2010**, *32*, 569–575. [CrossRef]
13. Abraham, M.T.; Satyam, N.; Rosi, A.; Pradhan, B.; Segoni, S. The selection of rain gauges and rainfall parameters in estimating intensity-duration thresholds for landslide occurrence: Case study from Wayanad (India). *Water* **2020**, *12*, 1000. [CrossRef]
14. Abraham, M.T.; Satyam, N.; Rosi, A. Empirical rainfall thresholds for occurrence of landslides in Wayanad, India. *EGU Gen. Assem.* **2020**, 5194. [CrossRef]
15. Department of Mining and Geology Kerala. *District Survey Report of Minor. Minerals*; Department of Mining and Geology Kerala: Thiruvananthapuram, India, 2016.
16. Abraham, M.T.; Satyam, N.; Reddy, S.K.P.; Pradhan, B. Runout modeling and calibration of friction parameters of Kurichermala debris flow, India. *Landslides* **2021**, *18*, 737–754. [CrossRef]
17. United Nations Development Programme. *Kerala Post Disaster Needs Assessment Floods and Landslides-August 2018*; United Nations Development Programme: Thiruvananthapuram, India, 2018.
18. Hao, L.; van Westen, C.; Martha, T.R.; Jaiswal, P.; McAdoo, B.G. Constructing a complete landslide inventory dataset for the 2018 monsoon disaster in Kerala, India, for land use change analysis. *Earth Syst. Sci. Data* **2020**, *12*, 2899–2918. [CrossRef]
19. Dou, J.; Yunus, A.P.; Bui, D.T.; Merghadi, A.; Sahana, M.; Zhu, Z.; Chen, C.W.; Khosravi, K.; Yang, Y.; Pham, B.T. Assessment of advanced random forest and decision tree algorithms for modeling rainfall-induced landslide susceptibility in the Izu-Oshima Volcanic Island, Japan. *Sci. Total Environ.* **2019**, *662*, 332–346. [CrossRef] [PubMed]
20. Pedregosa, F.; Varoquaux, G.; Gramfort, A.; Michel, V.; Thirion, B.; Grisel, O.; Blondel, M.; Prettenhofer, P.; Weiss, R.; Dubourg, V.; et al. Scikit-learn: Machine learning in Python. *J. Mach. Learn. Res.* **2011**, *12*, 2825–2830.
21. Miner, A.; Vamplew, P.; Windle, D.J.; Flentje, P.; Warner, P. A Comparative study of various data mining techniques as applied to the modeling of landslide susceptibility on the Bellarine Peninsula, Victoria, Australia. In Proceedings of the 11th IAEG Congress of the International Association of Engineering Geology and the Environment, Auckland, New Zealand, 5–10 September 2010; pp. 1327–1336.
22. Cabrera, A.F. Logistic regression analysis in higher education: An applied perspective. *High. Educ. Handb. Theory Res.* **1994**, *10*, 225–256.
23. Huang, X.; Wu, W.; Shen, T.; Xie, L.; Qin, Y.; Peng, S.; Zhou, X.; Fu, X.; Li, J.; Zhang, Z.; et al. Estimating forest canopy cover by multiscale remote sensing in northeast Jiangxi, China. *Land* **2021**, *10*, 433. [CrossRef]
24. Omohundro, S.M. *Five Balltree Construction Algorithms*; Tech. Rep. TR-89-063; International Computer Science Institute (ICSI): Berkeley, CA, USA, 1947.
25. Marjanovic, M.; Bajat, B.; Kovacevic, M. Landslide susceptibility assessment with machine learning algorithms. In Proceedings of the 2009 International Conference on Intelligent Networking and Collaborative Systems, IEEE, Barcelona, Spain, 4–6 November 2009; pp. 273–278.
26. Bröcker, J.; Smith, L.A. Increasing the reliability of reliability diagrams. *Weather Forecast.* **2007**, *22*, 651–661. [CrossRef]
27. Ho, T.K. Random decision forests. In Proceedings of the International Conference on Document Analysis and Recognition, ICDAR, Montreal, QC, Canada, 14–16 August 1995; Volume 1, pp. 278–282.
28. Chen, W.; Sun, Z.; Zhao, X.; Lei, X.; Shirzadi, A.; Shahabi, H. Performance evaluation and comparison of bivariate statistical-based artificial intelligence algorithms for spatial prediction of landslides. *ISPRS Int. J. Geo Inf.* **2020**, *9*, 696. [CrossRef]
29. Zhou, W.; Wu, W.; Lin, Z.; Zhang, G.; Chen, R.; Song, Y.; Wang, Z.; Lang, T.; Qin, Y.; Ou, P.; et al. Zonation of landslide susceptibility in Ruijin, Jiangxi, China. *Int. J. Environ. Res. Public Health* **2021**, *18*, 5906. [CrossRef] [PubMed]
30. Zhang, Y.; Wu, W.; Qin, Y.; Lin, Z.; Zhang, G.; Chen, R.; Song, Y.; Lang, T.; Zhou, X.; Huangfu, W.; et al. Mapping landslide hazard risk using random forest algorithm in Guixi, Jiangxi, China. *ISPRS Int. J. Geo-Inf.* **2020**, *9*, 695. [CrossRef]
31. Cortes, C.; Vapnik, V. Suppport vector networks. *Mach. Learn.* **1995**, *20*, 273–297. [CrossRef]
32. Vapnik, V.N. *The Nature of Statistical Learning Theory*; Springer: New York, NY, USA, 1995.
33. Vapnik, V.; Lerner, A.Y. Recognition of patterns with help of generalized portraits. *Avtomat. Telemekh* **1963**, *24*, 774–780.
34. Yao, X.; Dai, F.C. Support vector machine modeling of landslide susceptibility using a GIS: A case study. *IAEG2006* **2006**, *793*, 1–12.
35. Gao, R.; Wang, C.; Liang, Z.; Han, S.; Li, B. A research on susceptibility mapping of multiple geological hazards in Yanzi river basin, China. *ISPRS Int. J. Geo Inf.* **2021**, *10*, 218. [CrossRef]
36. Alaska Satellite Facility Distributed Active Archive Center (ASF DAAC) Dataset: ASF DAAC 2015, ALOS PALSAR Radiometric Terrain Corrected high res; Includes Material© JAXA/METI 2007. Available online: https://asf.alaska.edu/data-sets/derived-data-sets/alos-palsar-rtc/alos-palsar-radiometric-terrain-correction/ (accessed on 13 December 2020).

37. Capitani, M.; Ribolini, A.; Bini, M. The slope aspect: A predisposing factor for landsliding? *Comptes Rendus Geosci.* **2013**, *345*, 427–438. [CrossRef]
38. Zhang, T.; Han, L.; Zhang, H.; Zhao, Y.; Li, X.; Zhao, L. GIS-based landslide susceptibility mapping using hybrid integration approaches of fractal dimension with index of entropy and support vector machine. *J. Mt. Sci.* **2019**, *16*, 1275–1288. [CrossRef]
39. Achour, Y.; Pourghasemi, H.R. How do machine learning techniques help in increasing accuracy of landslide susceptibility maps? *Geosci. Front.* **2020**, *11*, 871–883. [CrossRef]
40. Ray, R.L.; Jacobs, J.M.; Cosh, M.H. Landslide susceptibility mapping using downscaled AMSR-E soil moisture: A case study from Cleveland Corral, California, US. *Remote Sens. Environ.* **2010**, *114*, 2624–2636. [CrossRef]
41. Department of Economics and Statistics Government of Kerala Official website of Department of Economics & Statistics, Government of Kerala. Available online: http://www.ecostat.kerala.gov.in/index.php/agri-state-wyd (accessed on 5 September 2021).
42. Fiorucci, F.; Ardizzone, F.; Mondini, A.C.; Viero, A.; Guzzetti, F. Visual interpretation of stereoscopic NDVI satellite images to map rainfall-induced landslides. *Landslides* **2019**, *16*, 165–174. [CrossRef]
43. India Meteorological Department (IMD) Data Supply Portal. Available online: http://dsp.imdpune.gov.in/ (accessed on 3 May 2019).
44. Sun, D.; Xu, J.; Wen, H.; Wang, Y. An optimized random forest model and its generalization ability in landslide susceptibility mapping: Application in two areas of Three Gorges Reservoir, China. *J. Earth Sci.* **2020**, *31*, 1068–1086. [CrossRef]
45. Ou, P.; Wu, W.; Qin, Y.; Zhou, X.; Huangfu, W.; Zhang, Y.; Xie, L.; Huang, X.; Fu, X.; Li, J.; et al. Assessment of landslide hazard in Jiangxi using geo-information technology. *Front. Earth Sci.* **2021**, *9*, 178. [CrossRef]
46. Chalkias, C.; Ferentinou, M.; Polykretis, C. GIS-based landslide susceptibility mapping on the Peloponnese Peninsula, Greece. *Geosciences* **2014**, *4*, 176–190. [CrossRef]
47. El-Fengour, M.; El Motaki, H.; El Bouzidi, A. Landslides susceptibility modelling using multivariate logistic regression model in the Sahla Watershed in northern Morocco. *Soc. Nat.* **2021**, *33*. [CrossRef]
48. Sharma, S.; Mahajan, A.K. Information value based landslide susceptibility zonation of Dharamshala region, northwestern Himalaya, India. *Spat. Inf. Res.* **2019**, *27*, 553–564. [CrossRef]
49. Frattini, P.; Crosta, G.; Carrara, A. Techniques for evaluating the performance of landslide susceptibility models. *Eng. Geol.* **2010**, *111*, 62–72. [CrossRef]

Article

Definition of Environmental Indicators for a Fast Estimation of Landslide Risk at National Scale

Samuele Segoni * and Francesco Caleca

Department of Earth Sciences, University of Florence, Via La Pira 4, 50121 Florence, Italy; francesco.caleca@unifi.it
* Correspondence: samuele.segoni@unifi.it; Tel.: +39-055-275-5975

Abstract: The purpose of this paper is to propose a new set of environmental indicators for the fast estimation of landslide risk over very wide areas. Using Italy (301,340 km^2) as a test case, landslide susceptibility maps and soil sealing/land consumption maps were combined to derive a spatially distributed indicator (LRI—landslide risk index), then an aggregation was performed using Italian municipalities as basic spatial units. Two indicators were defined, namely ALR (averaged landslide risk) and TLR (total landslide risk). All data were processed using GIS programs. Conceptually, landslide susceptibility maps account for landslide hazard while soil sealing maps account for the spatial distribution of anthropic elements exposed to risk (including buildings, infrastructure, and services). The indexes quantify how much the two issues overlap, producing a relevant risk and can be used to evaluate how each municipality has been prudent in planning sustainable urban growth to cope with landslide risk. The proposed indexes are indicators that are simple to understand, can be adapted to various contexts and at various scales, and could be periodically updated, with very low effort, making use of the products of ongoing governmental monitoring programs of Italian environment. Of course, the indicators represent an oversimplification of the complexity of landslide risk, but this is the first time that a landslide risk indicator has been defined in Italy at the national scale, starting from landslide susceptibility maps (although Italy is one of the European countries most affected by hydro-geological hazards) and, more in general, the first time that land consumption maps are integrated into a landslide risk assessment.

Keywords: landslide; Italy; risk; soil sealing

Citation: Segoni, S.; Caleca, F. Definition of Environmental Indicators for a Fast Estimation of Landslide Risk at National Scale. *Land* **2021**, *10*, 621. https://doi.org/10.3390/land10060621

Academic Editors: Enrico Miccadei, Cristiano Carabella, Giorgio Paglia and Paul Aplin

Received: 7 May 2021
Accepted: 8 June 2021
Published: 9 June 2021

Publisher's Note: MDPI stays neutral with regard to jurisdictional claims in published maps and institutional affiliations.

Copyright: © 2021 by the authors. Licensee MDPI, Basel, Switzerland. This article is an open access article distributed under the terms and conditions of the Creative Commons Attribution (CC BY) license (https://creativecommons.org/licenses/by/4.0/).

1. Introduction

Landslide risk is the possibility that a landslide occurs in a specific area and in a specific period of time, causing damages to population, buildings, infrastructure and services [1,2]. As a consequence, landslide risk is influenced by the overlapping in time and space of hazardous areas (where landslides are likely to occur) and potentially vulnerable exposed elements, resulting in an impact that could cause damages or losses. This has been traditionally translated into mathematical form by the classical equation [1]:

$$R = H \cdot V \cdot E \quad (1)$$

where R is the risk, H is the hazard (the probability for a dangerous event of a given intensity to happen in a certain place and time), V is the vulnerability (the degree of loss expected from the element impacted by the landslide) and E is exposition (the value of the elements exposed to the event).

Following this approach, quantitative risk analyses have been mainly published for small areas or, at the most, in regional scale applications [2–9].

A quantitative landslide risk assessment for very large areas (e.g., an entire nation) is still a very challenging objective, as it requires facing technical and scientific issues such as availability of complete, homogeneous and good quality input data of different natures (pertaining at least to the fields of geology, economy, demography and civil engineering) [10].

Italy is no exception to this, and to date the main strategies for assessing landslide risk at the national scale in Italy were based on different strategies. For instance, [11] proposed a statistical analysis on the spatial variability of recorded fatalities, to account for societal landslide risk in the whole Italian territory. Recently, [12] proposed a set of landslide risk indicators based on freely accessible data from an online governmental platform, including exposed population, number of buildings and landslide hazard zones as defined by the Italian regulation. Although representing a very complete overview of landslide risk in Italy, this approach has the drawback of presenting a spatial aggregation at the municipal level and leaving unexploited some scientific products that have a finer spatial resolution, such as landslide susceptibility maps, which have been proposed for several Italian regions [13–17] and for the whole Italian territory [18], or monitoring products of the artificialization of the territory such as soil sealing maps, which monitor the evolution of the processes of artificialization of the territory at high spatial resolution (10 m) at yearly time steps [19]. However, the approach of addressing national scale landslide risk problems with a set of simple indicators, rather than with a full QRA, seem promising and quite consolidated in landslide studies [10,20,21]. Undoubtedly, indicators are, by definition, simple means to describe and comprehend a complex phenomenon and are widely used in environmental studies by scientists and governmental agencies.

The purpose of this manuscript is to propose a new set of environmental indicators to characterize landslide risk over very wide areas and to apply it to characterize the Italian municipalities. The novelty in the proposed approach is to use advanced and high-resolution thematic layers: already existing landslide susceptibility maps [18] are used to identify hazardous areas, and soil sealing maps are used as they have a high resolution and constantly updated representation of the spatial distribution of the elements at risk (soil sealing maps are released on a yearly basis to monitor the expansion of urban fabric [19,22]). At the same time, the general objective is keeping the resulting indexes easy to understand, quick to update and flexible enough to be adapted at varying spatial units. In its basic formulation, a spatially distributed Landslide Risk Index (LRI) is defined on a pixel basis at 50 m resolution. Afterwards, we show an application to the whole Italian territory, in which the LRI is aggregated at the municipal level following two different approaches, generating two additional indexes that can be used to gain useful understanding on the interferences between geomorphological slope dynamics and urban expansion, which give birth to landslide risk.

2. Materials and Methods
2.1. Test Site

The study area considered for this work is the whole Italian territory (301,340 km^2) (Figure 1a). Italy is a peninsula located in Southern Europe and extending into the Mediterranean Sea. It is characterized by two main mountain ranges: the Alps, to the north, which separate Italy from the rest of Europe, and the Apennines, forming the backbone of the peninsula and running from NW to SE.

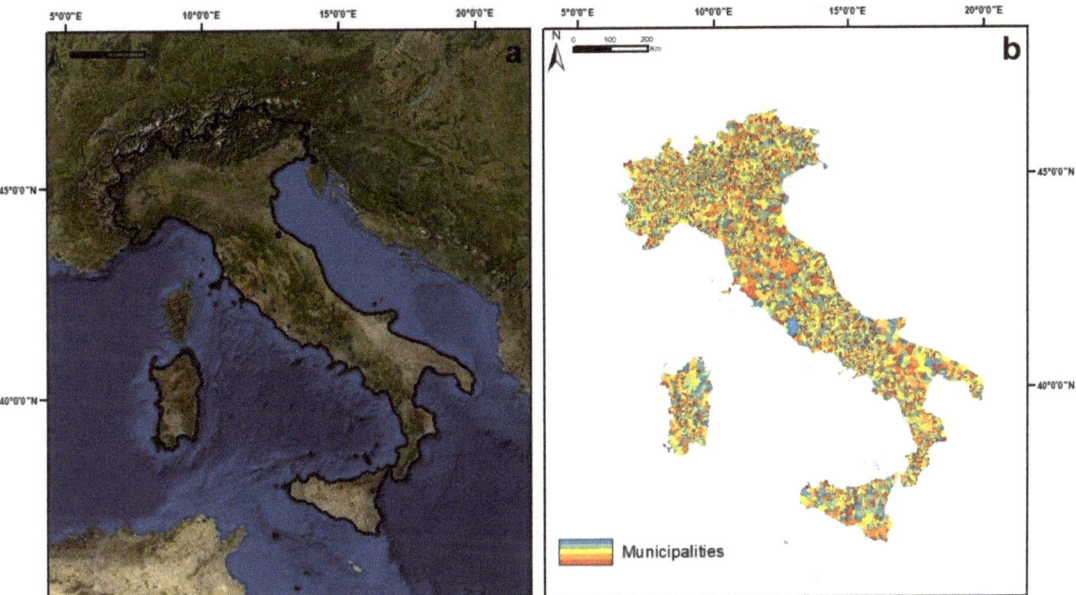

Figure 1. (a) Overview of Italy; (b) administrative subdivision into 7904 municipalities.

The geological setting and morphological features of the Italian peninsula are the result of a still active geological process that led to the formation of the two mountain chains [23]. The Alps are the typical example of a collisional belt: it was generated during the Cretaceous period by the convergence of the Adriatic continental upper plate (Argand's African promontory) and a subducting lower plate including the Mesozoic ocean and the European passive continental margin. In the Eocene, a complete closure of the ocean marked the onset of the Adria/Europe collision. The collisional zone is represented by the Austroalpine-Penninic wedge, a fossil subduction complex, showing that even coherent fragments of light continental crust may be deeply subducted in spite of their natural buoyancy [24]. The Apennines extend from the northwest part of the peninsula to the isle of Sicily, and link the western Alps with the Magrebian chain of North Africa [25]. The Apennines are a NW–SE oriented fold-and-thrust belt formed during the Oligocene period by the closure (started during the Cretaceous period), of the Mesozoic Tethys Ocean and following the collision between the European (Corso-Sardinian block) and African plates [26,27].

From a geomorphological point of view, Italy has a marked energy of relief: mountains are present in every Italian region and occupy more than the 35.2% of the territory. The greatest part of Italy, however, is characterized by hills, representing the 41.6% of the land surface. This juvenile morphological setting, in a still tectonically active territory, brings the consequence that landslide hazard is widespread in every part of Italy, excluding flat alluvial and coastal plains.

Landslide hazard is further exacerbated by climatic and meteorological constraints. Due to the large latitude range covered by Italy, the climate varies largely: from the cold climate of the north, EFH according to Koppen classification, typical of the highest mountain peaks, with annual precipitation higher than 2000 mm, to the Subtropical climate (BS in Koppen classification) of the southernmost coastal areas of Sicily, Apulia, Sardinia and Calabria, with long, hot, dry summers and precipitation less than 400 mm in Sicily [28]. Recently, due to the effects of climate change, periods of precipitation are becoming shorter and more intense in many parts of Italy [29,30], causing an increase in landslide activity and in the number of harmful landslide events per year [31,32].

For a full understanding of the application reported in this study, it is worth noting that Italy is subdivided into 7904 municipalities (Figure 1b), which represent the smallest administrative subdivisions of the territory and that have important responsibilities in territorial planning, urban design and risk management.

2.2. Landslides in Italy: National Inventory and Existing Susceptibility Maps

For the reasons explained above, each year hundreds to thousands of landslides affect Italy, causing victims and damages to buildings, infrastructure and cultural heritage [12,32–34]. An official landslide database exists at the national scale that is managed by ISPRA (National Institute for Research and Environmental Protection). The database is called IFFI (Italian National Landslide Inventory) and maps all known landslides (both active and inactive), mapped at the 1:10,000 scale by means of field surveys, remote sensing techniques and collection of ancillary data. According to IFFI, 620,808 landslides are present, covering about the 7.9% of the Italian territory. IFFI is openly accessible via an online platform [12] and it is acknowledged to be one of the most complete and homogeneous national-scale inventories in Europe [35–37]. IFFI is widely used as a base for landslide hazard and risk assessments at various scales [12,15,38–40].

In particular, in Italy, an overwhelming literature exists about landslide susceptibility studies. Landslide susceptibility maps (LSMs) represent, over appropriate spatial units, the spatial probability of the occurrence of landslides, and they are usually obtained by a statistical analysis of the spatial distribution of a set of predisposing factors [41]. Although LSMs do not contain temporal predictions, they are usually considered the starting point for landslide hazard and risk assessment. This is also the approach used for this work, but a literature review showed that most of the published LSMs refer to basin-scale studies [42–46]. Some examples of regional-scale susceptibility assessments also are present [13–16,47], but the use of a combination of regional maps obtained with different approaches to compose a nation-wide mosaic of landslide susceptibility would pose huge problems of consistency of the data. To our knowledge, the only LSM at the Italian scale available to be used as input data for this work is the national scale susceptibility assessment performed by [18]. The susceptibility assessment was performed separately for three different landslide typologies (rockfalls, rapid shallow slides, slow deep slides), producing three susceptibility maps at 50 m resolution. A Random Forest algorithm [48], which is a machine learning technique widely consolidated in LSM studies [49–51], was calibrated with the IFFI landslide inventory and a set of environmental variables including lithology, land cover, morphometric parameters (elevation, slope gradient, aspect, curvature), and hydrological parameters (topographic wetness index, stream power index, upslope contributing area). Overall, 196,087 sample points (50% randomly sampled inside landslides and 50% randomly sampled outside the mapped landslides) were used to train the Random Forest model and 84,641 independent points were used to quantify its accuracy in terms of AUC (area under receiver-operator characteristic curve), which is reported as 0.85.

2.3. Soil Sealing in Italy

In addition to the natural physical features (such as geological and climatic settings), anthropogenic dynamics are also deeply involved in landslide risk in Italy. On one hand, urban elements (such as buildings and infrastructure) may contribute to destabilizing slopes, acting as predisposing factors for landslide hazard. On the other hand, the ongoing expansion of urban fabric and infrastructure generates, at an alarming rate, new elements that are exposed to hazard, determining a relevant degree of landslide risk.

Since 2015, ISPRA has undertaken a nation-wide monitoring program of soil sealing. Soil sealing is the most intense form of artificial land take and it can be defined as the removal or covering of soil by buildings, constructions or other totally or partly impermeable artificial material [52]. Since then, every year, a national cartography of soil sealing is produced by remote sensing techniques and it is released as a raster map (pixel size 10*10 m), in which the whole Italian territory is classified into two classes: sealed soil/not sealed soil [53,54]. Sealed soil includes built-up areas, paved areas, railways, airports, ports and even reversible land consumption such as dirt roads [54]. Although all those elements are not distinguished from each other, the information conveyed by the soil sealing maps is very useful for the aim of this study because it provides useful information (updated on a yearly basis) about all anthropic elements exposed to risk, with a relatively very high spatial resolution.

To this regard, it should be stressed that, typically, urban areas in Italy are not clustered and are characterized by a peculiar diffuse pattern (referred to as "sprawl" and "sprinkling") [19]. As a consequence, other land cover/land use monitoring products (such as Corine Land Cover) are not able to adequately capture the spatial and temporal evolution of this phenomenon [54]. Moreover, the remote sensing techniques developed by ISPRA are specifically conceived and calibrated to detect the diffuse and scattered patterns of Italian urban fabric [19].

2.4. Methodology

Italian regulation (D. P. C. M. 27/12/1998) dictates that environmental assessment should be performed by subdividing the environment into environmental components, each of them described and characterized by indicators, which are parameters used to describe a given phenomenon and that should have the following characteristics: being concise, easy to understand and easy to measure and update.

In this framework, the objective of this study is to propose a set of indictors at a national scale to characterize landslide risk by depicting how urban expansion interferes with geomorphological slope processes. To this end, we started with some input data that consists of the outputs of bigger ongoing or concluded research activities, and we combined them by means of GIS analyses.

Input data are:

- Susceptibility maps of Italy at 50 m spatial resolution (as described in Section 2.2) [18]. Three separate maps exist, each focusing on a peculiar kind of landslides typically affecting Italian territory: rockfalls, shallow rapid slides, and deep-seated slow slides. Each map is in raster format and each raster cell expresses, with a numerical susceptibility index ranging from 0 to 100, the spatial probability of occurrence of a landslide of that typology.
- Soil sealing map of Italy, which identifies in the Italian territory the soil sealed or consumed by anthropic activities. In its basic form, the map can be used to subdivide the territory into (semi)natural soil cover and artificially covered soil, but the latter category is not further subdivided into sub-classes and the elements contributing to soil sealing cannot be assessed. Considering the scale of application, the scarce thematic accuracy is compensated by a high spatial and temporal accuracy: the map is in raster format, at 10 m pixel size, and is updated yearly. In this work, the most recent update available was used (monitoring of the reference year 2019, officially released in 2020). The map can be visualized as a binary raster assuming value 1 where sealed soil has been detected and 2 where it has not.
- Shapefile of municipalities borders, with reference coordinate system WGS84.

In short, the procedure consists of identifying a landslide risk following a revised and simplified version of Equation (1). For our purposes, hazard is considered equal to the spatial probability of occurrence (thus, equal to susceptibility). Over the susceptibility we superimpose the spatial distribution of anthropic elements (depicted by the soil sealing map), in order to consider elements at risk only on a presence/absence basis. Vulnerability is neglected (mathematically it is considered equal to 1 in Equation (1)) for different reasons: first, it would be nearly impossible to assess separately the physical vulnerability of each element (e.g., buildings) at national scale (and, to our knowledge this is a still unattempted task); and second, the soil sealing map does not effectively allow for distinguishing between different typologies of buildings or infrastructure. Moreover, in national scale studies, the approach of considering vulnerability as equal to 1 (the maximum possible degree) is considered a viable and cautionary approach [12]. The resulting index is then aggregated at the municipality basis.

The first step of the proposed procedure consists of blending the three susceptibility maps into a single information. It can be considered quite unlikely that, in a single spatial unit of the susceptibility map (pixel with 50 m size), two or more landslides of different typologies could be contemporarily present. Indeed, every predictive landslide model should first make a typological prediction, trying to predict what kind of landslide will take place [55]. As a consequence, the three susceptibility maps were imported into ArcGis software, and the "cell statistics" operation was performed to assess the "maximum" value. In this way, the output is a raster map in which the susceptibility index associated to each cell is the highest value found in the three input maps. This is equivalent to considering the landslide type with the highest susceptible value as the most probable to occur in a given location, and surmising that this landslide typology is the one that will be most likely affecting that area, controlling the related hazard. The resulting raster will be called "hazard index map" henceforth (Figure 2).

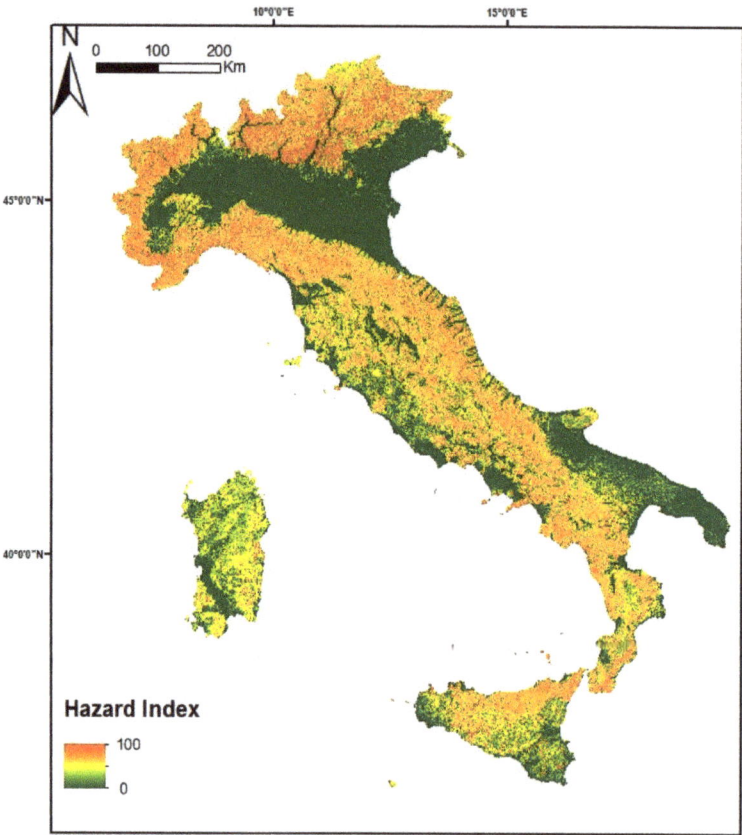

Figure 2. Hazard index map.

Before overlying the soil sealing map to the hazard index map, a procedure of homogenization is needed as the two raster maps have different cell sizes. Using ArcGis "block statistics" function, the resolution of the soil sealing raster was changed from 10 m to 50 m. Despite the loss of spatial resolution, this operation was necessary for the perfect match of soil sealing map with landslide hazard map, and some authors demonstrated that this spatial resolution is a good compromise in wide-area landslide hazard assessment studies [49]. The "minimum" statistics type was used: in this way the resulting raster obtains the value 1 (sealed soil) if at least one 10 m cell of sealed soil is present in each 50 m block. This choice determines a small expansion of the sealed soil that could be considered precautionary, and that is more desirable than alternate approaches. For instance, we verified that using the "majority" operator, many small infrastructure are completely neglected (e.g., roads cutting rural or mountain areas usually represent a small fraction of the 10 m pixels inside the 50 m block, and a relevant source of landslide risk would be completely ignored). In addition, it should be noted that the original soil sealing map represents the presence of sealed soil, but it is widely acknowledged [19] that the effects of the sealing may extend also to the surrounding areas (e.g., concerning hillslope hydrology, small surficial drainage systems connected to infrastructure could have discharge outlets a few meters away from the sealed area).

The resulting raster was reclassified, assuming a value of 1 in soil-sealed 50 m pixels and "no data" elsewhere. From a mathematical point of view, the reclassified soil sealing map and the hazard index raster were combined with a multiplication by means of the "raster calculator" tool of ArcMap. From the point of view of spatial information, the values "1" and "no data" in a multiplication act as a filter that maintains unaltered the input value of spatial probability of occurrence only in correspondence of anthropic elements, while far from them the index is not defined (conceptually, it is similar to assuming a risk equal to zero). This output raster was named Landslide Risk Index (LRI), because it accounts for the interaction between hazard and anthropic elements, giving a spatially distributed picture of how much they are exposed to landslide risk (Figure 3). It should be observed that a thorough assessment of the interaction between landslides and elements at risk would require accounting for the propagation of mass movements (for which run-out models would be necessary). This element is rarely encompassed in landslide susceptibility assessments, especially in wide-area applications; this shortcoming will be further investigated in the discussion of the results.

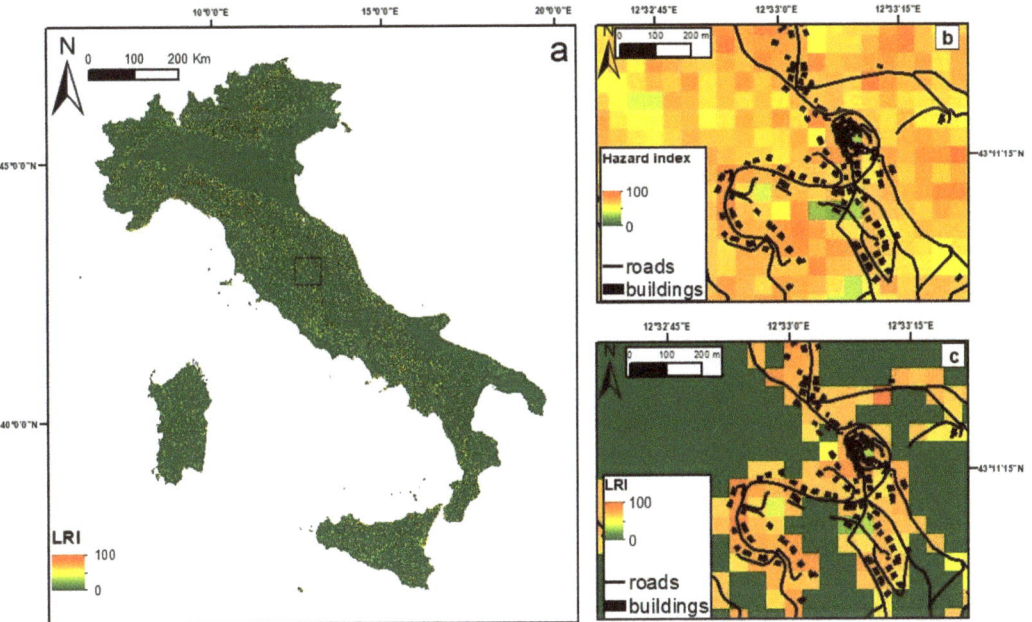

Figure 3. (**a**) Landslide Risk Index (LRI) map for the whole Italian territory; (**b**) Focus on hazard index map; (**c**) Focus on LRI map. Roads and buildings are from OpenStreetMap dataset.

LRI ranges from 0 to 100 and represents a spatially distributed indicator, which can be considered a basic element to be aggregated over larger spatial units in order to characterize them with respect to landslide risk. In this work, we derived from LRI two more indexes at municipal scale. The LRI raster and the shapefile of the borders of the Italian municipalities were overlaid in ArcMap and a "zonal statistics" was performed twice, using "mean" and "sum" to characterize each municipality with respect to two indexes named Average Landslide Risk (ALR) and Total Landslide Risk (TLR), respectively. The outcome of this operation represents the last step of the proposed procedure: the resulting indexes and a discussion about their interpretation are contained in the next section.

3. Results and Discussion

The TLR index (Figure 4) expresses for each mapping unit (municipalities in this study) the sum of the susceptibility values of all the cells with urbanized soil. Basically, this index cumulates for each administration the situations of interaction between spatial hazard and urbanized areas, expressing how much the development of the municipality has let hazardous areas to be "invaded" by constructions, infrastructure and services. In this regard, TLR could be used to describe the attention of an administration to harmonize the urban development with the main geomorphological hazard affecting its territory. Figure 4 shows that the Italian areas characterized by the highest TLR values are the Apennines (mainly the northern and central sectors), the isle of Sicily and, to a lesser extent, the eastern Alps. The drawback of this index is that it is sensitive to the extension of each aggregation unit: large municipalities have a greater chance than small ones to have a high TLR value, because of the higher number of pixels. For this reason, the value of the index does not have a fixed upper limit, and the value could theoretically tend to infinite, requiring particular attention for a correct interpretation. Indeed, when comparing different municipalities, a similarly high value of the index could be determined by many pixels with mid LRI values or by fewer pixels with higher LRI values. For this reason, the municipalities with the higher TLR index are large and densely urbanized municipalities. This result is not an artifact or a bias: the index effectively describes a recurring situation in some of the largest and most densely urbanized municipalities, which are exposed to a very high landslide risk in their territory because, during their urban expansion, they have had to cope with more hazardous areas than small municipalities. The highest values are found in the cities of Rome and Genova (both characterized by a very wide territory, densely populated and almost completely urbanized), and in the municipalities of Perugia, Gubbio and Messina, which are less populated but still have large portions of territory urbanized in hazardous areas (Figure 4). Nevertheless, TLR seems effective in highlighting the municipalities most affected by landslide risk, as the aforementioned territories correspond to areas where news about landslides continuously appear in newspapers and online blogs, as reported by [32]. In the last ten years, 4% of the landslide news catalogued and geotagged by their semantic engine is located in the aforementioned five municipalities with higher TLR values: in particular, 600 online news providers talked about landslides in Genova, 533 in Rome, and 235 in Messina.

Our results are further supported by the governmental data coming from ItaliaSicura web platform (http://mappa.italiasicura.gov.it/ last accessed on 31 May 2021), which collects the number of interventions and the economic resources allocated to mitigate hydrogeological risk in Italy. Rome is the Italian municipality with the highest number of interventions (64), likewise Genova has the highest total cost (about 378 m €) (however, it should be noted that data also include interventions for flood risk mitigation).

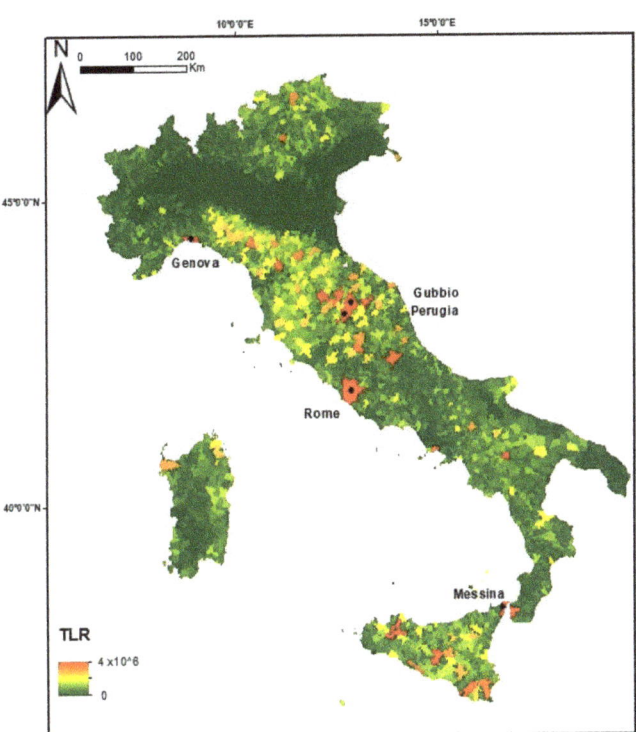

Figure 4. Characterization of the Italian municipalities with the Total Landslide Risk (TLR) index.

ALR index characterizes each municipality with the mean value of hazard found in correspondence of anthropic elements (Figure 5a). This index expresses, for each municipality, how hazardous is the portion of the territory where buildings, infrastructure and other services have been located. The values of the index range from 0 (minimum value) to 100 (maximum value): low values mean that the local administration has been cautious in planning urban development avoiding landslide risk, while high values are associated with municipalities where a consistent percentage of the urban structure has been built in hazardous areas, resulting in a relevant level of risk. It should be stressed that this does not necessarily mean that urban expansion has been recklessly planned: landslide hazard is so widespread in Italy that sometimes a municipality could be almost entirely interested by a relevant level of hazard posed by landslides or other geohazards (e.g., flood or volcanic activity). Nevertheless, also in such cases, ALR is an indicator that can be used to highlight situations where landslide risk is a very serious issue and should be carefully evaluated before further planning activities, or in the perspective of considering mitigation strategies. From a mathematical point of view, the value of ALR is independent from the areal extension of each municipality. However, a close investigation on the distribution of the values (Figure 5b) reveals that the highest values are found in small municipalities, most of them renowned international holiday destinations located by the sea, in rocky coasts (Positano, Amalfi, Capri, and Portofino, to name a few). We do not consider this outcome as a bias, and we explain it with a concurrence of factors of different nature. Firstly, in correspondence of many rocky and high-cliff coasts, the susceptibility to rockfalls presents very high values. Secondly, the territory of these municipalities is very steep and traditionally managed with the terracing method. This could be an effective method to cope with landslide hazard, but several studies highlighted that currently the loss of farmed land and the lack of maintenance seem to have recently increased the landslide hazard in these areas [56–58]. Thirdly, in the touristic locations with very high real estate value, the building of houses, accommodation facilities, infrastructure and services has been more intense than elsewhere. It has been driven mainly by market law and, especially in the last decades of the last century, not adequately counterbalanced by countermeasures concerning landslide hazard or environmental protection. This effect is particularly exacerbated in small municipalities, because the territory that can be used for urban expansion is limited and causes a severe competition between economic interests (urban expansion to support tourism and investments on the real estate market) and geomorphological processes. This is particularly alarming because small municipalities usually have scarce resources (both in terms of funds and manpower) to effectively face emergencies or to manage in-house risk mitigation strategies.

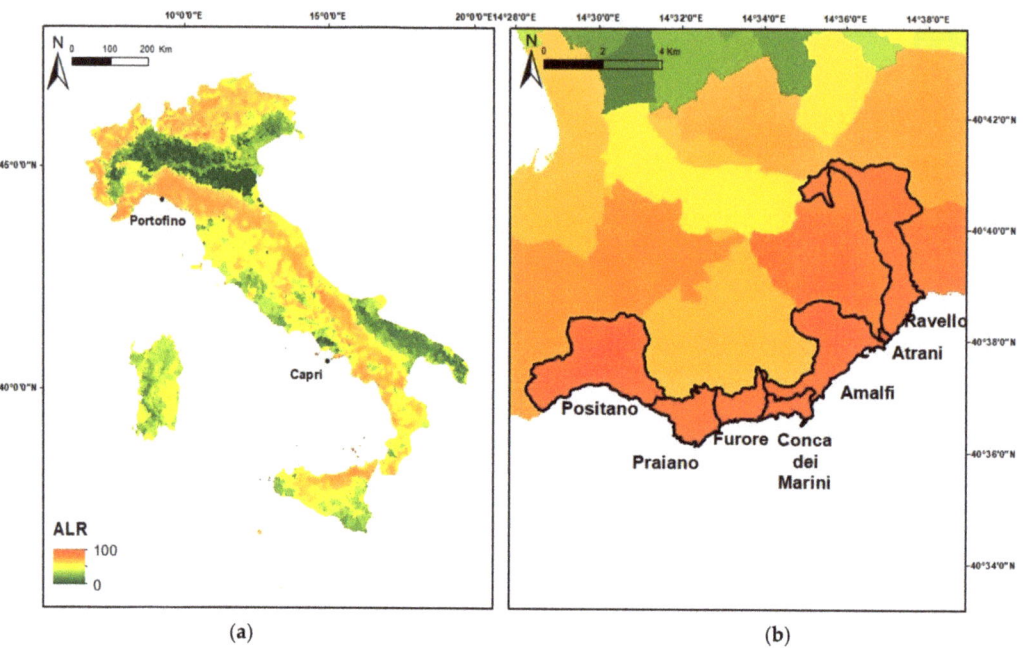

Figure 5. (a) Characterization of the Italian municipalities with the Average Landslide Risk (ALR) index; (b) Focus on the Amalfi Coast, where seven municipalities are ranked among the 10 Italian municipalities with the highest ALR value.

Our findings are in accordance with the evidence resulting from the governmental WebGIS platform presented by [12]: in most of the high-ALR municipalities highlighted in Figure 5 (especially Positano, Amalfi, Conca dei Marini), a high number of buildings are built in areas classified as landslide hazard areas according to Italian laws. For example, at least 90% of Amalfi buildings are located in hazardous areas.

The combined use of ALR and TLR indexes can be useful in gaining preliminary insights on the landslide risk of municipalities. Starting from the LRI index, which is defined at the pixel level, the same principle could be applied to other spatial units and ALR and TLR could be calculated for administrative subdivision of different level (e.g., provinces or districts) or for geographical areas (e.g., basins). It should be stressed that the proposed indexes are environmental indicators and, by definition, are conceived to simplify a complex phenomenon to aid an easy understanding also for non-experts. As a consequence, we acknowledge that the proposed indexes are an oversimplification of reality and cannot substitute a thorough quantitative risk assessment. The main utility of the indexes lies in the fact that a nation-wide quantitative landslide risk assessment is still far from being accomplished for Italy; thus, the proposed indexes can be used to explain, at scales ranging from the local to the national, the severity of the phenomenon, and to evaluate how the administrations have dealt with landslide hazards when planning urban expansion and associated services.

One of the most important requirements for indicators is the possibility to be easily updated. Concerning LRI, the updating procedure can be accomplished in GIS environment whenever updated input data (susceptibility and soil sealing maps) are available. Soil sealing is a dynamic anthropogenic process, and an updated nation-wide map is officially released every year, thus allowing for a yearly update of LRI to account for variations in urban expansion. Conversely, susceptibility is traditionally considered a quasi-static element and a constant update of this element is not expected. However, the index could be updated if a nation-wide susceptibility map is released and deemed more accurate than the one used in this work. E.g., a susceptibility map considering also the runout of landslides would be particularly indicated to thoroughly consider the interactions between hillslope processes and elements at risk. Indeed, we acknowledge that one of the main limitations of the present work is the absence of a method to explicitly include the landslide runout in the model. Unfortunately, complex modeling techniques are required to assess the post-failure displacement of landslides [59,60] and the travel distance is correlated to lithological and morphological factors [61]. For these reasons, a model accounting for landslide runout at the scale of this work ($3*10^6$ km^2) has not been proposed yet; even the latest attempts to include landslide runout in susceptibility assessments are limited to few case studies with limited extension [62,63].

Once LRI is updated, the derivation of TLR and ALR at municipal level can be also accomplished easily in a GIS system. This procedure could be carried out using the last update of the shapefile representing the Italian municipalities, which is also updated every year to account for small variations mainly consisting of the merging of very small and scarcely populated municipalities.

4. Conclusions

A nation-wide quantitative landslide risk assessment is not yet feasible in Italy (301,304 km^2) because of the lack of homogeneous, complete and detailed data. In this work we partially fill this gap by proposing a set of indicators to characterize landslide risk. Indicators are simple numerical indexes widely used in environmental studies by scientists and governmental agencies to simplify and describe complex phenomena. By definition, indicators should be simple and easy to measure, update and understand.

Firstly, a spatially distributed landslide risk index (LRI) was obtained combining state-of-the-art nation-wide susceptibility maps and a soil sealing map released in the framework of a governmental monitoring program of the urban expansion. While the former account for hazardous areas, the latter indicates if anthropic elements could be exposed; their spatial overlapping defines the relevance of the degree of risk.

LRI was then aggregated at the municipal scale to define the average landslide risk index (ALR) and the total landslide risk index (TLR), expressing respectively how hazardous the areas occupied by settlements or infrastructure are, and how serious the overall risk level in each municipality is. ALR and TLR proposed in this work are simple to update and can be adapted to various contexts and scales; in this test they were applied at the municipal level because municipalities are the key administrative subdivisions involved in urban development, land planning and risk mitigation strategies. The proposed indexes cannot substitute a detailed quantitative risk assessment, nevertheless they can provide a preliminary outlook on the spatial distribution of landslide risk at a national scale, and they can be used to evaluate how cautionary each municipality has been in planning its development to deal with the geomorphological hazards threatening its territory.

Author Contributions: Conceptualization, S.S.; methodology, S.S.; investigation, F.C., S.S.; data curation, S.S.; writing—original draft preparation, S.S., F.C.; writing—review and editing, S.S.; visualization, F.C.; supervision, S.S. Both authors have read and agreed to the published version of the manuscript.

Funding: This research received no external funding.

Institutional Review Board Statement: Not applicable.

Informed Consent Statement: Not applicable.

Data Availability Statement: Part of the input data used for this study are accessible online at https://www.istat.it/it/archivio/222527 and http://groupware.sinanet.isprambiente.it/uso-copertura-e-consumo-di-suolo/library/consumo-di-suolo (last accessed on 31 May 2021). The new data presented in this study are available on request from the corresponding author.

Conflicts of Interest: The authors declare no conflict of interest.

References

1. Varnes, D. *Landslide Hazard Zonation: A Review of Principles and Practice*; UNESCO: Paris, France, 1984.
2. Fell, R.; Ho, K.K.S.; Lacasse, S.; Leroi, E. A framework for landslide risk assessment and management. *Int. Conf. Landslide Risk Manag. Vanc. Can.* **2005**, *31*, 3–25.
3. Van Westen, C.J.; van Asch, T.W.J.; Soeters, R. Landslide hazard and risk zonation—Why is it still so difficult? *Bull. Eng. Geol. Environ.* **2006**, *65*, 167–184. [CrossRef]
4. Remondo, J.; Bonachea, J.; Cendrero, A. A statistical approach to landslide risk modelling at basin scale: From landslide susceptibility to quantitative risk assessment. *Landslides* **2005**, *2*, 321–328. [CrossRef]
5. Hungr, O. A Review of Landslide Hazard and Risk Assessment Methodology. In *Landslides and Engineered Slopes. Experience, Theory and Practice*; Aversa, S., Cascini, L., Picarelli, L., Scavia, C., Eds.; CRC Press: Boca Raton, FL, USA, 2018; pp. 3–27, ISBN 978-1-315-37500-7.
6. Huang, J.; Griffiths, D.V. Gordon Fenton Quantitative Risk Assessment of Individual Landslides. In Proceedings of the 7th International Symposium on Geotechnical Safety and Risk (ISGSR), Taipei, Taiwan, 11–13 December 2019; pp. 45–54.
7. Guo, Z.; Chen, L.; Yin, K.; Shrestha, D.P.; Zhang, L. Quantitative risk assessment of slow-moving landslides from the viewpoint of decision-making: A case study of the Three Gorges Reservoir in China. *Eng. Geol.* **2020**, *273*, 105667. [CrossRef]
8. Catani, F.; Casagli, N.; Ermini, L.; Righini, G.; Menduni, G. Landslide hazard and risk mapping at catchment scale in the Arno River basin. *Landslides* **2005**, *2*, 329–342. [CrossRef]
9. Lu, P.; Catani, F.; Tofani, V.; Casagli, N. Quantitative hazard and risk assessment for slow-moving landslides from Persistent Scatterer Interferometry. *Landslides* **2014**, *11*, 685–696. [CrossRef]
10. Pereira, S.; Santos, P.P.; Zêzere, J.L.; Tavares, A.O.; Garcia, R.A.C.; Oliveira, S.C. A landslide risk index for municipal land use planning in Portugal. *Sci. Total Environ.* **2020**, *735*, 139463. [CrossRef]
11. Dilley, M.; Chen, R.S.; Deichmann, U.; Lerner-Lam, A.; Arnold, M.; Agwe, J.; Buys, P.; Kjekstad, O.; Lyon, B.; Yetman, G. *Natural Disaster Hotspots: A Global Risk Analysis*; Disaster Risk Management Series; World Bank Publications: Washington, DC, USA, 2005; Volume 5, pp. 1–132.
12. Iadanza, C.; Trigila, A.; Starace, P.; Dragoni, A.; Biondo, T.; Roccisano, M. IdroGEO: A Collaborative Web Mapping Application Based on REST API Services and Open Data on Landslides and Floods in Italy. *ISPRS Int. J. Geo-Inf.* **2021**, *10*, 89. [CrossRef]
13. Tiranti, D.; Nicolò, G.; Gaeta, A.R. Shallow landslides predisposing and triggering factors in developing a regional early warning system. *Landslides* **2019**, *16*, 235–251. [CrossRef]
14. Donnini, M.; Modica, M.; Salvati, P.; Marchesini, I.; Rossi, M.; Guzzetti, F.; Zoboli, R. Economic landslide susceptibility under a socio-economic perspective: An application to Umbria Region (Central Italy). *Rev. Reg. Res.* **2020**, *40*, 159–188. [CrossRef]
15. Manzo, G.; Tofani, V.; Segoni, S.; Battistini, A.; Catani, F. GIS techniques for regional-scale landslide susceptibility assessment: The Sicily (Italy) case study. *Int. J. Geogr. Inf. Sci.* **2013**, *27*, 1433–1452. [CrossRef]
16. Segoni, S.; Lagomarsino, D.; Fanti, R.; Moretti, S.; Casagli, N. Integration of rainfall thresholds and susceptibility maps in the Emilia Romagna (Italy) regional-scale landslide warning system. *Landslides* **2015**, *12*, 773–785. [CrossRef]
17. Piacentini, D.; Troiani, F.; Soldati, M.; Notarnicola, C.; Savelli, D.; Schneiderbauer, S.; Strada, C. Statistical analysis for assessing shallow-landslide susceptibility in South Tyrol (south-eastern Alps, Italy). *Geomorphology* **2012**, *151–152*, 196–206. [CrossRef]
18. Trigila, A.; Frattini, P.; Casagli, N.; Catani, F.; Crosta, G.; Esposito, C.; Iadanza, C.; Lagomarsino, D.; Mugnozza, G.S.; Segoni, S.; et al. Landslide Susceptibility Mapping at National Scale: The Italian Case Study. In *Landslide Science and Practice: Volume 1: Landslide Inventory and Susceptibility and Hazard Zoning*; Margottini, C., Canuti, P., Sassa, K., Eds.; Springer: Berlin/Heidelberg, Germany, 2013; pp. 287–295, ISBN 978-3-642-31325-7.
19. Munafò, M. *Consumo di Suolo, Dinamiche Territoriali e Servizi Ecosistemici*; SNPA: Rome, Italy, 2019; p. 224.
20. Guillard-Gonçalves, C.; Cutter, S.L.; Emrich, C.T.; Zêzere, J.L. Application of Social Vulnerability Index (SoVI) and delineation of natural risk zones in Greater Lisbon, Portugal. *J. Risk Res.* **2015**, *18*, 651–674. [CrossRef]

21. de Almeida, L.Q.; Welle, T.; Birkmann, J. Disaster risk indicators in Brazil: A proposal based on the world risk index. *Int. J. Disaster Risk Reduct.* **2016**, *17*, 251–272. [CrossRef]
22. Munafò, M.; Salvati, L.; Zitti, M. Estimating soil sealing rate at national level—Italy as a case study. *Ecol. Indic.* **2013**, *26*, 137–140. [CrossRef]
23. Bosellini, A. Outline of the Geology of Italy. In *Landscapes and Landforms of Italy*; Soldati, M., Marchetti, M., Eds.; Springer International Publishing: Cham, Switzerland, 2017; pp. 21–27.
24. Dal Piaz, G.V.; Bistacchi, A.; Massironi, M. Geological outline of the Alps. *Episodes* **2003**, *26*, 175–180. [CrossRef]
25. Vezzani, L.; Festa, A.; Ghisetti, F.C. *Geology and Tectonic Evolution of the Central-Southern Apennines, Italy*; Geological Society of America: Boulder, CO, USA, 2010. [CrossRef]
26. Scisciani, V.; Tavarnelli, E.; Calamita, F. The interaction of extensional and contractional deformations in the outer zones of the Central Apennines, Italy. *J. Struct. Geol.* **2002**, *24*, 1647–1658. [CrossRef]
27. Boccaletti, M.; Corti, G.; Martelli, L. Recent and active tectonics of the external zone of the Northern Apennines (Italy). *Int. J. Earth Sci.* **2011**, *100*, 1331–1348. [CrossRef]
28. Pinna, M. Contributo alla classificazione del clima d'Italia. *Riv. Geogr. Ital.* **1970**, *77*, 129–152.
29. Alpert, P.; Ben-Gai, T.; Baharad, A.; Benjamini, Y.; Yekutieli, D.; Colacino, M.; Diodato, L.; Ramis, C.; Homar, V.; Romero, R.; et al. The paradoxical increase of Mediterranean extreme daily rainfall in spite of decrease in total values. *Geophys. Res. Lett.* **2002**, *29*, 31-1–31-4. [CrossRef]
30. Libertino, A.; Ganora, D.; Claps, P. Technical note: Space–time analysis of rainfall extremes in Italy: Clues from a reconciled dataset. *Hydrol. Earth Syst. Sci.* **2018**, *22*, 2705–2715. [CrossRef]
31. Gariano, S.L.; Guzzetti, F. Landslides in a changing climate. *Earth-Sci. Rev.* **2016**, *162*, 227–252. [CrossRef]
32. Battistini, A.; Segoni, S.; Manzo, G.; Catani, F.; Casagli, N. Web data mining for automatic inventory of geohazards at national scale. *Appl. Geogr.* **2013**, *43*, 147–158. [CrossRef]
33. Battistini, A.; Rosi, A.; Segoni, S.; Lagomarsino, D.; Catani, F.; Casagli, N. Validation of landslide hazard models using a semantic engine on online news. *Appl. Geogr.* **2017**, *82*, 59–65. [CrossRef]
34. Calvello, M.; Pecoraro, G. FraneItalia: A catalog of recent Italian landslides. *Geoenviron. Disasters* **2018**, *5*, 13. [CrossRef]
35. Trigila, A. *Rapporto Sulle Frane in Italia: Il Progetto IFFI: Metodologia, Risultati e Rapporti Regionali*; APAT: Rome, Italy, 2007; ISBN 88-448-0310-0.
36. Trigila, A.; Iadanza, C.; Spizzichino, D. Quality assessment of the Italian Landslide Inventory using GIS processing. *Landslides* **2010**, *7*, 455–470. [CrossRef]
37. Herrera, G.; Mateos, R.M.; García-Davalillo, J.C.; Grandjean, G.; Poyiadji, E.; Maftei, R.; Filipciuc, T.-C.; Jemec Auflič, M.; Jež, J.; Podolszki, L.; et al. Landslide databases in the Geological Surveys of Europe. *Landslides* **2018**, *15*, 359–379. [CrossRef]
38. Budetta, P. Landslide hazard assessment of the Cilento rocky coasts (Southern Italy). *Int. J. Geol.* **2013**, *7*, 1–8.
39. Sacchini, A.; Faccini, F.; Ferraris, F.; Firpo, M.; Angelini, S. Large-scale landslide and deep-seated gravitational slope deformation of the Upper Scrivia Valley (Northern Apennine, Italy). *J. Maps* **2016**, *12*, 344–358. [CrossRef]
40. Pellicani, R.; Argentiero, I.; Spilotro, G. GIS-based predictive models for regional-scale landslide susceptibility assessment and risk mapping along road corridors. *Nat. Hazards Risk* **2017**, *8*, 1012–1033. [CrossRef]
41. Fell, R.; Corominas, J.; Bonnard, C.; Cascini, L.; Leroi, E.; Savage, W.Z. Guidelines for landslide susceptibility, hazard and risk zoning for land-use planning. *Eng. Geol.* **2008**, *102*, 99–111. [CrossRef]
42. Cervi, F.; Berti, M.; Borgatti, L.; Ronchetti, F.; Manenti, F.; Corsini, A. Comparing predictive capability of statistical and deterministic methods for landslide susceptibility mapping: A case study in the northern Apennines (Reggio Emilia Province, Italy). *Landslides* **2010**, *7*, 433–444. [CrossRef]
43. Conforti, M.; Robustelli, G.; Muto, F.; Critelli, S. Application and validation of bivariate GIS-based landslide susceptibility assessment for the Vitravo river catchment (Calabria, south Italy). *Nat. Hazards* **2012**, *61*, 127–141. [CrossRef]
44. Zizioli, D.; Meisina, C.; Valentino, R.; Montrasio, L. Comparison between different approaches to modeling shallow landslide susceptibility: A case history in Oltrepo Pavese, Northern Italy. *Nat. Hazards Earth Syst. Sci.* **2013**, *13*, 559–573. [CrossRef]
45. Segoni, S.; Tofani, V.; Lagomarsino, D.; Moretti, S. Landslide susceptibility of the Prato–Pistoia–Lucca provinces, Tuscany, Italy. *J. Maps* **2016**, *12*, 401–406. [CrossRef]
46. Segoni, S.; Pappafico, G.; Luti, T.; Catani, F. Landslide susceptibility assessment in complex geological settings: Sensitivity to geological information and insights on its parameterization. *Landslides* **2020**, *17*, 2443–2453. [CrossRef]
47. Esposito, G.; Carabella, C.; Paglia, G.; Miccadei, E. Relationships between Morphostructural/Geological Framework and Landslide Types: Historical Landslides in the Hilly Piedmont Area of Abruzzo Region (Central Italy). *Land* **2021**, *10*, 287. [CrossRef]
48. Lagomarsino, D.; Tofani, V.; Segoni, S.; Catani, F.; Casagli, N. A tool for classification and regression using random forest methodology: Applications to landslide susceptibility mapping and soil thickness modeling. *Environ. Modeling Assess.* **2017**, *22*, 201–214. [CrossRef]
49. Catani, F.; Lagomarsino, D.; Segoni, S.; Tofani, V. Landslide susceptibility estimation by random forests technique: Sensitivity and scaling issues. *Nat. Hazards Earth Syst. Sci.* **2013**, *13*, 2815–2831. [CrossRef]
50. Lee, S. Current and future status of GIS-based landslide susceptibility mapping: A literature review. *Korean J. Remote Sens.* **2019**, *35*, 179–193.

51. Shano, L.; Raghuvanshi, T.K.; Meten, M. Landslide susceptibility evaluation and hazard zonation techniques—A review. *Geoenviron. Disasters* **2020**, *7*, 1–19. [CrossRef]
52. Prokop, G.; Jobstmann, H.; Schönbauer, A. *Overview on Best Practices for Limiting Soil Sealing and Mitigating Its Effects in EU-27*; European Communities. Brussels, Belgium, 2011.
53. Munafò, M.; Assennato, F.; Congedo, L.; Luti, T.; Marinosci, I.; Monti, G.; Riitano, N.; Sallustio, L.; Strollo, A.; Tombolini, I. *Il Consumo di Suolo in Italia*; Rapporti ISPRA n.218/2015; ISPRA: Roma, Italy, 2015; p. 90.
54. Luti, T.; Segoni, S.; Catani, F.; Munafò, M.; Casagli, N. Integration of Remotely Sensed Soil Sealing Data in Landslide Susceptibility Mapping. *Remote Sens.* **2020**, *12*, 1486. [CrossRef]
55. Hartlen, J.; Viberg, L. General report: Evaluation of landslide hazard. In Proceedings of the International Symposium on Landslides, Lausanne, Switzerland, 10–15 July 1988; pp. 1037–1057.
56. Di Napoli, M.; Carotenuto, F.; Cevasco, A.; Confuorto, P.; Di Martire, D.; Firpo, M.; Pepe, G.; Raso, E.; Calcaterra, D. Machine learning ensemble modelling as a tool to improve landslide susceptibility mapping reliability. *Landslides* **2020**, *17*, 1897–1914. [CrossRef]
57. Tarolli, P.; Preti, F.; Romano, N. Terraced landscapes: From an old best practice to a potential hazard for soil degradation due to land abandonment. *Anthropocene* **2014**, *6*, 10–25. [CrossRef]
58. Savo, V.; Salvati, L.; Caneva, G. In-between soil erosion and sustainable land management: Climate aridity and vegetation in a traditional agro-forest system (Costiera Amalfitana, Southern Italy). *Int. J. Sustain. Dev. World Ecol.* **2016**, *23*, 423–432. [CrossRef]
59. Stamatopoulos, C.A.; Di, B. Analytical and approximate expressions predicting post-failure landslide displacement using the multi-block model and energy methods. *Landslides* **2015**, *12*, 1207–1213. [CrossRef]
60. Firmansyah, S.; Feranie, S.; Tohari, A.; Latief, F.D.E. Prediction of landslide run-out distance based on slope stability analysis and center of mass approach. In Proceedings of the International Symposium on Geophysical Issues PEDISGI, Badung, Indonesia, 8–10 June 2015; Volume 29.
61. Guo, D.; Hamada, M.; He, C.; Wang, Y.; Zou, Y. An empirical model for landslide travel distance prediction in Wenchuan earthquake area. *Landslides* **2014**, *11*, 281–291. [CrossRef]
62. Mergili, M.; Schwarz, L.; Kociu, A. Combining release and runout in statistical landslide susceptibility modeling. *Landslides* **2019**, *16*, 2151–2165. [CrossRef]
63. Napoli, M.D.; Martire, D.D.; Bausilio, G.; Calcaterra, D.; Confuorto, P.; Firpo, M.; Pepe, G.; Cevasco, A. Rainfall-induced shallow landslide detachment, transit and runout susceptibility mapping by integrating machine learning techniques and GIS-based approaches. *Water* **2021**, *13*, 488. [CrossRef]

Article

Combining Site Characterization, Monitoring and Hydromechanical Modeling for Assessing Slope Stability

Shirin Moradi [1], Thomas Heinze [2,†], Jasmin Budler [2,†], Thanushika Gunatilake [2,‡], Andreas Kemna [2] and Johan Alexander Huisman [1,*]

1 Agrosphere Institute (IBG 3), Forschungszentrum Jülich GmbH, 52425 Jülich, Germany; mrd.shir@gmail.com
2 Geophysics, Institute of Geosciences, University of Bonn, 53113 Bonn, Germany; thomas.heinze@rub.de (T.H.); jasmin.budler@rub.de (J.B.); thanushika.gunatilake@unine.ch (T.G.); kemna@geo.uni-bonn.de (A.K.)
* Correspondence: s.huisman@fz-juelich.de
† Current address: Applied Geology, Institute of Geology, Mineralogy and Geophysics, Ruhr-University Bochum, 44801 Bochum, Germany.
‡ Current address: Centre for Hydrogeology and Geothermics (CHYN), University of Neuchâtel, 2000 Neuchâtel, Switzerland.

Citation: Moradi, S.; Heinze, T.; Budler, J.; Gunatilake, T.; Kemna, A.; Huisman, J.A. Combining Site Characterization, Monitoring and Hydromechanical Modeling for Assessing Slope Stability. *Land* **2021**, *10*, 423. https://doi.org/10.3390/land10040423

Academic Editor: Enrico Miccadei

Received: 29 March 2021
Accepted: 14 April 2021
Published: 15 April 2021

Publisher's Note: MDPI stays neutral with regard to jurisdictional claims in published maps and institutional affiliations.

Copyright: © 2021 by the authors. Licensee MDPI, Basel, Switzerland. This article is an open access article distributed under the terms and conditions of the Creative Commons Attribution (CC BY) license (https://creativecommons.org/licenses/by/4.0/).

Abstract: Rainfall-induced landslides are a disastrous natural hazard causing loss of life and significant damage to infrastructure, farmland and housing. Hydromechanical models are one way to assess the slope stability and to predict critical combinations of groundwater levels, soil water content and precipitation. However, hydromechanical models for slope stability evaluation require knowledge about mechanical and hydraulic parameters of the soils, lithostratigraphy and morphology. In this work, we present a multi-method approach of site characterization and investigation in combination with a hydromechanical model for a landslide-prone hillslope near Bonn, Germany. The field investigation was used to construct a three-dimensional slope model with major geological units derived from drilling and refraction seismic surveys. Mechanical and hydraulic soil parameters were obtained from previously published values for the study site based on laboratory analysis. Water dynamics were monitored through geoelectrical monitoring, a soil water content sensor network and groundwater stations. Historical data were used for calibration and validation of the hydromechanical model. The well-constrained model was then used to calculate potentially hazardous precipitation events to derive critical thresholds for monitored variables, such as soil water content and precipitation. This work introduces a potential workflow to improve numerical slope stability analysis through multiple data sources from field investigations and outlines the usage of such a system with respect to a site-specific early-warning system.

Keywords: landslide; hydromechanical modeling; early-warning; slope stability; rainfall-induced landslides; local factor of safety; SoilNet; geophysical characterization; water content distribution; bedrock topography

1. Introduction

Landslides are common natural hazards in many parts of the world, often triggered by increased pore pressure after heavy rainfall. Due to climate change, more intense rainfall events are expected and the frequency of destructive landslides may increase [1–3]. In order to evaluate hazards associated with landslide-prone hillslopes, several modeling approaches can be used to determine critical precipitation events that may lead to slope failure. Common limit-equilibrium models tend to overestimate the factor of safety as a measure of slope stability in complex geometrical setups [4]. The concept of a local factor of safety in Coulomb stress-field based finite element models is one approach used to overcome this limitation [5]. Studies applying limit-equilibrium and continuous finite element models have partly found good agreement between the results for stability analysis e.g., [6]. However, variation between models depends strongly on the methods used and the application scenarios [7,8].

The incorporation of spatial variability of soil and rock types into models for slope stability evaluation is crucial to assess the structural and hydrological state of a hillslope e.g., [9]. Slope morphology and spatial distribution of the material properties affect the slope failure potential e.g., [10,11]. Especially bedrock topography and soil depth were identified as important factors with respect to slope hydrology and stability [9,12,13]. Geophysical characterization e.g., [14–16] and monitoring e.g., [17,18] are increasingly used for structural and hydrological assessment of landslide-prone hillslopes. Such geophysical studies often combine multiple methods to study the subsurface structures and seismics and electric resistivity tomography (ERT) are the most commonly used e.g., [19,20]. Seismic refraction is particularly useful to identify lithological layers and slip surfaces e.g., [21,22], whereas ERT is able to provide information on the water content distribution in the subsurface e.g., [23,24] by using the correlation between bulk electrical resistivity and saturation e.g., [25,26]. Compared to drilling methods or point sensors, geophysical measurements usually provide information with a higher spatial resolution at lower cost and are only minimally invasive [27]. The combination of geophysical methods allows collection of complementary information and can be analyzed using data fusion methods [28]. Typically, supporting laboratory studies are used to improve geophysical monitoring concepts and to evaluate innovative geophysical methods, such as self-potential measurements, for detecting critical hydrological conditions [29,30]. An extensive review of geophysical monitoring methods for failure-prone hillslopes is given in Whiteley et al. [31]. In addition, the development of cost-effective sensor networks for monitoring soil water content and slope movement has gained momentum in recent years, which allows us to bridge the gap between costly boreholes and extensive geophysical monitoring and surveying e.g., [32,33].

A few case studies have combined hydrogeological and geomorphological site characterization, geophysical monitoring and hydromechanical modeling. For shallow landslides in pyroclastic soils, water content, derived from electric resistivity profiles, were combined with statistical modeling using a cellular automaton to derive a relationship between the factor of safety of a hillslope with in situ measurable quantities [34]. For the La Clapiere landslide in France, vertical electrical sounding was used to obtain the underground structure used in a geomechanical model [35]. An extensive ERT survey was combined with groundwater measurements and meteorological data to study the groundwater dynamics of a translational landslide and to develop a conceptual model [36]. To assess the slope stability of the Brzozowka landslide near Cracow (Poland), ERT monitoring was combined with drilling and laboratory tests to improve a stability analysis in which critical conditions for slope stability were derived from simulations for extreme precipitation events [37].

In this study, we present a slope stability analysis of a failure-prone hillslope in Germany based on the combination of thorough site investigation and monitoring with a hydromechanical finite element model. The site investigation includes a geological, geomorphological, hydrological, and geophysical characterization and soil water content and groundwater monitoring using a sensor network as well as geoelectric measurements. We introduce a work flow incorporating the different data sources into the model setup and for validating the model results. The data used in this study are partly obtained from former studies at the study site, and include (i) groundwater level [38], (ii) borehole logs [38], (iii) laboratory tests for soil hydraulic and mechanical properties [38,39], (iv) slope movement and (v) precipitation [40], as well as recent (vi) seismic refraction surveys, (vii) electric resistivity tomography, (viii) soil water content monitoring network and (ix) a digital elevation model. The parameterized, constrained and verified hydromechanical model was subsequently used to study potential hazardous precipitation events. We compared the results to the observations in the field and discuss how the developed approach could be extended towards a site-specific early warning system.

2. Study Area

The study site is a failure-prone hillslope at the Dollendorfer Hardt located 16 km south-east of the city of Bonn in the Siebengebirge, Germany (Figure 1). This site has been investigated in several previous studies since the 1990s [39–41]. At least two major landslides occurred at the study site in the past 100 years (1958 and 1972), and minor movements have been observed since the last failure event [39,40]. There is anecdotal evidence that attributes the initiation of the first major landslide to the construction of a trail at the upper part of the current scarp area. However, intense rainfall is suspected to be the triggering factor of both major events [38].

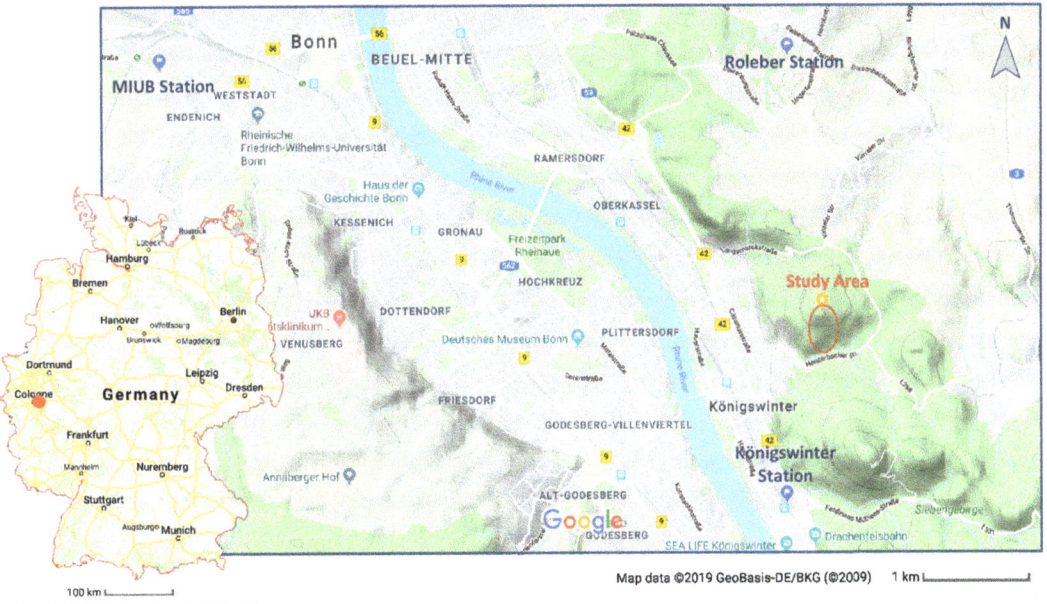

Figure 1. Location of the study site near Bonn, Germany (source: Google Earth).

2.1. Site Characterization

The Dollendorfer Hardt is a horst structure with a height of roughly 250 m above sea level, representing the most northern peak of the Siebengebirge of volcanic origin. Towards the north, there is the lower Rhine Bay. The lithology of the area is characterized by Devonian shales, consisting of interchanging bedding of slate and graywacke, on which Tertiary sediments, mainly clay and quartz sand, are deposited. These lithological layers are partly overlain by basaltic and trachytic tuffs as a result of volcanic eruptions in this region [39].

In order to obtain subsurface information for the study site, drillings were performed between March and August of 1998 [38]. The techniques included closed core percussion drilling, open core percussion drilling and manual auger and percussion drilling (Figure 2). The soil of the drill cores was used to determine hydraulic and mechanical properties of the different identified layers according to the guidelines of the German Industrial Standard of that time (DIN 1993) [38]. In particular, the following parameters were determined: particle size distribution, soil water content, consistency limits, particle density and maximum soil water content but were not completed for all soil types due to the limited availability of samples. Further, shear strength was determined by direct shear tests and triaxial tests. Samples from outcrops were used to determine the hydraulic conductivity [38].

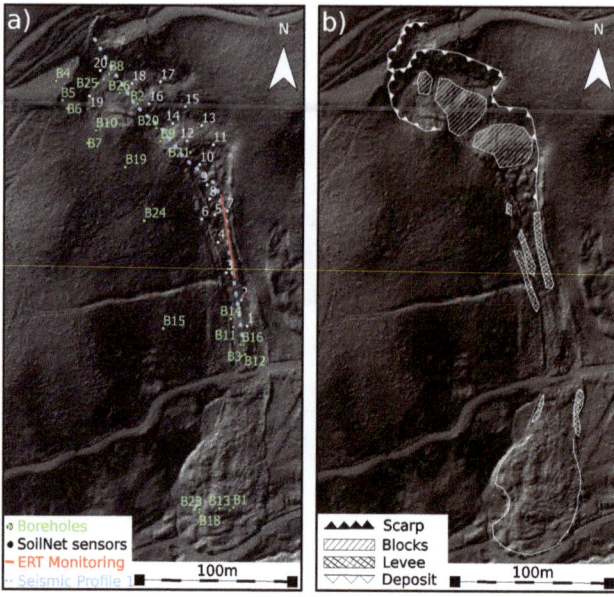

Figure 2. (**a**) Digital elevation model of the study area provided by the Geobasisdatendienst NRW with geoelectric monitoring profile line, seismic profile line, borehole locations and water content sensors, (**b**) as well as a simplified geomorphological map.

The scarps of the two landslides at the study site cover an area of almost 30,000 m^2 located at a steep (up to 35–40°) south-facing hillslope with a river at the valley bottom. The area is approximately 330 m long with a variable width ranging from 17 m at the narrowest passage, up to 65 m close to the scarp (Figure 2). The well defined scarp is the northern boundary of the landslide zone and is a result of a rotational movement. Relatively undisturbed rotational blocks can be identified in the landslide mass in the upper part of the slope [38], which then convert to a mass flow in the middle part of the slope (transport zone). The transport zone is the narrowest part of the landslide and over 140 m long. The earthflow is constrained by one, sometimes two sets of levees at both sides, originating from the two landslide events. The translational mass has been deposited in the debris zone in the lower part of the slope up to a small river with little to no inclination [39,42]. The landslide mass consists mostly of trachytic tuff and clayey sediments from the Tertiary. The first landslide mass additionally contains loess from the Quaternary loess cover. The slope is naturally covered by a forest consisting mainly of beech trees.

Previous studies provided evidence of regular slope movement in the middle part of the scarp area using inclinometers and tiltmeters. Movement of ±3 cm year^{-1} was observed for the transport zone in 'extraordinary wet conditions' in spring when heavy rainfall coincided with high groundwater levels [38]. In addition, continuous soil creep of small magnitude on the order of a few millimeters per year was observed in the transport zone [40]. Schmidt [38] attributed the vulnerability of the site for slope instability to the specific geological setting of the area with an abundance of clay-rich soil layers. This also explains the reported elastic swelling/shrinking of the lower rotational block of the scarp zone and the elastic movement associated with groundwater level changes in the debris zone [38]. Ruptures and breakups of tuff blocks of ≈0.3 m can be observed regularly along the scarp.

To investigate the extent of the lithological layers in the study area, ten seismic refraction profiles were acquired between September 2015 and January 2017. The measurement profiles had lengths between 50 to 150 m with a geophone spacing varying between 1 and 2.5 m for the different profiles. The refracted seismic waves were generated by strikes with a 5 kg hammer on a metal plate close to each geophone. For a good signal to noise ratio, ten strikes per geophone were stacked. Signals were recorded with a SUMMIT II Compact (DMT-Group, Essen, Germany). The maximum recording time was 350 ms. The data were processed and analyzed using the software ReflexW (J. Sandmeier, Karlsruhe, Germany). A bandpass filter was applied to cut frequencies below 10 Hz and above 150 Hz. First arrivals were picked and an inversion was performed using regular grids with a grid spacing of one quarter or less of the geophone distance.

The results for the first arrival of the refracted seismic waves indicated a three layer case (Figure 3). The inversion results suggest that the seismic wave velocity range associated with the three layers are <300 m s^{-1}, 400–600 m s^{-1}, and >800 m s^{-1}. These layer velocities were consistently found in all measured profiles. Based on the seismic data and additional core drilling data from [38] and literature values for the rock and mineral types [43,44], the three different layers are interpreted as follows:

- a top layer consisting of clayey sediments, trachyte tuff and loess, transported and mixed by the landslides;
- an intermediate layer of Tertiary sediments mainly consisting of silt and clay;
- a base layer of Devonian bedrock, strongly weathered at the top.

The seismic measurements also indicated isolated sand structures within the Tertiary sediments. Furthermore, the thickness of the Tertiary sediments was found to decrease downhill and to disappear completely within the transport zone. The abrasive character of the former landslides eroded the originally deposited sediments and replaced them with the deposited landslide mass. In the transport zone, the sliding surface corresponds to a lithological boundary and was identified in the seismic refraction. In the upper part of the rotational landslide, the sliding surface lies within the Tertiary sediments and therefore cannot be determined by seismic refraction [38,39]. Based on previous studies, the upper layer is assumed to be temporally unstable and thus prone to landslides [41]. A similar layering was found on the adjacent slopes. This supports the assumption that only the upper layer of tuff and parts of the Tertiary sediments were transported and mixed by the landslides.

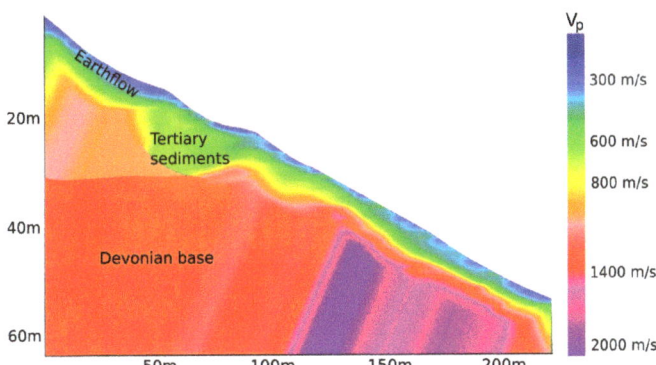

Figure 3. Tomogram of p-wave velocity derived for seismic profile 1 (see Figure 2a) and its subsequent interpretation.

2.2. Meteorological Data

Precipitation was continuously measured at the meteorological station at Königswinter-Heiderhof as well as at the station of the Department of Meteorology of the University of Bonn (MIUB). The Königswinter-Heiderhof station is located 3 km southwest of the test site, whereas the MIUB station is approximately 10 km northwest of the study site (Figure 1). At the Königswinter-Heiderhof station, precipitation and temperature data are recorded at a daily resolution since 1990 and 2000 onward, respectively. For the MIUB station, daily precipitation data are available between 1999 and 2001, and 10-min resolution precipitation data are available since 2010. Temperature is recorded with a 10 min resolution since 1995. The mean annual precipitation for the time period of 1995–2017 varied between 650–950 mm with an average of 769 mm. For the same period, the mean annual temperature varied between 8 °C and 13 °C with an average of 9 °C (Figure 4). Since temperature data were missing at the Königswinter-Heiderhof station for the period 1995–1999, temperature data from the MIUB station were used in this period. As common in the temperate climate zone, the strongest rainfall events were associated with thunderstorms in summer. The maximum monthly precipitation was 235 mm in July 2014.

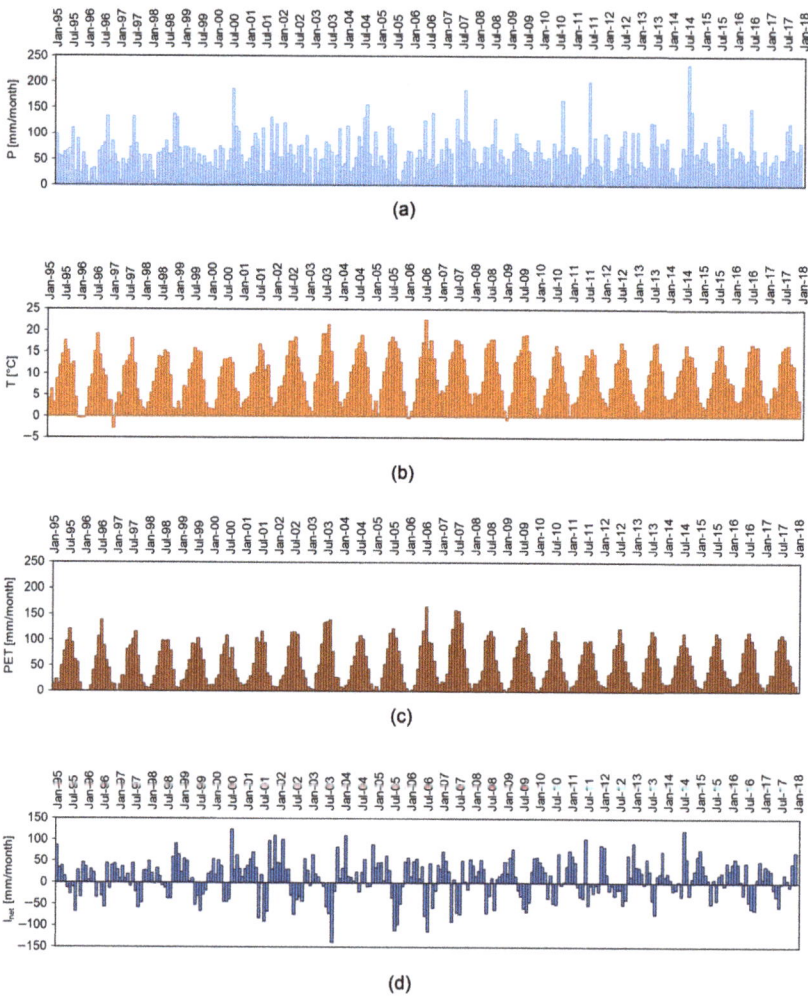

Figure 4. (**a**) Monthly precipitation, (**b**) temperature, (**c**) potential evapotranspiration, and (**d**) net infiltration for the Dollendorfer Hardt test site based on the data of Königswinter-Heiderhof station from 1995–2017.

The net infiltration, I_{net} [mm month^{-1}], is the difference between precipitation P [mm month^{-1}] and actual evapotranspiration ET [mm month^{-1}]:

$$I_{net} = P - ET \qquad (1)$$

To obtain net infiltration from precipitation, an estimate of the actual evapotranspiration is required. To minimize the number of required parameters, the temperature-based approach of Thornthwaite [45] was used here. In this approach, the monthly potential evapotranspiration was calculated using

$$PET = 16 \left(\frac{L_d}{12}\right)\left(\frac{N}{30}\right)\left(10\frac{T_a}{I}\right)^a \qquad (2)$$

$$a = (6.75 \times 10^{-7})I^3 - (7.71 \times 10^{-5})I^2 + (1.792 \times 10^{-2})I + 0.49239 \qquad (3)$$

$$I = \sum_{i=1}^{12}\left(\frac{T_{ai}}{5}\right)^{1.514} \qquad (4)$$

where PET [mm month^{-1}] is the estimated potential evapotranspiration, L_d [h] is the average daylength of the month, N [-] is the number of days of the month, T_a [°C] is the mean daily temperature of the month and a [-] is an exponent that is a function of the annual Thornthwaite heat index, I [-]. Based on this approach, the mean annual PET was estimated to be 644 mm for the Dollendorfer Hardt. This value is in good agreement with PET values provided for the Roleber station (8 km north of the study site) by the Deutscher Wetterdienst (DWD, German Weather Service), which were calculated based on the Haude method [46].

The resulting monthly PET for the study area is shown in Figure 4. It can be seen that monthly PET occasionally exceeded precipitation in the summer months when temperature is relatively high. In these cases, net infiltration was considered to be negative, which was modeled as outflow. The mean annual precipitation, PET, and net infiltration for the simulation period of 1995–2017 were 769, 644, and 125 mm, respectively. It should be noted that it was assumed here that PET was equal to actual evapotranspiration. This assumption is certainly not valid in all conditions [47]. However, the Dollendorfer Hardt is experiencing precipitation all over the year and has a relatively high groundwater level. Therefore, it was assumed that the trees did not experience water stress and were able to transpire with rates dictated by the atmospheric conditions.

Slope instability at the Dollendorfer Hardt site is suspected to be triggered by intensive rainfalls. In order to identify the potential magnitude of high-intensity rainfall events both the daily rainfall data from the Königswinter station and the high-resolution data from the MIUB station were analyzed. The maximum daily rainfall amount at the Königswinter and MIUB stations are 88 mm and 45 mm, respectively. The maximum hourly precipitation per year was derived from the MIUB station with 35 mm h^{-1} in 2013. The mean value of maximum hourly precipitation over the years 2010 to 2017 was 15.7 mm h^{-1}. Based on this analysis, a rainfall intensity of 20 mm h^{-1} was selected to represent an intensive rainfall event at the study site which is also not too rare.

2.3. Monitoring of Ground Water Level and Soil Water Content

Following the drilling in 1998, 26 standpipe groundwater gauges were installed [38]. For long-term groundwater monitoring, an electronic water level indicator was used. Twelve tubes showed strong variations in groundwater level and were monitored with hourly resolution with pressure transducers [38]. Groundwater monitoring data are available from April 1999 to May 2001 [38].

To characterize spatial variability of near-surface soil water content, the soil water content sensor network SoilNet [32] was installed at the Dollendorfer Hardt test site in August 2016. The installed network consisted of 20 SoilNet nodes. The locations of the nodes were chosen to achieve a homogeneous distribution of sensors across the slope. At each location, six SMT100 sensors (Truebner GmbH, Neustadt, Germany) were installed horizontally at depths of −0.05, −0.20 and −0.50 m. Two sensors at each depth were used to increase the measurement volume and to provide a control of the measurement quality. The operating principle of the SMT100 sensor and the calibration approach for determining the relative dielectric permittivity from the sensor response are described in Bogena et al. [48]. To link the measured soil permittivity to soil water content using petrophysical relationships, 12 undisturbed soil samples were taken at 4 locations along the slope at the depths of −0.10, −0.30 and −0.50 m. The soil samples were saturated in the laboratory and the soil permittivity as well as the weight was measured daily during a drying period of 42 days at room temperature. Subtracting the dry weight of the samples from the measured weight, the water content was determined and linked to the soil permittivity. The samples were roughly categorized into two classes based on their grain size distribution. The coarse-grained samples were located in the rotated blocks while the fine-grained material was found at all other locations. No difference was found for the petrophysical relationship for clay and tuff samples. For the coarse-grained soil samples, the petrophysical relationship of Roth et al. [49] showed the best agreement and was used for the conversion of the soil permittivity to water content for the SoilNet sensors 11–14 and 16–18. For the fine-grained samples, the petrophysical relationship of Robinson et al. [50] showed the best agreement with the laboratory data and was subsequently used for the remaining SoilNet sensors. Soil water content measurements were taken from August 2016 to July 2018.

To assess the water dynamics at greater depth and to achieve a higher spatial resolution than the SoilNet sensor network, an electric resistivity monitoring system was installed in the transport zone of the landslide. Measurements were conducted between March 2016 and May 2018 with various time intervals from daily to monthly measurements. An ABEM Terrameter LS (Guideline Geo AB, Sundbyberg, Sweden) with 96 electrodes was used with an electrode spacing of 0.5 m. A dipole–dipole configuration with skips of 0, 2, 4 and 6 electrodes was combined with multiple gradient measurements. In the data processing prior to the inversion, data points were removed due to systematic errors, such as bad electrode connections, problems with power supply or high current strength (>1 A). To account for temperature effects, the electric conductivities were corrected to the mean subsurface temperature of 10 °C following the procedure of Hayley et al. [51]. Due to the lack of temperature measurements at depth, the model of Brunet et al. [52] was used to calculate the required temperature information $T(t,z)$ [°C] over the given time and depth range. The preprocessed electrical data were inverted using the finite element based inversion code CRTomo [53]. We used a resistance error model with parameters a and b for absolute and relative resistance error contributions, resulting in a resistance error ΔR [Ω] for the measured resistance R [Ω] [54]. The absolute error was set to $a = 0.001\,\Omega$ and the relative one to $b = 3\%$. To improve the resistivity estimation, the inversion was performed with the seismic layer boundary between bedrock and landslide as a structural constraint e.g., [18,55,56]. To better uncover temporal changes, a difference inversion was also performed [57].

In the ERT inversion results, the three layers found with seismic refraction could not be identified, as the difference in electric conductivity of the two upper layers was too small. The inversion results only show the electrically conductive and heterogeneous landslide mass and the more resistive bedrock. From these results, water content was estimated using the relationship as described in Waxman and Smits [26]:

$$\sigma_b = \frac{S^i}{F_f}\left(\sigma_w + \frac{\sigma_s}{S}\right) \quad (5)$$

with saturation, S [-], water conductivity, σ_w [S m^{-1}], surface conductivity, σ_s [S m^{-1}], formation factor, F_f [-] and saturation exponent, i [-]. Water conductivity, $\sigma_w = 0.1$ [S m^{-1}], was gained from in situ measurements in the boreholes. As surface conductivity depends on the clay content, the empirical relationship between clay content, cc [%] and the surface conductivity, σ_s [S m^{-1}], in the work of Rhoades et al. [58] was used:

$$\sigma_s = (2.3 \times cc - 2.1) \times 10^{-3} \quad (6)$$

where cc is the clay content (cc_1 = 57% and cc_2 = 40%, where the index 1 denotes the upper layer and 2 the lower layer) obtained from Schmidt [38]. The formation factor $F_f = \varphi^{-j}$ was calculated based on the porosity φ [-] and the cementation exponent $j = 2$, where the porosity values (φ_1 = 55%, φ_2 = 40%) were also taken from Schmidt [38] and the cementation exponent as well as the saturation exponent $i = 2$ were based on literature values e.g., [59].

The derived water content values from ERT (Figure 5) were in agreement with values measured by the SoilNet sensor network, which were limited to a depth of -0.50 m of the soil. In general, observed water dynamics were low during the monitoring period, as only the first two meters showed a response to precipitation events and a decrease in soil water content during dry periods. However, the change in volumetric water content below the top layer was less than $2\,\text{cm}^3\,\text{cm}^{-3}$ and declined strongly with depth over a 7 week dry period from 5 April to 24 May 2017. Continuous low-intensity precipitation with less than $10\,\text{mm}\,\text{d}^{-1}$ over several days resulted in increasing saturation in the upper 0.50 m of the soil.

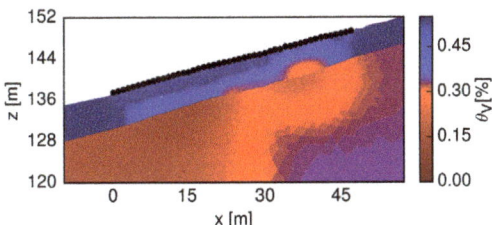

Figure 5. Water content derived from ERT measurements taken at 31 May 2017.

3. Hydromechanical Model

The applied hydromechanical model is based on the approach of Lu et al. [5] solving the Richard's equation to describe the transient water flow in the subsurface. In the mixed form, the Richard's equation is given as

$$\nabla K(h) \nabla H + W = \frac{\partial \theta(h)}{\partial t}, \tag{7}$$

with the hydraulic head, H [m], pressure head, h [m], hydraulic conductivity, $K(h)$ [m s^{-1}], volumetric water content, $\theta(h)$ [m^3 m^{-3}] and possible source/sink terms, W [s^{-1}]. Following the model of van Genuchten [60], the water retention curve is given by

$$S_e = \frac{\theta(h) - \theta_r}{\theta_s - \theta_r} = (1 + (\alpha|h|)^n)^{-m} \tag{8}$$

with effective saturation, S_e [-], residual and saturated water content, $\theta_{r/s}$ [m^3 m^{-3}], and soil specific parameters, α [m^{-1}], n and $m = 1 - 1/n$ [-]. The hydraulic conductivity is based on the effective saturation

$$K(h) = K_s \sqrt{S_e} \left(1 - (1 - S_e^{1/m})^m\right)^2 \tag{9}$$

with saturated hydraulic conductivity, K_s [m s^{-1}].

The mechanical response of the soil is considered to be linear elastic with no poroelasticity and described by the Young's modulus, E [Pa] and Poisson's ratio, ν [-]. In this case, momentum equilibrium is given by the total stress tensor, σ [Pa], the unit weight, γ [N m^{-3}] and the body force vector, F_v [-], as

$$\nabla \cdot \sigma + \gamma F_v = 0 \tag{10}$$

From the total stress tensor, the effective stress tensor, σ' [Pa], can be calculated based on the pore water pressure, P_w [Pa], [61,62]:

$$\sigma' = \sigma - S_e P_w \mathbf{I} \tag{11}$$

with the unit vector, $\mathbf{I}$[-]. To assess the stability of a hillslope, the local factor of safety, LFS [-], is calculated based on the Mohr–Coulomb failure criteria

$$LFS = \frac{\cos\phi'(c' + \sigma_m \tan\phi')}{R} \tag{12}$$

with effective friction angle, ϕ' [°], effective cohesion, c' [Pa], mean effective stress, $\sigma_m = 0.5 \times (\sigma_1 + \sigma_3) - S_e P_w$ [Pa] and the radius of the Mohr circle, $R = 0.5 \times (\sigma_1 - \sigma_3)$ [Pa]. Here, $\sigma_{1,3}$ [Pa] are the major and minor principal stresses. A slope is considered to be prone to failure for $LFS < 1$ as the restricting forces are smaller than the downhill forces in this case.

4. Combination of Field Observations and Hydromechanical Modeling

In this study, the hydromechanical model was used to assess slope stability through the calculation of the local factor of safety. Site characterization, laboratory tests and continuous monitoring of soil water content and precipitation are input for the model to increase the quality of the assessment (Figure 6). The input can be separated into one-time constraints and dynamic information. The base for the numerical model is the physical/mathematical model as described in Section 3. To solve the governing equations, the parameters, the geometry of the domain as well as boundary and initial conditions need to be specified.

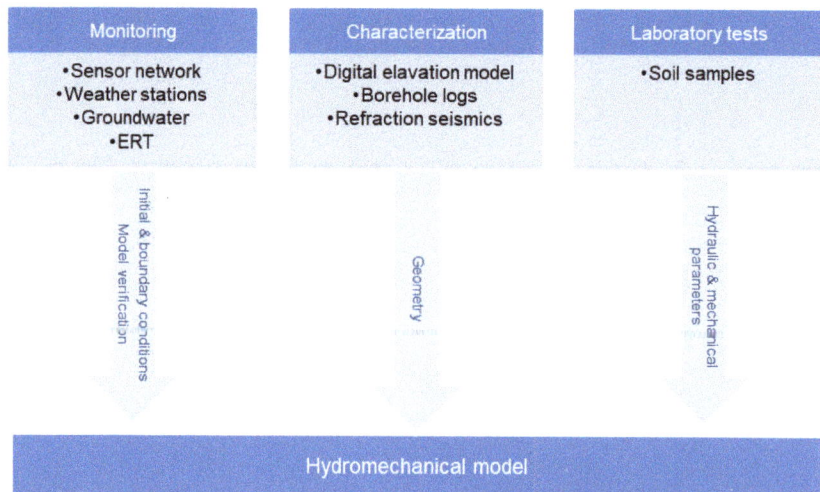

Figure 6. Schematic sketch of the different input streams into the hydromechanical model for stability analysis.

4.1. Model Setup

Surface topography was taken from a digital terrain model with a spatial resolution of 1 m^{-2} derived from remote sensing and provided by the Geodatenbasisdienst NRW. Bedrock topography as well as the varying thickness of soil layers was obtained from the seismic refraction and the drilling logs. For this purpose, the wave velocities of 300 ms^{-1}, 600 ms^{-1} and 800 ms^{-1} were selected as limit values for the lithological layers in agreement with the lithological units identified in the borehole logs. The position of the interfaces was digitized in the seismic tomograms, georeferenced and converted to absolute depths using the digital terrain model. The information was entered into the open source software GrassGis (GrassGis Development Team, 2017) and interpolated to surfaces using inverse distance weighting. Hence, a volume model of the investigated area was created based on the generated surfaces (Figure 7). The geophysically-derived volume model was used to generate a 3D geometrical model of the study site in Comsol Multiphysics [COMSOL Inc, Stockholm, Sweden].

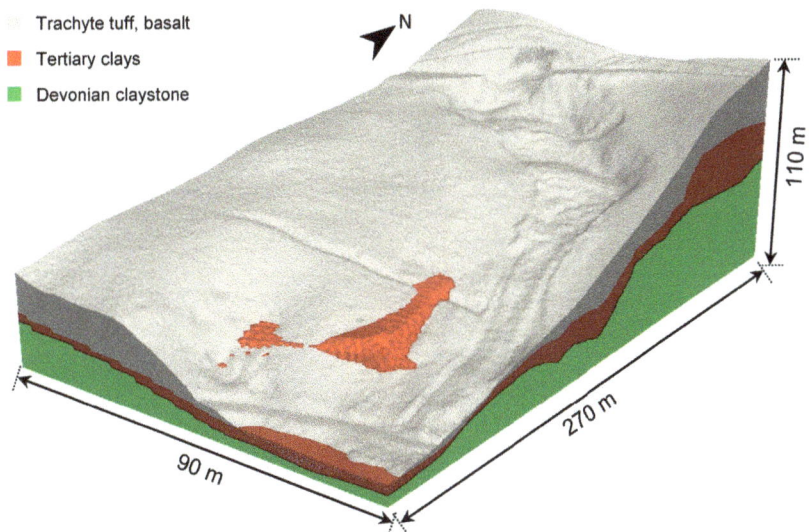

Figure 7. Derived 3D geological model based on refraction seismic survey and borehole logs. Digital elevation model provided by Geobasisdatendienst NRW.

Despite the complex geometry of the study site that can influence subsurface flow and the resulting stability, the hydromechanical model was applied to the mid-cross-section of the landslide area only for computational reasons. The transition from the 3D to 2D geometrical model and the hydromechanical simulations were performed by COMSOL Multiphysics. The 2D domain was discretized using an unstructured triangular mesh with an increasing mesh size from surface to bottom, so that the highly dynamic hydrological conditions in the top soil are captured with a reasonable computational efficiency and accuracy. The mesh size near the slope surface was ≈0.05 m. The maximum mesh size in the top layer was ≈0.2 m. The maximum mesh size increased to 0.3 m in the mid layer and increased further towards the deeper part of the bottom layer. The modeling domain was defined to be substantially larger than the region of interest to reduce the impact of the boundary conditions.

The groundwater monitoring data from the boreholes was used to define the hydraulic boundary conditions (Figure 8). It has been observed that the groundwater level can occasionally reach the surface in the lower part of the slope. Accordingly, the slope surface is defined as a mixed boundary. The inflow and outflow rate depend on pressure, net infiltration, storage capacity of the soil and also its hydraulic conductivity. The bottom boundary was defined as a no-flow boundary. A fixed pressure head boundary at the lower right lateral boundary of the domain was defined using the river level at the slope toe. Here, it was assumed that the river level was constant and at a height of 100 m above the bottom of the modeling domain. The top 25 m of this boundary was a seepage boundary in which water is free to exit from the saturated subsurface. On the left side of the domain, a pressure head boundary was defined using the minimum level of the measured groundwater level at the closest borehole to this boundary. The upper 38 m of the left lateral boundary was a no-flow boundary. From the mechanical point of view, the ground surface is a free boundary with no external loads and constraints, whereas the lateral boundaries were defined as Roller boundaries and the bottom boundary was fixed (Figure 8).

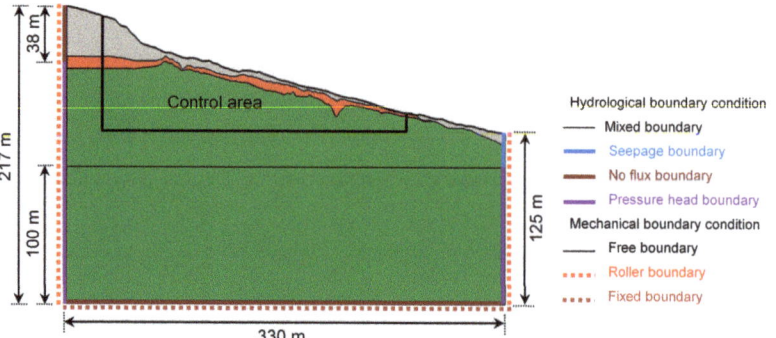

Figure 8. The 2D cross-section used for hydromechanical modeling including boundary conditions for the hydrological and mechanical model components.

In order to parameterize the three layers of the subsurface model, each layer was considered as a homogeneous and isotropic medium. Estimates for bulk, dry and saturated soil density, particle size distribution, porosity, soil cohesion, friction angle and saturated hydraulic conductivity were given by Schmidt [38] and are summarized in Tables 1 and 2. The bulk and dry density of the soil were determined by oven drying following the German Industrial Standard (DIN) of that time, DIN [63]. The maximum moisture content was determined using the method described in DIN [64]. The saturated conductivity was determined using a constant head as well as a falling head permeameter based on DIN [65]. The soil particle distribution were defined by analyzing the samples taken from the drilling cores along the landslide zone and by lithological interpretations of the borehole logs [38]. Soil cohesion and friction angle were determined by a shear box and a triaxial test based on Schmidt [38], DIN [66]. The laboratory tests showed good repeatability with a statistical variance between repeated measurements of less than 10% [38]. The Tertiary clay is the material with the highest density but the lowest cohesion (Table 2). The Devonian clay/silt has the highest cohesion of around 30 kPa but the Trachyte tuff has the highest friction angle of all tested materials. The derived maximum water content is around 35 to 40% for all materials, while the residual water content is around a 6%. The hydraulic conductivity for the Tertiary clay and the Devonian clay/silt layers were provided as a range of possible values and constrained through a calibration of the model with regard to the measured mean groundwater levels of the period 1999 to 2001 by assuming a net infiltration of 125 mm year^{-1}. Using the available information on bulk density and particle size distribution (Table 1), estimates of the soil hydraulic parameters of each layer were obtained using the Rosetta pedotransfer function (Rosetta Lite v 1.1) [67] (Table 2). The soil elastic moduli (E, ν) were estimated from typical ranges provided in the literature [68,69] (Table 3). Here, we have considered a stiff clay-rich Trachyte tuff and Tertiary clay layers with medium to high plasticity (E = 15 MPa) and a harder Devonian clay/silt layer with relatively lower plasticity (E = 30 MPa). The typical Poisson's ratio, ν [-], for silty soils is given as 0.3–0.35 and for unsaturated to saturated clay as 0.1–0.5. Accordingly, for the unsaturated Trachyte tuff and Tertiary clay layers and saturated Devonian clay/silt layer, a Poisson' ratio of 0.35 is considered. All parameters required for the simulations are listed in Table 2. While all parameters may be subject to changes over time in principle, e.g., due to ongoing compaction or root growth [70], they were considered to be constant in our simulations because the test site showed little dynamics in recent monitoring periods.

134

Table 1. Particle size distribution and bulk density of each soil layer of the Dollendorfer Hardt test site (derived from Schmidt [38]).

Description	Unit	Trachyte Tuff	Tertiary Clay	Devonian Clay/Silt
Sand	%	26	11	3
Silt	%	40	41	64
Clay	%	34	48	33
ρ_b	kg m^{-3}	1900	2000	1900

Table 2. Soil model parameters for the Dollendorfer Hardt test site.

Symbol	Unit	Trachyte Tuff	Tertiary Clay	Devonian Clay/Silt
θ_s	-	0.40	0.35	0.40
θ_r *	-	0.06	0.07	0.065
α *	m^{-1}	1.9	2.0	1.1
n *	-	1.22	1.18	1.31
K_s	m s^{-1}	10^{-6}	10^{-9}–10^{-7}	10^{-10}–10^{-6} **
K_s ***	m s^{-1}	10^{-6}	10^{-7}	8.0×10^{-9}
ρ_b	kg m^{-3}	1900	2000	1900
E **	MPa	15	15	30
ν **	-	0.35	0.35	0.35
ϕ'	°	34	32	30
c'	kPa	20	10	30

* Calculated value by the Rosetta pedotransfer function. ** Values derived from literature. *** Obtained by model calibration using mean groundwater level in the period of 1999–2001 using a net infiltration of 125 mm year^{-1}.

Table 3. Typical values of Young's modulus for cohesive material (MPa) [68,69].

Soil Type	Very Soft to Soft	Medium	Stiff to Very Stiff	Hard
Silts with slight plasticity	2.5–8	10–15	15–40	40–80
Silts with low plasticity	1.5–6	6–10	10–30	30–60
Clays with low-medium plasticity	0.5–5	5–8	8–30	30–70
Clays with high plasticity	0.35–4	4–7	7–20	20–32

4.2. Model Calibration and Validation

The model was initialized by simulating the long-term average state of the slope using a spin-up period of 300 years with a constant mean annual net infiltration of 125 mm. This long spin-up period was used to ensure that the model reaches steady-state flow condition. The hydraulic conductivity of the three soil layers was calibrated with regard to the measured mean groundwater level for the period of 1999–2001 using data from three boreholes along the mid cross-section of the test site within the value range given above. The mean groundwater depth from the surface at the boreholes B8, B14 and B21 in the measurement period of 1999–2001 are −7.4 m, −2.5 m, and −2.8 m, respectively (Figure 2). For every set of hydraulic conductivity values, the 300-year model initialization was repeated. For the final choice of parameters, the measured and simulated values were highly correlated ($R^2 > 0.9$) but the simulated mean groundwater level was 0.66 m lower than the measured level. After calibration and model spin-up for 300 years, the groundwater level is lower in the upper part of the slope and closer to the surface in the middle part of the slope, which is in agreement with the measured values at the site.

The calculated steady-state pressure distribution was considered as the initial condition for the next simulation step. The mean monthly conditions for the test site were simulated using the mean monthly net infiltration for the period of 1995–2017. These simulation results with the calibrated model were verified using two series of measured data. First, the simulated groundwater level was compared to the mean monthly groundwater level (Figure 9) by averaging daily measured groundwater levels. Second, simulated soil water content for the top soil was compared to mean monthly measured soil water content obtained from the SoilNet data at the site. Here, the measured soil water content from the twelve sensors nodes that are located along the mid-cross-section were used (sensor nodes 1, 2, 3, 4, 5, 8, 9, 10, 12, 14, 18 and 20 in Figure 2). It should be noted that unreliable parts of the measured data were not considered.

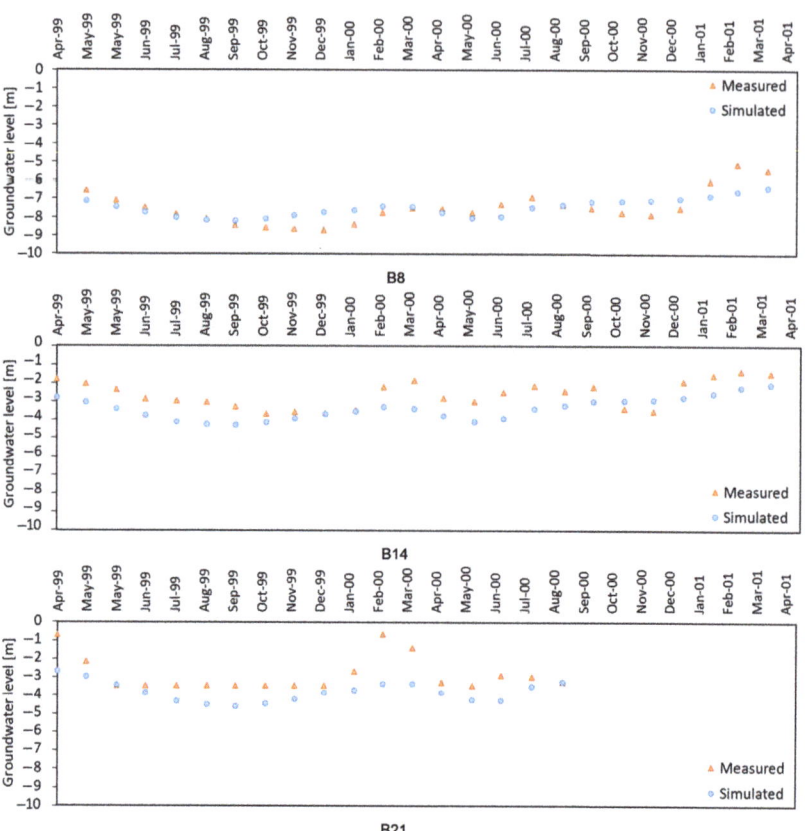

Figure 9. Time series of measured and simulated mean monthly groundwater level for the three boreholes used for model calibration and validation.

This simplified model could capture some of the key features that are deemed important for slope stability assessment. The seasonal pattern of the simulated groundwater levels correlated reasonably well with the measured values (R^2-values ranging from 0.49 to 0.59). However, the simulation shows less pronounced groundwater variations. This is attributed to the use of the mean monthly net infiltration that ignores daily variations associated with high-intensity rainfall, root water uptake and actual evapotranspiration at the site. Accordingly, there are few fluctuations in the precipitation and little variation between wet and dry conditions, which is directly reflected in the lower dynamics of the simulated groundwater level. Moreover, the use of low-intensity mean monthly net infiltration results in the absence of infiltration fronts as well as a lack of water perching due to soil layering, which also contributes to the reduced fluctuation of the simulated groundwater level. In addition to the overall decreasing groundwater levels from top to mid-part of the slope, the simulation captures the remarkably different dynamics of the groundwater level for nearby cross-sections as a result of variation in depth and topography of soil layers. This is consistent with the measured values at the site. Further, the maximum mean groundwater level was observed in January, February and March. This is in agreement with the findings of Schmidt [38], who found that most slope movements occurred during the rainy spring.

In the next step, the simulated soil water content of the top 0.5 m for the period from September 2016 until May 2017 was compared to the measured values from SoilNet. The results show that the seasonal variation as well as the variation with depth is much smaller for the simulated soil water content. The low variation in the simulated water content with depth is mainly attributed to the implementation of the low-intensity mean monthly net infiltration in which the fluctuation in precipitation and the extremely rainy and dry periods are moderated. Notably, the use of a mean monthly low-intensity net infiltration results in the absence of infiltration fronts. In combination with the homogeneous soil hydraulic properties within each layer and the lack of depth-dependent root water uptake, this results in highly simplified and incorrect water content distributions with depth. Therefore, it was concluded that the simulation results do not seem to capture relevant features of the measured near-surface water content distribution, which is the area of interest for slope stability evaluation. Thus, it is evident that the evaluation of slope stability in response to intensive rainfall events cannot be based on monthly net infiltration. Therefore, the measured soil water content data will be considered directly for slope stability evaluation. As the ERT monitoring showed very little dynamics in the deeper layers, soil water content below the top 0.5 m was taken from the simulations using low-intensity mean monthly net infiltration.

5. Model Results for Precipitation Events

To study the slope stability in response to precipitation, two days were selected from the available SoilNet data as the initial conditions for the simulation of a hypothetical rainfall event. These two days were chosen so that both dry and wet initial conditions are considered. According to the values of I_{net} [mm month^{-1}] within the SoilNet measurement period, the driest day occurred at the end of September 2016 after a three-month period with negative net infiltration. The wettest initial conditions occurred at the end of March 2017 after an extended period with positive net infiltration starting in October 2016. Accordingly, the data from 30 September 2016 and 31 March 2017 were used to define dry and wet initial conditions, respectively. The measured soil water content was horizontally quite variable due to the heterogeneity of the soil hydraulic properties within each soil layer. Since the soil layers were assumed to be homogeneous in the simulations, some data points exceeded the saturated water content for the wet conditions. Therefore, the measured soil water content was normalized with 0.395 cm^{3}cm^{-3} as the maximum water content because a full saturation at a water content of 0.40 cm^{3}cm^{-3} resulted in numerical issues. In order to ensure numerical stability, it was also required that the water content distribution with depth varied by at least 0.02 cm^{3}cm^{-3} between depth. Therefore, some normalized soil water content values were manually adjusted. The measurements in dry conditions (30 September 2016) were not normalized and used as is. The measured values were linearly interpolated between the sensor locations by Comsol Multiphysics.

The saturation and simulated LFS distributions before the start of the event rainfall on 30 September 2016 and 31 March 2017 are shown in Figure 10. As expected, the most failure-prone areas besides the scarp area appeared in the middle part of the slope (i.e., locations B and C), where the groundwater was close to the surface and the water content was higher. The LFS at position A in the scarp area was low and close to 1 due to the geometry of the slope with a high inclination in this area. In the following, the simulation results for the two most failure-prone locations B and C will be discussed, because the slope instability at those spots is attributed to water level changes. The initial water content distribution along the depth for locations B and C of the mentioned days is shown in Figure 11. For both days, the groundwater level and the soil water content are higher at location B than at location C. In case of the wet conditions, the groundwater level is similar at both locations. The sharp changes in water content below -2 m are related to soil layering.

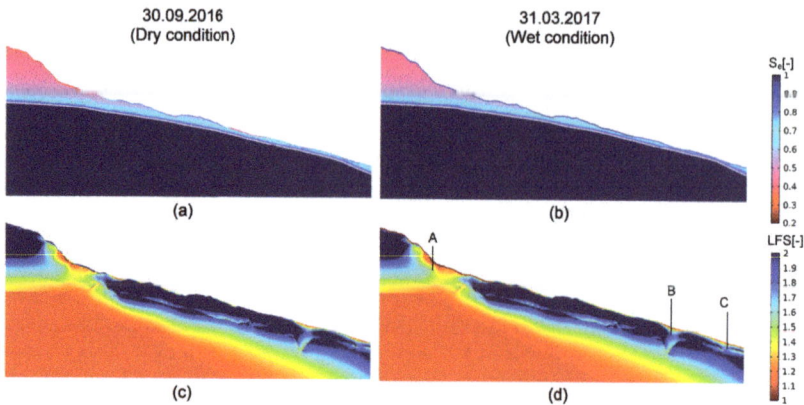

Figure 10. The initial saturation (**a,b**) and LFS distribution (**c,d**) for the rainfall simulations on 30 September 2016 and 31 March 2017.

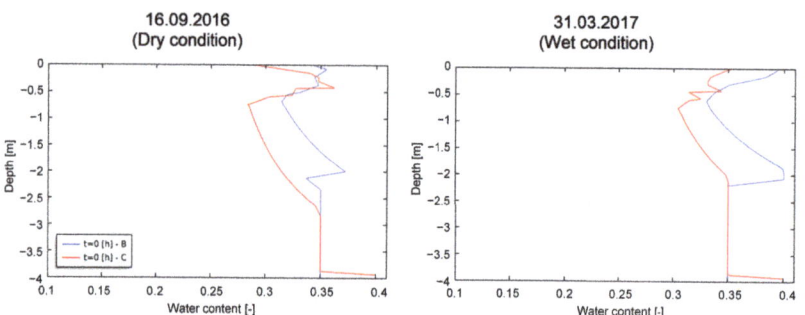

Figure 11. Initial conditions for water content for the event rainfall on 30 September 2016 and 31 March 2017 at the two most vulnerable locations B and C indicated in Figure 10.

In the final step, simulated slope stability in response to an event rainfall of 20 mm h^{-1} is presented for the two selected days. The simulated LFS profiles at location B and C for the dry initial conditions have a maximum difference at the soil surface because of the difference in water content (Figure 12). Below a depth of −0.5 m, the differences in LFS are associated with the differences in the mean monthly conditions of the soil. Accordingly, water content and pressure decreased uniformly from the groundwater table upwards below the depth of −0.5 m. As discussed before, this is attributed to the homogeneous soil layers with no root water uptake and the implemented low-intensity mean monthly net infiltration to obtain the initial conditions below the depth of −0.5 m. The groundwater level and the soil water content below the depth of −0.5 m in March 2017 was higher than in September 2016, which resulted in a relatively lower LFS in March 2017 for the equivalent locations.

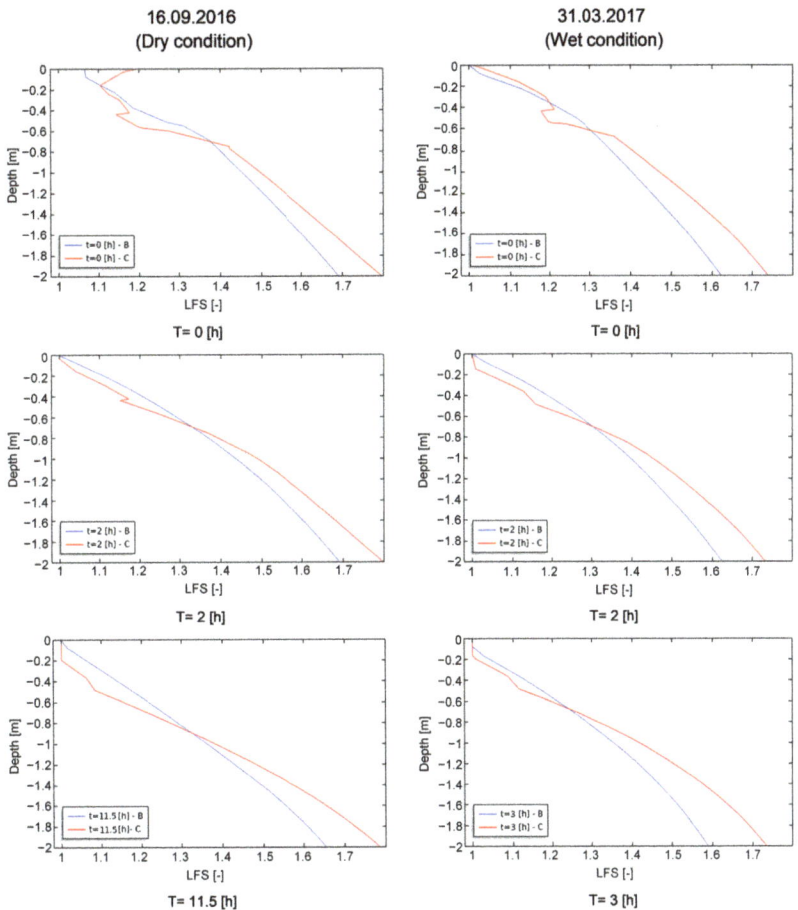

Figure 12. Variation in LFS during the event rainfall for two days (30 September 2016 and 31 March 2017) with different initial conditions, comparably dry and wet soil, for the two most vulnerable locations B and C indicated in Figure 10.

Taking these two sets of initial conditions, the response to an event rainfall of 20 mm h^{-1} was simulated. Figure 12 shows the temporal development of the LFS for the two locations during the rainfall until a LFS of 1 is reached. The critical amounts of rainfall needed are 60 mm and 230 mm for the wet and dry initial conditions, respectively. It can be seen that significantly less rainfall is required on 31 March 2017 with the overall wetter initial condition compared to 16 September 2016. These results are consistent with previous studies [71,72] that have shown that initial hydrological conditions play an important role in the timing of failure initiation. It is interesting to note that the instability threshold was reached first in location C for both wet and dry initial conditions, although the top 0.5 m of the soil at the beginning of the rainfall was drier at location C for the both dry and wet initial conditions. Within a short time (<2 h) after the start of the rainfall, the near surface water content at location C becomes higher than that at location B. Consequently, location C reaches the failure threshold earlier. This is attributed to the difference in bedrock topography and soil depth at these two locations. In particular, the depth of the less permeable mid- and base layer is only 0.4 m from the surface at location C compared to 2.2 m at location B. Accordingly, the local pore pressure and saturation increase faster at this location, which means that a potential failure state is reached after less rainfall. This is in agreement with the findings of Moradi et al. [13] that specifically highlighted the importance of bedrock topography and soil layering for slope stability evaluation.

6. Discussion

The coupled hydromechanical model of Lu et al. [5] was applied for slope stability assessment of the hillslope at the Dollendorfer Hardt (Bonn, Germany) with a relatively complex geometry and heterogeneity in material properties. The test site has a long history of investigations and slope analysis. Drilling, groundwater monitoring, laboratory tests, geophysical, and geomorphological surveys as well as a soil water content sensor network were used to design and parameterize the hydromechanical model. The soil water content obtained from the sensor network was successfully combined with the hydromechanical model to provide realistic initial conditions for the near-surface water content distribution. The ERT monitoring was only applied on a segment of the hillslope and was not used to define initial conditions for the simulations. As the ERT revealed very little changes in the water content distribution at greater depth even during summer time or heavy rainfall events, the sensor network covers all relevant shallow water dynamics. The results of the simulations show that both the lithological layers and the initial conditions play an important role in the redistribution of the pore water pressure and thus determine the position of the potentially unstable locations. The initially wetter conditions require significantly less rainfall than the drier initial conditions to reach potentially unstable conditions. Instabilities develop at locations that facilitate the accumulation of water due to the subsurface topography of less permeable layers. The obtained model allows us to study the influence of initial conditions and precipitation events on the slope stability.

For the test site itself, a significant mass movement seems unlikely as only potentially unstable locations close to the surface were identified in the model. However, the modeling results clearly identify the regions of most significant movement during long-lasting rainfall events at reasonable precipitation rates for the region in agreement with previous studies [38,40]. The hillslope is possibly moving in response to heavy precipitation but at a low rate of approximately 1–3 mm year^{-1} [40]. Modeling of more severe precipitation events would be required for a complete hazard assessment, also incorporating expected future changes in precipitation patterns due to climate change. Further, the model results indicate that the scarp is continuously unstable due to its high inclination. This is consistent with field observations of ruptures of small sizes that occur regularly. A slow extension of the scarp towards the top of the hill seems likely, potentially compromising hiking tracks.

In principle, the presented workflow for slope stability assessment can be extended towards a model-based early-warning system as almost all available data sources were incorporated into the model. So far, site-specific early warning systems are predominantly sensor-based with warning thresholds derived from conventional stability analysis e.g., [73–75]. Incorporating precipitation forecasts, a potentially critical state of a hillslope could be calculated based on the current water content distribution using the current state as initial conditions in the numerical model. However, computational demands are often the bottleneck to include near-real time complex numerical simulations into early warning systems. Therefore, it is challenging to calculate slope stability for a predicted rainfall based on the most recent soil water content measurements. A well-constrained model as presented in this work can be used to derive thresholds for implemented sensor networks or geophysical monitoring based on pre-calculated scenarios. Besides such a model-centered early-warning system, multiple data sources could be used for a more robust early-warning system by acknowledging the value of the in situ measurements itself. Data robustness could be improved through the combination of multiple different types of sensors [70]. In this work, soil water content sensors and ERT were combined to monitor the water content distribution in the subsurface. Through redundancy, erroneous measurements of single sensors or electrodes can be detected in order to reduce the risk of false alarms. Additionally, extensometers or GPS sensors could be used to capture slope movement [75]. By combining precipitation forecasts, soil water content sensor networks and geophysical monitoring to capture water content dynamics and redistribution, and model-derived thresholds, it may be possible to establish a multi-level early warning system with high accuracy.

This study showed the rich possibilities that arise through the combination of various survey and monitoring methods with a hydromechanical model for slope stability. The complementary data sources allow for constraining the model in most aspects and reduce uncertainty in the model design. However, as seen in the comparison with the soil water content sensor network, there is a small scale heterogeneity that can not be resolved using geophysics as the contrast in geophysical parameters, e.g., resistivity or wave velocity, is too small. The high clay content in the soil made the application of ground penetrating radar, as another good methodology to detect near surface heterogeneity, impossible. On other hillslopes, this may be a useful addition to reveal near-surface heterogeneity. With additional near-surface soil sampling and extensive laboratory testing, it may be possible to describe the variability in parameters of the upper soil layers and to include the variability in the numerical model using geostatistical methods. For the studied test site, the assumption of homogeneous soil layers was considered to be adequate as slope topography and layering were the dominant factors controlling slope stability. The added value of ERT monitoring was not high in this study, but this method is expected to be more useful for slopes with stronger water content dynamics at greater depth. This would require an elaborate ERT setup with multiple electrodes and various electrode spacings. In more dynamic scenarios, material parameters and also soil layering may also change within an observation period. In the introduced framework, this could be included through an updated model.

In the future, an approach considering data assimilation seems desirable, especially in the context of a continuous monitoring of the test site. This should preferably be explored with a more sophisticated numerical model for slope stability analysis. In particular, a more advanced coupling of surface flow and infiltration into the soil would significantly improve the model. Field observations showed that strong precipitation events with more than 20 mm h^{-1} only resulted in small increases in soil water content, since a large amount of the water ran off as surface water. Surface morphology supported this observation with visible runoff channels at the surface. Modeling surface runoff and ponding of water based on the surface morphology might alter infiltration dynamics along the slope during those rare heavy precipitation events. Daily water content dynamics along the hillslope are also altered by root-water uptake. While in the current study, net infiltration was considered, a more complex model for root-water uptake would change the water content distribution with depth and increase the heterogeneity of the water content distribution in the subsurface. If this process is adequately considered, the hydromechanical model could be operated with a daily or even higher resolution for the meteorological boundary conditions. In addition, differences between overgrown and bare parts of the slope could be incorporated into the model and possibly linked to the effective cohesion, as roots are known to contribute to the slope stability. With growing computational power, the full three-dimensional geometry of the hillslope and adjacent slopes could be modeled to eliminate simplifying assumptions and effects resulting from the geometrical reduction. An extension towards the calculation of plastic deformation would allow to determine slope movement and water dynamics for situations with a factor of safety larger than one, which not necessarily result in immediate rapid slope failure. Additional model verification could be achieved through measurements of slope movement.

7. Conclusions

A workflow to design, parameterize and validate a hydromechanical model using various data sources from site characterization, monitoring and laboratory tests was presented. Through the combination of complementary data sources, a detailed numerical model was implemented despite the site heterogeneity. Monitoring with point-scale sensors revealed small scale heterogeneity in the hydraulic properties of the upper soil layer, which was not resolved during site characterization. This heterogeneity had minor influence on the slope stability analysis as it is dominated by slope and bedrock topography as well as initial conditions. The multiple data sources and the presented model can be extended towards a model-centered early warning system as well as towards an early warning system based on multiple data streams with thresholds defined through the model.

Author Contributions: Conceptualization, S.M., T.H., A.K. and J.A.H.; methodology, S.M., T.H., J.B., T.G., A.K. and J.A.H.; software, S.M., T.H., J.B., T.G.; validation, S.M., T.H., J.B., T.G., A.K. and J.A.H.; formal analysis, S.M., T.H., J.B., T.G., A.K. and J.A.H.; investigation, S.M., T.H., J.B., T.G., A.K. and J.A.H.; resources, A.K. and J.A.H.; data curation, S.M., T.H., J.B., and T.G.; writing—original draft preparation, S.M., T.H., J.B., and T.G.; writing—review and editing, S.M., T.H., A.K., and J.A.H.; visualization, S.M., T.H., J.B. and T.G.; supervision, A.K. and J.A.H.; project administration, A.K. and J.A.H.; funding acquisition, A.K. and J.A.H. All authors have read and agreed to the published version of the manuscript.

Funding: We thank the Federal Ministry for Science and Education of Germany for funding of the Project CMM-Slide through the GEOTECHNOLOGIEN initiative under contract 03G08498.

Institutional Review Board Statement: Not applicable.

Informed Consent Statement: Not applicable.

Data Availability Statement: The data presented in this study are available on reasonable request from the authors.

Acknowledgments: We thank the Geobasis NRW for providing us with the digital elevation model. We thank Matthias Leschinski and Constanze Reinken for their support with the seismic analysis and Ansgar Weuthen and Bernd Schilling for the installation and maintenance of the soil water content sensor network.

Conflicts of Interest: The authors declare no conflict of interest. The funders had no role in the design of the study; in the collection, analyses, or interpretation of data; in the writing of the manuscript, or in the decision to publish the results.

Abbreviations

The following abbreviations are used in this manuscript:

LFS	Local factor of safety
ERT	Electrical resistivity tomography
MIUB	Department of Meteorology of the University of Bonn
PET	Potential evapotranspiration

References

1. Jakob, M.; Lambert, S. Climate change effects on landslides along the southwest coast of British Columbia. *Geomorphology* **2009**, *107*, 275–284. [CrossRef]
2. Petley, D.N. On the impact of climate change and population growth on the occurrence of fatal landslides in South, East and SE Asia. *Q. J. Eng. Geol. Hydrogeol.* **2010**, *43*, 487–496. [CrossRef]
3. Saez, J.L.; Corona, C.; Stoffel, M.; Berger, F. Climate change increases frequency of shallow spring landslides in the French Alps. *Geology* **2013**, *41*, 619–622. [CrossRef]
4. Lu, N.; Godt, J. Infinite slope stability under steady unsaturated seepage conditions. *Water Resour. Res.* **2008**, *44*. [CrossRef]
5. Lu, N.; Sener-Kaya, B.; Wayllace, A.; Godt, J.W. Analysis of rainfall-induced slope instability using a field of local factor of safety. *Water Resour. Res.* **2012**, *48*, 1–14. [CrossRef]
6. Liu, Q.Q.; Li, J.C. Effects of Water Seepage on the Stability of Soil-slopes. *Procedia IUTAM* **2015**, *17*, 29–39. doi:10.1016/j.piutam.2015.06.006. [CrossRef]
7. Rabie, M. Comparison study between traditional and finite element methods for slopes under heavy rainfall. *HBRC J.* **2014**, *10*, 160–168. [CrossRef]
8. Stianson, J.; Chan, D.; Fredlund, D. Role of Admissibility Criteria in Limit Equilibrium Slope Stability Methods Based on Finite Element Stresses. *Comput. Geotech.* **2015**, *66*, 113–125. [CrossRef]
9. Lanni, C.; McDonnell, J.; Hopp, L.; Rigon, R. Simulated effect of soil depth and bedrock topography on near-surface hydrologic response and slope stability. *Earth Surf. Process. Landforms* **2013**, *38*, 146–159. [CrossRef]

10. Reid, M.E.; Iverson, R.M. Gravity-driven groundwater-flow and slope failure potential. 2. Effects of slope morphology, material properies, and hydraulic heterogeneity. *Water Resour. Res.* **1992**, *28*, 939–950. [CrossRef]
11. Kim, M.S.; Onda, Y.; Kim, J.K.; Kim, S.W. Effect of topography and soil parameterisation representing soil thicknesses on shallow landslide modelling. *Quat. Int.* **2015**, *384*, 91–106. [CrossRef]
12. Tromp-van Meerveld, H.J.; McDonnell, J.J. On the interrelations between topography, soil depth, soil moisture, transpiration rates and species distribution at the hillslope scale. *Adv. Water Resour.* **2006**, *29*, 293–310. [CrossRef]
13. Moradi, S.; Huisman, J.; Class, H.; Vereecken, H. The effect of bedrock topography on timing and location of landslide initiation using the local factor of safety concept. *Water* **2018**, *10*, 1290. [CrossRef]
14. Lapenna, V.; Lorenzo, P.; Perrone, A.; Piscitelli, S.; Sdao, F.; Rizzo, E. High-resolution geoelectrical tomographies in the study of Giarrossa landslide (Southern Italy). *Bull. Eng. Geol. Environ.* **2003**, *62*, 259–268. [CrossRef]
15. Sass, O.; Bell, R.; Glade, T. Comparison of GPR, 2D-resistivity and traditional techniques for the subsurface exploration of the Öschingen landslide, Swabian Alb (Germany). *Geomorphology* **2008**, *93*, 89–103. [CrossRef]
16. Jongmans, D.; Bievre, G.; Renalier, F.; Schwartz, S.; Beaurez, N.; Orengo, Y. Geophysical investigation of a large landslide in glaciolacustrine clays in the Trieves area (French Alps). *Eng. Geol.* **2009**, *109*, 45–56. [CrossRef]
17. Gance, J.; Malet, J.P.; Supper, R.; Sailhac, P.; Ottowitz, D.; Jochum, B. Permanent electrical resistivity measurements for monitoring water circulation in clayey landslides. *J. Appl. Geophys.* **2016**, *126*, 98–115. [CrossRef]
18. Uhlemann, S.; Chambers, J.; Wilkinson, P.; Maurer, H.; Merritt, A.; Meldrum, P.; Kuras, O.; Gunn, D.; Smith, A.; Dijkstra, T. 4D imaging of moisture dynamics during landslide reactivation. *J. Geophys. Res. Earth Surf.* **2016**. [CrossRef]
19. Göktürkler, G.; Balkaya, C.; Erhan, Z. Geophysical investigation of a landslide: The Altindag landslide site, Izmir (Western Turkey). *J. Appl. Geophys.* **2008**, *65*, 84–96. [CrossRef]
20. Chambers, J.E.; Wilkinson, P.B.; Kuras, O.; Ford, J.R.; Gunn, D.A.; Meldrum, P.I.; Pennington, C.V.L.; Weller, A.L.; Hobbs, P.R.N.; Ogilvy, R.D. Three-dimensional geophysical anatomy of an active landslide in Lias Group mudrocks, Cleveland Basin, UK. *Geomorphology* **2011**, *125*, 472–484. [CrossRef]
21. Mauritsch, H.J.; Seiberl, W.; Arndt, R.; Römer, A.; Schneiderbauer, K.; Sendlhofer, G.P. Geophysical investigations of large landslides in the Carnic region of southern Austria. *Eng. Geol.* **2000**, *56*, 373–388. [CrossRef]
22. Glade, T.; Stark, P.; Dikau, R. Determination of potential landslide shear plane depth using seismic refraction—A case study in Rheinhessen, Germany. *Bull. Eng. Geol. Environ.* **2005**, *64*, 151–158. [CrossRef]
23. Baron, I.; Supper, R. Application and reliability of techniques for landslide site investigation, monitoring and early warning-Outcomes from a questionnaire study. *Nat. Hazards Earth Syst. Sci.* **2013**, *13*, 3157–3168. [CrossRef]
24. Lehmann, P.; Gambazzi, F.; Suski, B.; Baron, L.; Askarinejad, A.; Springman, S.M.; Holliger, K.; Or, D. Evolution of soil wetting patterns preceding a hydrologically induced landslide inferred from electrical resistivity survey and point measurements of volumetric water content and pore water pressure. *Water Resour. Res.* **2013**, *49*, 7992–8004. [CrossRef]
25. Archie, G.E. The electrical resistivity log as an aid in determining some reservoir characteristics. *Trans. AIME* **1942**, *146*, 54–62. [CrossRef]
26. Waxman, M.H.; Smits, L.J.M. Electrical conductivities in oil-bearing shaly sands. *Soc. Pet. Eng. J.* **1968**, *8*, 107–122. [CrossRef]
27. Hübner, R.; Heller, K.; Günther, T.; Kleber, A. Monitoring Hillslope Moisture Dynamics with Surface ERT for Enhancing Spatial Significance of Hydrometric Point Measurements. *Hydrol. Earth Syst. Sci.* **2015**, *19*, 225–240. [CrossRef]
28. Hibert, C.; Grandjean, G.; Bitri, A.; Travelletti, J.; Malet, J.P. Characterizing Landslides through Geophysical Data Fusion: Example of the La Valette Landslide (France). *Eng. Geol.* **2012**, *128*, 23–29. [CrossRef]
29. Hojat, A.; Arosio, D.; Ivanov, V.I.; Longoni, L.; Papini, M.; Scaioni, M.; Tresoldi, G.; Zanzi, L. Geoelectrical Characterization and Monitoring of Slopes on a Rainfall-Triggered Landslide Simulator. *J. Appl. Geophys.* **2019**, *170*, 103844. [CrossRef]
30. Heinze, T.; Limbrock, J.; Pudasaini, S.; Kemna, A. Relating Mass Movement with Electrical Self-Potential Signals. *Geophys. J. Int.* **2019**, *216*, 55–60. [CrossRef]
31. Whiteley, J.S.; Chambers, J.E.; Uhlemann, S.; Wilkinson, P.B.; Kendall, J.M. Geophysical Monitoring of Moisture-Induced Landslides: A Review. *Rev. Geophys.* **2019**, *57*, 106–145. [CrossRef]
32. Bogena, H.R.; Herbst, M.; Huisman, J.A.; Rosenbaum, U.; Weuthen, A.; Vereecken, H. Potential of wireless sensor networks for measuring soil water content variability. *Vadose Zone J.* **2010**, *9*, 1002–1013. [CrossRef]
33. Srinivasan, K.; Howell, B.; Anderson, E.; Flores, A. A low cost wireless sensor network for landslide hazard monitoring. In Proceedings of the 2012 IEEE International Geoscience and Remote Sensing Symposium (IGARSS), Munich, Germany, 22–27 July 2012.
34. Piegari, E.; Cataudella, V.; Di Maio, R.; Milano, L.; Nicodemi, M.; Soldovieri, M.G. Electrical resistivity tomography and statistical analysis in landslide modelling: A conceptual approach. *J. Appl. Geophys.* **2009**, *68*, 151–158. [CrossRef]
35. Tric, E.; Lebourg, T.; Jomard, H.; Le Cossec, J. Study of Large-Scale Deformation Induced by Gravity on the La Clapière Landslide (Saint-Etienne de Tinée, France) Using Numerical and Geophysical Approaches. *J. Appl. Geophys.* **2010**, *70*, 206–215. [CrossRef]
36. Ling, C.; Xu, Q.; Zhang, Q.; Ran, J.; Lv, H. Application of electrical resistivity tomography for investigating the internal structure of a translational landslide and characterizing its groundwater circulation (Kualiangzi landslide, Southwest China). *J. Appl. Geophys.* **2016**, *131*, 154–162. [CrossRef]
37. Pasierb, B.; Grodecki, M.; Gwóźdź, R. Geophysical and Geotechnical Approach to a Landslide Stability Assessment: A Case Study. *Acta Geophys.* **2019**, *67*, 1823–1834. [CrossRef]

38. Schmidt, J. The Role of Mass Movements for Slope Evolution. Ph.D. Thesis, Universitäts-und Landesbibliothek Bonn, Bonn, Germany, 2001.
39. Hardenbicker, U. *Hangrutschungen im Bonner Raum-Naturraeumliche Einordnung und ihre Anthropogenen Ursachen*; Ferdinand Dümmlers Verlag: Bonn, Germany, 1994.
40. García, A.; Hördt, A.; Fabian, M. Landslide monitoring with high resolution tilt measurements at the Dollendorfer Hardt landslide, Germany. *Geomorphology* **2010**, *120*, 16–25. [CrossRef]
41. Schmidt, J.; Dikau, R. Preparatory and triggering factors for slope failure: Analyses of two landslides near Bonn, Germany. *Z. Geomorphol.* **2005**, *49*, 121–138.
42. Weber, M. Welchen Beitrag kann die Luftbildinterpretation zur Erfassung und Datierung von Hangrutschungen leisten – Erste Ergebnisse aus dem Bonner Raum. *Arb. Zur Rheinischen Landeskd.* **1991**, *60*, 19–30.
43. Reynolds, J.M. *An Introduction to Applied and Environmental Geophysics*; Wiley: Chichester, UK, 1997.
44. Schön, J.H. *Physical Propoperties of Rocks*; Elsevier: Amsterdam, The Netherlands, 2011.
45. Thornthwaite, C.W. An Approach toward a Rational Classification of Climate. *Geogr. Rev.* **1948**, *38*, 55–94. [CrossRef]
46. Schrödter, H. *Verdunstung: Anwendungsorientierte Meßverfahren und Bestimmungsmethoden*; Springer: Berlin, Germany, 1985.
47. Xiang, K.Y.; Li, Y.; Horton, R.; Feng, H. Similarity and difference of potential evapotranspiration and reference crop evapotranspiration—A review. *Agric. Water Manag.* **2020**, *232*, 1–16. [CrossRef]
48. Bogena, H.R.; Huisman, J.A.; Schilling, B.; Weuthen, A.; Vereecken, H. Effective calibration of low-cost soil water content sensors. *Sensors* **2017**, *17*, 208. [CrossRef]
49. Roth, K.; Schulin, R.; Fluhler, H.; Attinger, W. Calibration of time domain reflectometry for water content measurement using a composite dielectric approach. *Water Resour. Res.* **1990**, *26*, 2267–2273. [CrossRef]
50. Robinson, D.A.; Jones, S.B.; Blonquist, J.M.; Friedman, S.P. A physically derived water content/permittivity calibration model for coarse-textured, layered soils. *Soil Sci. Soc. Am. J.* **2005**, *69*, 1372–1378. [CrossRef]
51. Hayley, K.; Bentley, L.R.; Gharibi, M.; Nightingale, M. Low temperature dependence of electrical resistivity: Implications for near surface geophysical monitoring. *Geophys. Res. Lett.* **2007**, *34*, L18402. [CrossRef]
52. Brunet, P.; Clément, R.; Bouvier, C. Monitoring soil water content and deficit using Electrical Resistivity Tomography (ERT)—A case study in the Cevennes area, France. *J. Hydrol.* **2010**, *380*, 146–153. [CrossRef]
53. Kemna, A. Tomographic Inversion of Complex Resistivity. Ph.D. Thesis, Ruhr-Universität Bochum, Bochum, Germany, 2000.
54. Slater, L.; Binley, A.; Daily, W.; Johnson, R. Cross-hole electrical imaging of a controlled saline tracer injection. *J. Appl. Geophys.* **2000**, *44*, 85–102. [CrossRef]
55. Günther, T.; Rücker, C. A New Joint Inversion Approach Applied to the Combined Tomography of DC Resistivity and Seismic Refraction Data. In *Symposium on the Application of Geophysics to Engineering and Environmental Problems 2006*; Society of Exploration Geophysicists: Tulsa, OK, USA, 2006; pp. 1196–1202. [CrossRef]
56. Doetsch, J.; Linde, N.; Pessognelli, M.; Green, A.G.; Günther, T. Constraining 3-D electrical resistance tomography with GPR reflection data for improved aquifer characterization. *J. Appl. Geophys.* **2012**, *78*, 68–76. [CrossRef]
57. Nguyen, F.; Kemna, A. Strategies for Time-Lapse Electrical Resistivity Inversion. In *Near Surface Geoscience 2005 11th European Meeting of Environmental and Engineering Geophysics*; European Association of Geoscientists and Engineers: Houten, The Netherlands, 2005; p. A005. [CrossRef]
58. Rhoades, J.D.; Manteghi, N.A.; Shouse, P.J.; Alves, W.J. Soil Electrical Conductivity and Soil Salinity: New Formulations and Calibrations. *Soil Sci. Soc. Am. J.* **1989**, *53*, 433. [CrossRef]
59. Gillis, G.; Pirie, G. Schlumberger Oilfield Glossary—Geophysics Module. 2018. Available online: https://www.glossary.oilfield.slb.com (accessed on 15 October 2018).
60. Van Genuchten, M.T. A Closed-form Equation for Predicting the Hydraulic Conductivity of Unsaturated Soils1. *Soil Sci. Soc. Am. J.* **1980**, *44*, 892. [CrossRef]
61. Lu, N.; Likos, W.J. Suction stress characteristic curve for unsaturated soil. *J. Geotech. Geoenviron. Eng.* **2006**, *132*, 131–142.:2(131). [CrossRef]
62. Lu, N.; Godt, J.W.; Wu, D.T. A closed-form equation for effective stress in unsaturated soil. *Water Resour. Res.* **2010**, *46*, 1–14. [CrossRef]
63. DIN. *Baugrund, Untersuchung von Bodenproben—Bestimmung der Dichte des Bodens (DIN 18125)*; Deutsches Institut für Normung e.V. (DIN): Berlin, Germany, 1993.
64. DIN. *Baugrund, Versuche und Versuchsgeräte—Bestimmung des Wasseraufnahmevermögens (DIN 18132)*; Deutsches Institut für Normung e.V. (DIN): Berlin, Germany, 1993.
65. DIN. *Baugrund, Untersuchung von Bodenproben—Bestimmung des Wasserdurchlässigkeitsbeiwerts (DIN 18130)*; Deutsches Institut für Normung e.V. (DIN): Berlin, Germany, 1993.
66. DIN. *Baugrund, Untersuchung von Bodenproben—Bestimmung der Scherfestigkeit (DIN 18137)*; Deutsches Institut für Normung e.V. (DIN): Berlin, Germany, 1993.
67. Schaap, M.G.; Leij, F.J.; van Genuchten, M.T. ROSETTA: A computer program for estimating soil hydraulic parameters with hierarchical pedotransfer functions. *J. Hydrol.* **2001**, *251*, 163–176. [CrossRef]
68. Kezdi, A. *Handbook of Soil Mechanics*; Elsevier: Amsterdam, The Netherlands, 1980.

69. Obrzud, R.; Truty, A. *The Hardening Soil Model—A Practical Guidbook*; Z_Soil.PC 100701 Report; Zace Services: Lausanne, Switzerland, 2018.
70. Stahli, M.; Sattele, M.; Huggel, C.; McArdell, B.W.; Lehmann, P.; Van Herwijnen, A.; Berne, A.; Schleiss, M.; Ferrari, A.; Kos, A.; et al. Monitoring and prediction in early warning systems for rapid mass movements. *Nat. Hazards Earth Syst. Sci.* **2015**, *15*, 905–917. [CrossRef]
71. Greco, R.; Pagano, L. Basic features of the predictive tools of early warning systems for water-related natural hazards: Examples for shallow landslides. *Nat. Hazards Earth Syst. Sci.* **2017**, *17*, 2213–2227. [CrossRef]
72. Montrasio, L.; Schilirò, L.; Terrone, A. Physical and numerical modelling of shallow landslides. *Landslides* **2015**. [CrossRef]
73. Intrieri, E.; Gigli, G.; Mugnai, F.; Fanti, R.; Casagli, N. Design and implementation of a landslide early warning system. *Eng. Geol.* **2012**, *147*, 124–136. [CrossRef]
74. Michoud, C.; Bazin, S.; Blikra, L.H.; Derron, M.H.; Jaboyedoff, M. Experiences from site-specific landslide early warning systems. *Nat. Hazards Earth Syst. Sci.* **2013**, *13*, 2659–2673. [CrossRef]
75. Sättele, M.; Krautblatter, M.; Bründl, M.; Straub, D. Forecasting rock slope failure: How reliable and effective are warning systems. *Landslides* **2016**, *13*, 737–750. [CrossRef]

Article

Evaluation of the Effect of Hydroseeded Vegetation for Slope Reinforcement

Okoli Jude Emeka [1], Haslinda Nahazanan [1,*], Bahareh Kalantar [2], Zailani Khuzaimah [3] and Ojogbane Success Sani [1]

1. Department of Civil Engineering, Faculty of Engineering, Universiti Putra Malaysia, Serdang 43400, Malaysia; gs43137@student.upm.edu.my (O.J.E.); gs47871@student.upm.edu.my (O.S.S.)
2. RIKEN Center for Advanced Intelligence Project, Goal-Oriented Technology Research Group, Disaster Resilience Science Team, Tokyo 103-0027, Japan; Bahareh.kalantar@riken.jp
3. Institute of Plantation Studies, Universiti Putra Malaysia, Serdang 43400, Malaysia; zailani@upm.edu.my
* Correspondence: n_haslinda@upm.edu.my

Citation: Emeka, O.J.; Nahazanan, H.; Kalantar, B.; Khuzaimah, Z.; Sani, O.S. Evaluation of the Effect of Hydroseeded Vegetation for Slope Reinforcement. *Land* **2021**, *10*, 995. https://doi.org/10.3390/land10100995

Academic Editors: Enrico Miccadei, Cristiano Carabella and Giorgio Paglia

Received: 21 July 2021
Accepted: 21 August 2021
Published: 22 September 2021

Publisher's Note: MDPI stays neutral with regard to jurisdictional claims in published maps and institutional affiliations.

Copyright: © 2021 by the authors. Licensee MDPI, Basel, Switzerland. This article is an open access article distributed under the terms and conditions of the Creative Commons Attribution (CC BY) license (https://creativecommons.org/licenses/by/4.0/).

Abstract: A landslide is a significant environmental hazard that results in an enormous loss of lives and properties. Studies have revealed that rainfall, soil characteristics, and human errors, such as deforestation, are the leading causes of landslides, reducing soil water infiltration and increasing the water runoff of a slope. This paper introduces vegetation establishment as a low-cost, practical measure for slope reinforcement through the ground cover and the root of the vegetation. This study reveals the level of complexity of the terrain with regards to the evaluation of high and low stability areas and has produced a landslide susceptibility map. For this purpose, 12 conditioning factors, namely slope, aspect, elevation, curvature, hill shade, stream power index (SPI), topographic wetness index (TWI), terrain roughness index (TRI), distances to roads, distance to lakes, distance to trees, and build-up, were used through the analytic hierarchy process (AHP) model to produce landslide susceptibility map. Receiver operating characteristics (ROC) was used for validation of the results. The area under the curve (AUC) values obtained from the ROC method for the AHP model was 0.865. Four seed samples, namely ryegrass, rye corn, signal grass, and couch, were hydroseeded to determine the vegetation root and ground cover's effectiveness on stabilization and reinforcement on a high-risk susceptible 65° slope between August and December 2020. The observed monthly vegetation root of couch grass gave the most acceptable result. With a spreading and creeping vegetation ground cover characteristic, ryegrass showed the most acceptable monthly result for vegetation ground cover effectiveness. The findings suggest that the selection of couch species over other species is justified based on landslide control benefits.

Keywords: hazard; landslide; hydroseeding; slope; vegetation; AHP

1. Introduction

Landslide hazard management, prevention, and control are critical to preventing loss of lives and properties [1–3]. For every hazard management, mitigation is the final stage, and it provides the methodology of controlling any form of natural hazard [4–6]. Mitigating an existing landslide hazard or preventing future landslides is an element of a dangerous decrease in the causative components or an increase in the opposing forces [4,7,8]. Vegetation influence in slope stability analysis reveals a significant and still ongoing challenge for research. Few research studies within the mitigation framework of slope stability have increasingly leaned towards the influence exerted by either vegetation canopy [9,10] or vegetation root mechanisms [11,12], with more emphasis on the surface runoff on vegetation cover. The present study focuses on the effective means of reducing the rate of surface runoff, which practically reduces the effect of landslides through their combined vegetation cover and vegetation root system. Bare slopes are prone to landslides as a result of the lack of surface cover.

Moreover, due to the tropical weather conditions of Malaysia, the soils go through intense soil weathering, which may elevate soil erosion and surface runoff. Landslide control measures are essential to address the issues of slope failures due to their impact on lives and the economy. So far, there has not been much available information or research on vegetation slope stabilization methods in Malaysia. According to Popescu [13], a list of approaches to control and prevent landslides was organized into four experimental groups by the International Union of Geological Sciences Working Group on Landslides (IUGS WG/L), which includes: (i) Slope geometry modification (ii) Drainage (iii) Retaining structures (iv) Internal slope reinforcement. Moreover, Hutchinson [14] listed drainage as the most crucial methodology to reduce landslides, followed by slope geometry modification as the second most applied approach. The study also suggested that although a single mitigation strategy may prevail, most slope failure mitigation strategies combine some groups.

Over a decade, there has been an apparent shift to "soft engineering", non-structural preventive measures such as drainage and slope geometry modification, as well as some novel approaches such as stabilization using lime/cement, soil nailing, or grouting [15]. Non-structural solutions are less expensive compared to the cost of structural solutions [16]. On the other hand, structural measures such as retaining walls include opening the slope during the construction process and sometimes needing steeper temporary cuts [17]. Both techniques increase the risk of landslide during construction or increased precipitation infiltration [18]. The mitigation approach should be designed to significantly fit the condition of the specific slope under investigation [19]. In high-risk terrains where landslides pose a danger to lives or primarily affect properties, a landslide expert, a geotechnical, or civil engineer should be consulted before carrying out any stabilization work [20]. However, these methods have their limitations. Thus, to address this, we looked into applying a soil bioengineering methodology, especially on plant species that have never been used in Malaysia. For effective landslide management, one must identify the most critical conditioning factors that affect the slope's stability, determine landslide-susceptible areas, then select the appropriate and cost-effective method to be sufficiently utilized to minimize the possibility of landslide [21]. Here, we analyze the effects of different plant species used for slope stabilization in other parts of the world. Furthermore, the focus was on species selection and the vegetation engineering properties, especially their root architecture and ground cover. The present study also considers runoff, rainfall, soil type, vegetation, and slope. Thus, knowledge of the relationships between the landslide susceptibility mapping and an effective landslide control mechanism for effective landslide hazard management is fundamental [8].

Different landslide susceptibility models with a considerable level of accuracy have been developed and broadly classified into data-driven (quantitative) and expert opinion (qualitative) categories, both of which have their respective advantages and limitations. The choice of mapping techniques depends on the structure of the terrain, landslide triggering factors, data availability, and landslide types [22,23]. Data-driven methods are divided into statistical and deterministic approaches, according to Mantovani et al. [24]. A statistical method is an indirect approach that uses statistical analysis to obtain landslide predictions from several parameters. The statistical method is further divided into bivariate and multivariate models. Statistical models include frequency ratio (FR) [25,26], information value method (IVM) [27–29], weights of evidence method (WOE) [30–33], and machine learning algorithms, such as logistic regression (LR) [25,34], artificial neural network (ANN) [35,36], support vector machine (SVM) [37,38], and deep learning (DL) [39,40] are based on data collected from previous landslides and their spatial coverage. Integration between event data and targets makes landslide mapping easier in a geographic information system (GIS) environment [41]. These methods require less human knowledge and experience to produce and utilize as susceptibility models [36]. The machine learning algorithms effectively analyze large datasets with higher precision than statistical methods [42]. Generally, these data-driven strategies have their limitations, such as the application of a relative generalization approach for landslide parameters and applications over a large area, a lack of knowledge of the correlation between landslides and the conditioning parameters, and a lack of understanding of relevant expert ideas in empirical modeling [33]. In recent years, machine learning algorithms have become a more robust approach in landslide research [43], but the models require managing uncertainties. These uncertainties could result from errors and model variability [44], difficulties in the selection of parameters [45], system understanding [46], the weighting of parameters [47], and human judgment [48]. Moreover, machine learning may encounter prediction errors if trained with a small data set [43]. Besides errors from model building, the selection of input variables also impacts the prediction accuracy in machine learning [49]. Uncertainties resulting from these landslide susceptibility models are inevitable [48] and can threaten the selection of the most suitable landslide susceptibility approach [50].

In qualitative methodologies divided into geomorphic and heuristic methods, such as AHP [51–53] and weighted linear combination [8,54,55], weights and ranks are assigned by the experts regardless of any existing landslide inventory maps and terrain variations [28]. The deterministic approach includes static and dynamic methods that evaluate landslide hazards using slope stability models, which results in the calculation of safety factors. Contrary to data-driven models, expert knowledge has been projected as outstanding, more reliable, consistent, and generally applicable when the knowledge is formalized, particularly for a large-scale mapping [56]. Another advantage of these approaches is that each polygon on the map can be evaluated separately, according to its unique set of conditions [57]. The main limitations of these methods are their subjectivity in factor selection, mapping, and the weighting of the parameters, but avoid generalization, as is often used in data-driven strategies [58]. Due to these different landslide susceptibility approaches, they may produce varying results.

Although these methods may yield accurate results in most cases, there is also a certain degree of uncertainty, sometimes leading to inaccurate results [59]. The weighting of criteria can result in many uncertainties in expert opinion methods [60]. The user's preferred data input choices are seen as the actual rules in the spatial decision-making process and but may often be erroneous. These uncertainties can unfavorably influence the accuracy of the landslide susceptibility results if ignored [61]. Attempts have been made to improve the accuracy of these models, as they are valuable tools for solving a wide range of spatial anomalies [62]. Therefore, the authenticity of any spatial simulation models and expert opinion methods depends largely on calculating the relative importance of each parameter [63]. Moreover, the model validation process must be reliable, robust, and have a certain degree of fitting and prediction skill [64]. However, the performance evaluation of most of the landslide susceptibility maps (LSMs) can be based on the testing datasets [65].

AHP, first developed by Saaty [62], is categorized under this qualitative approach and applied by several researchers for landslide susceptibility assessment [51,52,66,67]. AHP uses a hierarchical process of landslide parameters to compare possible pairs and assign weights and a consistency ratio [8]. The AHP method is based on three principles: Decomposition, comparative judgment, and assigning priorities [8]. AHP enables the experts to derive significant parameters from a pair of criteria for multi-criteria decision-making. It decides the parameters based on the objective and knowledge of the problem [68].

Hydroseeding is an efficient method of plant establishment on a land surface [69] and involves applying seed under pressure using a water carrier [70]. The basic hydroseeding concept sprays seed mixed with water or dried onto an already prepared surface [71]. Hydroseeding is a mechanism that involves the application of a mix of seeds, adhesives, mulch, fertilizers, and water on soils, using an appropriate hydroseeding tool [70]. This experiment evaluates the reinforcement effect of four seed samples on a failing terrain within the Universiti Putra Malaysia (UPM) by applying the same mixtures and spray methodology, comparing their vegetation root system and ground cover, and evaluating their vegetation landslide control strength. This approach is one of the techniques of ground revegetation used to stabilize bare soil surface to control landslide hazards [72].

Cellulose mulch mixed with tackifier as a binder is also applied [73]. The cellulose mulch, combined with the seeds, germinator, fertilizer, are mixed with water in the hydroseeder, forming a homogeneous slurry and uniformly sprayed on the soil [74]. The fertilizer-mixed cellulose fiber mulch and the seed act as an absorbent mat and hold a large water capacity that helps seed germination and forms a stable blanket cover on the surface before the seed germination period. Hydroseeding is a widely used method of landslide or soil erosion control [75]. The United States and Australia are among the countries that use hydroseeding because it is considered the fastest, most efficient, and most economical method of landslide control [76]. Studies revealed the following observations and conclusions from hydroseeding application: (a) The hydroseeding process is the fastest means of landslide and soil erosion prevention [74]; (b) Some seeds germinate within two days, which enables the topsoil of the embankment to be already 100% stabilized right before the development of the vegetation ground cover [77]; (c) Watering is required less as soon as the ground cover grass seed has been established [78]; (d) The mulch serves as water retention and absorbing mat, and reduces the development of the unwanted weeds [74]; (d) The end product of hydroseeding requires a very minimal maintenance policy as soon as the permanent ground cover is purely developed [74]. In summary, the contribution of this study is to compare the slope reinforcement effectiveness of the four seed samples (rye corn, ryegrass, signal grass, and couch) used in different parts of the world in Malaysia in relation to the soil conditions.

2. Study Area and Materials

2.1. Study Area

The study was conducted within the UPM, a government tertiary institution in Serdang, Selangor, Peninsular Malaysia (Figure 1a), a land surface of about 1108 hectares made up of different natural and artificial facilities in buildings, agricultural lands, trees, and lakes. The entire area is within latitude: 2°59′34.19″ N and longitude: 101°42′16.79″ E. It is mostly warm with a sunny tropical rainforest climate and abundant rainfall, particularly in the northeast monsoon season between October and March. It has a constant annual temperature with a maximum between 31 and 33 °C, while minimums are usually between 22 and 23.5 °C. The average yearly rainfall is about 2400 mm, with relative dryness between June and July. The geology of Serdang consists of three different rock formations, which include the Kajang, Kuala Lumpur, and Kenny hills formations. Kajang formation consists of schist and some intercalations of limestone and phyllites. On the other hand, the Kuala Lumpur formation consists of limestones with intercalation of phyllite, while quartzes and phyllite make up the Kenny hills formation [79]. The experiment was carried out on the 11 August 2020 on a 65° slope gradient within 2°59′34.3″ N 101°43′30.1″ E. The area's soil is a mix of sandy, clay, and loam collectively identified under Serdang and Malacca. The surrounding vegetation in the site consisted of native and introduced lower grasslands species. The land is irregular, highly degraded with a uniform slope subject to severe erosion, and prone to landslide, with no vegetation. Before carrying out the field tests, a calibration test was conducted to identify the machine calibrations and settings. This test showed that the spray mechanism was reliable and ready to go with the tests.

The site was cleared of stumps, rock debris, stones, and unwanted materials of more than 4–5 cm (Figure 1c). A hand rake was used to rough grade the entire area to uncover debris and level the site. Initial tiling to a depth of 5 cm was done to reduce earth compaction, allow bonding of the topsoil and the subsoil, and improve vegetation root penetration and groundwater movement. We also studied the soil's physical and chemical properties to reveal the soil nutrient level, pH, acidity, and alkalinity level. The site was measured and marked and ready for hydroseeding. Table 1 presents the description of the study area.

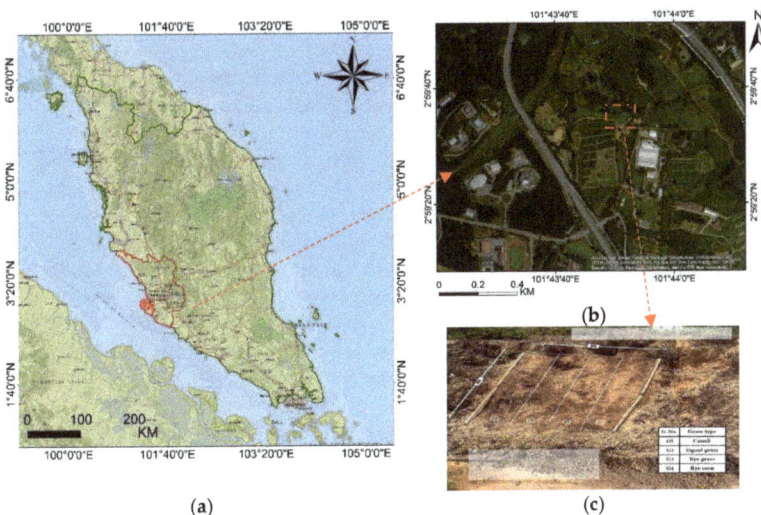

Figure 1. Study area (**a**) map of Malaysia; (**b**) LiDAR image of the study area; (**c**) hydroseeded site.

Table 1. Site description of the study area.

Properties	Description
Location	UPM, Serdang
Slope angle	65 degrees
Soil type	Serdang (Typic Kandiudults) and Malacca Series (Typic Hapludox)
Precipitation	2400 mm annually
Humidity	The average annual percentage of humidity is: 80.0%
Temperature	22 to 33 °C

Physical and Chemical Properties of Serdang and Malacca Series

The soil of the study area was predominantly sand. The soil textures were primarily sandy clay loam to sandy clay. The clayey property of the soil increased slightly with depth. This is one of the characteristics of the Serdang series, developed from sandstone parent material [80]. The Serdang and Malacca Series consisted of topsoil at 0.1 m and 1.5 m subsurface depth. At a depth of 1.5 m, the value of the soils of the Serdang series was about 57.21% sand, 10.21% silt, and 32.53% clay. For the Malacca series, at between 1.5 m, the values of the soils samples collected were about 41.58% sand, 2.94% silt, and 55.49% clay.

The physical and chemical properties of soils depend primarily on soil texture, which gives more ideas on soil classification (a nutrient required by plants), crop suitability, and soil interpretation [81]. The particle size distribution of the Malacca soil series observed in the study area reveals that the Malacca series's soil is dominated by clay. Sand content decreases with depth while clay content increases with depth. High clay content on the B horizon shows that illuviation has occurred and that the Malacca series contains more ferromagnesium minerals, which produces more clay [82]. The soil had a lower bulk density on the topsoil, which means that the soil organic matter content was lesser on topsoil with values between 1.47 and 1.42 grcm^{-3} and from 1.48 to 1.49 grcm^{-3} and 1.44 to 1.45 grcm^{-3} on the subsoil. The bulk density of the Serdang series was moderate and can improve crop production. The value of bulk density of the Malacca series was 1.36 and 1.27 grcm^{-3} at the topsoil, and between 1.37 to 1.41 grcm^{-3} and 1.30 to 1.32 grcm^{-3} on subsoil. The bulk density was higher on the subsoil. This is due to the compaction of soil at the oxic horizon as a result of the illuviation process. It could also be a result of the decrease in organic matter content with depth.

The water holding capacity of the topsoil and subsoil at 33 kPa ranged from 23.02 to 16.61 v/v and 21.16 to 17.01 v/v while the water holding capacity of the topsoil and subsoil at 1500 kPa ranged from 12.59 to 7.58 v/v and 11.97 to 8.36 v/v. The value of soil moisture content at the potential 33 kPa (FC) and 1500 kPa (PWP) was higher on topsoil than subsoil. More organic matter content in the topsoil increases the soil's ability to hold water. At 33 kPa, the water holding capacity of the Malacca series' was 23.83% to 21.51% v/v on the topsoil and 23.72% to 21.81% v/v on subsoil, while at 1500 kPa, water was held from 10.75% to 10.77% v/v on topsoil and from 10.29% to 13.23% v/v on the subsoil. The low available water holding capacity is due to low organic matter, which reduces the specific surface area of the soil. Both soil series show very strong to moderate pH values from 4.99 to 4.90 and 5.68 to 5.22. The pH value of the topsoil is higher than the subsoil, probably as a result of the H^+ supply to the soil by organic matter. The pH values were between 4 and 5 and are considered low. This value is typical for tropical soil, where soil erosion, weathering, and leaching are considered very high. The pH of Ultisols in Malaysia is acidic below 7, which ranges from 4 to 5 in the B horizon due to climatic conditions that wash away the soil's cations such as Ca^{2+} Ma^{2+}, P, K, and Na, and causes the accumulation of sesquioxide and has an impact on low productivity [83]. Table 2 summarizes the properties of the soil of the study location.

Table 2. The soil physical and chemical properties of the study area.

Properties		Serdang Series	Malacca Series
Depth (m) Soil texture		Topsoil 0.1 Subsoil 1.5 Sandy clay loam to sandy clay	0.1 1.5 Clay and Sand
Bulk density (grcm^{-3})	Topsoil Subsoil	1.47 and 1.42 1.48 to 1.49 and 1.44 to 1.45	1.36 and 1.27 1.37 to 1.41 to 1.30 to 1.32
Porosity (%)	Topsoil Subsoil	36.8 and 38.26 30.15 to 21.87 and 344 to 23.15	41.37 and 47.08 29.15 to 34.12 and 42.22 to 34.51
Water holding capacity (%)	Topsoil 33 kPa Subsoil 33 kPa	23.02 to 16.61 21.16 to 17.01	23.83 to 21.51 23.72 to 21.81
Soil pH (%)	Topsoil 1500 kPa Subsoil 1500 kPa Topsoil Subsoil	12.59 to 7.58 11.97 to 8.36 4.99 to 4.90 5.68 to 5.22	10.75 to 10.77 10.29 to 13.23 4.47 and 5.30

2.2. Data Description

2.2.1. Hydroseeding Mixture

After studying the nature of the slope, climate, and soil condition, the experiment was carried out on a steep slope within the study area. We prepared the land and divided the layout into four segments measuring 8 ft in length and 2 ft wide, creating a square-shaped plot of 32 ft on all sides—a rigid distinction made due to a slope gradient and separate subplots. The seeds were separately mixed with fertilizer and liquid seed germinator in the exact quantities and sprayed accordingly. The following samples and mixtures were applied:

G1: Rye grass (Lolium perenne L.). Ryegrass originated from Europe, Asia, and North Africa and is mainly cultivated and naturalized in Australia, America, and some islands in Oceania. Characterized by bunch-like growth habits, they are perennial. Perennial ryegrass is significant in forage/livestock systems. The high palatability and quick digestibility characteristics make it highly valuable for dairy and sheep foraging systems. In temperate regions, ryegrass is sometimes called forage grass. Its main features include increased yield potential, faster establishment, appropriate for reduced-tillage renovation, and application on heavy and waterlogged soils (Figure 2a).

Figure 2. Seed samples: (**a**) Ryegrass; (**b**) rye corn; (**c**) signal grass; (**d**) couch.

G2: Rye corn (Secale cereale) originates from Turkey. Rye corn is a species of cereal that has been commonly grown on the sandy outlays of the Mallee regions of South Australia and Victoria. A versatile species that tolerate arid conditions and grows well on sandy soil types, it performs best when grown on good fertile soils that respond well to nitrogen applications to gain the most from the shorter growing season soils (Figure 2b).

G3: Signal grass (Brachiaria decumbens/Urochioa Decumbens): It is originally from Uganda and widely grown in tropical and subtropical countries. It forms a thick, high-yielding sward that reacts very well to increased nitrogen. It is also a perennial grass with a solid stoloniferous root system and extended training stems that roots down from their nodes. It is recommended for shallow slope erosion control and is predominantly grown on most road cuts in Malaysian highways. Signal grass was initially used in the wet, humid tropics, but in recent years it has been grown over much broader climatic conditions (Figure 2c).

G4: Couch Bermuda grass (Cynodon dactylon) is native to most eastern hemispheres (Afro-Eurasia and Australia). Moreover, it is perennial and has both stolon and rhizomes. The couch is used both as a cover crop and for erosion control. This grass produces good quality hay and grazing. This grass adapts mainly in areas where the annual rainfall varies between 600 mm and 1750 mm (Figure 2d). Table 3 presents the seed sample's origin, quantity, and price.

Bio Green is a concentrated, highly effective, water-soluble, granular fertilizer with the optimum ratio of nutrients. It is highly suitable for grass, greens, and leafy plants. Bio green contains three significant elements, namely Nitrogen, Phosphorus, and Potassium (NPK), which enhance the growth of plants. Nitrogen (N) aids in leaf growth, which then forms proteins and chlorophyll. Phosphorus (P) aids in root development, while Potassium (K) helps in stem and root growth and protein synthesis. Bio-green is a Malaysian formulated product that has undergone extensive research and testing to be equivalent in performance to fertilizers' NPK 15.15.15 range. It is a product of composted organic materials with micro-nutrients naturally present. It contains natural humic and amino compounds and is highly water-soluble, with nutrient leaching being significantly reduced. It also improves soil structure, and its slow-release effect lasts longer compared to other fertilizers of a similar range.

Table 3. The quantity, origin, and prices of the seed samples.

Seeds	Quantity (kg)	Origin	Price per kg (AUD)
Rye grass (*Lolium perenne* L.)	1 kg	Europe, Asia and northern Africa	5.10
Rye corn (Secale cereale)	1 kg	Turkey	1.90
Signal grass (Brachiaria Decumbens/Urochioa Decumbens)	1 kg	Uganda	16.50
Couch Bermuda grass (Cynodon dactylon)	1 kg	Afro-Eurasia and Australia	24.00

2.2.2. Landslide Inventories and Conditioning Factors

A landslide inventory map with 202 landslide sites was obtained from an aerial photo from the UPM's Geospatial Information Science Research Centre (GISRC) database supported with field observation. Furthermore, the landslide inventory was divided into two datasets (ground truth and classified); 70% (141 points) were used as ground truth data to train the models, while 30% (61 points) were used as classified data in a confusion matrix to validate the models.

For this study, LiDAR data were analyzed by a qualitative method. The LiDAR data was captured in 2015 by Ground Data Solution Bhd using a Riegl scanner aboard an EC-120 Helicopter flown over the University Putra Malaysia at an altitude of about 600 m above the terrain surface. The acquired point cloud averages 6 points per square meter, with a 15 cm vertical accuracy on non-vegetated terrain and a 25 cm horizontal accuracy. Based on expert's opinion, literature, and significance to the study area, twelve landslide conditioning factors within and around the study area, namely the elevation, slope, aspect, curvature, hill shade, land use, distance to trees, distance to road, distance to urban, distance to lake, stream power index (SPI), terrain roughness index (TRI), and topographic wetness index (TWI). Slope, aspect, elevation, curvature, and hill shade were directly created from the digital elevation model (DEM) derived from LiDAR data using their layer toolboxes. Stream power index (SPI), topographic wetness index (TWI), and terrain roughness index (TRI) were created from spatial layers such as slope, flow direction, and flow accumulation. Shapefiles of distances to roads, lakes, trees, and build-up were digitized as land-use/landcover from the LiDAR image and produced using the Euclidean distance method in ArcGIS with 10×10 cell sizes.

2.2.3. Effects of Climate on Vegetation

Changes in climatic conditions could reduce crop yield [84]. The increase in temperature results in enhanced evapotranspiration, decreasing water availability, and further exacerbating dry months [85]. High storms associated with heavy rainfall increase flood frequency, which negatively impacts vegetation growth. Moreover, an increase in air and the temperature of water reduces the efficiency of plants. Low rainfall and high temperature reduces soil moisture content, water availability for irrigation and impair crop growth in non-irrigated areas. We studied the climatic conditions of the study area (temperature and precipitation) before and after the test (from August to December 2020) to assess its relationship to seed germination, growth, and development. The climate data obtained from the Malaysian Meteorological Department were examined critically and implemented in the study. The study area has a constant annual temperature with a maximum between 31 and 33 °C, and minimums usually between 22 and 23.5 °C. The average annual rainfall is also about 2400 mm. Table 4 below shows the average climate condition of the study area.

Table 4. The annual climatic condition of the study area in 2020 (from climate data).

	Jan.	Feb.	Mar.	Apr.	May	Jun.	Jul.	Aug.	Sept.	Oct.	Nov.	Dec.
Avg. Temperature °C (°F)	25.1 (77.2)	25.7 (78.3)	26 (78.8)	26.1 (78.9)	26.3 (79.3)	26.2 (79.2)	26.1 (78.9)	26 (78.8)	25.8 (78.5)	25.7 (78.2)	25.2 (77.4)	25.1 (77.2)
Min Temperature °C (°F)	21.9 (71.4)	21.9 (71.4)	22.7 (72.8)	23.1 (73.6)	23.4 (74.1)	23.1 (73.5)	22.8 (73.1)	22.8 (73.1)	22.8 (73)	22.8 (73)	22.7 (72.8)	22.3 (72.2)
Max Temperature °C (°F)	209 (8.2)	174 (6.9)	268 (10.6)	300 (11.8)	246 (9.7)	174 (6.9)	183 (7.2)	219 (8.6)	243 (9.6)	308 (12.1)	373 (14.7)	284 (11.2)
Precipitation/Rainfall mm (in)	209 (8.2)	174 (6.9)	268 (10.6)	300 (11.8)	246 (9.7)	174 (6.9)	183 (7.2)	219 (8.6)	243 (9.6)	308 (12.1)	373 (14.7)	284 (11.2)
Humidity	85%	82%	85%	87%	87%	85%	84%	84%	85%	87%	90%	88%
Rainy days	20	18	24	27	25	22	24	24	25	26	26	24

3. Methodology

3.1. AHP for Landslide Susceptibility Mapping

The production of a landslide susceptibility map of the whole area to predict the possibility of landslide occurrence was carried out to determine landslide high and low-risk areas using the AHP method. Professor Thomas L. Saaty originally developed AHP as a multi-criteria decision-making (MCDM) approach [66]. Applying AHP in this study aims to identify, correlate, weigh, and rank different parameters that determine slope susceptibility to landslide [86]. The method assigned weights to the conditioning factors by pairwise comparison. This pairwise comparison creates judgments between pairs of the set of variables instead of prioritizing them. The conditioning parameters associated with the study area and essential to assess landslide susceptibility are significant in achieving this judgment. The conditioning factors were ranked between 1 to 9 according to their level of importance. AHP is one of the most widely successful GIS-based methodologies for landslide susceptibility mapping over the past decades [52]. This is due to its ease of use and high capabilities in providing prediction maps [87]. It uses the consistency ratio (CR) to identify consistencies by comparing the priorities of each criterion using the CR equation, according to Saaty [62].

$$CR = \frac{CI}{RI} \quad (1)$$

where RI the random consistency index and CI represents the consistency index expressed as

$$CI = \frac{\lambda max - 1}{n - 1} \quad (2)$$

where λmax represents the principal value of the matrix, n is the order of the matrix. Moreover, the weights of each criterion were integrated into a single landslide susceptibility index by applying the equation:

$$LSI = \sum_{i=1}^{n} Ri \times wi \quad (3)$$

This decision-making tool specifies the value of each factor. Where Ri represents the rating classes of each layer, and wi represents the weight of each of the landslide conditioning factors. Each factor's weight from the matrix class was multiplied by the weight class. The local representation of elements in the slope area where landslide is predicated determines the susceptibility of map results.

3.2. Field Experimental Design

We carried out the hydroseeding experiment on the 11 August 2020 within the study area on a terrain sloping at 65° with a total horizontal surface area of 304 cm and a total vertical area of 244 cm wide and divided into four equal parts of 61 cm each, as shown in Figure 3. The site was leveled and tilled with a rake to remove large soil particles. We studied the nature of the soil and climatic condition and did not modify the earth as this may alter the experiment results. Seed samples of 1 kg each were separately mixed with 500 g of fertilizer and 50 mL of seed germinator mixtures before the hydroseeding experiment. The mixtures were sprayed with an Ozito cordless hand spreader.

Figure 3. The spraying process on the study area.

3.2.1. Vegetation Ground Cover

Vegetation ground cover has long been adequate, especially in reducing landslide and erosion on the roadside slopes [88]. Vegetation cover promotes infiltration and provides resistance to topsoil by stabilizing the soil structure and intercepting rainfall and runoff, thereby playing a vital role in soil and water conservation [89]. A related approach to stabilize disturbed slopes is by hydroseeding. In this research, we used the line-based method to measure the vegetation ground cover. Geometrically, it is a single-dimensional distance line measurement that measures the distance of the first contact to the last touch of the species. It also determines the percentage cover for each line then averages the lines together to estimate vegetation cover. The formula of the line-based method is as shown below.

$$\% ground\ cover = \frac{Total\ distance\ of\ specie\ A}{Distance\ of\ all\ specie\ along\ line} \times 100 \quad (4)$$

When the vegetation ground cover is denser and complex, the surface is shielded from wind, and direct rainfall is intercepted and redirected by the canopy.

3.2.2. Vegetation Root

According to structural stability and vegetation ecology theory, if the root systems are more complicated and profound, the soil is more stable, and infiltration is higher [90]. Therefore, well-developed root systems can help reduce runoff by directing rainfall into groundwater storage more quickly, helping to reduce the slope strain and stabilizing the slope. The vegetation root architecture was determined using Yen's [91] pull-out tests. Root samples from each of the four species were manually uprooted from the field, washed, and the root diameters were measured using a measuring tape.

3.2.3. Estimation of Surface Runoff

Rainfall surface runoff is an indicator for determining the water loss of a slope [92]. It occurs when the intensity of rainfall is greater than the intensity of infiltration, leading to the failure of excess water to infiltrate [93]. To determine the rainfall rate, surface runoff, rainfall duration, and rainfall intensity were recorded. The rainfall events were based on the following criteria: (i) Erosive rainfall with daily rainfall amount greater than 12 mm (ii) Similar rainfall intervals (iii) Rainfall duration of no less than 60 min. These factors proved to be effective in rainfall interception for surface runoff [94]. Duration is the extent of the rainfall, and intensity is the rate at which it rains, mathematically expressed by the height of the rainfall layer per minute (mm/min) [95]. Rainfall intensity is the ratio of the total rainfall amount in a given period to the duration of the period [96]. The present study focuses on the relationship among runoff, rainfall, soil type, vegetation, slope, with the primary focus of finding an effective means of reducing the rate of surface runoff, which will practically reduce the effect of landslides. Different methodologies exist for computing and estimating surface runoff [97]. However, this research applies the rational method to determine the surface runoff of the site experiment.

Rational Method

It is considered the most widely employed and easy-to-use practical method to estimate and determine the surface runoff of any specific rainfall. This method is used primarily on small scales. The rational approach is expressed mathematically as

$$Q = CIA \qquad (5)$$

where C = coefficient of runoff (runoff volume/rainfall), A = area of the catchment, and I = intensity of the precipitation.

We made sizeable holes and installed plastic containers with rigid channels that channel water into the collection bottles at the bottom of each plot to measure runoff volume (Figure 4). The surface runoff volume of the plants was collected in this plastic container at the same rainfall intensity and duration during the rainfall events and transferred to a measuring cylinder to get the readings. The surface runoff volume of each vegetation was therefore determined by dividing volume by the period.

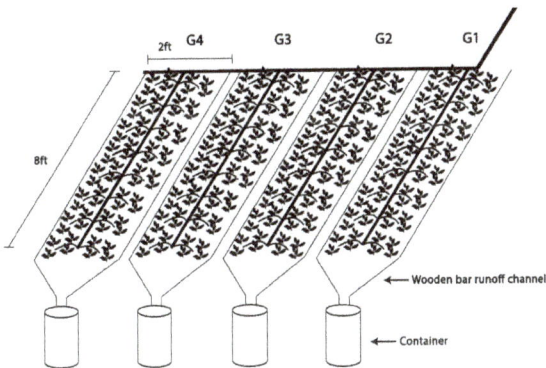

Figure 4. Experimental surface runoff plot.

This experiment shows how vegetation characteristics such as the vegetation roots and vegetation ground cover reduce surface runoff. This experiment only considers surface runoff from September to December 2020. We collected four rainfall events for this study. These event dates were chosen based on a weather forecast predicting rainfall of >12 mm, which matched the experimental criteria. The daily rainfall data was obtained from the National Hydraulic Research Institute of Malaysia (NAHRIM).

4. Results

4.1. Landslide Susceptibility Assessment

The landslide susceptibility result using the AHP model includes the weights of the conditioning factors, class weights, and a CR [98]. Applying the AHP, a decision hierarchy was built through a pairwise comparison of each conditioning factor. The AHP prioritized the effective criteria and variables by using pairwise comparisons to create a matrix. Ratings of the parameters were provided on a 9-point scale. The values of 1/9 represent the least important, 1 represents equal importance, and 9 represents the most important. The degree of consistency used in developing the rating was also determined. Moreover, a procedure through which an index of consistency, known as a CR, was produced. The CR indicates the probability that the matrix judgments were generated randomly [99]. Spatial Analyst Tools ArcGIS 10.7 software was used to reclassify each cell of the final map into five categories and values assigned ranging from 1 to 5, representing very low to very high in the LSM. The landslide susceptibility map was classified into five classes using the natural break classifier. Figure 5 shows the LSM produced by AHP, which reveals 20.65% of very low, 20.18% low, 20.37% moderate, 19.45% high, and 19.35% very high susceptible areas.

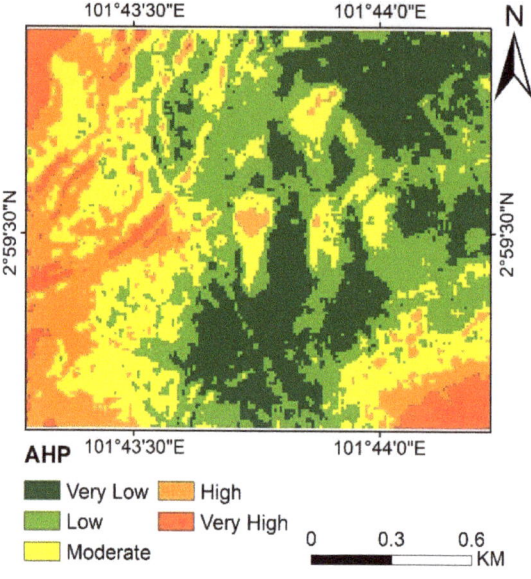

Figure 5. Landslide Susceptibility map by AHP.

Validation

The landslide susceptibility model was evaluated using the receiver operating characteristic curve (ROC curve), a non-dependent threshold approach. The ROC curve shows the validity of the diagnostic ability of a binary classifier system as its discrimination threshold [100]. The ROC curve is created by plotting the values of the true positive rate (TPR) against the values of the false-positive rate (FPR) at different threshold settings [101]. This curve validates the model's accuracy regardless of the prediction model since it compares random landslide points and a separated dataset of landslides [102,103]. A synthetic index was calculated for the ROC, utilizing the area under the curve (AUC), which has generally been applied in past studies to evaluate the accuracy of the landslide susceptibility map. A higher AUC value indicates a higher accuracy of the susceptibility map [98]. The AUC value obtained from the AHP model revealed a 0.865 accuracy (Figure 6). Therefore, the results of the model indicate an accurate susceptibility map. A precise LSM model highly depends on conditional factors. It thus assists decision-makers, landscapers, and urban planners in identifying hazard-prone areas for early mitigation. After studying the nature and environmental conditions of the study area, a hydroseeding experiment was introduced and carried out on a high-risk slope within the study location to control and prevent potential landslides.

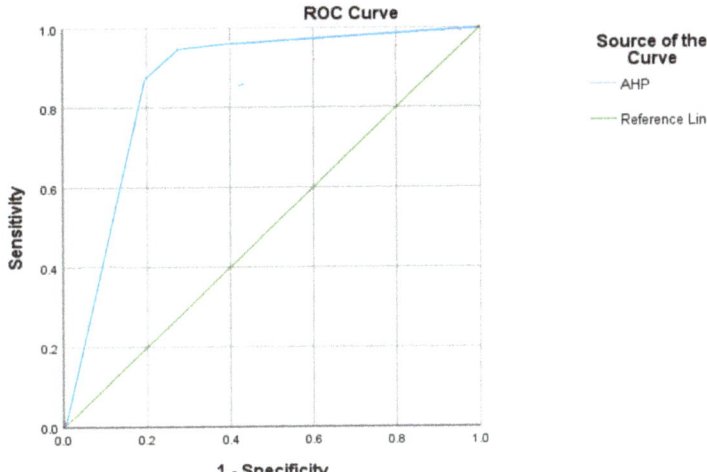

Figure 6. ROC curve.

4.2. Experimental Observations, Monitoring, and the Result of Hydroseeding

In this study, we evaluated and recorded the germination performance of the four hydroseeded seed species. The main goal was to determine which of these seeds, currently used to hydroseed slopes in different parts of the world, is more effective in the study area. Moreover, the vegetation ground cover and the vegetation root were studied to test their effectiveness in controlling landslides (Figure 7).

Figure 7. Hydroseeded vegetation.

4.2.1. Germination Rate

Before the experiment, we studied environmental factors such as rainfall and temperature. The surveys were conducted at the end of each month from August to December 2020. In each subplot, the species were identified and recorded. Couch (G4) seed germinated between the 2nd and 3rd day and showed an approximate 92% germination rate, followed by signal grass (G3) which grew on the 9th day with an 88% germination rate. The survey also recorded a delay in germination on both rye corn (G2) and ryegrass (G1), which germinated between day 15 and day 17, with 84% and 60% germination rates, respectively.

4.2.2. Vegetation Root Length

The vegetation root is the most crucial aspect of the plant for slope stabilization. Adequate subsurface drainage is essential to reduce the pore-water pressure of the subsoil. Over the past decades, several types of research have revealed significant roles played by plant roots to minimize the detachment rates of the soil as a result of concentrated flows and are therefore very effective in controlling landslides [104,105]. Vegetation roots support the slope drainage system and act as a scale preferential flow direction on the hillslope and drain the subsoil's water content from unstable terrain. When vegetation root systems converge, or the subsurface flow ends abruptly in the slope, it may lead to a concentration of water pressure in a critical region of the hill, thereby leading to instability. Flow direction may occur, resulting in both positive and negative outcomes on slope stability. Practical and precise knowledge of the disposition of vegetation roots in the slope is necessary to get it right. In this study, the vegetation root architecture was determined using the pull-out tests introduced by Yen [91]. Root samples from each of the four species were uprooted from the field, washed, and the root length was measured using a measuring tape, and the values were recorded and plotted in Figure 8.

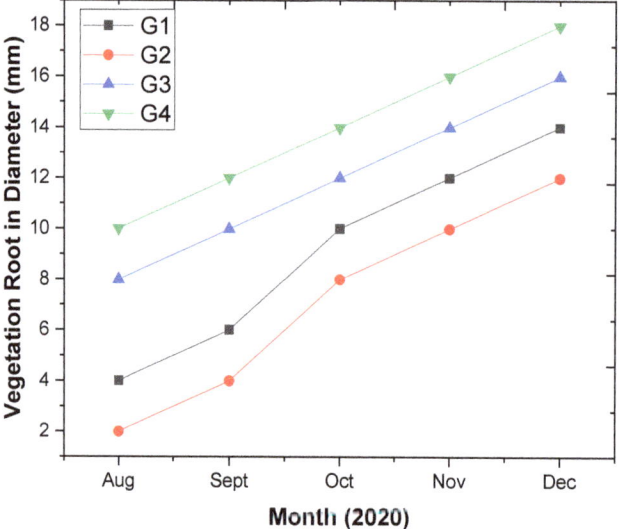

Figure 8. Graph plot of the monthly vegetation root length.

The root architecture of the four seed species in Figure 9a–d shows the typical distribution of the root system. It offers a general idea of how the roots developed and indicates the localization of deep-rooted and fibrous roots within the root system. Ryegrass has a fibrous root system with thick primary roots and thinner lateral branches and showed a poor result. The roots of ryegrass are usually arbuscular mycorrhizal. Flawed due to the environmental and climatic conditions, ryegrass, if under suitable conditions, may germinate faster than some other grass seeds. However, the roots spread slowly and grow naturally into clumps that spread their shoots vertically, called tillers. The shallow roots of ryegrass limit its tolerance for heat and drought. It adapts well to a wide variety of improved soil conditions, such as acidic and alkaline soils. It also thrives better under soil pH. Rye corn showed an extensive, fibrous root system that may expand.

Figure 9. The different vegetation root samples: (**a**) Ryegrass (G1); (**b**) rye corn (G2); (**c**) signal grass (G3); (**d**) couch (G4).

On the other hand, signal grass showed a better root development in the soil with a deep taproot system that can effectively grow deeper on a wide range of soils and adapt to different environmental conditions. The experiment also reveals that the couch has the most extended root system, with a shallow taproot system that can spread, anchoring the shallower soil depth and around the subsoil. The experiment so far compared the effectiveness of the four seed plant roots in stabilizing the soil against possible landslides. However susceptible to incisive landslide occurrence, no research has been done to compare the effectiveness of these four plants in controlling landslides in this study location. Hence, this study introduced the landslide-reducing potential of these species with both fibrous and tap roots systems on the slope of the study area. The experimental results showed that plant roots with taproot systems were more adaptive and efficient in the soil type than the seed with a fibrous root system. According to [106], tap-rooted plant species penetrate more into thick soils than fibrous-rooted plant species. Therefore it is well adapted for use in landslide mitigation and control.

The different vegetation roots showed a significant variation regarding the site characteristics; with an increase in rainfall, the roots considerably showed their best tolerance following an increasing monthly pattern, with the highest length observed from G4 in December, followed by G3. Both G1 and G2, with a fibrous root system, also showed a significant monthly root development. The vegetation root was plotted using an origin software.

4.3. Result of the Hydroseeded Vegetation Ground Cover

Hydroseeded vegetation ground cover is the percentage of the ground surface covered by vegetation. It controls landslides by anchoring the soil against rainfall and other landslide causative factors. The plants were monitored, and the vegetation ground cover was recorded from August to December.

The species with the highest values was ryegrass (G1). It showed leaf spreading and creeping characteristics. From August to September 2020, the experimental result recorded 40% vegetation cover due to its prolonged germination rate. A significant 180% of coverage was recorded between November and December. Rye corn (G2) was second to ryegrass and reached 160% coverage between November and December. However, the values sharply increased from 60% to 120% between August and October 2020. Signal grass (G3) showed the lowest result with 40% to 80% between August and October and 95% between November and December. Moreover, couch (G4) gave values between 80% and 120% from August to October and remained unchanged until December with 140%.

Vegetation ground cover can help to prevent landslides by protecting the soil surface against the impact of rainfall and surface runoff. It also reduces runoff volume, increases surface roughness, and reduces sediment traps and transportation [105,107]. The result was plotted using the Origin software, as shown (Figure 10) below.

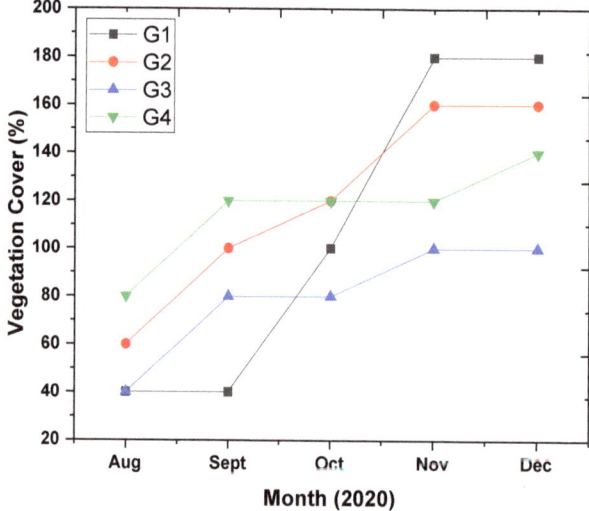

Figure 10. A plot of the vegetation ground cover.

At the start of the monitoring period, August 2020, G1 and G3 showed the lowest vegetation cover while the vegetation cover reached 80% in G4. Between September and December, the vegetation covers of G1 significantly increased, topping other species and showing the best choice for control of surface landslide. Test G3 showed the least acceptable result with a 95% vegetation cover.

4.4. Vegetation Surface Runoff

The experiment focused on analyzing the overall influence of the vegetation root and vegetation ground cover to reduce surface runoff, reducing the effects of landslides on the study area. The four different vegetation species were subjected to rainfall intensities of 35 mm/h in September, 48 mm/h in October, 33 mm/h in November, and 34 mm/h in December on a Malacca and Serdang soil series sloping at 65°, as shown in Table 5 below. The area's soil type is a mix of sandy, clay, and loam, collectively identified under the

Serdang and Malacca series. In line with past literature, the runoff rate increased due to the percentage of vegetation ground cover and vegetation root characteristics [108].

Table 5. Showing selected daily runoff under different rainfall intensities.

Date	Rainfall (mm/h)	G1 Runoff (mm)	G2 Runoff (mm)	G3 Runoff (mm)	G4 Runoff (mm)
September	35	9.03	8.53	6.77	5.52
October	48	4.62	4.32	3.73	3.05
November	33	8.25	6.91	7.36	5.13
December	34	7.81	8.28	6.15	5.68

The results show that G2 has the highest surface runoff rate followed by the G1 plot, then G3, with G4 having the least runoff amount under different rainfall intensities. The surface vegetation runoff of the vegetation is shown in Figure 11 below.

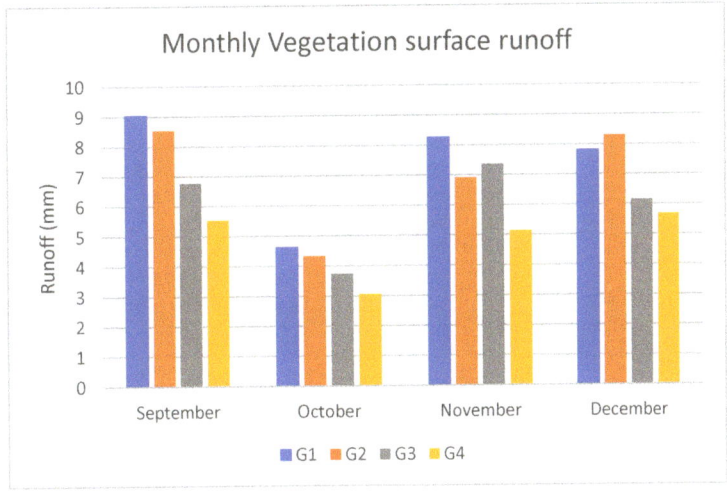

Figure 11. Monthly vegetation surface runoff.

It shows that the runoff rate changes due to their vegetation characteristics under the same rainfall intensity. One can observe from Figure 11 that the amount of surface runoff in the G1 and G2 plots was significantly higher than the G3 and G4 plots under the same rainfall intensity.

The average surface runoff of G4 is 4.85 mm, while G3 comes behind with an average surface runoff value of 5.89 m^3. G2 and G1 also have surface runoff values of 7.01 mm and 7.43 mm, respectively. Couch vegetation offers the most acceptable landslide control benefits to signal grass, rye corn, and ryegrass vegetation. The findings suggest that the selection of couch species over other species is justified based on landslide control benefits.

5. Discussion

The AHP method was applied in this study to identify, correlate, weigh, and rank different landslide parameters that determine slope susceptibility. Using the systematic analysis of a pairwise comparison matrix of a list of the conditioning factors by expert's opinion scores between one and nine indicates the relative importance that was used. The susceptibility map helped determine high-risk, landslide susceptible areas for the mitigation process. The output map of the landslide susceptibility model was evaluated qualitatively, which is essential in selecting the most suitable site to carry out the landslide mitigation experiment. Applying the AHP approach, very low susceptibility areas were 20.65% low, 20.18% low, 20.37% moderate, 19.45% high, and 19.35% very highly susceptible areas.

From the AHP susceptibility map, a hydroseeding experiment was carried out on the high-risk visibly sliding slope, with the result revealing that hydroseeding has great potential on degraded terrains in a similar environment. Results also showed that the grasses thrived more during rainfall season than dry season with daily watering. The results of the vegetation root system appear to show that vegetation has the best potential for long-term slope stabilization. Comparing the development of this study with other researchers that successfully carried out a similar experiment in a laboratory using rainfall simulations, the experimental results are very similar with natural precipitation and daily watering. Improved knowledge on this research is equally expected to support and enhance the selection of hydroseeding components and improve the soil nutrient and native species in future restoration projects.

Several authors have argued that climatic conditions and soil types are the main factors limiting the vegetation reinforcement of soil, especially in semiarid environments, resulting in low vegetation cover and distorted roots [109]. Hence, this experiment went through rainy and dry seasons and different soil types to prove the effectiveness of the selected seed with good results.

According to Garcia-Palacios et al. [110], vegetation covering more than 50% of semi-arid terrain requires no further mitigation to prevent landslide or soil erosion. In this research, the vegetation ground cover for ryegrass and signal grass was less than 50% in August. Ryegrass was also less than 40% in September, with both plants maintaining soil stability. Risse et al. [111] noted that the rate of soil loss was about 10 to 20 times higher in construction sites than in agricultural lands. Moreover, Wang et al. [112] stated that several studies have revealed that soil properties such as moisture content and nutrients could affect their root morphology as vegetation matures. Thus, in our experiment, the vegetation root system increased with time, though evidence of nutrient loss could be seen from the plant leaves.

6. Conclusions

After studying the nature and environmental conditions of the study area, a hydroseeding experiment was introduced and carried out on the high-risk, steep slope within the study location to control and prevent potential landslides. Four seed samples were separately hydroseeded, and their success rate was compared in relation to the soil condition of the study area. The vegetation was tested under four different rainfall intensities. The result showed that couch (G4) has the least average surface runoff of 4.85 mm. It also has the fastest estimated germination rate of 92% in 3 days and showed densely populated, strong stemmed vegetation cover, with a tall height and a deep, long shallow taproot system. It showed the most reliable and most effective for landslide control in the study area, followed by G3 with a value of 5.89 mm, and an estimated 88% germination rate on day nine, with thick, stemmed solid vegetation ground cover and also has the tallest vegetation height, as well as a deep-rooted taproot system. G2 has an average surface runoff of 7.01 mm with an estimated 84% germination rate on day fifteen, with creeping and sparsely populated vegetation cover, short vegetation height, and an extensive fibrous root system. More so, G1, with the highest average runoff value of 7.43 mm, germinated on day 17, with an estimated 92% germination rate. It showed similar results and characteristics to G2 with a creeping and sparsely distributed vegetation cover and short vegetation height with fibrous, thick, and thinner lateral root branches. Based on the results of this investigation, couch vegetation offers the best landslide control benefits to that of signal grass, rye corn, and ryegrass vegetation. The findings suggest that the selection of the couch species over other species is justified based on landslide control benefits.

Further studies are recommended to use different satellite sensors for landslides susceptibility mapping of the study area. Moreover, other types of slope seeds should be experimented with. Ryegrass and rye corn developed poorly under the soil and climatic condition of this study area. They could also thrive well under improved soil and favorable climate conditions.

Author Contributions: O.J.E. performed the experiments and data acquisition; O.J.E. wrote the manuscript, discussion, and analyzed the data. H.N. and Z.K. supervised; B.K. and O.S.S. edited, restructured, and optimized the manuscript professionally. All authors have read and agreed to the published version of the manuscript.

Funding: The research is supported by the Universiti Putra Malaysia.

Institutional Review Board Statement: Not applicable.

Informed Consent Statement: Not applicable.

Data Availability Statement: Available on request.

Acknowledgments: The authors would like to thank the Universiti Putra Malaysia for providing all facilities during this research. We are also thankful to RIKEN Center for Advanced Intelligence Project, Japan.

Conflicts of Interest: The authors declare no conflict of interest.

References

1. Al-Najjar, H.A.H.A.; Pradhan, B.; Kalantar, B.; Sameen, M.I.; Santosh, M.; Alamri, A. Landslide Susceptibility Modeling: An Integrated Novel Method Based on Machine Learning Feature Transformation. *Remote Sens.* **2021**, *13*, 3281. [CrossRef]
2. Kalantar, B.; Ueda, N.; Al-Najjar, H.A.H.; Idrees, M.O.; Motevalli, A.; Pradhan, B. Landslide susceptibility mapping at Dodangeh watershed, Iran using LR and ANN models in GIS. In Proceedings of the SPIE Remote Sensing, Berlin, Germany, 10–13 September 2018; Volume 10790.
3. Al-Najjar, H.A.H.; Pradhan, B. Spatial landslide susceptibility assessment using machine learning techniques assisted by additional data created with generative adversarial networks. *Geosci. Front.* **2021**, *12*, 625–637. [CrossRef]
4. Kalantar, B.; Ueda, N.; Al-Najjar, H.A.H.; Saeidi, V.; Gibril, M.B.A.; Halin, A.A. A comparison between three conditioning factors dataset for landslide prediction in the sajadrood catchment of iran. *ISPRS Ann. Photogramm. Remote Sens. Spat. Inf. Sci.* **2020**, *5*, 625–632. [CrossRef]
5. Al-Najjar, H.A.H.; Kalantar, B.; Pradhan, B.; Saeidi, V. Conditioning factor determination for mapping and prediction of landslide susceptibility using machine learning algorithms. In Proceedings of the SPIE 11156 Earth Resources Environmental Remote Sensing/GIS Appllications X, Strasbourg, France, 10–12 September 2019; Volume 19. [CrossRef]
6. Kalantar, B.; Ueda, N.; Lay, U.S.; Al-Najjar, H.A.H.; Halin, A.A. Conditioning factors determination for landslide susceptibility mapping using support vector machine learning. In Proceedings of the IGARSS 2019-2019 IEEE International Geoscience and Remote Sensing Symposium, Yokohama, Japan, 28 July–2 August 2019; pp. 9626–9629.
7. Kalantar, B.; Ueda, N.; Saeidi, V.; Ahmadi, K.; Halin, A.A.; Shabani, F. Landslide susceptibility mapping: Machine and ensemble learning based on remote sensing big data. *Remote Sens.* **2020**, *12*, 1737. [CrossRef]
8. Psomiadis, E.; Papazachariou, A.; Soulis, K.X.; Alexiou, D.S.; Charalampopoulos, I. Landslide mapping and susceptibility assessment using geospatial analysis and earth observation data. *Land* **2020**, *9*, 133. [CrossRef]
9. Osiński, P.; Rickson, R.J.; Hann, M.J.; Koda, E. Assessment of slope stability influenced by vegetation cover and additional loads applied. *Ann. Warsaw Univ. Life Sci. Land Reclam.* **2018**, *46*, 81–91. [CrossRef]
10. Eugeniusz, K.; Piotr, O. Improvement of Slope Stability as a Result combining diverse reinforcement methods. *Architectura* **2012**, *11*, 3–14.
11. Wang, S.; Zhao, M.; Meng, X.; Chen, G.; Zeng, R.; Yang, Q.; Liu, Y.; Wang, B. Evaluation of the effects of forest on slope stability and its implications for forest management: A case study of Bailong River Basin, China. *Sustainability* **2020**, *12*, 6655. [CrossRef]
12. Ali, N.; Farshchi, I.; Mu'azu, M.A.; Rees, S.W. Soil-root interaction and effects on slope stability analysis. *Electron. J. Geotech. Eng.* **2012**, *17C*, 319–328.
13. Popescu, M. A suggested method for reporting landslide remedial measures. *Bull. Eng. Geol. Environ.* **2001**, *60*, 69–74. [CrossRef]
14. Hutchinson, J.N. Some Aspect of morphological and Geotechnical parameters of Landslides, with examples drwan from Italy and elsewhere. *Geol. Romana* **1994**, *30*, 5–13.
15. Popescu, M.E.; Sasahara, K. Engineering Measures for Landslide Disaster Mitigation BT. In *Landslides—Disaster Risk Reduction*; Sassa, K., Canuti, P., Eds.; Springer: Berlin/Heidelberg, Germany, 2009; pp. 609–631. ISBN 978-3-540-69970-5.
16. Bakri, N.N.O.; Mydin, M.A.O. General Building Defects: Causes, Symptoms and Remedial Work. *Eur. J. Technol. Des.* **2014**, *3*, 4–17. [CrossRef]
17. Choi, K.Y.; Cheung, R.W.M. Landslide disaster prevention and mitigation through works in Hong Kong. *J. Rock Mech. Geotech. Eng.* **2013**, *5*, 354–365. [CrossRef]
18. Sun, J.; Liu, Q.; Li, J.; An, Y. Effects of rainfall infiltration on deep slope failure. *Sci. China Ser. G Phys. Mech. Astron.* **2009**, *52*, 108–114. [CrossRef]

19. Nelson, M.; Saftner, D.; Carranza-Torres, C. Slope Stabilization for Local Government Engineers in Minnesota. In Proceedings of the Congress on Technical Advancement 2017: Construction and Forensic Engineering, Duluth, Minnesota, 10–13 September 2017; pp. 127–138. [CrossRef]
20. Lacasse, S.; Nadim, F.; Kalsnes, B. Living with landslide risk. *Geotech. Eng. J. SEAGS AGSSEA* **2010**, *41*, 4–14.
21. Noroozi, A.G.; Hajiannia, A. The Effects of Various Factors on Slope Stability. *Int. J. Sci. Eng. Investig.* **2015**, *4*, 44–48.
22. Pham, B.T.; Tien Bui, D.; Pourghasemi, H.R.; Indra, P.; Dholakia, M.B. Landslide susceptibility assesssment in the Uttarakhand area (India) using GIS: A comparison study of prediction capability of naïve bayes, multilayer perceptron neural networks, and functional trees methods. *Theor. Appl. Climatol.* **2017**, *128*, 255–273. [CrossRef]
23. Althuwaynee, O.F.; Pradhan, B. Semi-quantitative landslide risk assessment using GIS-based exposure analysis in Kuala Lumpur City. *Geomat. Nat. Hazards Risk* **2017**, *8*, 706–732. [CrossRef]
24. Lacerda, W.; Ehrlich, M.; Fontoura, S.; Sayão, A.; Van Westen, C. Geo-Information tools for landslide risk assessment: An overview of recent developments. *Landslides Eval. Stab. Terrain Eval. Stabilisation Set 2 Vol.* **2004**, 39–56. [CrossRef]
25. Rasyid, A.R.; Bhandary, N.P.; Yatabe, R. Performance of frequency ratio and logistic regression model in creating GIS based landslides susceptibility map at Lompobattang Mountain, Indonesia. *Geoenviron. Disasters* **2016**, *3*, 19. [CrossRef]
26. Javad, M.; Baharin, A.; Barat, M.; Farshid, S. Using frequency ratio method for spatial landslide prediction. *Res. J. Appl. Sci. Eng. Technol.* **2014**, *7*, 3174–3180. [CrossRef]
27. Wubalem, A. Landslide Susceptibility Mapping Using Statistical Methods in Uatzau Catchment Area, Northwestern Ethiopia. *Geoenviron. Disasters* **2020**, *8*, 1. [CrossRef]
28. Ciurleo, M.; Cascini, L.; Calvello, M. A comparison of statistical and deterministic methods for shallow landslide susceptibility zoning in clayey soils. *Eng. Geol.* **2017**, *223*, 71–81. [CrossRef]
29. Gordo, C.; Zêzere, J.L.; Marques, R. Landslide susceptibility assessment at the basin scale for rainfall- and earthquake-triggered shallow slides. *Geosciences* **2019**, *9*, 268. [CrossRef]
30. Rahman, M.S.; Ahmed, B.; Di, L. Landslide initiation and runout susceptibility modeling in the context of hill cutting and rapid urbanization: A combined approach of weights of evidence and spatial multi-criteria. *J. Mt. Sci.* **2017**, *14*, 1919–1937. [CrossRef]
31. Elmoulat, M.; Brahim, L.A. Landslides susceptibility mapping using GIS and weights of evidence model in Tetouan-ras-Mazari area (Northern Morocco). *Geomat. Nat. Hazards Risk* **2018**, *9*, 1306–1325. [CrossRef]
32. Thiery, Y.; Malet, J.P.; Sterlacchini, S.; Puissant, A.; Maquaire, O. Landslide susceptibility assessment by bivariate methods at large scales: Application to a complex mountainous environment. *Geomorphology* **2007**, *92*, 38–59. [CrossRef]
33. Ghosh, S.; Das, R.; Goswami, B. Developing GIS-based techniques for application of knowledge and data-driven methods of landslide susceptibility mapping. *Indian J. Geosci.* **2013**, *67*, 249–272.
34. Maalouf, M. Logistic regression in data analysis: An overview. *Int. J. Data Anal. Tech. Strateg.* **2011**, *3*, 281–299. [CrossRef]
35. Qiao, G.; Lu, P.; Scaioni, M.; Xu, S.; Tong, X.; Feng, T.; Wu, H.; Chen, W.; Tian, Y.; Wang, W.; et al. Landslide investigation with remote sensing and sensor network: From susceptibility mapping and scaled-down simulation towards in situ sensor network design. *Remote Sens.* **2013**, *5*, 4319–4346. [CrossRef]
36. Polykretis, C.; Ferentinou, M.; Chalkias, C. A comparative study of landslide susceptibility mapping using landslide susceptibility index and artificial neural networks in the Krios River and Krathis River catchments (northern Peloponnesus, Greece). *Bull. Eng. Geol. Environ.* **2014**, *74*, 27–45. [CrossRef]
37. Vakhshoori, V.; Pourghasemi, H.R.; Zare, M.; Blaschke, T. Landslide susceptibility mapping using GIS-based data mining algorithms. *Water* **2019**, *11*, 2292. [CrossRef]
38. Ramachandra, T.V.; Aithal, B.H.; Kumar, U.; Joshi, N.V. Prediction of shallow landslide prone regions in undulating terrains. *Disaster Adv.* **2013**, *6*, 54–64.
39. Al-Najjar, H.A.H.; Kalantar, B.; Pradhan, B.; Saeidi, V.; Halin, A.A.; Ueda, N.; Mansor, S. Land cover classification from fused DSM and UAV images using convolutional neural networks. *Remote Sens.* **2019**, *11*, 1461. [CrossRef]
40. Kalantar, B.; Ueda, N.; Saeidi, V.; Janizadeh, S.; Shabani, F. Deep Neural Network Utilizing Remote Sensing Datasets for Flood Hazard Susceptibility Mapping in Brisbane, Australia. *Remote Sens.* **2021**, *13*, 2638. [CrossRef]
41. Shahabi, H.; Hashim, M. Landslide susceptibility mapping using GIS-based statistical models and Remote sensing data in tropical environment. *Sci. Rep.* **2015**, *5*, 9899. [CrossRef]
42. Nefeslioglu, H.A.; Sezer, E.; Gokceoglu, C.; Bozkir, A.S.; Duman, T.Y. Assessment of Landslide Susceptibility by Decision Trees in the Metropolitan Area of Istanbul, Turkey. *Math. Probl. Eng.* **2010**, *2010*, 901095. [CrossRef]
43. Sarker, I.H. Machine Learning: Algorithms, Real-World Applications and Research Directions. *SN Comput. Sci.* **2021**, *2*, 160. [CrossRef]
44. Tharwat, A. Classification error: Bias and variance, Underfitting and Overfitting. *Mach. Learn. Present.* **2018**. [CrossRef]
45. Cao, Q.; Banerjee, R.; Gupta, S.; Li, J.; Zhou, W.; Jeyachandra, B. Data Driven Production Forecasting Using Machine Learning. In Proceedings of the SPE Argentina Exploration and Production of Unconventional Resources Symposium, Buenos Aires, Argentina, 1–3 June 2016.
46. Abdar, M.; Pourpanah, F.; Hussain, S.; Rezazadegan, D.; Liu, L.; Ghavamzadeh, M.; Fieguth, P.; Cao, X.; Khosravi, A.; Acharya, U.R.; et al. A review of uncertainty quantification in deep learning: Techniques, applications and challenges. *Inf. Fusion* **2021**, *76*, 243–297. [CrossRef]

47. Blundell, C.; Cornebise, J.; Kavukcuoglu, K.; Wierstra, D. Weight uncertainty in neural networks. In Proceedings of the 32nd International Conference on Machine Learning, Lille, France, 6–11 July 2015; Volume 2, pp. 1613–1622.
48. Feizizadeh, B.; Jankowski, P.; Blaschke, T. *A Spatially Explicit Approach for Sensitivity and Uncertainty Analysis of GIS-Multicriteria Landslide Susceptibility Mapping*; Verlag der Österreichischen Akademie der Wissenschaften: Vienna, Austria, 2013; pp. 157–164. [CrossRef]
49. Han, B.; Yao, Q.; Yu, X.; Niu, G.; Xu, M.; Hu, W.; Tsang, I.W.; Sugiyama, M. Co-teaching: Robust training of deep neural networks with extremely noisy labels. *Adv. Neural Inf. Process. Syst.* **2018**, *2018*, 8527–8537.
50. Adnan, M.S.G.; Rahman, M.S.; Ahmed, N.; Ahmed, B.; Rabbi, M.F.; Rahman, R.M. Improving spatial agreement in machine learning-based landslide susceptibility mapping. *Remote Sens.* **2020**, *12*, 3347. [CrossRef]
51. Ait Brahim, L.; Bousta, M.; Jemmah, I.A.; El Hamdouni, I.; ElMahsani, A.; Abdelouafi, A.; Sossey alaoui, F.; Lallout, I. Landslide susceptibility mapping using AHP method and GIS in the peninsula of Tangier (Rif-northern morocco). In Proceedings of the MATEC Web Conference, Rabat, Morocco, 22–25 November 2017; Volume 149, p. 02084. [CrossRef]
52. Moradi, M.; Bazyar, M.H.; Mohammadi, Z. GIS-based landslide susceptibility mapping by AHP method, a case study, Dena City, Iran. *J. Basic Appl. Sci. Res.* **2012**, *2*, 6715–6723.
53. Nguyen, T.T.N.; Liu, C.C. A new approach using AHP to generate landslide susceptibility maps in the chen-yu-lan watershed, Taiwan. *Sensors* **2019**, *19*, 505. [CrossRef]
54. Chou, J.R. A weighted linear combination ranking technique for multi-criteria decision analysis. *S. Afr. J. Econ. Manag. Sci.* **2013**, *16*, 28–41.
55. Malczewski, J. On the use of weighted linear combination method in GIS: Common and best practice approaches. *Trans. GIS* **2000**, *4*, 5–22. [CrossRef]
56. Syst, H.E.; Attribute, C.C. Interactive comment on "A Novel Strategy for landslide displacement and its direction monitoring" by Z.-W. Zhu et al. *Nat. Hazards Earth Syst. Sci. Discuss.* **2014**, *1*, C2961–C2962.
57. Stanley, T.; Kirschbaum, D.B. A heuristic approach to global landslide susceptibility mapping. *Nat. Hazards* **2017**, *87*, 145–164. [CrossRef] [PubMed]
58. Sun, D.; Xu, J.; Wen, H.; Wang, Y. An Optimized Random Forest Model and Its Generalization Ability in Landslide Susceptibility Mapping: Application in Two Areas of Three Gorges Reservoir, China. *J. Earth Sci.* **2020**, *31*, 1068–1086. [CrossRef]
59. Feizizadeh, B.; Shadman Roodposhti, M.; Jankowski, P.; Blaschke, T. A GIS-based extended fuzzy multi-criteria evaluation for landslide susceptibility mapping. *Comput. Geosci.* **2014**, *73*, 208–221. [CrossRef]
60. Loucks, D.P.; van Beek, E. System Sensitivity and Uncertainty Analysis. In *Water Resource Systems Planning and Management*; Springer: Cham, Switzerland, 2017. [CrossRef]
61. Ghorbanzadeh, O.; Feizizadeh, B.; Blaschke, T. Multi-criteria risk evaluation by integrating an analytical network process approach into GIS-based sensitivity and uncertainty analyses. *Geomat. Nat. Hazards Risk* **2018**, *9*, 127–151. [CrossRef]
62. Ghosh, S. *Knowledge Guided Empirical Prediction of Landslide Hazard*; ITC: Enschede, The Netherlands, 2011; ISBN 9789061643104.
63. Norton, J. An introduction to sensitivity assessment of simulation models. *Environ. Model. Softw.* **2015**, *69*, 166–174. [CrossRef]
64. Guzzetti, F.; Reichenbach, P.; Ardizzone, F.; Cardinali, M.; Galli, M. Estimating the quality of landslide susceptibility models. *Geomorphology* **2006**, *81*, 166–184. [CrossRef]
65. Sterlacchini, S.; Ballabio, C.; Blahut, J.; Masetti, M.; Sorichetta, A. Spatial agreement of predicted patterns in landslide susceptibility maps. *Geomorphology* **2011**, *125*, 51–61. [CrossRef]
66. Noorollahi, Y. Landslide modelling and susceptibility mapping using AHP and fuzzy approaches. *Int. J. Hydrol.* **2018**, *2*, 137–148. [CrossRef]
67. Grozavu, A.; Patriche, C.; Mihai, F. Application of ahp method for mapping slope geomorphic phenomena. *Int. Multidiscip. Sci. GeoConference Surv. Geol. Min. Ecol. Manag. SGEM* **2017**, *17*, 377–384. [CrossRef]
68. He, H.; Hu, D.; Sun, Q.; Zhu, L.; Liu, Y. A landslide susceptibility assessment method based on GIS technology and an AHP-weighted information content method: A case study of southern Anhui, China. *ISPRS Int. J. Geo-Inf.* **2019**, *8*, 266. [CrossRef]
69. Parsakhoo, A.; Jajouzadeh, M.; Motlagh, A.R. Effect of hydroseeding on grass yield and water use efficiency on forest road artificial soil slopes. *J. For. Sci.* **2018**, *64*, 157–163. [CrossRef]
70. Vallone, M.; Pipitone, F.; Alleri, M.; Febo, P.; Catania, P. Hydroseeding application on degraded slopes in the southern Mediterranean area (Sicily). *Appl. Eng. Agric.* **2013**, *29*, 309–319. [CrossRef]
71. Drake, D. Assessment of Hydroplanting Techniques and Herbicide. Master's Thesis, University of Hawai'i System, Honolulu, HI, USA, 2009.
72. Stokes, A.; Douglas, G.B.; Fourcaud, T.; Giadrossich, F.; Gillies, C.; Hubble, T.; Kim, J.H.; Loades, K.W.; Mao, Z.; Mcivor, I.R.; et al. Ecological mitigation of hillslope instability: Ten key issues facing researchers and practitioners. *Plant Soil* **2014**, *377*, 1–23. [CrossRef]
73. Kumarasinghe, U. A review on new technologies in soil erosion management. *J. Res. Technol. Eng.* **2021**, *2*, 120–127.
74. Cereno, M.M.; Tan, F.J.; Uy, F.A.A. Combined Hydroseeding and Coconet Reinforcement for Soil Erosion Control. *Soil Eros. Stud.* **2011**, 2–15. [CrossRef]
75. Faucette, L.B.; Risse, M.; Jordan, C.F.; Cabrera, M.; Coleman, D.C.; West, L.T. Vegetation and soil quality effects from hydroseed and compost blankets used for erosion control in construction activities. *J. Soil Water Conserv.* **2006**, *61*, 2–8. [CrossRef]

76. Blankenship, W.D.; Condon, L.A.; Pyke, D.A. Hydroseeding tackifiers and dryland moss restoration potential. *Restor. Ecol.* **2019**, *28*, S127–S138. [CrossRef]
77. Rivera, D.; Mejías, V.; Jáuregui, B.M.; Costa-Tenorio, M.; López-Archilla, A.I.; Peco, B. Spreading Topsoil Encourages Ecological Restoration on Embankments: Soil Fertility, Microbial Activity and Vegetation Cover. *PLoS ONE* **2014**, *9*, e101413. [CrossRef]
78. Heyes, S.; Butler, M.; Gartlan, C.; Ovington, A. Corangamite Seed Supply & Revegetation Project. 2008. Available online: https://www.ccmaknowledgebase.vic.gov.au/resources/Developing_seed_production_areas_for_native_plants.pdf (accessed on 24 August 2021).
79. Farhani, N.B.Y. Evaluation of Geological Formation for Potential Groundwater Aquifer by Integrated by Geophysical Technique. Available online: http://psasir.upm.edu.my/id/eprint/76090 (accessed on 5 September 2021).
80. Paramananthan, S.; Daud, N. Classification of Acid Sulfate Soils of Peninsular Malaysia. *Pertanika* **1986**, *9*, 323–330.
81. William, J. Chancellor Soil Physical Properties. *Adv. Soil Dyn.* **2013**, *1*, 21–254. [CrossRef]
82. Ghani, M.A.; Ghani, A. Petrology of granitic rocks along new Pos Selim to Kampung Raja highway (km 0 to km 22): Identification of different granitic bodies, its field and petrographic characteristics. *Bull. Geol. Soc. Malays.* **2003**, *46*, 35–40. [CrossRef]
83. Fatai, A.A.; Shamshuddin, J.; Fauziah, C.I.; Radziah, O.; Bohluli, M. Formation and characteristics of an Ultisol in Peninsular Malaysia utilized for oil palm production. *Solid Earth* **2017**, 1–21. [CrossRef]
84. Zhao, C.; Liu, B.; Piao, S.; Wang, X.; Lobell, D.B.; Huang, Y.; Huang, M.; Yao, Y.; Bassu, S.; Ciais, P.; et al. Temperature increase reduces global yields of major crops in four independent estimates. *Proc. Natl. Acad. Sci. USA* **2017**, *114*, 9326–9331. [CrossRef] [PubMed]
85. Chen, H.; Zhang, W.; Gao, H.; Nie, N. Climate change and anthropogenic impacts on wetland and agriculture in the Songnen and Sanjiang Plain, northeast China. *Remote Sens.* **2018**, *10*, 356. [CrossRef]
86. Grozavu, A.; Patriche, C.V. Mapping landslide susceptibility at national scale by spatial multi-criteria evaluation. *Geomat. Nat. Hazards Risk* **2021**, *12*, 1127–1152. [CrossRef]
87. Bahrami, Y.; Hassani, H.; Maghsoudi, A. Landslide susceptibility mapping using AHP and fuzzy methods in the Gilan province, Iran. *GeoJournal* **2021**, *86*, 1797–1816. [CrossRef]
88. Cerdà, A. Soil water erosion on road embankments in eastern Spain. *Sci. Total Environ.* **2007**, *378*, 151–155. [CrossRef] [PubMed]
89. Pan, C.; Shangguan, Z. Runoff hydraulic characteristics and sediment generation in sloped grassplots under simulated rainfall conditions. *J. Hydrol.* **2006**, *331*, 178–185. [CrossRef]
90. Gaurina-Medjimurec, N. *Handbook of Research on Advancements in Environmental Engineering*; IGI Global: Hershey, PA, USA, 2014; ISBN 9781466673373.
91. Yen, C. Study on the root system form and distribution habit of the ligneous plants for soil conservation in Taiwan. *J. Chin. Soil Water Conserv.* **1972**, *3*, 179–204.
92. Zhao, B.; Zhang, L.; Xia, Z.; Xu, W.; Xia, L.; Liang, Y.; Xia, D. Effects of Rainfall Intensity and Vegetation Cover on Erosion Characteristics of a Soil Containing Rock Fragments Slope. *Adv. Civ. Eng.* **2019**, *2019*, 7043428. [CrossRef]
93. Wu, L.; Peng, M.; Qiao, S.; Ma, X. Effects of rainfall intensity and slope gradient on runoff and sediment yield characteristics of bare loess soil. *Environ. Sci. Pollut. Res.* **2018**, *25*, 3480–3487. [CrossRef]
94. Li, X.; Niu, J.; Xie, B. The effect of leaf litter cover on surface runoff and Soil Erosion in Northern China. *PLoS ONE* **2014**, *9*, e107789. [CrossRef]
95. Repel, A. Developing of the rainfall intensity-duration-frequency curves for the Kosice region using multiple computational models. *IOP Conf. Ser. Mater. Sci. Eng.* **2019**, *566*, 012026. [CrossRef]
96. Gámez-Balmaceda, E.; López-Ramos, A.; Martínez-Acosta, L.; Medrano Barboza, J.P.; López, J.F.R.; Seingier, G.; Daesslé, L.W.; López-Lambraño, A.A. Rainfall intensity-duration-frequency relationship. Case study: Depth-duration ratio in a semi-arid zone in Mexico. *Hydrology* **2020**, *7*, 78. [CrossRef]
97. Fantina, D.E. A Comparison of Runoff Estimation Techniques. 2012. Available online: https://www.suncam.com/courses/100233-06.html (accessed on 24 August 2021).
98. Meena, S.R.; Ghorbanzadeh, O.; Blaschke, T. A comparative study of statistics-based landslide susceptibility models: A case study of the region affected by the Gorkha earthquake in Nepal. *ISPRS Int. J. Geo-Inf.* **2019**, *8*, 94. [CrossRef]
99. Al-Shabeeb, A.R.; Al-Adamat, R.; Mashagbah, A. AHP with GIS for a Preliminary Site Selection of Wind Turbines in the North West of Jordan. *Int. J. Geosci.* **2016**, *7*, 1208–1221. [CrossRef]
100. Rossi, M.; Reichenbach, P. LAND-SE: A software for statistically based landslide susceptibility zonation, version 1.0. *Geosci. Model Dev.* **2016**, *9*, 3533–3543. [CrossRef]
101. Tazik, E.; Jahantab, Z.; Bakhtiari, M.; Rezaei, A.; Alavipanah, S.K. Landslide susceptibility mapping by combining the three methods Fuzzy Logic, Frequency Ratio and Analytical Hierarchy Process in Dozain basin. *Int. Arch. Photogramm. Remote Sens. Spat. Inf. Sci.* **2014**, *40*, 267. [CrossRef]
102. Gigović, L.; Drobnjak, S.; Pamučar, D. The application of the hybrid GIS spatial multi-criteria decision analysis best–worst methodology for landslide susceptibility mapping. *ISPRS Int. J. Geo-Inf.* **2019**, *8*, 79. [CrossRef]
103. SafeLand Recommended Procedures for Validating Landslide Hazard and Risk Models and Maps, Report. 2011. Available online: https://www.ngi.no/download/file/5994 (accessed on 24 August 2021).
104. Baets, S.; Poesen, J. Empirical models for predicting the erosion-reducing effect of plant roots during concentrated flow. *Geomorphology* **2010**, *118*, 425–432. [CrossRef]

105. Vannoppen, W.; Vanmaercke, M.; De Baets, S.; Poesen, J. A review of the mechanical effects of plant roots on concentrated flow erosion rates. *Earth-Sci. Rev.* **2015**, *150*, 666–678. [CrossRef]
106. Chen, G.; Weil, R.R. Penetration of cover crop roots through compacted soils. *Plant Soil* **2010**, *331*, 31–43. [CrossRef]
107. Najafi, S.; Dragovich, D.; Heckmann, T.; Sadeghi, S.H. Sediment connectivity concepts and approaches. *Catena* **2021**, *196*, 104880. [CrossRef]
108. Luo, J.; Zhou, X.; Rubinato, M.; Li, G.; Tian, Y.; Zhou, J. Impact of multiple vegetation covers on surface runoff and sediment yield in the small basin of nverzhai, hunan province, China. *Forests* **2020**, *11*, 329. [CrossRef]
109. Bochet, E.; García-Fayos, P.; Alborch, B.; Tormo, J. Soil water availability effects on seed germination account for species segregation in semiarid roadslopes. *Plant Soil* **2007**, *295*, 179–191. [CrossRef]
110. García-Palacios, P.; Bowker, M.A.; Chapman, S.J.; Maestre, F.T.; Soliveres, S.; Gallardo, A.; Valladares, F.; Guerrero, C.; Escudero, A. Early-successional vegetation changes after roadside prairie restoration modify processes related with soil functioning by changing microbial functional diversity. *Soil Biol. Biochem.* **2011**, *43*, 1245–1253. [CrossRef]
111. Risse, M.; Cabrera, M.; Coleman, D.C.; West, L.T. Contaminated Soils, Sediments and Water. In *Contaminated Soils Sediments Water*; Springer: Berlin/Heidelberg, Germany, 2006. [CrossRef]
112. Kim, J.H.; Fourcaud, T.; Jourdan, C.; Maeght, J.L.; Mao, Z.; Metayer, J.; Meylan, L.; Pierret, A.; Rapidel, B.; Roupsard, O.; et al. Vegetation as a driver of temporal variations in slope stability: The impact of hydrological processes. *Geophys. Res. Lett.* **2017**, *44*, 4897–4907. [CrossRef]

Article

Snow Avalanche Assessment in Mass Movement-Prone Areas: Results from Climate Extremization in Relationship with Environmental Risk Reduction in the Prati di Tivo Area (Gran Sasso Massif, Central Italy)

Massimiliano Fazzini [1], Marco Cordeschi [2], Cristiano Carabella [1], Giorgio Paglia [1], Gianluca Esposito [1] and Enrico Miccadei [1,*]

[1] Department of Engineering and Geology, Università degli Studi "G. d'Annunzio" Chieti-Pescara, Via dei Vestini 31, 66100 Chieti Scalo, Italy; massimiliano.fazzini@unich.it (M.F.); cristiano.carabella@unich.it (C.C.); giorgio.paglia@unich.it (G.P.); gianluca.esposito@unich.it (G.E.)
[2] Altevie Engineering, Viale Francesco Crispi 19/b, 67100 L'Aquila, Italy; cordeschi@altevie.eu
* Correspondence: enrico.miccadei@unich.it

Abstract: Mass movements processes (i.e., landslides and snow avalanches) play an important role in landscape evolution and largely affect high mountain environments worldwide and in Italy. The increase in temperatures, the irregularity of intense weather events, and several heavy snowfall events increased mass movements' occurrence, especially in mountain regions with a high impact on settlements, infrastructures, and well-developed tourist facilities. In detail, the Prati di Tivo area, located on the northern slope of the Gran Sasso Massif (Central Italy), has been widely affected by mass movement phenomena. Following some recent damaging snow avalanches, a risk mitigation protocol has been activated to develop mitigation activities and land use policies. The main goal was to perform a multidisciplinary analysis of detailed climatic and geomorphological analysis, integrated with Geographic Information System (GIS) processing, to advance snow avalanche hazard assessment methodologies in mass movement-prone areas. Furthermore, this work could represent an operative tool for any geomorphological hazard studies in high mountainous environments, readily available to interested stakeholders. It could also provide a scientific basis for implementing sustainable territorial planning, emergency management, and loss-reduction measures.

Keywords: snow avalanche; mass movements-prone areas; hazard assessment; climate extremization; environmental risk; Gran Sasso Massif; Central Apennines

Citation: Fazzini, M.; Cordeschi, M.; Carabella, C.; Paglia, G.; Esposito, G.; Miccadei, E. Snow Avalanche Assessment in Mass Movement-Prone Areas: Results from Climate Extremization in Relationship with Environmental Risk Reduction in the Prati di Tivo Area (Gran Sasso Massif, Central Italy). *Land* **2021**, *10*, 1176. https://doi.org/10.3390/land10111176

Academic Editor: Giulio Iovine

Received: 8 October 2021
Accepted: 29 October 2021
Published: 2 November 2021

Publisher's Note: MDPI stays neutral with regard to jurisdictional claims in published maps and institutional affiliations.

Copyright: © 2021 by the authors. Licensee MDPI, Basel, Switzerland. This article is an open access article distributed under the terms and conditions of the Creative Commons Attribution (CC BY) license (https://creativecommons.org/licenses/by/4.0/).

1. Introduction

Mass movement phenomena (i.e., rockfalls, debris flows, shallow landslides, snow avalanches, etc.) play a significant role in the landscape evolution and occur in relation to physiographic, geomorphological, and climatic features and to triggering effects induced by human and/or seismic activity [1–12]. These phenomena cause significant disasters on a global scale every year, and the frequency of their occurrence seems to be on the rise. The expansion of urbanization and the tourism development in particular areas, such as mountainous regions, notably increased the environmental hazards and risks. Moreover, climate extremization and the potential for more severe weather conditions could also be acknowledged as contributing factors. Hence, these events can significantly impact mountain environments, residential areas in avalanche zones, ecosystems, and public infrastructures [13,14].

According to the Emergency Events Database—EMDAT [15], snowfall and snow avalanches are considered natural hazards belonging to hydrometeorological events. Snow avalanches are critical events connected to the sudden instability of snow-covered slopes in geodynamical active mountain regions. Moreover, they are undoubtedly one of the major denudational processes in cold and mountainous areas, representing a huge natural hazard with devastating socioeconomic and environmental impacts [16,17].

Mass movement triggering is linked to sudden changes in the geomorphological features of the slopes and the physical characteristics of the snow cover [18,19], resulting, in turn, from numerous variables in continuous changes, such as the geomorphological characteristics of the site, the static and dynamic climatological trends, the processes of metamorphism of the snowy mantle, and the effects of new snow overloading on a preexisting snow cover caused by the action of wind and seismic events of significant magnitude.

It is crucial to follow different approaches to map snow avalanches to provide correct and valuable hazard assessments. Hazard maps represent significant and essential tools needed to evaluate snow avalanche susceptibility of an area, such as a ski resort [20]. It is possible to distinguish between different types of avalanche hazard maps: inventory maps, such as France Carte de Localisation Probable des Avalanches CLPA, [21], depicting the maximum extends of known avalanches, usually compiled from literature, technical documents, and interviews and supported by air–photo interpretation and field investigations and hazard maps [22,23], outlining zones affected by different degrees of hazard, generally drawn based on known historical events, geomorphological studies, and statistical and/or dynamic computational models. In addition to these thematic maps, several techniques can be used to evaluate avalanche hazards and risks involving the implementation of defense structures, closures, and explosives [24,25]. Since the pioneering works in this research field [26,27], most studies were performed to evaluate the long-term risk on settlements and critical infrastructure. These authors all used solid explosives, investigated shock waves propagating through a snowpack, and showed the distinct damping effect of snow, e.g., [28–30]. According to the literature and technical reports [31–33], the techniques used to evaluate avalanche hazards and risks are different depending on the circumstances. The long-term risk affecting permanent settlements and critical infrastructure is typically managed by conducting hazard mapping during the main steps of the land planning process. On the other side, safety services for ski resorts, ski facilities, and temporary worksites are characterized by closures and explosives (i.e., Obellx® gas exploder) to manage short-term avalanche risk; guides adopt professional route selection to control the exposure of people, and public avalanche forecasters communicate regional avalanche danger to a direct stakeholder who manages their own risk [16].

Moreover, it must be considered that the devastating propagation of a snow avalanche may contribute to the mass wasting of rocks and vegetation being transported along the way and accumulated together with the snow avalanche debris. This induced mass wasting poses longer-lasting damages with more destructive effects [34]. As a result, to completely define the degree of hazard in mass movement-prone areas, dynamic computational models can help to estimate paths and impact pressures in the runout zone [35]. Modeling the avalanche triggering mechanisms is complicated, and this complexity has been widely described in many studies [36,37]. The morphological setting (i.e., terrain and slope), the snowpack, and the meteorological conditions contribute to the avalanche movement and propagation. Based on the interaction of these parameters, the avalanche formation and its propagation can eventually be modeled [38–42]. The models have been largely enhanced with the involvement of recent advanced technologies of Geographic Information Systems (GIS) [43–45], which have become powerful tools for the implementation of required databases to support decision-making activities in land planning, such as over hazardous regions posed as a threat by several geohazards (i.e., landslides and snow avalanches).

The mountain territories of the Abruzzo Region are not immune to the general phenomenon of increased tourists' fruition and related snow avalanche risk. Nevertheless, due to its geographical location and physiographic framework (Figure 1), the Abruzzo Region also shows peculiar meteorological and snow characteristics that differ from the rest of the Alps and Central Apennines [46].

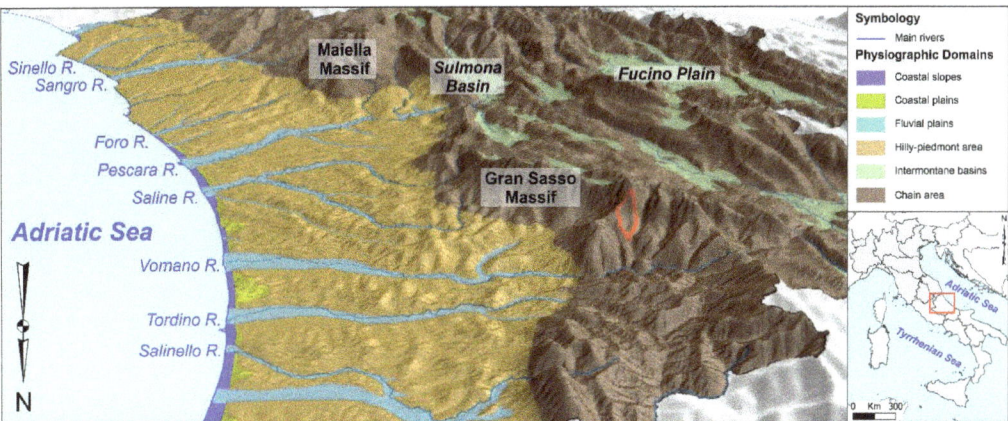

Figure 1. Three-dimensional view (from 20 m DTM, SINAnet) of the Abruzzo Region (Central Italy) and main physiographic domains. The red polygon indicates the study area.

The study area is located in the northeastern part of the Abruzzo Region within the Gran Sasso Massif (Figure 1). It is sited in the municipal territory of Pietracamela. It includes, to the south, a wide irregular mountainous landscape dominated by the Corno Grande (2912 m a.s.l.), featuring as the highest peak of the Apennines Chain.

To develop the present study, an integrated and multidisciplinary approach was followed to provide further advancement in snow avalanche hazard assessment methodologies. Combining and integrating morphometric, geomorphological, climatic, and nivological analyses, it was possible to better define the existing relationships between climate extremization and environmental risk reduction in a mass movement-prone area, such as the Prati di Tivo area. This paper focuses on the stepwise approach for a correct snow avalanche assessment by combining the patterns of snow avalanches and the main meteorological features of the study area. Morphometric and geomorphological analyses were carried out to evaluate landslide hazards in this mass movement-prone area, mainly focusing on the dynamic geomorphic action of snow avalanches. The role of the geomorphological and climatic features in the triggering of the avalanches was also evaluated. Furthermore, it describes the safety services and the risk mitigation protocol to perform over a ski facilities area in such a mass movement-prone setting. This work could represent an effective tool in geomorphological hazard studies for high mountainous environments readily available to interested stakeholders, which provides a scientific basis for territorial planning, emergency management, and mitigation measures.

2. Study Area
2.1. Geological and Geomorphological Setting

The study area is located in Central Italy within the northern sector of the Abruzzo Region, and it is strictly located in the Apennines Chain area, showing a high-relief mountainous landscape. The Central Apennines chain's morphology is characterized by the presence of a series of ridges trending from NW–SE to N–S (i.e., Gran Sasso Massif, 2912 m a.s.l.; Maiella Massif, 2793 m a.s.l.), separated by longitudinal and transversal valleys and broad intermontane basins (elevation 250–1000 m a.s.l.—i.e., Fucino Plain and Sulmona Basin) (Figure 1). The elevation abruptly drops down to the hilly piedmont area (ranging from ~800 m a.s.l. to the coastline), which features a mesa, cuesta, and plateau landscape [47–50]. The northeastern front of the asymmetric Abruzzo Apennines chain is characterized by a steep mountainside with large escarpments. The Gran Sasso Massif is the highest in the Central Apennines, with several peaks above 2500 m a.s.l. It features an arched shape, trending from W–E to N–S, and drops down to lower elevations (>1000 m a.s.l.), defining a large and steep mountain escarpment.

The chain is composed of pre-orogenic lithological sequences that belong to different Meso–Cenozoic paleogeographic domains (carbonate ramp and platform limestones and slope-to-pelagic limestones). The Neogene deformation of these sequences, along NW–SE to N–S-oriented (W-dipping) thrusts, determined the emplacement of the main mountain ridges, also including the Gran Sasso one (Figure 2) [51–58].

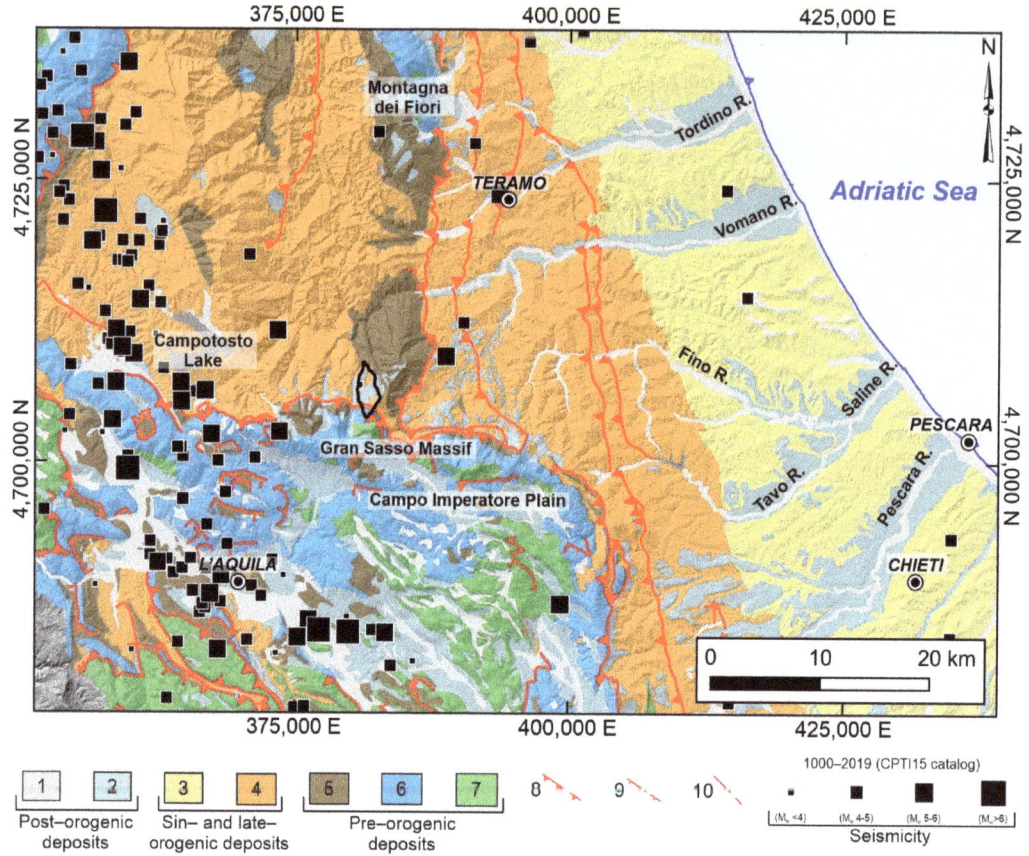

Figure 2. Geological map of NE Abruzzo (modified from [59]). Legend: post-orogenic deposits—(1) fluvial deposits (Holocene) and (2) fluvial and alluvial fan deposits (Middle-Late Pleistocene); sin- and late-orogenic deposits—(3) hemipelagic sequences with conglomerate levels (Late Pliocene–Early Pleistocene) and (4) turbiditic foredeep sequences (Late Miocene–Early Pliocene); pre-orogenic deposits—(5) carbonate ramp facies (Early Miocene-Early Pliocene), (6) slope and pelagic basin sequences (Cretaceous–Miocene), and (7) carbonate platform sequences (Jurassic–Miocene); (8) major thrust (dashed if buried); (9) major normal fault (dashed if buried); and (10) major fault with strike-slip or reverse component (dashed if buried). Seismicity derived from the CPTI15 catalog [60]. The black line indicates the location of the study area.

This compressional phase was followed by extensional and strike-slip tectonics along mostly the NW–SE to NNW–SSE-oriented faults, which define the present-day landscape configuration [47,61,62]. The hilly piedmont and coastal areas are made up of sin- and late-orogenic deposits (i.e., sandy-pelitic turbiditic foredeep sequences), largely covered and unconformably overlaid by Pleistocene hemipelagic sequences. The post-orogenic deposits mainly consist of fluvial and alluvial fan deposits, as well as glacial, travertine, slope, and eluvial–colluvial deposits (Figure 2).

The geomorphological framework is mainly related to mass wasting; gravity-induced (e.g., mostly rotational–translational slides, earth flows, rockfalls, and complex slides); and fluvial-related (e.g., debris flows, alluvial fans, etc.) processes. Ancient glacial processes are preserved as relict landforms [48,63–66].

According to the historical and instrumental data [67–69], Central Italy has been affected by many earthquakes, with recurrent seismic events of moderate-to-high intensity. The present-day regional tectonic setting is dominated by intense seismicity (up to Mw 7.0 [60]), with earthquakes mostly located in the chain sectors (i.e., 2009, L'Aquila and 2016–2017, Central Italy); moderate seismicity also affects the hilly piedmont and Adriatic areas.

2.2. Climatic Setting

The Abruzzo Region climate is affected by the physiographic and morphological setting of the Central Apennine Chain and its eastern front in the proximity of the upper watershed divide. This geographic position, located not far from the Adriatic Sea—approximately 40 km as the crow flies—largely influences the climate setting, varying from a Mediterranean type along the coasts and the hilly piedmont areas to a more temperate and continental type in the chain area [70,71]. The morphological arrangement also regulates the rainfall distribution; the highest annual rainfall values (up to 1500–2000 mm/y) occur along the main ridges and in the inland sectors, decreasing down to ~600 mm/year along the hilly piedmont and coastal areas. It is occasionally characterized by heavy rainfall events (>100 mm/d and 30–40 mm/h) [72–74]. The average temperature values range between 8 and 10 °C in the mountain sectors (average minimum values of 0–5 °C at high elevations) and between 16 and 18 °C along the coast. The winter temperature (average January values) shows low values in the inland areas (0–2 °C, with minimum values of approximately −5 to −10 °C at high elevations) and higher ones (8–10 °C) in the hilly piedmont sectors. Over the past two decades, the Abruzzo Region has been affected by some heavy rainfall events and snowstorms, generated by heavy rainfall ranging from 60 to 100 mm in a few hours to >200 mm per day and by snowfall up to >100 cm/day (e.g., January 2003, April 2004, October 2007, March 2011, September 2012, December 2013, February-March 2015, and January-February 2017 [75]).

More in detail, the mountainous landscape and the homogeneous aspect exposure distribution with north exposed slopes determines a harsh climate poorly mitigated by the maritime influence, as confirmed by the presence of the Calderone glacier—the southernmost one in Europe [64,76].

The Abruzzo Apennine chain sector represents an orographic barrier able to strongly diversify the effects of atmospheric currents on its slopes, with upwind (Stau) and downwind (Föhn or, locally, "Garbino") flows that rule and modify the spatial and altitudinal distributions of rainfall and snowfall events [77,78]. Inland sectors, according to their upwind exposure to cold polar currents moisture-laden after transit through the Adriatic and/or Tyrrhenian sides, are characterized by intense and frequent rainfall events [79]. Furthermore, even if, in such a climatically dynamic framework, it is not uncommon to detect minimum winter temperatures around −25 °C, the Central Apennines (i.e., Abruzzo and Molise regions) show geomorphological situations that determine the presence of a cold air pool. About this, the absolute minimum values were recorded in recent times in several intermontane plains at an elevation ranging from 1200 to 1500 m a.s.l., such as the Piane di Pezza Plain (−37.4 °C), Cinquemiglia Plain (−30 °C), Campo Felice Plain (−32 °C), and Marsia Plain (−36 °C) [80].

According to previous analyses and data [70,71,81,82], the study area is characterized by transitional thermal–meteoric features that largely influence the climate setting varying from continental sub-Apennine to sub-Mediterranean Apennine regimes, considering its southern latitudinal location and the relatively small distance from the Adriatic Sea, which exerts a strong maritime influence.

3. Materials and Methods

The study area was investigated through an integrated and multidisciplinary approach (Figure 3) based on (i) a morphometric analysis, (ii) geomorphological analysis, (iii) climatic analysis, (iv) nivological analysis, and (v) analysis for the assessment of snow avalanche hazard, supported by the combination of literature data and GIS-based techniques.

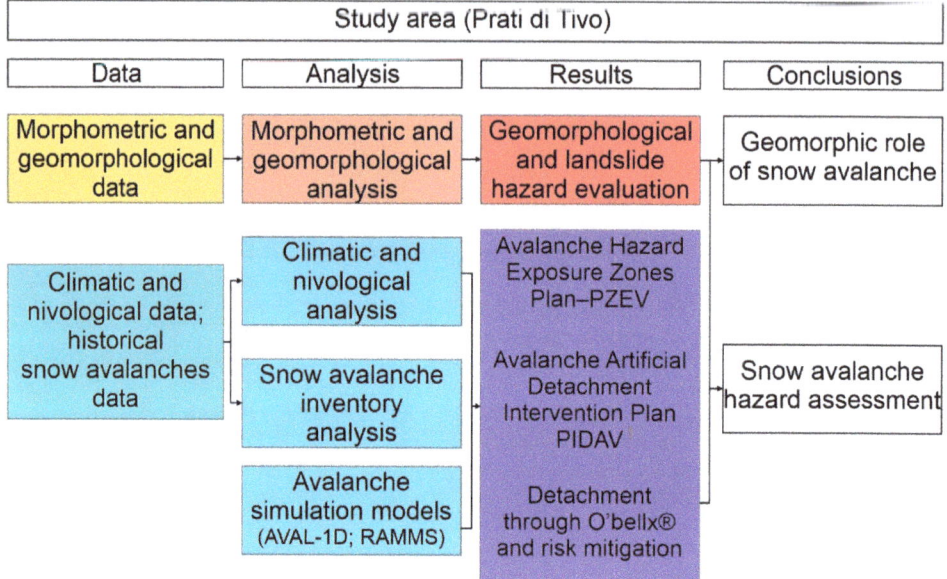

Figure 3. Schematic flowchart diagram showing the main methodological steps.

3.1. Morphometric Analysis

The analysis was performed using topographic maps (1:25,000–1:5000 scale) and supported by the creation of a Digital Elevation Model (5-m DEM) derived from 1:5000 scale regional technical maps, previously retrieved from Open Geodata Portal of the Abruzzo Region (http://opendata.regione.abruzzo.it/, accessed on 15 May 2021). It was carried out in Geographic Information System (GIS) software (QGIS 2020, version 3.16 "Hannover"). It was centered on the definition of the main physiographic features of the study area in order to highlight the morphological setting of this high mountainous environment quantitatively. In detail, the analysis was based on the computation of three main parameters: elevation, slope (first derivate of elevation [83]), and local relief. This latter was calculated as the elevation range within 1×1 km windows, according to Ahnert [84].

According to Schweizer et al. [85], snow avalanche formations result from the complex interaction between the topography, snowpack, and meteorological conditions. As a result, the morphometric characteristics (i.e., slope aspect, relative slope height, and slope inclination of the snow avalanche) are seen as most important in determining the spatial patterns of snow accumulation and, accordingly, the starting, transition, and runout zones [86]. Looking at the landscape parameters evaluated for the study area, the computed morphometric factors (elevation, slope, and local relief) appear to be the most relevant, which control the spatial distribution of snow avalanche activity.

The study area is strictly located to the main drainage basin, automatically extracted from the DEM using the Hydrological Tools in QGIS, whose closing point was located at

Pietracamela Village (Figure 4). This assumption was followed in order to have a basic unit to which to refer to in performing all the multidisciplinary analyses, revealing that the drainage basin scale may be the most convenient choice [87].

Figure 4. Spatial distribution of the weather gauges used in the present study. The red line indicates the location of the study area.

3.2. Geomorphological Analysis

This analysis involved preliminary storing and managing existing data, retrieved from public authorities' technical reports, databases, and the scientific literature. Specifically, geological–geomorphological data were supplied by the CARG Project-Sheet 349 "Gran Sasso d'Italia" [88], the Abruzzo-Sangro Basin Authority [89], the IFFI database [90], and scientific publications [47,65]. These data were integrated and verified through geomorphological field mapping, carried out at an appropriate scale (1:5000–1:10,000), and stereoscopic air photo interpretation using 1:33,000- and 1:10,000-scale stereoscopic air photos (Flight GAI 1954 and Flight Abruzzo Region 1981–1987), as well as an analysis of 1:5000-scale orthophoto color images (Flight Abruzzo Region 2010) and Google Earth® imagery (2019–2020). Field mapping was focused on the definitions of the lithological features and geomorphological landforms, with reference to the main mass movements affecting the study area. It was performed according to the Italian geomorphological guidelines [91], international guidelines [92], and thematic literature concerning geomorphological mapping and analysis in different geological and climatic contexts, as well as field-based and numerical analysis [93–97].

3.3. Climatic Analysis

Climatic data analysis was carried out to outline the distribution of the climatic parameters and conditions in the study area. The analysis was based on a dataset obtained from a network of 7 gauges (colored dots in Figure 4; data provided by the Functional Center and Hydrographic Office of the Abruzzo Region and the *amateur meteorological association* L'Aquila Caput Frigoris—https://www.caputfrigoris.it/, accessed on 12 January 2021). More in detail, according to the lack of historical thermo–pluviometric series suiting the World Meteorological Organization (WMO) directives [98], climatic data belonging to the Pietracamela gauge (1043 m a.s.l.; blue dot in Figure 4) were used to quantify the microclimatic setting of the study area properly. Its dataset gathers thermo–pluviometric series data covering a 50-year time record (1950–2004). Recently, three gauges, featuring technical elements in accordance with the WMO 1083 directives [98], were located at the Teramo (265 m a.s.l.; pink dot in Figure 4), near the tourist and ski facilities at Prati di Tivo (1450 m a.s.l.; green dot in Figure 4), and along the northeastern slope of Gran Sasso Massif at Rifugio Franchetti (2433 m a.s.l.; light blue dot in Figure 4). The northern exposure of these gauges provides a good representation of the climatic conditions occurring in correspondence with the detachment areas of snow avalanches, despite the lack in the snow datasets. Concerning the Rifugio Franchetti gauge, the available data covered different time records (1998–2003 and 2016–2018).

3.4. Nivological Analysis

The local nivological analysis was based on a detailed dataset manually collected at the nivo-meteorological station of the Meteomont service (https://www.sian.it/infoMeteo, accessed on 15 February 2021). It is located at the base of the slope in the Prati di Tivo area (Figure 3), at an elevation of 1450 m a.s.l. It features a northern exposure similar to avalanche-prone regions located at higher elevations. The available historical data for this station begins from the 1977/1978 winter season (from November to April) for 32 surveying seasons. The series is nearly uninterrupted, with a few gaps mostly occurring in correspondence of the beginning/end of seasons. However, data related to the 1992/1994 seasons are completely missing. More in detail, the considered dataset shows several temporal gaps, since it is deeply affected by the irregularity in the opening/closing dates of ski facilities—the former occurring after the first significant snowfall events and the latter during the spring period, usually in the presence of a thick snow cover. This condition was widely relevant before the 1986/1987 winter season and after the 2008/2009 one; consequently, the amounts of seasonal new snow were not correctly computed in these temporal intervals. To reduce this underestimation, we tried to derive good-quality data about the potential snowfall events by computing thermo–pluviometric records at gauges located at a comparable elevation not far from the Prati di Tivo area (e.g., Campotosto gauge, 1344 m a.s.l.—yellow dot in Figure 4). Nevertheless, to deduce a general nivometric trend and better define the nivometric regime of the study area, data belonging to a 20-year time period (1986/1987–2008/2009) were considered and thoroughly analyzed.

3.5. Snow Avalanche Hazard Assessment

This analysis was performed following a stepwise methodological approach that involved the snow avalanche inventory analysis, the analysis and mapping of snow avalanches' paths, the elaboration of a snow avalanche hazard map, and the definition of numerical models.

The snow avalanche inventory was retrieved from the State Forestry Corps of Italy and the Abruzzo Region (http://opendata.regione.abruzzo.it/content/carta-storica-della-valanghe, accessed on 15 May 2021) and allowed us to clearly describe the avalanches' spatial distribution over the study area. Moreover, it was integrated with information derived from the available literature and technical reports [44,75,99].

The analysis of snow avalanches' paths was achieved by combining the literature data, specific site investigations, investigations of the snow-covered ground, interviews of witnesses to past avalanche events, and studying of previous events recorded in various historical and technical archives [44,100].

The evaluation of the snow avalanche hazard map was carried out according to the Swiss mapping criteria [101,102] and thematic guidelines provided by AINEVA (Italian Interregional Association for Snow and Avalanche) [31,103]. Avalanche-exposed zones were defined and annexed within the Avalanche Hazard Exposure Zones Plan—PZEV (Piano delle Zone Esposte a Valanghe in Italian). Generally, this evaluation is fixed through mathematical parameters, which quantified the velocity and flow height, transmitted pressures, and stopping distances of the avalanches [31,102,104,105]. In the invasion zones, as reported in Table 1, some areas are identified and marked with different colors according to the estimated avalanche hazard—i.e., high hazard with red, moderate hazard with blue, and low hazard with yellow. Town planning and land use prescriptions are fixed for each of the identified zones.

Table 1. Synthesis of the AINEVA criteria [31] for the delimitation and the use of areas with different degrees of exposure to avalanche hazards (T = return time of the avalanche (years) and P_{imp} = impact pressure (kPa)).

Zone/Hazard Degree	Definition Land Use Restrictions
RED High Hazard	Areas affected either by avalanches with T = 30, even with low destructive power ($P_{imp} \geq 3$), or by highly destructive avalanches ($P_{imp} > 15$) with T = 100. New constructions are not allowed.
BLUE Moderate Hazard	Areas affected either by avalanches with T = 30 with low destructive power ($P_{imp} < 3$) or areas affected by rare events (T = 100) with a moderate destructive power ($3 < P_{imp} < 15$). New constructions are allowed but with strong restrictions (low building indexes, reinforced structures, etc.).
YELLOW Low Hazard	Areas affected either by events with a low destructive power ($P_{imp} < 3$) and T = 100 or by events with 100 < T < 300. New constructions are allowed, with minor restrictions (no public facilities, like schools, hotels, etc.).

In the PZEV's framework, morphometric and nivometric data are generally combined to define the degree of exposure of a specific area in terms of the frequency and intensity of avalanche events. This detailed analysis is usually expressed through:

- the avalanche return period—the average number of years between two events of the same intensity;
- the avalanche pressure—the forces per unit of surface exercised by the avalanche on a flat obstacle of big dimensions disposed perpendicularly to the trajectory of the advancing mass of snow. The pressure can be determined with reference to both the dynamic and static components of the solicitation.

The obtained maps effectively identify the avalanche sites and their expansion in the accumulation zones. This has proven to be most helpful in defining these zones in terms of avalanche frequency and dynamic pressure, thus determining the magnitude/frequency distribution in the runout zones [106–108].

The criteria established and reported in the Avalanche Artificial Detachment Intervention Plan—PIDAV (Piano di Intervento di Distacco Artificiale di Valanghe in Italian) [109] were followed to develop prevention and management activities in the study area. Generally, the main objective of these protection measures is to minimize negative consequences due to snow avalanche risk for people and goods in their settlements and along traffic lines, as well as for skiers [32,110]. The PIDAV plan is a tool, eventually complementary to the aforementioned PZEV, which refers to an area open to the public, clearly defined in space and time, where an artificial release of unstable snow masses is performed to reduce avalanche hazards and risks [109,111]. In case of an urban zone or a ski facility to be protected, as in the study area, it is necessary to define a management measures plan to protect the ski lift. It should include the plan for meteo-nivological conditions monitoring—which are in constant evolution during climatic events—and describes activities to be exerted to learn about this evolution at the meso- and microscale to evaluate snow cover stability conditions and their potential evolution.

In conclusion, this stepwise sequence was completed by avalanche simulation models. In detail, 1- and 2-dimensional avalanche simulation models (e.g., AVAL-1D and RAMMS [39,112]) were applied both to back-analyze documented avalanche events at a particular site, as well as to estimate the consequences of possible hazard scenarios. According to the literature and technical data [102,111,113], the main nivometric parameters required for dynamic avalanche modeling are represented by the maximum height of the snow cover (Hs) and the increase of the snow cover height over three consecutive days (Dh3gg). For developing the present study, an increase of 5 cm of new snow and a snow cover for every 100 m of elevation was proposed, taking into account the aforementioned literature data and the nivological expert judgment. These physical–mechanical characteristics, together with ancillary information concerning the physiography, steepness, and roughness of the ground, the presence of infrastructures s.l. were reported on a 5-m grid DTM base map and elaborated in a GIS environment.

AVAL-1D is a numerical avalanche dynamics program developed by the Swiss Federal Institute for Snow and Avalanche Research [112]. It allows the simulation of avalanches in one dimension from the starting zone to the runout one. It reproduces runout distances, flow velocities, and impact pressures of both flowing and powder snow avalanches along a specified avalanche track. It consists of two modules: FL-1D (dense flow model) for dense flow avalanches and SL-1D (powder snow model) for powder snow avalanches. It cannot reproduce the whole set of dynamical parameters, since it is a one-dimensional formulation that combines the internal distribution of flowing variables into basic ones controlled by two frictional parameters [114]. In order to supply this not accurately modeling, the RAMMS (RAapid Mass MovementS) code [115] was mainly used to calculate the pressure values on a site-specific avalanche path (such as Vallone della Giumenta) from initiation to runout in a three-dimensional terrain. It is a practical tool for avalanche practitioners, which requires a complete procedure to fulfill the morphological features and release parameters. Moreover, it can be used to estimate runout distances, flow velocities, flow heights, and impact forces [116–118].

4. Results

4.1. Morphometric Analysis

The study area reaches its maximum altitude on the peak of Corno Piccolo (2655 m a.s.l.) and is characterized by a morphology that gradually slopes down to a minimum of 1030 m a.s.l. in correspondence with Pietracamela Village. Based on the orography of the landscape, the area can be fairly divided into three different sectors: a northern one near Pietracamela village, a central one comprising the Prati di Tivo area, and a southern one corresponding to the northern slope of the Corno Piccolo ridge (Figure 5). The northern sector presents the lowest elevation, ranging approximately from 1100 to 1300 m a.s.l.; the slope values range from 0 to 40°, with the maximum values detected in correspondence with the N–S-oriented and, secondarily, W–E-oriented drainage lines; the energy of the relief ranges from 250 to 350 m, with the highest values along the Rio San Giacomo. The central sector is characterized by a flat and irregular morphology, featuring elevations ranging from 1300 to 1700 m a.s.l., and a homogeneous slope distribution (values between 5° and 20°); the energy of the relief, on the other side, shows heterogeneous values ranging from 250 m towards the western portion to 400 m towards the eastern one. The southern sector, finally, presents elevations ranging from 1700 up to 2500 m a.s.l.; the slope distribution is dominated by the highest values (between 60° and 80°), with peaks detected in correspondence with the N–S-oriented drainage lines and W–E-oriented steep scarps; the energy of the relief ranges from 500 up to 600 m, with the highest values along the northern escarpment of Corno Piccolo.

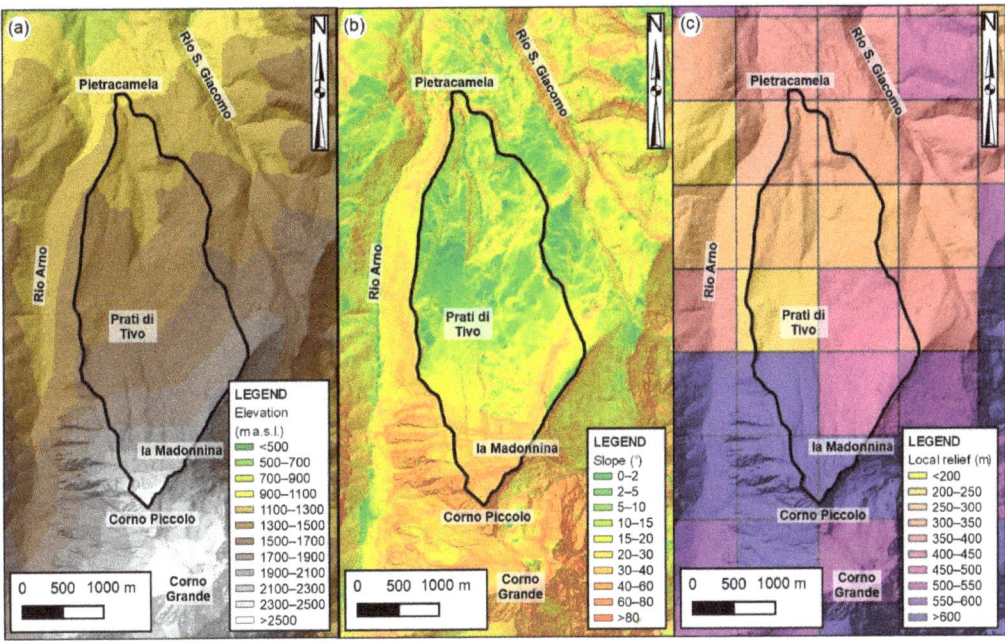

Figure 5. Physiographic features of the study area: (**a**) elevation map, (**b**) slope map, and (**c**) local relief map. The black line represents the study area.

4.2. Geomorphological Analysis

The study area is characterized by the outcropping of lithological sequences belonging to pre- to sin-orogenic deposits. In detail, it is characterized by calcareous and marly deposits outcropping in the southernmost sector, in correspondence with the Corno Piccolo ridge. Instead, the central sector is dominated by the presence of arenaceous-pelitic and pelitic-arenaceous deposits, mainly composed of turbiditic layers with fine sand or coarse silt and pelitic intercalations. The bedrock is widely covered by continental deposits (Figure 6). Scree slope deposits, mainly composed of cemented breccias, characterize the westernmost sector and, locally, the easternmost one, near la Madonnina.

In the Prati di Tivo area, recent glacial and alluvial fan deposits are present along the N–S-elongated outcrops; moving eastward, fluvio-glacial deposits, consisting of cemented breccias and largely marked by landslide bodies, alternate with recent scree slope deposits.

From a geomorphological standpoint, the most recurrent features are represented by structural, slope, fluvial, and glacial landforms (Figure 6). Concerning structural ones, in the southernmost sector, a W–E-oriented is detectable, overlapping calcareous deposits on overturned marly deposits. A second buried thrust is not clearly observable, but its existence can be inferred through minor in-field exposures that highlight the overlapping of marly deposits over sin-orogenic pelitic-arenaceous deposits. Slope landforms partly consist of active rockfalls and complex landslides in the northern sector near Pietracamela.

Large quiescent rotational and translational slides affect the central-eastern portion of the study area east of the Prati di Tivo area, together with localized quiescent earthflows. Smaller rotational and translational slides are found towards the north, mainly set on pelitic-arenaceous deposits. Finally, N–S-oriented rock gullies with debris discharges characterize the northern slop of the Corno Piccolo ridge. Landforms due to running water are mainly represented by a wide alluvial fan, as well as of several gullies; both are present in the western part of the study area, within the Prati di Tivo area, the former being set between glacial (to the West) and landslide (to the East) deposits, the latter with a general N–S direction and extending along the main drainage line. Finally, concerning glacial landforms, several scarps are preserved as relict landforms in the southern sector. Furthermore, several N–S-oriented avalanche tracks characterize the norther escarpment of the Corno Piccolo ridge, alternating themselves with the rock gullies.

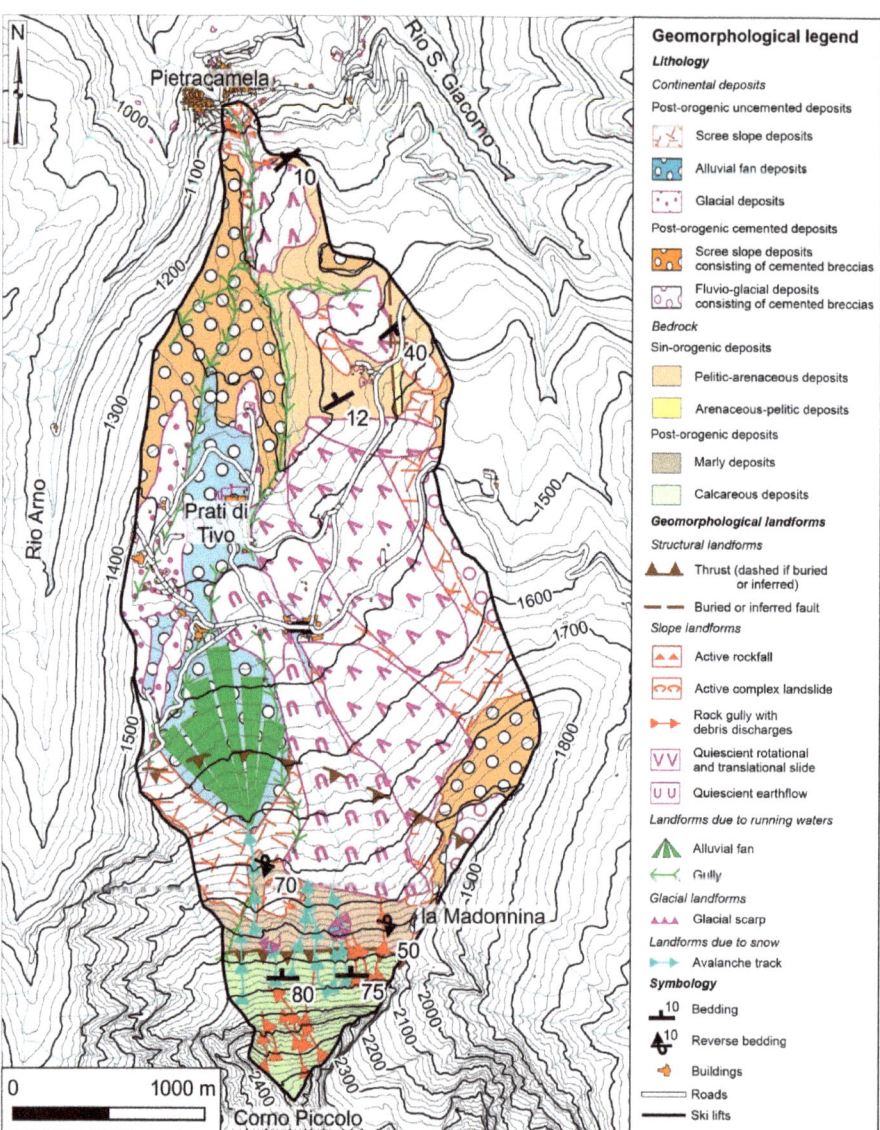

Figure 6. Simplified geomorphological map of the study area (modified and updated from [65,88,89]). The black line represents the study area.

4.3. Climatic Analysis

The microclimatic features of the study area were analyzed considering the historical thermo–pluviometric series available at the Pietracamela gauge (1043 m a.s.l.), covering a 50-year time period (1951–2004); nevertheless, the selected gauge is located approximately

400 m below the Prati di Tivo area and at least 1000 m downstream of the main avalanches and landslides detachment areas (Table 2). Nonetheless, for the study area, it was impossible to define spatial and altitudinal meteoric features since snowfall and rainfall events are often coupled with strong winds that can induce relevant rates of underestimations, especially at the highest elevations.

Table 2. Main values of the temperature and rainfall, resulting from the climatic analysis at the Pietracamela gauge (1043 m a.s.l.).

Yearly Average (1951–2004)			Monthly Average (1951–2004)										
Temperature (°C)			Temperature (°C)										
			Jan	Feb	Mar	Apr	May	Jun	Jul	Aug	Sep	Oct	Nov	Dec
Frost days	65	Frost days	16	15	12	4	0	0	0	0	1	5	12
Absolute maximum	36.8	Absolute maximum	19	20.7	25	24.5	31.5	35	35	36.8	34	27.8	25	22
Daily average	10.7	Daily average	3	3.4	5.5	8.4	12.8	16.7	19.7	19.8	16	11.3	7.2	4.2
Mean maximum	14.6	Mean maximum	6.5	7.2	9.4	12.4	17	21.2	24.5	24.6	20.2	14.9	10.4	7.4
Mean minimum	6.7	Mean minimum	−0.4	−0.5	1.5	4.4	8.7	12.3	14.9	14.9	11.8	7.7	4	1.1
Absolute minimum	−14	Absolute minimum	−14	−12.8	−12.1	−7	−0.5	3	4.5	4	−1.3	−7	−7	−13
Rainfall (mm)														
Total rainfall	1065.3													
Maximum in 1 h	57.8		Jan	Feb	Mar	Apr	May	Jun	Jul	Aug	Sep	Oct	Nov	Dec
Maximum in 24 h	268.6	Total rainfall	85.4	74.8	95.7	110.4	82.5	68.1	44.9	51.8	78.3	117.1	136.3	120.0
Rainy days	106	Rainy days	8.5	9.1	9.7	10.6	10.0	8.0	6.2	5.8	7.2	9.8	10.8	10.6

The average annual temperature is ~10.7 °C, with an average daily thermal excursion of ~7 °C; the maximum temperatures can eventually exceed 35 °C, while the minimum ones almost reach −15 °C. Frost days ($T_{min} < 0$ °C) are ~65 per year, while ice days ($T_{max} < 0$ °C) are no more than 10 per year.

The total rainfall is moderately abundant with respect to the Gran Sasso Massif geographic location, exposed to "Tramontana" and "Bora" dry and cold winds, as well as to "Scirocco" and "Libeccio" wetter ones, which often release the moisture taken in charge. During the summer, the ascent of convective cells from the near L'Aquila Basin and the middle Vomano River valley is common. The total annual rainfall is approximately 1100 mm, distributed along with an average of 106 rainy days; the hourly and daily maximum values, respectively, correspond to 58 mm and 267 mm. The meteoric regime shows peculiar features pertaining to the Apennine–Adriatic type [81,100], with a bimodal rainfall distribution characterized by a global maximum value in November with a secondary peak in April and a global minimum in July/August—months not in a drought, given the frequent occurrence of convective phenomena—with a secondary peak in February. It should also be stressed that the monthly rainfalls never drop below 50 mm.

Snowfalls are frequent during every winter season, with high amounts with respect to the geographic position of the study area, as previously reported in the thematic literature [100,119]. Considering the possible influence of disturbing fluxes coming from the south and associated with negative temperatures, the study area can present cumulative values among the highest of the Central Apennine area, as happened in February 2017 [119].

The performed climatic analysis confirms, even in this area, an increase of the temperature values of about 1.1 °C during the last 50 years; unfortunately, the weather station located at Pietracamela ceased its activity in 2004, thus making a more recent trend analysis impossible. The rainfall regimes, on the other side, do not show significant variations (−1.5 mm/year), but it is possible to observe a decrease of about 10% in the number of days with precipitation rates > 1 mm from 110 to 101; consequently, the average daily rainfall intensities have slightly increased. Considering the recorded datasets, homogeneous and complete data relating to short and intense precipitation events is unavailable. It was not possible to perform a comprehensive analysis for this specific climatic aspect.

The detailed climatic analysis shows that average annual temperatures for the year 2020, as observed for the Teramo gauge (blue line in Figure 7a), was approximately 1 °C higher than the conventional 30-year time period (1971–2000), known as CliNo (Climate Normal). Consequently, the elevations of the 0 and −1 °C isotherms correspond to 3099 and 3296 m a.s.l., far above those calculated by Dramis et al. [120], corresponding, respectively, to 2615 and 3028 m a.s.l. Considering that the average annual temperature recorded at the Pietracamela (1043 m a.s.l.) and Rifugio Franchetti (2433 m a.s.l.; green line in Figure 7a) gauges are, respectively, 10.7 and 2.6 °C, given a difference in elevation of about 1500 m, a vertical thermal gradient of approximately 6.1 °C/km can be estimated. Thus, an average annual temperature of 7.6 °C and 4.2 °C can be derived at, respectively, Prati di Tivo (orange line in Figure 7a) and the avalanche detachment areas located at ~2200 m a.s.l.

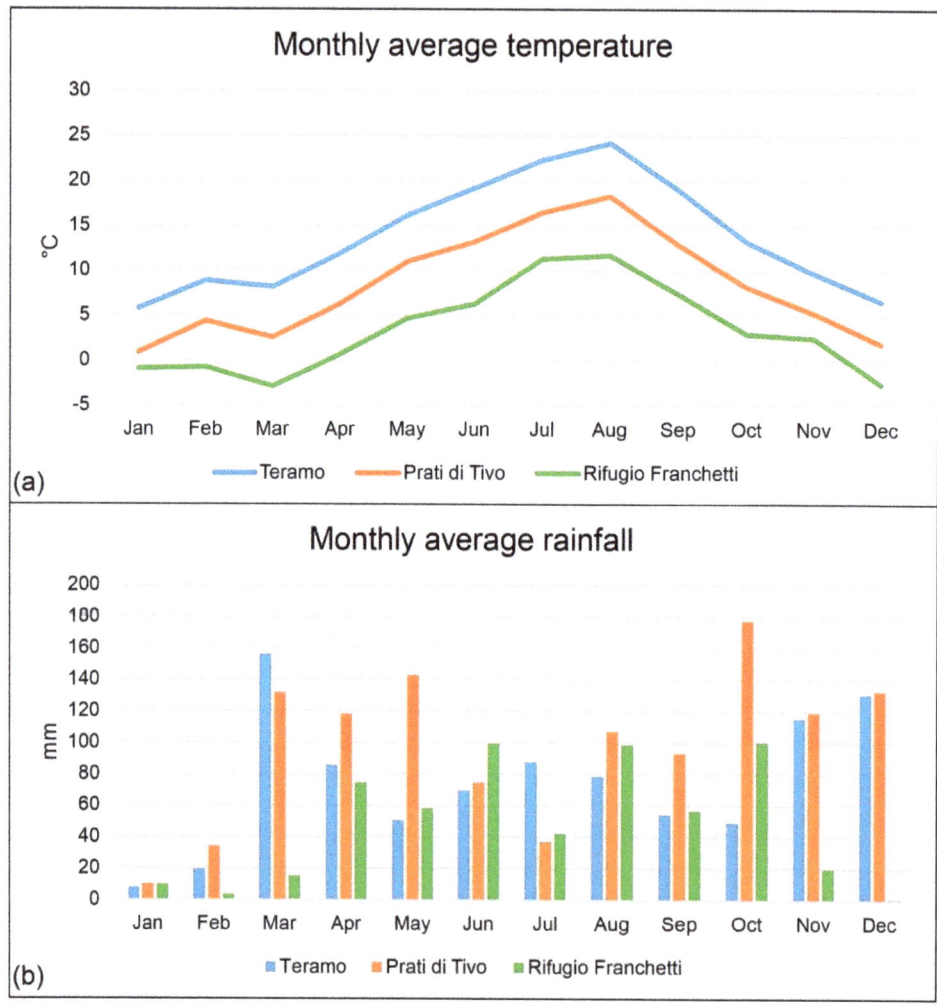

Figure 7. Monthly average temperature (**a**) and monthly average rainfall (**b**) trends for the year 2020.

The monthly average rainfalls (Figure 7b) show a noticeable growth with a direct relationship as the altitude increases due to an orographic effect, in accordance with previous estimates made for this sector of the Central Apennine Chain (30 mm/100 m) [100,119]. Nevertheless, a drastic decrease of these values occurs at higher elevations, which is typical of an arid boreal habitat. This substantial underestimation, up to 70%, occurs in areas exposed to powerful winds during rainfall and/or snowfall events [121].

The anemometric signal is significant in the whole area from 260 up to 2400 m a.s.l., thus promoting the accumulation of frames and lenses above all on the ridges of Gran Sasso Massif and in correspondence of steep channels and depressed morphologies downwind of the main flow during and after snowfalls (Figure 8a). In particular, at high elevations, the number of days with a maximum wind speed greater than 30 km/h, sufficient for inducing a reworking of the snow cover, is around 40. Main winds come from the second and the third quadrants during the winter, reaching speeds greater than 200 km/h (Figure 8b).

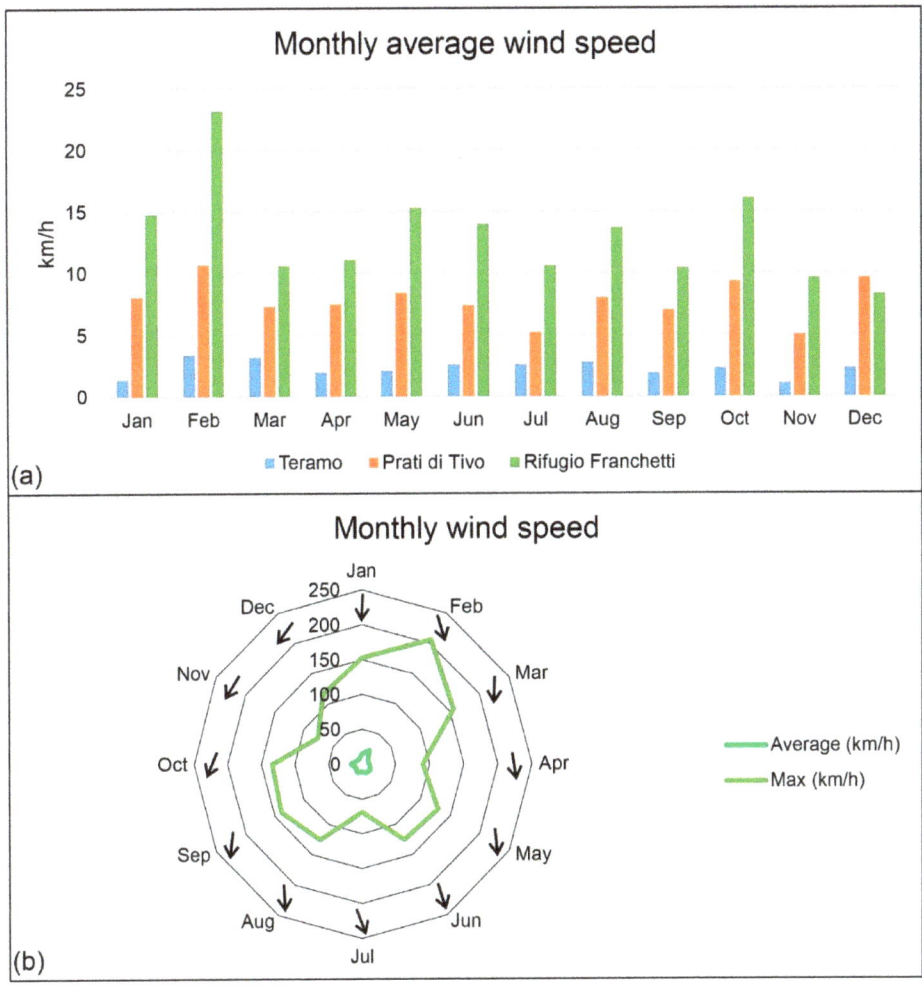

Figure 8. (a) Monthly average wind speed for the year 2020 and (b) monthly wind speed recorded at Rifugio Franchetti (2433 m a.s.l.). Black arrows indicate the monthly average direction of winds.

4.4. Nivological Analysis

By accounting for limits derived from the available nivological data, as previously reported in the Materials and Methods section, the following can be stated:

- The Prati di Tivo area shows a regular snow cover every year relatively abundant in certain winter seasons, such as 1994/1995, 1998/1999, and 2004/2005, with cumulative values greater than 400 cm. Only during the 1988/1989 winter season, the cumulative was less than 100 cm (Figure 9a). Recently, a more significant snowfall irregularity occurred, with long phases characterized by the absence of snow phenomena, alternating with short but intense heavy snowfalls events. Accumulations seem to have increased against a lower persistence of the snow cover. The trend analysis referring to the aforementioned period depicts a clear rise in the signal—over 3 cm per winter season—deriving from the highly irregular nivometric trend, with a hint of a ten years of periodicity and a more recent signal of about 3.8 cm per winter season detected on average for the Central Appennines Chain [100,119,122].
- Heavy snowfalls already occur from the middle of November. They are common throughout the winter and until the second half of March, becoming sporadic in April (Figure 9b). By accounting for the available datasets, the absolute monthly maximum values occurred in January 2017 at about 425 cm distributed in only seven days [119]. The winter's least snowy month is estimated to be December. The snowfall regime (Figure 8b) presents a unimodal distribution, with the maximum values detected in January and February. Arguably, for altitudes greater than 2000 m a.s.l., the trend tends to become fairly regular if not bimodal with a second peak during the spring, given the notable snowfall increase in March and April, as detected at the Campo Imperatore gauge (2137 m a.s.l.).
- The number of snowy days shows considerable intra-seasonal variations, strictly dependent on the synoptic seasonal evolution. This aspect was particularly evident in the last decade, albeit in a context of significant snowiness, with values ranging between 25 and 35 events per season, with peaks of about 40. During the last seasons, a general decrease of the phenomenology seemed to be occurring; these are increasingly concentrated in a few days and present a greater intensity, which underlines the climatic extremization in progress. Furthermore, a delay at the beginning of the snowy season seems evident, along with a greater frequency of events at the beginning of the spring season.
- Daily snowfall data (Figure 10a) highlight the possible occurrence of snowy events of high intensity and short-to-moderate durations. In particular, the maximum recorded daily amount of fresh snow is around 70 cm (13 February 1986 and 23 March 2009). Moreover, it is essential to consider unofficial recordings performed on 17–19 January 2017 (when abundant avalanche events occurred, reaching the Prati di Tivo area and causing considerable damages to infrastructures and ski facilities), which pointed out a daily maximum of 140 cm on 17 January and of 310 cm for the whole three-day period.
- Significant sudden temperature changes occurring more frequently after or during snowfalls generally disfavor the cohesion process between the strata composing the snow cover, thus causing a hypothetical increase of the avalanche hazard. Nevertheless, a clear Mediterranean type, the climatic extremization, and a not-excessively-high elevation determine an early beginning of the accelerated destructive metamorphism processes, with a subsequent quick decrease of the snow depth values on the ground up to the maximum elevation of avalanche-prone areas. Furthermore, close to detachments areas, a strong wind power occurring during and after snowfalls induces rapid mechanical metamorphism. In the case of intense snow events followed by exceptionally cold climatic phases and variable weather conditions, destructive metamorphism processes take place very slowly; constructive metamorphism is indeed established. The thickness of the snow cover remains relatively abundant for a long time (Figure 10b).

- Days with mixed snowfall and rainfall events or entirely rainy ones are also estimated to occur during the winter season. This is connected to the synoptic conditions inducing rainfall and to the eventual mixing within the frontal system. Field evidence and surveys in specific sites suggest that this has a significant repercussion on natural avalanche occurrences, especially below 1900 m a.s.l.
- From field surveys, as well as from the avalanche inventory and literature data (i.e., Meteomont service), it results that, in correspondence with a sudden temperature rise, avalanche events may occur with loose surface cohesion values already in the 24–36 h following the snowfall events, involving many buildings and anthropic structures present in the Prati di Tivo area.

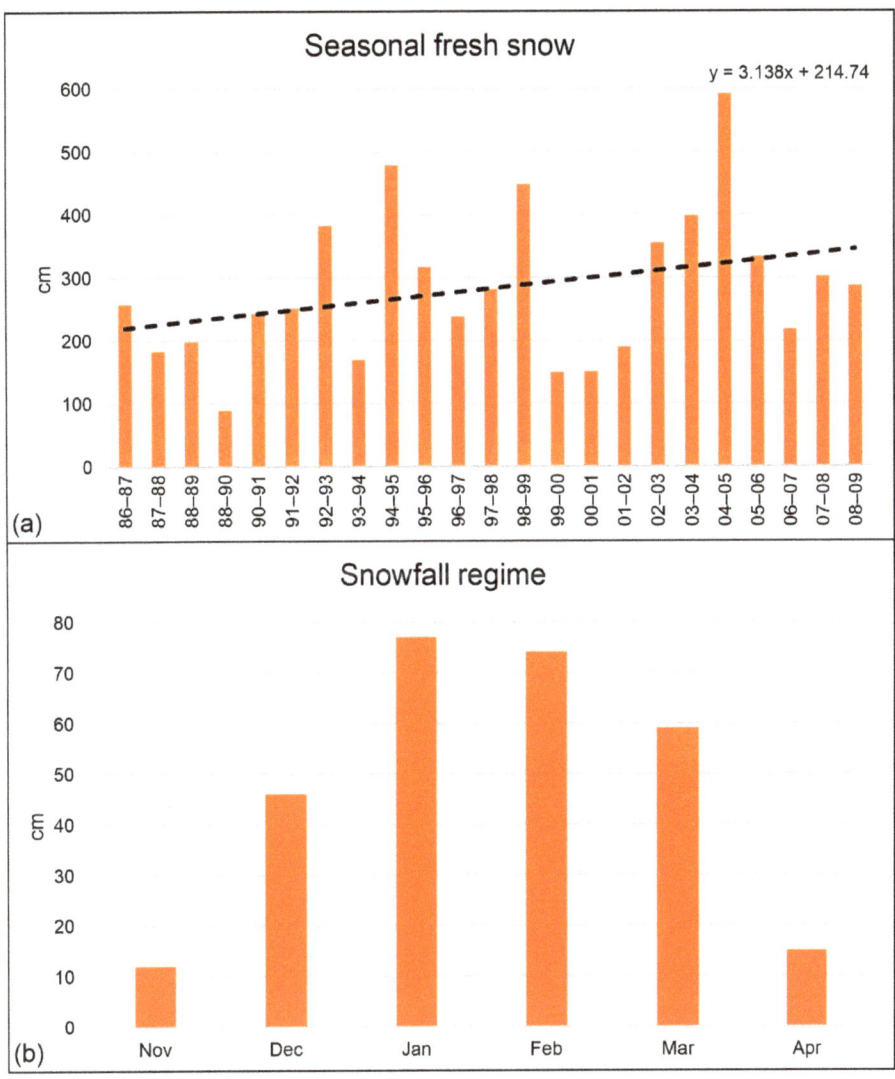

Figure 9. Seasonal fresh snow trend for the timespan 1986–2009 (**a**) and snowfall regime (**b**) at the Prati di Tivo gauge (1450 m a.s.l.).

In order to provide a complete and updated description of the nivometric trends, the historical series and datasets available at the Rifugio Il Ceppo gauge (1340 m a.s.l.) were analyzed. This weather station was taken into account since it shows snowmaking very similar to that of Prati di Tivo, even if it is located on the eastern side of the neighboring Laga Mountains at a distance of about 20 km from the study area.

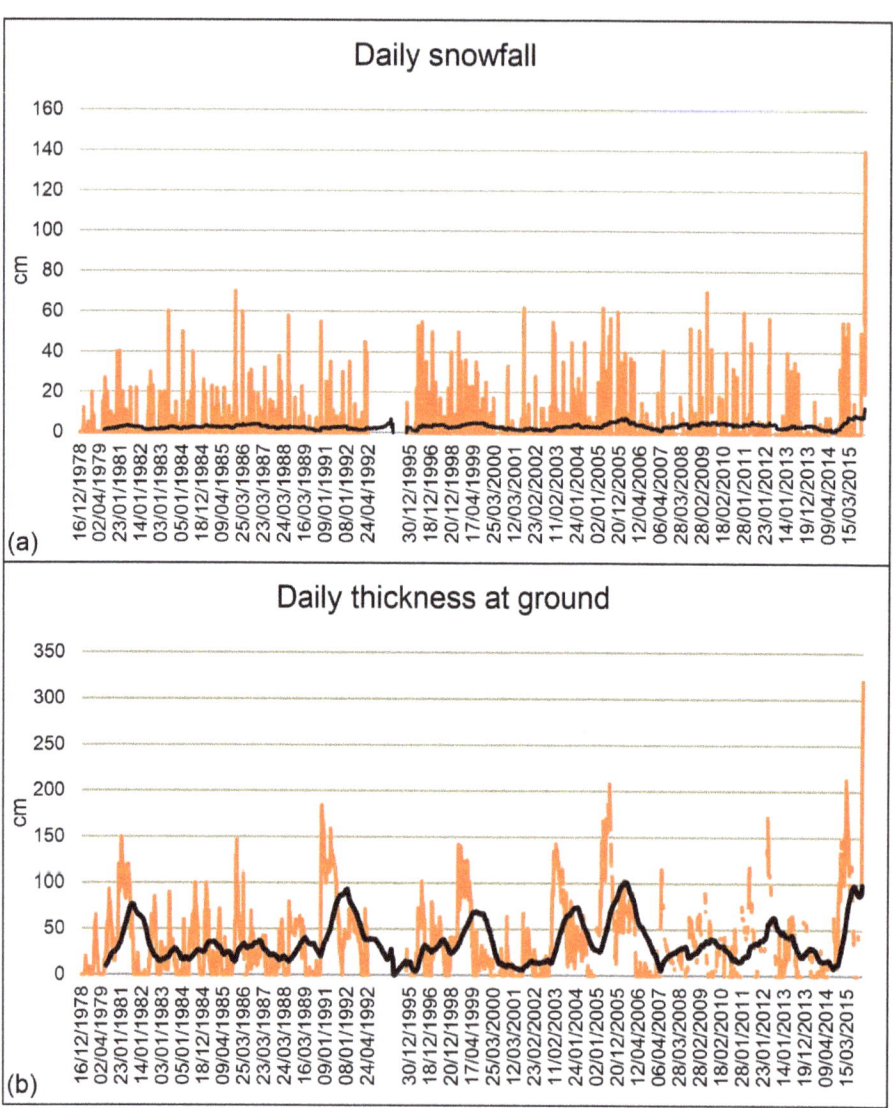

Figure 10. Daily snowfall (**a**) and snow cover thickness (**b**) at the Prati di Tivo gauge (1450 m a.s.l.).

As graphically reported in Figure 11, the nivometric trend spanning over a 40-year time period (1979–2019) confirms the increase of snow precipitation. However, it is less marked than that evident for the Prati di Tivo area (Figure 8a), with a minimal difference in the recent interseasonal variations (2.1 cm vs. 3.1 cm/season), whose trends are increasingly marked.

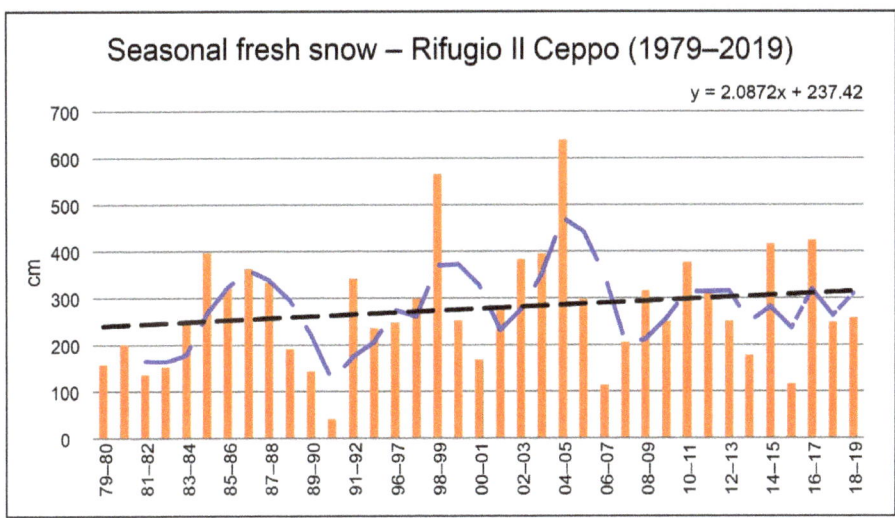

Figure 11. Seasonal fresh snow at Rifugio Il Ceppo (1340 m a.s.l.). The blue dashed line represents the 5-year moving average.

4.5. Snow Avalanche Hazard Assessment

A stepwise methodological approach allowed us to perform a complete snow avalanche hazard assessment, also taking into account the study area's main physiographic and geomorphological features. The avalanche-prone areas along the main N–S-oriented avalanche tracks characterize the northern escarpment of the Corno Piccolo ridge, which alternate with several rock gullies. The area shows elevation values ranging from 1320 to 2270 m a.s.l. As confirmed by the previous data and analysis [123,124], the avalanche paths largely affected residential structures and ski facilities, causing significant damages in recent times.

Firstly, a snow avalanche inventory analysis was carried out. The geodatabase retrieved from the State Forestry Corps of Italy and the Abruzzo Region stored and collected almost 800 avalanches over the whole Abruzzo Region from 1957 to 2013. The yearly number ranged up to 70 in the last decades, and about 40 events were recorded in the previous years covered by the catalog, with a poor direct correlation with the snow thickness. As graphically reported in Figure 12, for the Prati di Tivo area, the database reports 131 snow avalanches: 10 of which accounted as slab snow avalanches, 26 as glide snow avalanches, 54 as powder snow avalanches, 9 as loose snow avalanches, and 32 as mixed or not classified snow avalanche.

Additionally accounting for the main geomorphological features, a preliminary analysis of the spatial distribution of snow avalanches over the study area shows the northern escarpment of the Corno Piccolo ridge almost totally affected by avalanche phenomena whose detachment areas are located at elevations ranging from 1700 up to 2550 m a.s.l. These phenomena mainly involved several N–S-oriented rock gullies and trails, often anastomosed, allowing the snow movements to extend heterogeneously, depending on the type and amount of snow involved.

Moreover, to provide a complete inventory, official avalanche data of the Abruzzo Region were integrated with the literature data, local chronicles, eyewitness reports of past avalanche events, and studies of previous events recorded in various historical and technical archives.

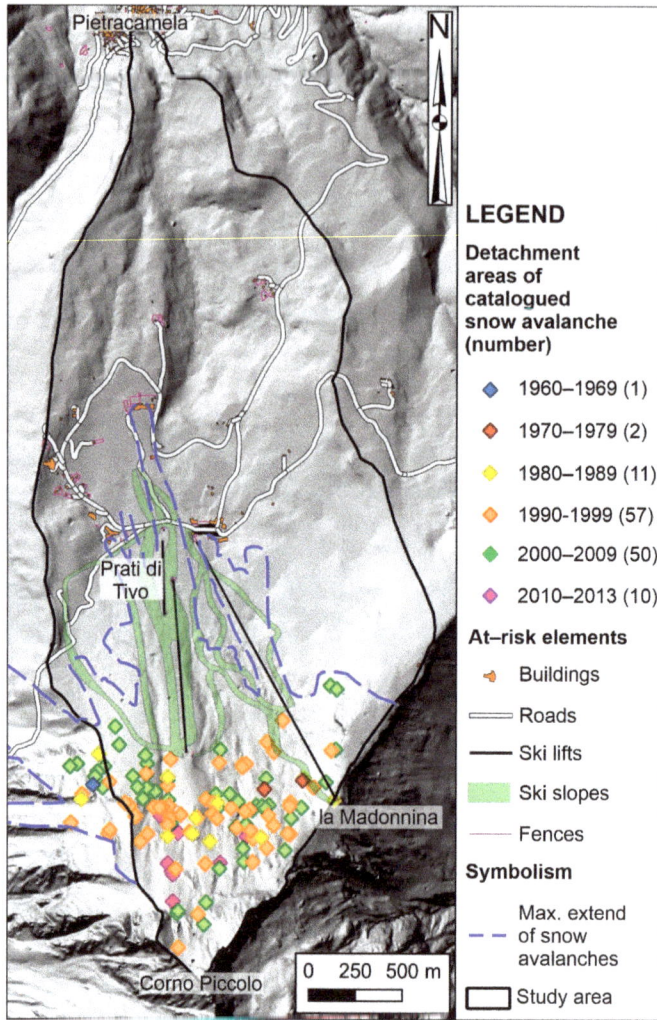

Figure 12. Historical snow avalanche map of the study area (1957–2013 period).

The study area has suffered a rapid and intense urban development for sport tourism purposes since 1965, although several avalanches have occurred in the past, according to local chronicles and eyewitness reports. These events were also integrated into the snow avalanche inventory analysis, enabling the mapping of the maximum extent of avalanches in the Prati di Tivo area. They are chronologically reported as follows:

- April 1929 (not reported in the official database of Abruzzo Region)—A large avalanche event reached the Guide Shelter (Rifugio delle Guide in Italian).
- 7 April 1978—A large avalanche, detached from the northern escarpment of the Corno Piccolo ridge, moved down through rock gullies and affected the ski lift and the other facilities downstream to the Madonnina location.
- 8 January 1981—An avalanche, detached from the northern slopes of the Corno Piccolo ridge, affected some houses and buildings located at Prati di Tivo.
- 2 March 1984—An avalanche affected the study area and posed in a threat the Madonnina chairlift bar.

- 3 March 1999—A large avalanche hit the study area, causing several damages to the ski facilities (e.g., pylons, intermediate station, and ticket office).

The resulting data highlight a scarce spatial characterization of snow avalanches along the study area. Given the proximity of different detachments sites, the reconstruction of past avalanche activity remains quite difficult, distinguishing events occurring from neighboring detachment areas. Consequently, considering the absence of an updated Probable Avalanche Location Map—CLPV (Carta di Localizzazione Probabile delle Valanghe in Italian) [44], an analysis of snow avalanches' paths was performed to identify areas likely to be exposed to avalanche hazard. This "static" approach [105,125] was based on the analysis of morphological features for delineating the predisposition to snow avalanche occurrence within the whole northern escarpment of the Corno Piccolo Ridge, also considering the spatial distribution and the recurrence of the main phenomena. Six main avalanche paths were selected (Figure 13) that are considered the most likely to occur and affect houses, roads, and sporting infrastructures. Moreover, most of the chosen sites were devoid of terminology, and according to their proximity to isolated buildings or specific sites/localities, it was decided to provide detailed descriptions for some of them.

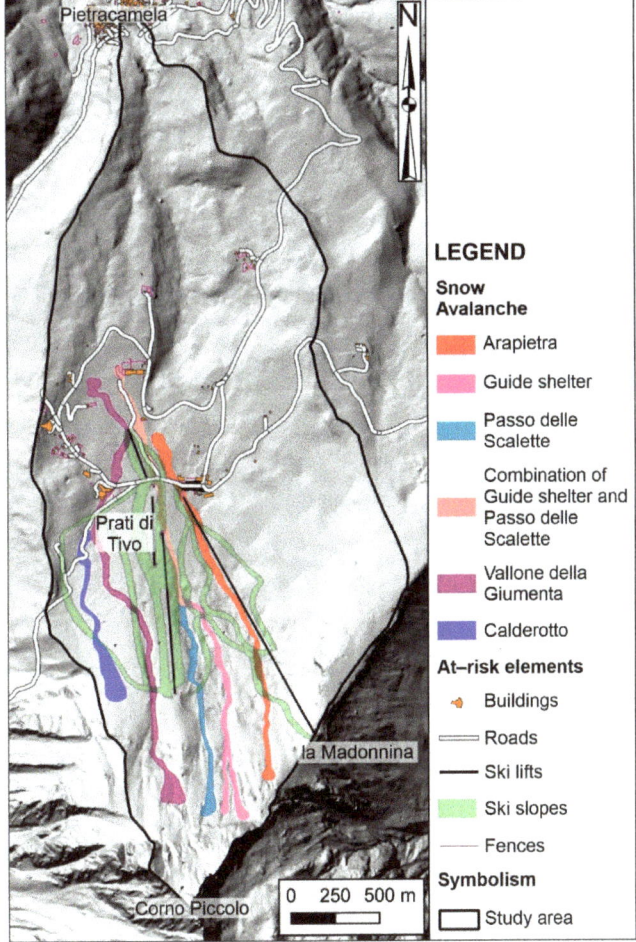

Figure 13. Snow avalanche paths recorded and mapped in the present study.

The exceptional snowfall events generating the snow avalanche disaster involving the Rigopiano Hotel in January 2017 [75,126,127] caused several collateral events in the surrounding areas, including the Prati di Tivo area. They generated a wide snow avalanche along the Vallone della Giumenta (Figure 13) and determined acute injuries to the Prati di Tivo residence (Figure 14).

Figure 14. Detail of damage caused by the Vallone della Giumenta avalanche to the Prati di Tivo residence on 18 January 2017 (source Il Martino, 2017). For the avalanche path's location, see Figure 12.

This event led to the definition and realization of the first Avalanche Hazard Map of the entire Central Appennines area within the Avalanche Hazard Exposure Zones Plan—PZEV (Piano delle Zone Esposte a Valanghe in Italian). At the same time, the PIDAV was presented to develop prevention, mitigation, and management activities of ski facilities directly exposed to avalanche dynamics [100]. In this context, the nivometric dataset available at Prati di Tivo gauge (1450 m a.s.l.) was widely analyzed to derive the required parameters for dynamic snow avalanche modeling (Table 3). This analysis allowed us to define the possible return times of snowfall events at different periods. In detail, the relationship between the maximum height of the snow cover (Hs) and the increase of the snow cover height over three consecutive days (Dh3gg) revealed that snow events (i.e., January 2017) that created a Hs > 3 m can be statistically considered as significant outliers (std > 3) with potential return times far exceeding 300 years.

Table 3. Main nivometric parameters required for the return time estimations, recorded at the Prati di Tivo gauge (1450 m a.s.l.). N.B.: Hs is the height of the snow cover and DH3gg is the increase of the snow cover height over three consecutive days.

	Return Time (Year)								
	5	10	15	30	50	100	150	200	300	500
Hs (t) cm	154	177	191	213	229	251	263	272	285	301
	Return Time (Year)								
	5	10	15	30	50	100	150	200	300	500
DH3gg (t) cm	64	78	86	99	109	122	130	135	143	153

A thematic map elaborated following the AINEVA criteria [31] and integrated with a detailed nivological analysis is graphically shown in Figure 15. This map clearly shows the main areas exposed to avalanche hazards. Considering the spatial distribution of the snow avalanche paths, as reported in Figure 12, it is possible to delineate the transit and invasion zones, marked with different colors according to the estimated avalanche hazard and the potential return periods (such as T = 30, 100, and 300 years). The analysis of the map highlights different scenarios: with a return time equal to 30 years, all tourist, sporting, and residential facilities can be affected by significant snow avalanche phenomena characterized by paths that reach the Prati di Tivo area at elevations of 1450 m a.s.l., while avalanches with a return time equal to 300 years can get a wider spatial extension (downward to 1370 m.a.s.l.) comparable to the maximum extent of the historical snow avalanches previously described (dashed blue line in Figure 12).

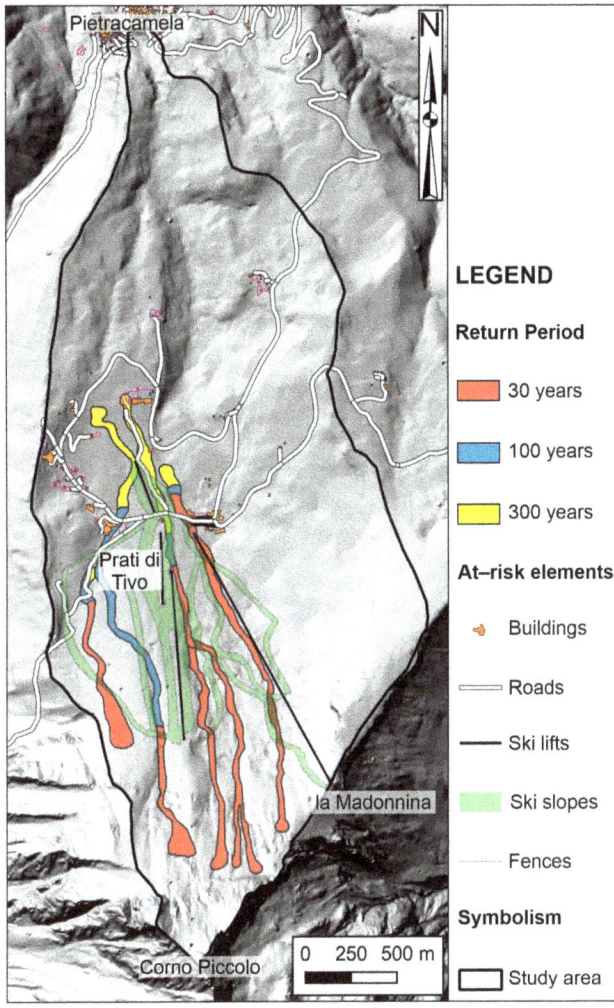

Figure 15. Avalanche hazard map of the Prati di Tivo area. Areas affected by snow avalanches with a return time equal to 30, 100, and 300 years are shown, respectively, in red, blue, and yellow.

5. Discussions

Snow avalanche hazards are computed to be increasing worldwide due to climate changes [120,129]. Among all the climatic contributors, climate extremization is identified as one of the factors influencing the behavior, irregularity, and frequency of snow avalanches [130,131]. In some areas, it causes the thinning and shortening of the duration of snow cover, contributing to an increased irregularity that raises the hazard. As a result, a correct climatic analysis involving investigations of changes in the snow cover and snow avalanche hazard assessment is vital for administering many crucial societal issues concerning territorial planning, risk mitigation, and resilience activities [25,132,133].

Exposure to this hazard may be voluntary, as is the case with skiing, or involuntary, such as on public transportation corridors and settlements. According to the literature and technical reports [31–33], the techniques used to evaluate avalanche hazards and risks are different depending on the circumstances.

Here, we attempted to understand the main interrelationships between climate extremization and environmental risk in a mass movement-prone area, such as the Prati di Tivo area. We discussed the stepwise approach to be followed for a correct snow avalanche assessment by combining the spatial distribution of the snow avalanches and the main climatic features of the study area. It was also essential to compare the findings with the detailed geomorphological features of the Vallone della Giumenta to outline the role of climate extremization in the triggering of the avalanches.

The combination of preliminary results and thematic maps allowed us to better characterize the study area from a morphometric, geomorphological, climatic, and nivological standpoint. In such a complex and mass movement-prone area, it was necessary to activate a risk mitigation protocol to develop land use policies and activities to define a significant snow avalanche assessment. According to the PIDAV project [99], the safety services for ski resorts and facilities at Prati di Tivo were updated by installing 12 Obellx® gas exploders [109,134] to manage short-term avalanche risks better. The installation was realized in correspondence with the main detachment areas at elevations ranging between 2100 and 2250 m a.s.l.

Moreover, as part of the increasingly more frequent processes of climate extremization, on 24–26 March 2020, a heavy snowfall event affected the study area. It was acknowledged as a prevalently stormy snowfall, which brought 90 cm of fresh snow (with a density of 140 kg/m^3) over the ski facilities located at Prati di Tivo at elevations of about 1400 m a.s.l. Given the high snow accumulation rates, explosive pitches were performed on 24 March immediately after the beginning of the snowfall event and on March 26 during the main event, inducing moderate detachments of fresh, humid, and low-cohesion snow. Even if the preventive activity of Obellx® gas exploders reasonably mitigated the snow dynamics, on the night of 27 March, around 4:20 a.m., two natural snow avalanche events occurred following new abundant snowfalls and affected the northern escarpment of the Corno Piccolo ridge (Figure 16). A detailed field survey and a specific site investigation were also performed in the early morning of 28 March, thanks to a clear weather improvement. Considering the information gathered from this survey, it was possible to make several essential deductions:

- Slightly downstream of the prominent peak (Corno Piccolo, 2655 m a.s.l.) at an elevation of about 2550 m a.s.l., a detachment area was visible, as graphically shown in Figure 16a. Moreover, according to no official local chronicles and eyewitness reports, it seemed to correspond with the site of an avalanche never reported and stored in the Geodatabase of the Abruzzo Region.
- The whole avalanche path mainly affected the Vallone della Giumenta (for the site's location, see Figure 13), with a clearly outlined detachment area at an elevation of 2300 m a.s.l. (Figure 16b).
- Significant snow accumulations generated by the snow mass releases produced by the Obellx® devices on the 25th and 26th of March were visible throughout the escarpment.

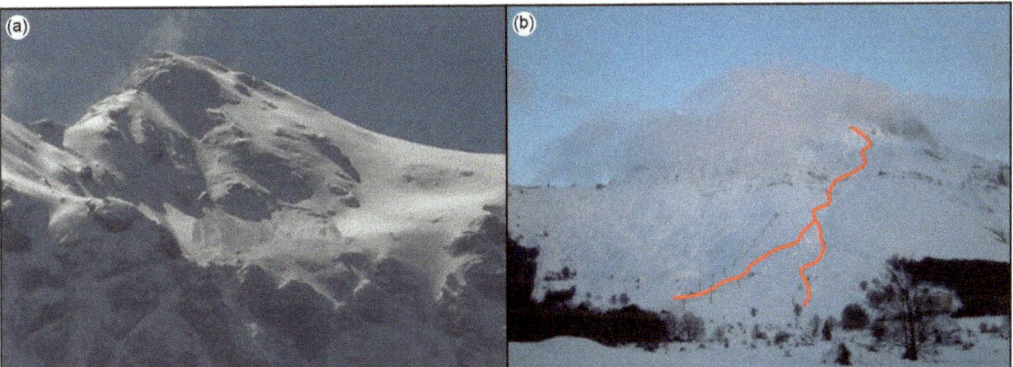

Figure 16. Photo documentation of the snow avalanches of March 2020. (**a**) Evidence of the summit detachment area at the base of Corno Piccolo with a clear surface slab and (**b**) a simplified snow avalanche path (in red) affecting the Vallone della Giumenta and involving the main ski facilities.

The presence of interdigitated snow mass accumulations belonging to the March 2020 avalanches, both partially converged into the Vallone della Giumenta, testified the dynamic avalanche framework of the study area. Unfortunately, this peculiar nivological setting makes it impossible to define the temporal evolution of the two different events. For these reasons, preliminary one- and two-dimensional avalanche simulation models (e.g., AVAL-1D and RAMMS [39,112]) were applied to better describe the possible evolution of the documented avalanche events at a particular site (such as the Vallone della Giumenta), as well as to calculate the consequences of possible hazard scenarios. The avalanche modeling was carried out by employing RAMMS software and implemented with GIS techniques. In detail, it was performed both by considering a scenario characterized by a limited thickness of the snow cover mitigated by the Obellx® devices' activity and a scenario in which the downstream slopes were totally covered by a thick snow cover (i.e., not secured by the preventive action of the Obellx® devices). The resulting data defined different stopping distances and paths of the selected avalanche under scenarios driven by other transmitted pressures, as graphically shown in Figure 16. It showed how, in the absence of the preventive action of the Obellx® devices (red stars in Figure 16), an avalanche event can predominantly occur along the Vallone della Giumenta. Moreover, transmitted pressures, which vary from 30 to 150 kPa, and different heights of snow cover (Hc = 0.5, 1.0, and 1.5 m) were accounted for. Under these scenarios, the avalanche path can widely reach the Prati di Tivo area, involving the residence (reported as the hotel in Figure 14) and the four-seat chairlift line (Figures 16b and 17).

Snow avalanches can generally act as geomorphic agents [135]. Snow avalanches can exert considerable erosive forces playing a significant role in landscape development. Evaluating the morphological features of the mass movement-prone area and the main avalanche features is essential to quantify the material entrained by the avalanche and transported to the deposition zone [136].

A semi-quantitative analysis was applied to the modeled avalanche path at Vallone della Giumenta (Figures 13 and 17) to better describe the geomorphic role of snow avalanches at the Prati di Tivo area. This specific site investigation presents peculiar morphometric features and snow avalanche pressures accounted as representative of the main avalanche events in the study area. The analysis focused on evaluating the pre- and post-avalanche setting, highlighting the variations in the contributing area caused by the snow avalanche along the Vallone della Giumenta. This evaluation showed a variation that increased by about two-fold (>50%), as the contribution of each avalanche track and rock gully was significant in the geomorphic action of the avalanche. It is graphically shown in Figure 18 and summarized in Table 4.

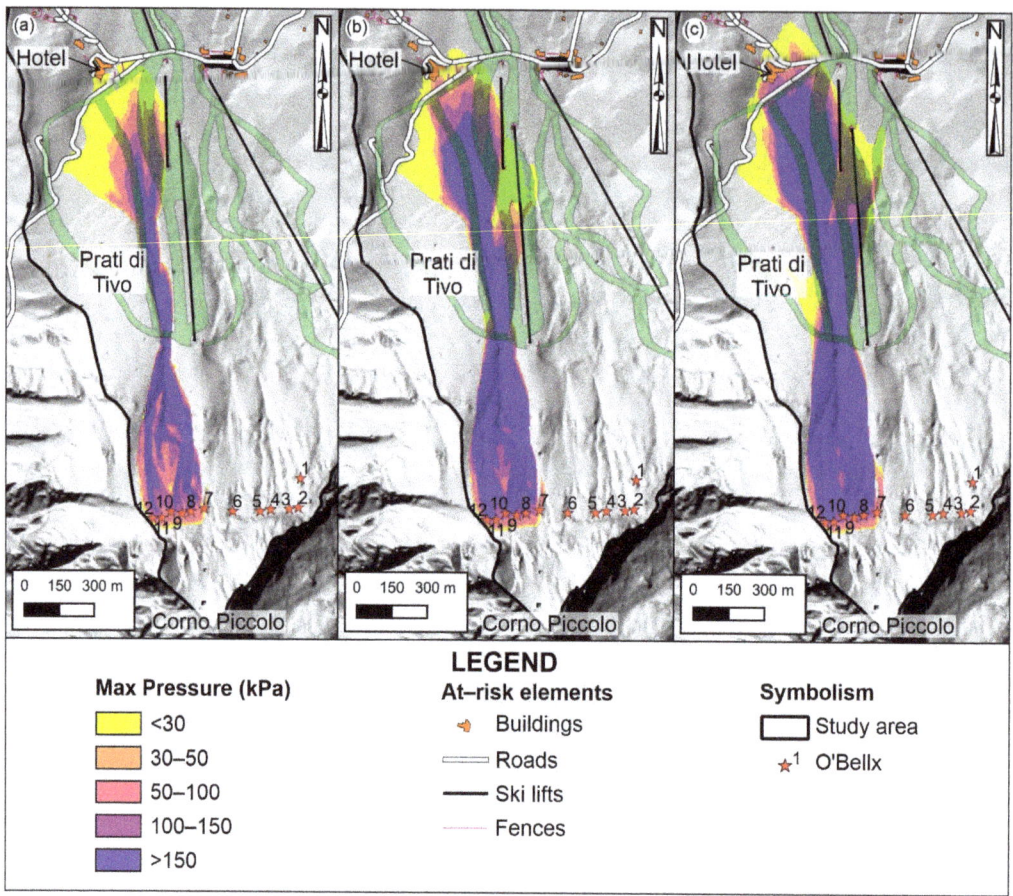

Figure 17. Results of the avalanche dynamics simulation under different transmitted pressures and heights of snow cover: (**a**) Hc = 0.5 m, (**b**) Hc = 1.0 m, and (**c**) Hc = 1.5 m.

In conclusion, the resulting data allowed us to properly define the main steps of the developed risk mitigation protocol. It was activated following some recent damaging snow avalanches affecting the Prati di Tivo area to better develop mitigation activities and land use policies needed for the management of permanent settlements, recreation infrastructures, and ski facilities. The relevance and the impact of the work are represented by: (1) the provision of new data on the physiography–geomorphology of the study area and the mass movement-prone areas, (2) the outline of a multidisciplinary methodological approach for the definition of snow avalanche critical areas and the configuration of hazard protocols not yet developed for the Central Apennines, and (3) a technical scientific basis to develop the civil protection plans required to increase the knowledge of citizens and interested stakeholders about proper land management considering multi-hazard scenarios (i.e., snow avalanches and landslides).

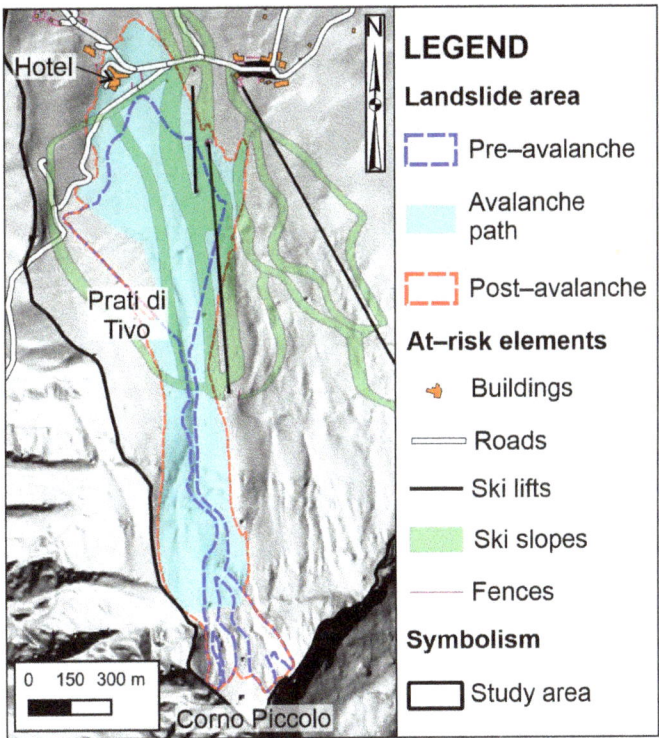

Figure 18. Pre- and post-avalanche landslide contributing area.

Table 4. Dimensions of the possible landslide area, modified by snow avalanche dynamics.

Pre-Avalanche Landslide Area (km²)	Snow Avalanche Area (km²)	Post-Avalanche Landslide Area (km²)
0.363	0.645	0.769

6. Conclusions

Snow avalanches are among the most destructive natural hazards threatening built structures, ski resorts, and landscapes in cold and mountainous regions. The Central Apennines high-mountain environment has been largely affected by different types of mass movements in recent years, accentuated in frequency and magnitude due to changes in the climate regime. The increase in temperatures, the irregularity of intense weather events, and several heavy snowfall events determined an increase in landslide and/or snow avalanche hazards, especially in areas with well-developed tourist facilities.

The Prati di Tivo area has been widely affected by several mass movement phenomena. Like other mountain territories of the Abruzzo Region, the study area is not immune to the general increased tourist fruition and related snow avalanche risk. It is located on the northern slope of the Gran Sasso Massif (Central Italy), showing peculiar meteorological and snow characteristics that differ from the rest of the Alps and Central Apennines. This work allowed us to better define and analyze the geomorphological and climatic features of the study area. The climate extremization results in relation to environmental risk reduction were evaluated by combining different thematic datasets (e.g., morphometric and geomorphological features, climatic and nivological data, technical information, and numerical modeling). In detail, we carried out a snow avalanche hazard assessment to outline a multidisciplinary methodological approach for defining snow avalanche-critical areas and a technical scientific basis to set up accurate civil protection plans and land management activities. The analysis was performed following a stepwise methodological approach, including the snow avalanche inventory analysis, the analysis and mapping of snow avalanches' paths, the elaboration of a snow avalanche hazard map, and the definition of numerical models.

Recent exceptional snowfall events in the Abruzzo Region (i.e., January 2017 [75,126,127]) caused several damages and injuries in the surrounding Prati di Tivo area. Consequently, the safety services for ski resorts and facilities were updated by installing 12 Obellx® gas exploders [109] to better manage short-term avalanche risks. However, despite the activation of this risk mitigation protocol, the recent snow avalanche event that occurred on 26 March 2020 testified that the local geomorphological dynamics amplified by the climatic extremization that could lead to approximate and insufficient results deriving from planned safety services. Understanding the likely scenarios and consequences of a changing climate on snow avalanche behavior is essential for planning and managing mountain developments. More specifically, the climatic evolution, characterized by further increases of the average winter temperatures and increasingly more irregular and intense snowfalls, could lead to avalanche events of even greater magnitudes compared to what was observed until now and, consequently, will determine a major need for constant updates of the calculations of the new snow avalanche paths.

Combining and integrating morphometric, geomorphological, climatic, and nivological analyses, it was possible to further advance the methodologies for a snow avalanche hazard assessment, defining the existing relationships between climate extremization and environmental risk in a mass-movement prone area, such as Prati di Tivo area. The resulting data also showed that, to perform a complete snow avalanche hazard assessment, it is necessary to consider the geomorphic role of snow avalanches, which can exert considerable erosive forces extending the contributing areas. Finally, a thorough expert-based study would be highly desirable to constantly evaluate the geomorphological and climatic dynamics of the study area. This kind of study can represent a valuable and operative tool for civil protection activities and territorial planning in relation to emergency management and mitigation measures by assuming the potential occurrence of extreme nivological and meteorological scenarios.

Author Contributions: Conceptualization, M.F. and E.M.; methodology, M.F. and E.M; software, C.C., M.C., G.E. and G.P.; validation, E.M. and M.F.; investigation, M.F. and E.M; resources, M.C. and M.F.; writing—original draft preparation, G.P., C.C., G.E. and M.F; writing—review and editing, G.P., M.F. and E.M.; supervision, E.M.; project administration, M.F. and E.M.; and funding acquisition, E.M. All authors have read and agreed to the published version of the manuscript.

Funding: This research and the APC were funded by Enrico Miccadei, the grant provided by Università degli Studi "G. d'Annunzio" Chieti-Pescara.

Institutional Review Board Statement: Not applicable.

Informed Consent Statement: Not applicable.

Data Availability Statement: The data presented in this study are available on request from the author. The data are not publicly available due to privacy. Images employed for the study will be available online for readers.

Acknowledgments: The authors would like to thank the anonymous reviewers that helped to improve the manuscript with their precious and constructive comments. The authors wish to thank the Cartographic Office of the Abruzzo Region by means of the Open Geodata Portal (http://opendata.regione.abruzzo.it/, accessed on 20 December 2020) and the Ministero dell'Ambiente e della Tutela del Territorio e del Mare (http://www.minambiente.it/, accessed on 20 December 2020) for providing the topographic and cartographic data used for this work. Climatic and nivological data were provided by the Functional Center and Hydrographic Office of the Abruzzo Region (Centro Funzionale e Ufficio Idrografico Regione Abruzzo) and the Meteomont service (https://www.sian.it/infoMeteo, accessed on 15 February 2021). The authors are also grateful to the amateur meteorological association L'Aquila Caput Frigoris (https://www.caputfrigoris.it/, accessed on 12 January 2021) for providing the updated climatic data (year 2020) used for the work.

Conflicts of Interest: The authors declare no conflict of interest.

References

1. Beniston, M.; Farinotti, D.; Stoffel, M.; Andreassen, L.M.; Coppola, E.; Eckert, N.; Fantini, A.; Giacona, F.; Hauck, C.; Huss, M.; et al. The European mountain cryosphere: A review of its current state, trends, and future challenges. *Cryosphere* **2018**, *12*, 759–794. [CrossRef]
2. Guerra, A.J.T.; Fullen, M.A.; Jorge, M.D.C.O.; Bezzerra, J.F.R.; Shork, M.S. Slope Processes, Mass Movement and Soil Erosion: A Review. *Pedosphere* **2017**, *27*, 27–41. [CrossRef]
3. Solari, L.; Del Soldato, M.; Raspini, F.; Barra, A.; Bianchini, S.; Confuorto, P.; Casagli, N.; Crosetto, M. Review of satellite interferometry for landslide detection in Italy. *Remote Sens.* **2020**, *12*, 1351. [CrossRef]
4. Leroueil, S.; Locat, J. Slope movements—Geotechnical characterization, risk assessment and mitigation. In *Geotechnical Hazards*; Maric, B., Lisac, L., Szavits-Nossan, A., Eds.; Balkema: Rotterdam, The Netherlands, 1998; pp. 95–106.
5. Martino, S.; Antonielli, B.; Bozzano, F.; Caprari, P.; Discenza, M.E.; Esposito, C.; Fiorucci, F.; Iannucci, R.; Marmoni, G.M.; Schilirò, L. Landslides triggered after the 16 August 2018 Mw 5.1 Molise earthquake (Italy) by a combination of intense rainfalls and seismic shaking. *Landslides* **2020**, *17*, 1177–1190. [CrossRef]
6. Aleotti, P.; Chowdhury, R. Landslide hazard assessment: Summary review and new perspectives. *Bull. Eng. Geol. Environ.* **1999**, *58*, 21–44. [CrossRef]
7. Glade, T.; Anderson, M.; Crozier, M.J. *Landslide Hazard and Risk*; John Wiley & Sons Ltd.: Chichester, UK, 2012; ISBN 9780470012659.
8. Marsala, V.; Galli, A.; Paglia, G.; Miccadei, E. Landslide susceptibility assessment of Mauritius Island (Indian ocean). *Geosci.* **2019**, *9*, 493. [CrossRef]
9. Peruccacci, S.; Brunetti, M.T.; Gariano, S.L.; Melillo, M.; Rossi, M.; Guzzetti, F. Rainfall thresholds for possible landslide occurrence in Italy. *Geomorphology* **2017**, *290*, 39–57. [CrossRef]
10. Quesada-Román, A.; Fallas-López, B.; Hernández-Espinoza, K.; Stoffel, M.; Ballesteros-Cánovas, J.A. Relationships between earthquakes, hurricanes, and landslides in Costa Rica. *Landslides* **2019**, *16*, 1539–1550. [CrossRef]
11. Tanyaş, H.; van Westen, C.J.; Allstadt, K.E.; Jessee, M.A.N.; Görüm, T.; Jibson, R.W.; Godt, J.W.; Sato, H.P.; Schmitt, R.G.; Marc, O.; et al. Presentation and Analysis of a Worldwide Database of Earthquake-Induced Landslide Inventories. *J. Geophys. Res. Earth Surf.* **2017**, *122*, 1991–2015. [CrossRef]
12. Calista, M.; Miccadei, E.; Piacentini, T.; Sciarra, N. Morphostructural, Meteorological and Seismic Factors Controlling Landslides in Weak Rocks: The Case Studies of Castelnuovo and Ponzano (North East Abruzzo, Central Italy). *Geosciences* **2019**, *9*, 122. [CrossRef]
13. Nadim, F.; Kjekstad, O.; Peduzzi, P.; Herold, C.; Jaedicke, C. Global landslide and avalanche hotspots. *Landslides* **2006**, *3*, 159–173. [CrossRef]
14. Rahmati, O.; Ghorbanzadeh, O.; Teimurian, T.; Mohammadi, F.; Tiefenbacher, J.P.; Falah, F.; Pirasteh, S.; Ngo, P.T.T.; Bui, D.T. Spatial modeling of snow avalanche using machine learning models and geo-environmental factors: Comparison of effectiveness in two mountain regions. *Remote Sens.* **2019**, *11*, 2995. [CrossRef]
15. CRED EM-DAT. The International Disaster Database. Available online: https://www.emdat.be (accessed on 14 February 2021).
16. Statham, G.; Haegeli, P.; Greene, E.; Birkeland, K.; Israelson, C.; Tremper, B.; Stethem, C.; McMahon, B.; White, B.; Kelly, J. A conceptual model of avalanche hazard. *Nat. Hazards* **2018**, *90*, 663–691. [CrossRef]
17. Voiculescu, M.; Ardelean, F.; Török-Oance, M.; Milian, N. Topographical factors, meteorological variables and human factors in the control of the main snow avalanche events in the fĂgĂraŞ massif (Southern carpathians—Romanian Carpathians): Case studies. *Geogr. Pol.* **2016**, *8*, 47–64. [CrossRef]
18. Navarre, J.P. The principles of snow mechanics: The mechanical properties of snow. In Proceedings of the Université Européenne d'été sur les Risques Naturels: Neige et Avalanches; Brugnot, G., Ed.; Cemagref Publications: Chamonix, France, 1992; pp. 75–86.
19. Fort, M.; Cossart, E.; Deline, P.; Dzikowski, M.; Nicoud, G.; Ravanel, L.; Schoeneich, P.; Wassmer, P. Geomorphic impacts of large and rapid mass movements: A review. *Geomorphol. Reli. Process. Environ.* **2009**, *15*, 47–64. [CrossRef]
20. Ghinoi, A.; Chung, C.J. STARTER: A statistical GIS-based model for the prediction of snow avalanche susceptibility using terrain features—Application to Alta Val Badia, Italian Dolomites. *Geomorphology* **2005**, *66*, 305–325. [CrossRef]
21. Borrel, G. La Carte de Localisation Probable des Avalanches. *Mappemonde* **1994**, *94*, 17–19.
22. Gruber, U.; Bartelt, P.; Haefher, H. Avalanche hazard mapping using numerical Voellmy-fluid models. *Publ. Norges Geotek. Inst.* **1998**, *203*, 117–121.
23. Sauermoser, S. Aavalanche hazard mapping—30 years' experience in Austria. In Proceedings of the 2006 International 901 Snow Science Workshop, Telluride, CO, USA, 1–6 October 2006; pp. 314–321.
24. Bründl, M.; Margreth, S. Integrative Risk Management: The Example of Snow Avalanches. In *Snow and Ice-Related Hazards, Risks, and Disasters*; Haeberli, W., Whiteman, C., Shroder, J.F., Eds.; Elsevier Inc.: Oxford, UK, 2015; pp. 263–301; ISBN 9780123964731.
25. Sinickas, A.; Jamieson, B.; Maes, M.A. Snow avalanches in western Canada: Investigating change in occurrence rates and implications for risk assessment and mitigation. *Struct. Infrastruct. Eng.* **2016**, *12*, 490–498. [CrossRef]

26. Mellor, M. Controlled release of avalanches by explosives. In *Advances in North American Avalanche Technology: 1972 Symposium*; Perla, R., Ed.; USDA Forest Service: Washington DC, USA, 1973; pp. 37–49.
27. Gubler, H. Artificial Release of Avalanches by Explosives. *J. Glaciol.* **1977**, *19*, 419–429. [CrossRef]
28. Binger, C.; Nelsen, J.; Olson, K.A. Explosive shock wave compression in snow: Effects of explosive orientation and snowpack compression. In Proceedings of the ISSW 2006. International Snow Science Workshop, Telluride, CO, USA, 1–6 October 2006; Gleason, J.A., Ed.; Montana State University Library: Bozeman, MT, USA, 2006; pp. 592–597.
29. Frigo, B.; Chiaia, B.; Cardu, M. Snowpack effects induced by blasts: Experimental measurements vs theoretical formulas. In Proceedings of the ISSW 2012. International Snow Science Workshop, Anchorage, AK, USA, 16–21 September 2012; Montana State University Library: Bozeman, MT, USA, 2012; pp. 943–947.
30. Johnson, J.B.; Solie, D.J.; Barrett, S.A. The response of a seasonal snow cover to explosive loading. *Ann. Glaciol.* **1994**, *19*, 49–54. [CrossRef]
31. Barbolini, M.; Cordola, M.; Natale, L.; Tecilla, G. Hazard mapping and land use regulation in avalanche prone areas: Recent developments in Italy. In *Recommendations to Deal with Snow Avalanches in Europe*; Hervas, J., Ed.; ISPRA: Rome, Italy, 2003.
32. Gubler, H.; Wyssen, S.; Kogelnig, A. *Guidelines for Artificial Release of Avalanches*; Wyssen Avalanche Control AG: Reichenbach, Germany, 2012; p. 48.
33. Canadian Avalanche Association (CAA). *Technical Aspects of Snow Avalanche Risk Management—Resources and Guidelines for Avalanche Practitioners in Canada*; Campbell, C., Conger, S., Gould, B., Haegeli, P., Jamieson, B., Statham, G., Eds.; Canadian Avalanche Association: Revelstoke, BC, Canada, 2016; p. 126.
34. Choubin, B.; Borji, M.; Hosseini, F.S.; Mosavi, A.; Dineva, A.A. Mass wasting susceptibility assessment of snow avalanches using machine learning models. *Sci. Rep.* **2020**, *10*, 18363. [CrossRef] [PubMed]
35. Jamieson, B.; Margreth, S.; Jones, A. Application and Limitations of Dynamic Models for Snow Avalanche Hazard Mapping. In Proceedings of the International snow science workshop proceedings 2008, Whistler, BC, Canada, 22 September 2008; pp. 730–739.
36. McClung, D.M.; Lied, K. Statistical and geometrical definition of snow avalanche runout. *Cold Reg. Sci. Technol.* **1987**, *13*, 107–119. [CrossRef]
37. Voiculescu, M. Patterns of the dynamics of human-triggered snow avalanches at the Făgăras massif (Southern Carpathians), Romanian Carpathians. *Area* **2014**, *46*, 328–336. [CrossRef]
38. Sharma, S.; Ganju, A. Complexities of avalanche forecasting in western Himalayas—An overview. *Cold Reg. Sci. Technol.* **2000**, *31*, 95–102. [CrossRef]
39. Gruber, U.; Bartelt, P. Snow avalanche hazard modelling of large areas using shallow water numerical methods and GIS. *Environ. Model. Softw.* **2007**, *22*, 1472–1481. [CrossRef]
40. Valero, C.V.; Wever, N.; Bühler, Y.; Stoffel, L.; Margreth, S.; Bartelt, P. Modelling wet snow avalanche runout to assess road safety at a high-Altitude mine in the central Andes. *Nat. Hazards Earth Syst. Sci.* **2016**, *16*, 2303–2323. [CrossRef]
41. Choubin, B.; Borji, M.; Mosavi, A.; Sajedi-Hosseini, F.; Singh, V.P.; Shamshirband, S. Snow avalanche hazard prediction using machine learning methods. *J. Hydrol.* **2019**, *577*, 123929. [CrossRef]
42. Gaume, J.; van Herwijnen, A.; Gast, T.; Teran, J.; Jiang, C. Investigating the release and flow of snow avalanches at the slope-scale using a unified model based on the material point method. *Cold Reg. Sci. Technol.* **2019**, *168*, 102847. [CrossRef]
43. Bühler, Y.; Kumar, S.; Veitinger, J.; Christen, M.; Stoffel, A. Snehmani Automated identification of potential snow avalanche release areas based on digital elevation models. *Nat. Hazards Earth Syst. Sci.* **2013**, *13*, 1321–1335. [CrossRef]
44. Pecci, M.; D'Aquila, P. Zonazione delle aree valanghive a partire dalla suscettibilità al distacco di valanghe. *Neve e Valanghe* **2010**, *69*, 36–47.
45. Yariyan, P.; Avand, M.; Abbaspour, R.A.; Karami, M.; Tiefenbacher, J.P. GIS-based spatial modeling of snow avalanches using four novel ensemble models. *Sci. Total Environ.* **2020**, *745*, 141008. [CrossRef]
46. Romeo, V.; Fazzini, M. La neve in Appennino. Prime analisi su 30 anni di dati nivometeorologici. *Neve e Valanghe* **2008**, *60*, 58–67.
47. D'Alessandro, L.; De Sisti, G.; D'Orefice, M.; Pecci, M.; Ventura, M. Geomorphology of the summit area of the Gran Sasso d'Italia. *Geogr. Fis. Din. Quat.* **2003**, *26*, 126–141.
48. Miccadei, E.; Piacentini, T.; Buccolini, M. Long-term geomorphological evolution in the Abruzzo area, Central Italy: Twenty years of research. *Geol. Carpathica* **2017**, *68*, 19–28. [CrossRef]
49. Piacentini, T.; Miccadei, E. The role of drainage systems and intermontane basins in the Quaternary landscape of the Central Apennines chain (Italy). *Rend. Lincei* **2014**, *25*, 139–150. [CrossRef]
50. Cavinato, G.P.; Miccadei, E. Sintesi preliminare delle caratteristiche tettoniche e sedimentarie dei depositi quaternari della Conca di Sulmona (L'Aquila). *Alp. Mediterr. Quat.* **1995**, *8*, 129–140.
51. Speranza, F.; Adamoli, L.; Maniscalco, R.; Florindo, F. Genesis and evolution of a curved mountain front: Paleomagnetic and geological evidence from the Gran Sasso range (Central Apennines, Italy). *Tectonophysics* **2003**, *362*, 183–197. [CrossRef]
52. Calamita, F.; Satolli, S.; Scisciani, V.; Esestime, P.; Pace, P. Contrasting styles of fault reactivation in curved orogenic belts: Examples from the central Apennines (Italy). *Bull. Geol. Soc. Am.* **2011**, *123*, 1097–1111. [CrossRef]
53. Vezzani, L.; Festa, A.; Ghisetti, F.C. Geology and tectonic evolution of the Central-Southern Apennines, Italy. *Spec. Pap. Geol. Soc. Am.* **2010**, *469*, 1–58. [CrossRef]
54. Cardello, G.L.; Doglioni, C. From Mesozoic rifting to Apennine orogeny: The Gran Sasso range (Italy). *Gondwana Res.* **2015**, *27*, 1307–1334. [CrossRef]

55. Pace, P.; Di Domenica, A.; Calamita, F. Summit low-angle faults in the Central Apennines of Italy: Younger-on-older thrusts or rotated normal faults? Constraints for defining the tectonic style of thrust belts. *Tectonics* **2014**, *33*, 756–785. [CrossRef]
56. Adamoli, L.; Bertini, T.; Chiocchini, M.; Deiana, G.; Mancinelli, A.; Pieruccini, U.; Romano, A. Ricerche geologiche sul Mesozoico del Gran Sasso d'Italia (Abruzzo). II. Evoluzione tettonico-sedimentaria dal Trias superiore al Cretaceo inferiore dell'area compresa tra il Corno Grande e S. Stefano di Sessanio (F. 140 Teramo). *Stud. Geol. Camerti* **1978**, *4*, 7–17.
57. Adamoli, L.; Bertini, T.; Deiana, G.; Pieruccini, U.; Romano, A. Ricerche geologiche sul Gran Sasso d'Italia (Abruzzo). VI. Primi risultati dello studio strutturale della catena del Gran Sasso d'Italia. *Stud. Geol. Camerti* **1982**, *7*, 97–103.
58. Adamoli, L. Evidenze di tettonica di inversione nell'area Corno Grande—Corno Piccolo (Gran Sasso d'Italia). *Boll. Della Soc. Geol. Ital.* **1992**, *111*, 53–66.
59. Carabella, C.; Buccolini, M.; Galli, L.; Miccadei, E.; Paglia, G.; Piacentini, T. Geomorphological analysis of drainage changes in the NE Apennines piedmont area: The case of the middle Tavo River bend (Abruzzo, Central Italy). *J. Maps* **2020**, *16*, 222–235. [CrossRef]
60. Rovida, A.; Locati, M.; Camassi, R.; Lolli, B.; Gasperini, P.; Antonucci, A. *The Italian Earthquake Catalogue CPTI15—Version 3.0*; Istituto Nazionale di Geofisica e Vulcanologia (INGV): Rome, Italy, 2021. [CrossRef]
61. Geurts, A.H.; Whittaker, A.C.; Gawthorpe, R.L.; Cowie, P.A. Transient landscape and stratigraphic responses to drainage integration in the actively extending central Italian Apennines. *Geomorphology* **2020**, *353*, 107013. [CrossRef]
62. Ciccacci, S.; D'Alessandro, L.; Dramis, F.; Miccadei, E. Geomorphologic evolution and neotectonics of the Sulmona intramontane basin (Abruzzi Apennine, Central Italy). *Z. Fur. Geomorphol. Suppl.* **1999**, *118*, 27–40.
63. Giraudi, C. I rock glacier tardo-pleistocenici e olocenici dell'Appennino—Età, distribuzione, significato paleoclimatico. *Quat. Ital. J. Quat. Sci.* **2002**, *15*, 45–52.
64. Pecci, M.; D'Agata, C.; Smiraglia, C. Ghiacciaio del calderone (Apennines, Italy): The mass balance of a shrinking mediterranean glacier. *Geogr. Fis. Din. Quat.* **2008**, *31*, 55–62.
65. Pecci, M.; D'Aquila, P. Geomorphological features and cartography of the Gran Sasso d'Italia massif between Corno Grande-Corno Piccolo and Pizzo Intermesoli. *Geogr. Fis. Din. Quat.* **2011**, *34*, 127–143.
66. Bianchi-Fasani, G.; Esposito, C.; Lenti, L.; Martino, S.; Pecci, M.; Scarascia-Mugnozza, G. Seismic analysis of the Gran Sasso catastrophic rockfall (Central Italy). In Proceedings of the Landslide Science and Practice: Risk Assessment, Management and Mitigation; Mrgottini, C., Ed.; Springer: Berlin, Germany, 2013; pp. 263–267.
67. Gruppo di lavoro. *Gruppo di Lavoro CPTI Catalogo Parametrico dei Terremoti Italiani, 2004 (CPTI04)*; Editrice Compositori: Bologna, Italy, 2004.
68. Chiarabba, C.; Jovane, L.; DiStefano, R. A new view of Italian seismicity using 20 years of instrumental recordings. *Tectonophysics* **2005**, *395*, 251–268. [CrossRef]
69. ISIDe Working Group. *Italian Seismological Instrumental and Parametric Database (ISIDe)*; Istituto Nazionale di Geofisica e Vulcanologia (INGV): Rome, Italy, 2007. [CrossRef]
70. Peel, M.C.; Finlayson, B.L.; McMahon, T.A. Updated world map of the Köppen-Geiger climate classification. Spatial Data Access Tool (SDAT)OGC Standards-based Geospatial Data Visualization/Download. *Hydrol. Earth Syst.* **2007**, *11*, 1633–1644. [CrossRef]
71. Di Lena, B.; Antenucci, F.; Mariani, L. Space and time evolution of the Abruzzo precipitation. *Ital. J. Agrometeorol.* **2012**, *1*, 5–20.
72. Miccadei, E.; Piacentini, T.; Daverio, F.; Di, R. Geomorphological Instability Triggered by Heavy Rainfall: Examples in the Abruzzi Region (Central Italy). In *Studies on Environmental and Applied Geomorphology*; Piacentini, T., Miccadei, E., Eds.; IntechOpen: London, UK, 2012; pp. 45–62.
73. Chiaudani, A.; Antenucci, F.; Di Lena, B. Historical analysis of maximum intensity precipitation(1, 3, 6, 12 hours) in Abruzzo Region (Italy)—Period 1951–2012. In Proceedings of the Agrometeorologia per la Sicurezza Ambientale ed Alimentare; 2013. Available online: http://agrometeorologia.it/documenti/AIAM2013/113-114_Chiaudani.pdf (accessed on 24 October 2020).
74. Piacentini, T.; Galli, A.; Marsala, V.; Miccadei, E. Analysis of soil erosion induced by heavy rainfall: A case study from the NE Abruzzo Hills Area in Central Italy. *Water* **2018**, *10*, 1314. [CrossRef]
75. Piacentini, T.; Calista, M.; Crescenti, U.; Miccadei, E.; Sciarra, N. Seismically induced snow avalanches: The central Italy case. *Front. Earth Sci.* **2020**, *8*, 507. [CrossRef]
76. Smiraglia, C.; Azzoni, R.S.; D'agata, C.; Maragno, D.; Fugazza, D.; Diolaiuti, G.A. The evolution of the Italian glaciers from the previous data base to the new Italian inventory. Preliminary considerations and results. *Geogr. Fis. Din. Quat.* **2015**, *38*, 79–87. [CrossRef]
77. Vergni, L.; Todisco, F.; Di Lena, B.; Mannocchi, F. Effect of the North Atlantic Oscillation on winter daily rainfall and runoff in the Abruzzo region (Central Italy). *Stoch. Environ. Res. Risk Assess.* **2016**, *30*, 1901–1915. [CrossRef]
78. Vergni, L.; Di Lena, B.; Todisco, F.; Mannocchi, F. Uncertainty in drought monitoring by the Standardized Precipitation Index: The case study of the Abruzzo region (central Italy). *Theor. Appl. Climatol.* **2017**, *128*, 13–26. [CrossRef]
79. Di Lena, B.; Curci, G.; Vergni, L. Analysis of rainfall erosivity trends 1980–2018 in a complex terrain region (Abruzzo, central italy) from rain gauges and gridded datasets. *Atmosphere* **2021**, *12*, 657. [CrossRef]
80. Fazzini, M.; Cardillo, A.; Di Fiore, T.; Lucentini, L.; Scozzafava, M. Extreme temperatures in the cold air pool of the central Apennines (Italy): Comparison with those of the Veneto Pre-Alps during winter 2016–2017. In Proceedings of the 34th International Conference on Alpine Meteorology, Reykjavík, Iceland, 19–23 June 2017; pp. 42–47.

81. Fazzini, M.; Giuffrida, A. Une nouvelle proposition quantitative des régimes pluviométriques dans le territoire de Italie: Premiers résultats. In Proceedings of the Climat Urbain, Ville et Architecture—Actes XVIII Colloque Internationale de Climatologie, Genova, Italy, 7–11 September 2005; pp. 361–364.
82. Fazzini, M.; Magagnini, L.; Giuffrida, A.; Frustaci, G.; Di Lisciandro, M.; Gaddo, M. Nevosità in Italia negli ultimi 20 anni. *Neve Valanghe* **2006**, *58*, 24–35.
83. Strahler, A.N. Dynamic basis of geomorphology. *Bull. Geol. Soc. Am.* **1952**, *63*, 923–938. [CrossRef]
84. Ahnert, F. Local relief and the height limits of mountain ranges. *Am. J. Sci.* **1984**, *284*, 1035–1055. [CrossRef]
85. Schweizer, J.; Jamieson, J.B.; Schneebeli, M. Snow avalanche formation. *Rev. Geophys.* **2003**, *41*, 1–25. [CrossRef]
86. Laute, K.; Beylich, A.A. Morphometric and meteorological controls on recent snow avalanche distribution and activity at hillslopes in steep mountain valleys in western Norway. *Geomorphology* **2014**, *218*, 16–34. [CrossRef]
87. Romshoo, S.A.; Bhat, S.A.; Rashid, I. Geoinformatics for assessing the morphometric control on hydrological response at watershed scale in the upper Indus Basin. *J. Earth Syst. Sci.* **2012**, *121*, 659–686. [CrossRef]
88. ISPRA. Geological Map of Italy, Scale 1:50,000, Sheet 349 "Gran Sasso d'Italia". Available online: https://www.isprambiente.gov.it/Media/carg/349_GRANSASSO/Foglio.html (accessed on 28 March 2021).
89. Abruzzo-Sangro Basin Authority. *Geomorphological Map, Scale 1:25,000. Piano Stralcio di Bacino per l'Assetto 1035 Idrogeologico dei Bacini di Rilievo Regionale Abruzzesi e del Bacino del Fiume Sangro.* (L.R. 18.05 1989 n.81 e L. 24.08.2001); Abruzzo Region: L'Aquila, Italy, 2005.
90. ISPRA IFFI Project—Italian Landslide Inventory. Available online: https://idrogeo.isprambiente.it/app/iffi?@=41.55172525894153,12.57350148381829,1 (accessed on 24 October 2020).
91. ISPRA. AIGEO Aggiornamento ed Integrazione delle Linee Guida della Carta Geomorfologica D'italia in Scala 1:50.000. In *Quaderni Serie III*; Servizio Geologico d'Italia: Rome, Italy, 2018.
92. Smith, M.J.; Paron, P.; Griffiths, J. *Geomorphological Mapping, Methods and Applications*; Elsevier Science: Oxford, UK, 2011; ISBN 9780444534460.
93. Miccadei, E.; Mascioli, F.; Ricci, F.; Piacentini, T. Geomorphology of soft clastic rock coasts in the mid-western Adriatic Sea (Abruzzo, Italy). *Geomorphology* **2019**, *324*, 72–94. [CrossRef]
94. Pasculli, A.; Palermi, S.; Sarra, A.; Piacentini, T.; Miccadei, E. A modelling methodology for the analysis of radon potential based on environmental geology and geographically weighted regression. *Environ. Model. Softw.* **2014**, *54*, 165–181. [CrossRef]
95. Carabella, C.; Miccadei, E.; Paglia, G.; Sciarra, N. Post-Wildfire Landslide Hazard Assessment: The Case of The 2017 Montagna Del Morrone Fire (Central Apennines, Italy). *Geosciences* **2019**, *9*, 175. [CrossRef]
96. Patton, A.I.; Rathburn, S.L.; Capps, D.M. Landslide response to climate change in permafrost regions. *Geomorphology* **2019**, *340*, 116–128. [CrossRef]
97. Gustavsson, M.; Kolstrup, E.; Seijmonsbergen, A.C. A new symbol-and-GIS based detailed geomorphological mapping system: Renewal of a scientific discipline for understanding landscape development. *Geomorphology* **2006**, *77*, 90–111. [CrossRef]
98. World Meteorological Organization (WMO). *Guide to the Implementation of Education and Training Standards in Meteorology and Hydrology*; WMO: Geneve, Switzerland, 2015.
99. Altevie. *Avalanche Risk Defense Zone of Vena Rossa—Gran Sasso d'Italia*; Altevie: L'Aquila, Italy, 2020.
100. Fazzini, M.; Bisci, C.; De Luca, E. Clima e neve sul massiccio del Gran Sasso. *Neve Valanghe* **1999**, *36*, 36–45.
101. Swiss Federal Institute for Snow and Avalanche Research (SLF). *Direttive per la Considerazione del Pericolo di Valanghe Nelle Attività di Incidenza Territoriale*; Swiss Federal Institute for Snow and Avalanche Research SLF: Davos, Switzerland, 1984; p. 22.
102. Salm, B.; Burkard, A.; Gubler, H.U. *Berechnung von Fliesslawinen; eine Anleitung für Praktiker mit Beispielen*; Mitteilungen des Eidgenössischen Institutes für Schnee und Lawinenforschung: Davos, Switzerland, 1990; p. 37.
103. Percitti, G. Avalanche study in Italy. In Proceedings of the European Summer University on Snow and Avalanches, Chamonix, France, 14–25 September 1992; Cemagref Publications: Beaucouze, France, 1992.
104. Chrustek, P.; Kolecka, N.; Bühler, Y. Snow avalanches mapping—Evaluation of a new approach. In Proceedings of the International Snow Science Workshop, Grenoble, France, 7–11 October 2013; pp. 750–755.
105. Brandolini, P.; Faccini, F.; Fratianni, S.; Freppaz, M.; Giardino, M.; Maggioni, M.; Perotti, L.; Romeo, V. Snow-avalanche and climatic conditions in the Ligurian ski resorts (NW-Italy). *Geogr. Fis. Din. Quat.* **2017**, *40*, 41–52. [CrossRef]
106. Barbolini, M.; Pagliardi, M. Analisi costi-benefici applicata alla gestione del problema valanghe: Applicazione ad un caso di studio in Alta Valbrembana (BG). In Proceedings of the Internation Symposion Interpraevent, Riva del Garda, Italy, 24–27 May 2004; pp. 13–24.
107. Cappabianca, F.; Barbolini, M.; Natale, L. Snow avalanche risk assessment and mapping: A new method based on a combination of statistical analysis, avalanche dynamics simulation and empirically-based vulnerability relations integrated in a GIS platform. *Cold Reg. Sci. Technol.* **2008**, *54*, 193–205. [CrossRef]
108. Maggioni, M.; Gruber, U.; Purves, R.S.; Freppaz, M. Potential release areas and return period of avalanches: Is there a relation? In Proceedings of the International Snow Science Workshop, Telluride, CO, USA, 1–6 October 2006; pp. 566–571.
109. Bruno, E.; Maggioni, M.; Freppaz, M.; Zanini, E. *Distacco Artificiale di Valanghe: Linee Guida per la Procedura Operativa Metodi e Normativa*; Regione Autonoma Valle d'Aosta—Région Autonome Vallée d'Aoste: Aosta, Italy, 2012; p. 130.

110. Boccardo, P.; Fissore, V.; Morreale, S.; Ilardi, E.; Baldo, M. Aerial Lidar technology in support to avalanches prevention and risk mitigation: An operative application at "Colle della Maddalena" (Italy). *ISPRS Ann. Photogramm. Remote Sens. Spat. Inf. Sci.* **2020**, *VI-3/W1-20*, 11–17. [CrossRef]
111. Vagliasindi, M.; Theodule, A.; Maggioni, M.; Levera, E. Artificial avalanche release as a protection measure for major roads: The case study of road S.S. 21 "Colle della Maddalena" (CN, Western Italian Alps). In Proceedings of the International Snow Science Workshop, Telluride, CO, USA, 7–11 October 2013; pp. 875–882.
112. Christen, M.; Bartelt, P.; Gruber, U. *Numerical Calculation of Dense Flow and Powder Snow Avalanches*; Swiss Federal Institute for Snow and Avalanche Research (SLF): Davos, Switzerland, 2010; p. 136.
113. Margreth, S. *Costruzione di Opere di Premunizione Contro le Valanghe Nella Zona di Distacco. Direttiva Tecnica: Aiuto all'Esecuzione*; Ufficio federale dell'ambiente (WSL); Istituto Federale per lo Studio della Neve e delle Valanghe (SNV): Davos, Switzerland, 2007; p. 139.
114. Oller, P.; Janeras, M.; de Buen, H.; Arnó, G.; Christen, M.; García, C.; Martínez, P. Using AVAL-1D to simulate avalanches in the eastern Pyrenees. *Cold Reg. Sci. Technol.* **2010**, *64*, 190–198. [CrossRef]
115. Bartelt, P.; Bühler, Y.; Christen, M.; Deubelbeiss, Y.; Salz, M.; Schneider, M.; Schumacher, L. *RAMMS. Avalanche Numerical Model for Snow Avalanches in Research and Practice. User Manual*; Swiss Federal Institute for Snow and Avalanche Research (SLF): Davos, Switzerland, 2017; p. 104.
116. Fischer, J.T.; Kowalski, J.; Pudasaini, S.P. Topographic curvature effects in applied avalanche modeling. *Cold Reg. Sci. Technol.* **2012**, *74–75*, 21–30. [CrossRef]
117. Christen, M.; Bartelt, P.; Gruber, U. AVAL-1D: An avalanche dynamics program for the practice. In Proceedings of the Protection of Habitat against Floods, Debris Flows and Avalanches, Matsumoto, Japan, 14–18 October 2002; pp. 715–725.
118. Christen, M.; Kowalski, J.; Bartelt, P. RAMMS: Numerical simulation of dense snow avalanches in three-dimensional terrain. *Cold Reg. Sci. Technol.* **2010**, *63*, 1–14. [CrossRef]
119. Bisci, C.; Fazzini, M.; Romeo, V.; Cardillo, A. Intense snowfalls of January 2017 along the central-southern Apennines (Italy), in comparisons with the 2015, 2012 and 2005 events. In Proceedings of the 34th International Conference on Alpine Meteorology, Reykjavík, Iceland, 18–23 June 2017; pp. 48–50.
120. Dramis, F.; Fazzini, M.; Pecci, M.; Smiraglia, C. The effects of Global Warming onto the Meditearranean high altitudes: The naturaly laboratory of Calderone Glacier (Central Apennines Italy). In Proceedings of the 32th International Geological Congress (IGC), Florence, Italy, 20–28 August 2004; Abbate, E., Ed.; IUGS, International Union of Geological Sciences: Paris, France; EC, European Commission: Brussels Belgium; pp. 111–117.
121. Bisci, C.; Fazzini, M. *Studio Idraulico-Ambientale Mediante L'analisi dei Processi Geomorfologici in Atto per la Caratterizzazione dei Bacini Idrografici Principali della Regione Marche—Analisi Climatologica*; Consorzio di Bonifica delle Marche: Ancona, Italy, 2019; p. 34.
122. Fazzini, M.; Romeo, V. L'enneigement dans les Apennins durant les derniers 30 ans. In Proceedings of the Actes XXIV Colloque AIC "Climat montagnard et risques"; Stamperia Romana: Roma, Italy, 2011; pp. 249–254.
123. D'Alessandro, L.; Pecci, M. Valanghe sul Gran Sasso d'Italia: Nota preliminare. *Mem. della Soc. Geol. Ital.* **2001**, *56*, 315–320.
124. De Sisti, G.; Monopoli, S.; Pecci, M. Valanghe sul Gran Sasso d'Italia: Analisi delle condizioni meteoclimatiche e implicazioni dell'assetto geomorfologico con particolare riferimento all'attività valanghiva dell'inverno 2002–2003. *Neve Valanghe* **2004**, *52*, 20–33.
125. Barbolini, M.; Pagliardi, M.; Ferro, F.; Corradeghini, P. Avalanche hazard mapping over large undocumented areas. *Nat. Hazards* **2011**, *56*, 451–464. [CrossRef]
126. Issler, D. The 2017 Rigopiano avalanche—dynamics inferred from field observations. *Geosciences* **2020**, *10*, 446. [CrossRef]
127. Braun, T.; Frigo, B.; Chiaia, B.; Bartelt, P.; Famiani, D.; Wassermann, J. Seismic signature of the deadly snow avalanche of January 18, 2017, at Rigopiano (Italy). *Sci. Rep.* **2020**, *10*, 18563. [CrossRef]
128. Ballesteros-Cánovas, J.A.; Trappmann, D.; Madrigal-González, J.; Eckert, N.; Stoffel, M. Climate warming enhances snow avalanche risk in the Western Himalayas. *Proc. Natl. Acad. Sci. USA* **2018**, *115*, 3410–3415. [CrossRef]
129. Strapazzon, G.; Schweizer, J.; Chiambretti, I.; Brodmann Maeder, M.; Brugger, H.; Zafren, K. Effects of Climate Change on Avalanche Accidents and Survival. *Front. Physiol.* **2021**, *12*, 450. [CrossRef]
130. Martin, E.; Giraud, G.; Lejeune, Y.; Boudart, G. Impact of a climate change on avalanche hazard. *Ann. Glaciol.* **2001**, *32*, 163–167. [CrossRef]
131. Castebrunet, H.; Eckert, N.; Giraud, G.; Durand, Y.; Morin, S. Projected changes of snow conditions and avalanche activity in a warming climate: The French Alps over the 2020–2050 and 2070-2100 periods. *Cryosphere* **2014**, *8*, 1673–1697. [CrossRef]
132. Komarov, A.; Seliverstov, Y.; Sokratov, S.; Glazovskaya, T.; Turchaniniva, A. Avalanche risk assessment in Russia. In Proceedings of the Geophysical Research Abstracts, EGU General Assembly, Vienna, Austria, 23–28 April 2017; p. 1.
133. Hovelsrud, G.K.; Karlsson, M.; Olsen, J. Prepared and flexible: Local adaptation strategies for avalanche risk. *Cogent Soc. Sci.* **2018**, *4*, 1460899. [CrossRef]
134. Yount, J.M.; Gorsage, B.R. Evolution of an avalanche program: From artillery to infrastructure. In Proceedings of the International Snow Science Workshop; 2016; pp. 442–449. Available online: https://www.slf.ch/fileadmin/user_upload/WSL/Mitarbeitende/schweizj/vanHerwijnen_etal_PTV_ISSW2016.pdf (accessed on 24 October 2020).
135. Luckman, B.H. The Geomorphic Activity of Snow Avalanches. *Geogr. Ann. Ser. A Phys. Geogr.* **1977**, *59*, 31–48. [CrossRef]
136. Freppaz, M.; Godone, D.; Filippa, G.; Maggioni, M.; Lunardi, S.; Williams, M.W.; Zanini, E. Soil erosion caused by snow avalanches: A case study in the Aosta Valley (NW Italy). *Arctic Antarct. Alp. Res.* **2010**, *42*, 412–421. [CrossRef]

Article

Evolution of Deep-Seated Gravitational Slope Deformations in Relation with Uplift and Fluvial Capture Processes in Central Eastern Sardinia (Italy)

Valentino Demurtas *, Paolo Emanuele Orrù and Giacomo Deiana

Department of Chemical and Geological Sciences, University of Cagliari, 09042 Monserrato, Italy; orrup@unica.it (P.E.O.); giacomo.deiana@unica.it (G.D.)
* Correspondence: valentino.demurtas@unica.it

Abstract: Connections between Plio-Pleistocenic tectonic activity and geomorphological evolution were studied in the Pardu Valley and Quirra Valley (Ogliastra, East Sardinia). The intensive Quaternary tectonic activity in Sardinia linked to the opening of the Tyrrhenian Basin is known. In Eastern Sardinia, it manifests with an uplift that is recorded by geomorphological indicators, such as deep-seated gravitational slope deformation, fluvial captures, engraved valleys, waterfalls, and heterogeneous water drainage. The Pardu River flows from the NW toward the SE and then abruptly changes direction toward the NE. At this point, a capture elbow adjacent to the current head of the Quirra River is well developed. The Quirra River, in its upstream part, flows at altitudes approximately 200 m higher than the Pardu River. It also shows an oversized and over-flooded valley with respect to the catchment area upstream. This setting indicates that the Pardu River, which previously flowed south along the Quirra River, was captured by the Pelau River. We analyzed long-term landslides with lateral spreading and sackung characteristics, which involve giant carbonate blocks and underlying foliated metamorphites in both valleys. The use of LiDAR, high-resolution uncrewed aerial vehicle digital photogrammetry (UAV-DP), and geological, structural, and geomorphological surveys enabled a depth morphometric analysis and the creation of interpretative 3D models of DGSDs. Space-borne interferometric synthetic aperture radar (InSAR) data using ERS and Sentinel-1 satellites identified downslope movement of up to 20 mm per year in both Pardu Valley flanks. Multi-source and multi-scale data showed that the state of activity of the DGSDs is closely linked to the geomorphological evolution of the catchment areas of the Rio Pardu and Rio Quirra. The intense post-capture erosion acted in the Rio Pardu Valley, giving it morphometric characteristics that were favorable to the current evolution of the DGSDs, while the Rio Quirra Valley presents paleo-DGSDs that have been fossilized by pre-capture terraced alluvial deposits.

Keywords: morphotectonic; morphostratigraphy; DGSDs; river capture; fluvial terraces; Sardinia; Italy

Citation: Demurtas, V.; Orrù, P.E.; Deiana, G. Evolution of Deep-Seated Gravitational Slope Deformations in Relation with Uplift and Fluvial Capture Processes in Central Eastern Sardinia (Italy). *Land* **2021**, *10*, 1193. https://doi.org/10.3390/land10111193

Academic Editor: Giulio Iovine

Received: 15 October 2021
Accepted: 3 November 2021
Published: 5 November 2021

Publisher's Note: MDPI stays neutral with regard to jurisdictional claims in published maps and institutional affiliations.

Copyright: © 2021 by the authors. Licensee MDPI, Basel, Switzerland. This article is an open access article distributed under the terms and conditions of the Creative Commons Attribution (CC BY) license (https:// creativecommons.org/licenses/by/ 4.0/).

1. Introduction

The Pliocene and Quaternary geodynamic processes related to the Tyrrhenian basin opening led an uplift in Sardinia [1–3]. This is evidenced by a morphotectonic setting linked to fluvial and gravitative morphologies [4,5]. Therefore, the hydrographic basins of the Rio Quirra and the Rio Pardu have been studied in detail in order to analyze their evolutionary scenarios in relation to a river capture.

Rio Pardu and Rio Quirra are two of the most important rivers in central eastern Sardinia. The two basins are separated by a river capture caused by the Rio Pelau, which isolated Rio Pardu, the catchment area of Rio Quirra. Rio Pardu flows from northwest to southeast and then flows towards the northeast through a river capture elbow with the name of Rio Pelau. Rio Quirra flows from the north to the south parallel to the coast and then abruptly bends towards the Tyrrhenian Sea near the mouth. The flow directions are closely related to the structural conditioning of the main alpine structural setting and are linked to the opening of the Tyrrhenian basin [6,7] (Figure 1).

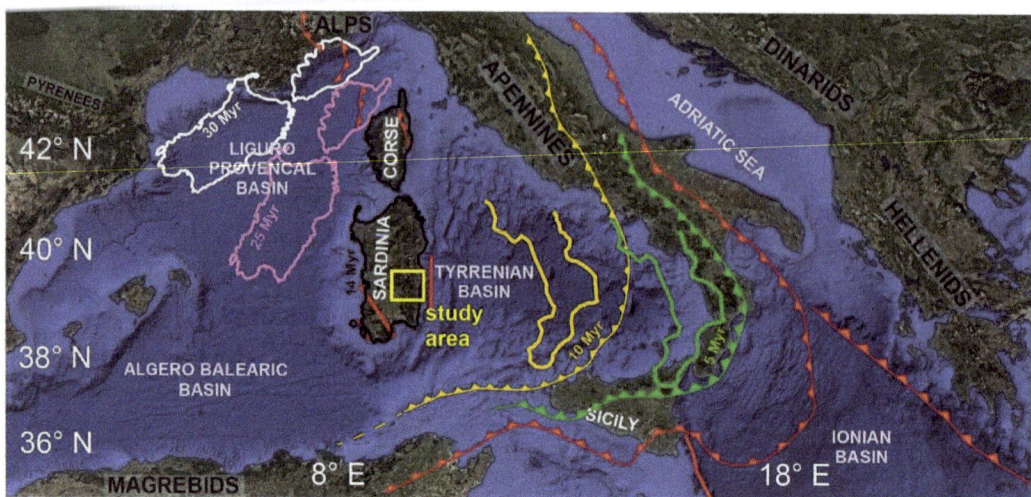

Figure 1. Geographical location and structural features of the study area, modified after [8]; red lines represent thrust fronts; white lines are the Sardinian–Corse Block translation at 30 Ma; the pink line represents the Sardinian–Corse Block translation at 25 Ma; the yellow line represents the Calabrian block translation at 10 Ma; the green line represents the Calabrian block translation at 5 Ma [2].

River drainage systems are very dynamic features of the landscape. Geological changes can cause fluvial captures, leading to abnormal large-scale river networks [7,9–16]. The main geological changes that cause river captures are glaciation and tectonic movements associated with earthquakes and faults [17]. Tectonic movements, especially landscape uplift, are much slower than glacial processes. Therefore, the development of tectonic river capture normally requires hundreds of thousands of years, or even millions of years [18,19].

The particular evolutionary characteristics of the Pardu and Quirra valleys in relation to slope instability dynamics have been the subject of various studies [4–7]. This sector of Sardinia represents one of the most susceptible areas to landslides in the region. This high hazard is closely linked to the particular vulnerability to important weather events, especially rainstorms. Rainfall-induced landslides represent a relevant threat to the population, infrastructure, buildings, and cultural heritage [20–24]. Among the most important catastrophic geological events in Sardinia are those that occurred in the Rio Pardu valley, which involved the inhabited centers of Gairo, Osini, Ulassai, and Jerzu. Between 15 and 17 October, 1951, extreme rainfall of about 1000 mm involved this area, triggering mudflows and landslides. This catastrophic event caused abandonment of the villages of Osini and Gairo [5,7,25]. These settlements have been rebuilt at least in part, sometimes with transfer to another area on the same slope. However, these measures proved useless, as the new sites present the same geo-hydrological risks as the previous ones [26,27]. Landslides affected schistose Paleozoic metamorphites on the left slope, while on the right, there was also widespread rockfall. Recent studies have highlighted the presence of deep-seated gravitational slope deformations with sacking [28–32] and lateral spread [33–35] characteristics that affect the sub-horizontal carbonate succession and the underlying metamorphites [6].

Deep-seated gravitational slope deformation (DGSD, [36]) is a complex type of rock slope failure characterized by large dimensions generated in stone rocks [37]. DGSDs are characterized by slow movements that can suddenly accelerate and cause catastrophic collapse of sections of the deformed slopes [30,38–41]. Therefore, this phenomenon represents an important geo-hazard in relation to the deformation of large infrastructures and secondary collateral landslides. Although DGSDs play an important role in slope evolution and geo-hydrological risk, knowledge about them was scarce for a long time [42]. They are characterized by very slow deformation rates [34], landform assemblages (such as double-crested ridges, trenches, synthetic and antithetic scarps, tension cracks, and convex bulged toes), and deep basal shear zones [43–47]. Often, shear zones present characteristics of cataclastic breccias with an abundant fine matrix [48] and thicknesses up to tens of meters [41]. DGSD is a common phenomenon in the relief of the Mediterranean Sea in relation to the particular geodynamic context that characterizes the region and to the widespread orogenetic chains. In this context, DGSDs play an important role in slope relief evolution, showing at least geometric analogies with gravity-accommodated structural wedges. Often, DGSD phenomena are influenced by the scale structural context of the slope and use pre-existing tectonic structures (fault and thrust) to guide their evolution, which is also in relation to a reactivation linked to a slope stress field variation [49,50].

In Sardinia, the studies and evidence of DGSDs are quite scarce, but the distensive tectonics and the Plio-Quaternary uplift could justify the favorable conditions for the development of DGSDs, which could also be due to local reactivation of Hercynian and Alpine tectonic structures. In this context, the slope evolutionary characteristics are analyzed—in particular, the DGSDs and the evolution of watercourses in relation to the uplift. The aim is to correlate these different aspects through geomorphological analysis with both field surveys and remote sensing techniques. Furthermore, the choice to analyze these basins takes on a particular characteristic due to the economic and social repercussions that the conditions of instability of the slopes determine in the populations of the towns of Ulassai, Osini, Jerzu, and Gairo. In fact, as is well known, these inhabited centers are continually threatened by disasters. Different types of interventions were carried out to protect inhabited centers and infrastructures, but they were carried out without a global study of the problem and, therefore, without real knowledge of the evolutionary modalities of the valley and the real gravitational dynamics of the slopes. Understanding the kinematics and temporal behavior of DGSDs and landslides is important for designing monitoring systems based on strong process knowledge. In some cases, continuous monitoring is the only way to reduce risk [51–55].

We hypothesize that the Plio-Quaternary tectonics and uplift in the Ogliastra area are the main forcing mechanisms for sustaining the necessary gravitational forces of DGSDs.

Here, we present an innovative approach for analysis of DGSDs and fluvial dynamics by using morphotectonic, morphostratigraphic, and geomorphic data and time-series InSAR data in the Pardu and Quirra rivers. We also integrated stratigraphic and morphotectonic data of the drainage basin scale to support our observations and analyses about the relation between DGSD activity and fluvial capture.

2. Geological Setting

East-central Sardinia (Italy) is characterized by widespread Jurassic dolomitic plateaus—called "Tacchi" in Sardinia—overlying a Paleozoic basement (Figures 2 and 3a,b) [56,57].

The area is characterized by the Pardu River Valley in the north, the Quirra River Valley in the south, and the Rio Pelau toward the east (Figure 2). The geological basement primarily comprises low-grade Paleozoic metamorphites affected by complex plicative structures, while in the coastal sector, there are widespread outcrops of carboniferous granites placed in the terminal phases of the Hercynian orogeny [1,59–61]. The major metamorphic Paleozoic units are the Filladi del Gennargentu Formation and Monte Santa Vittoria Formation, which are constituted by metasandstones, quartzites, phyllites, and metavolcanites (Middle Cambrian–Middle Ordovician) [57,62,63]. The summit of the metamorphic basement has suffered chemical alteration associated with a warm humid climate during the Permian and Triassic periods [64,65].

The marine and transitional Mesozoic sedimentary succession rests on the metamorphic basement in angular unconformity. These Mesozoic deposits are extensive and decipherable from their plateau morphology and are clearly visible along the right slope of the Rio Pardu and the Rio Quirra (Figure 2). The basal layers are primarily fluvial sediments of the Genna Selole Formation (Middle Jurassic) (Figure 3d), which are overlain by dolomitic limestones of the Dorgali Formation (Middle–Upper Jurassic) (Figure 3e). [57,64,66,67].

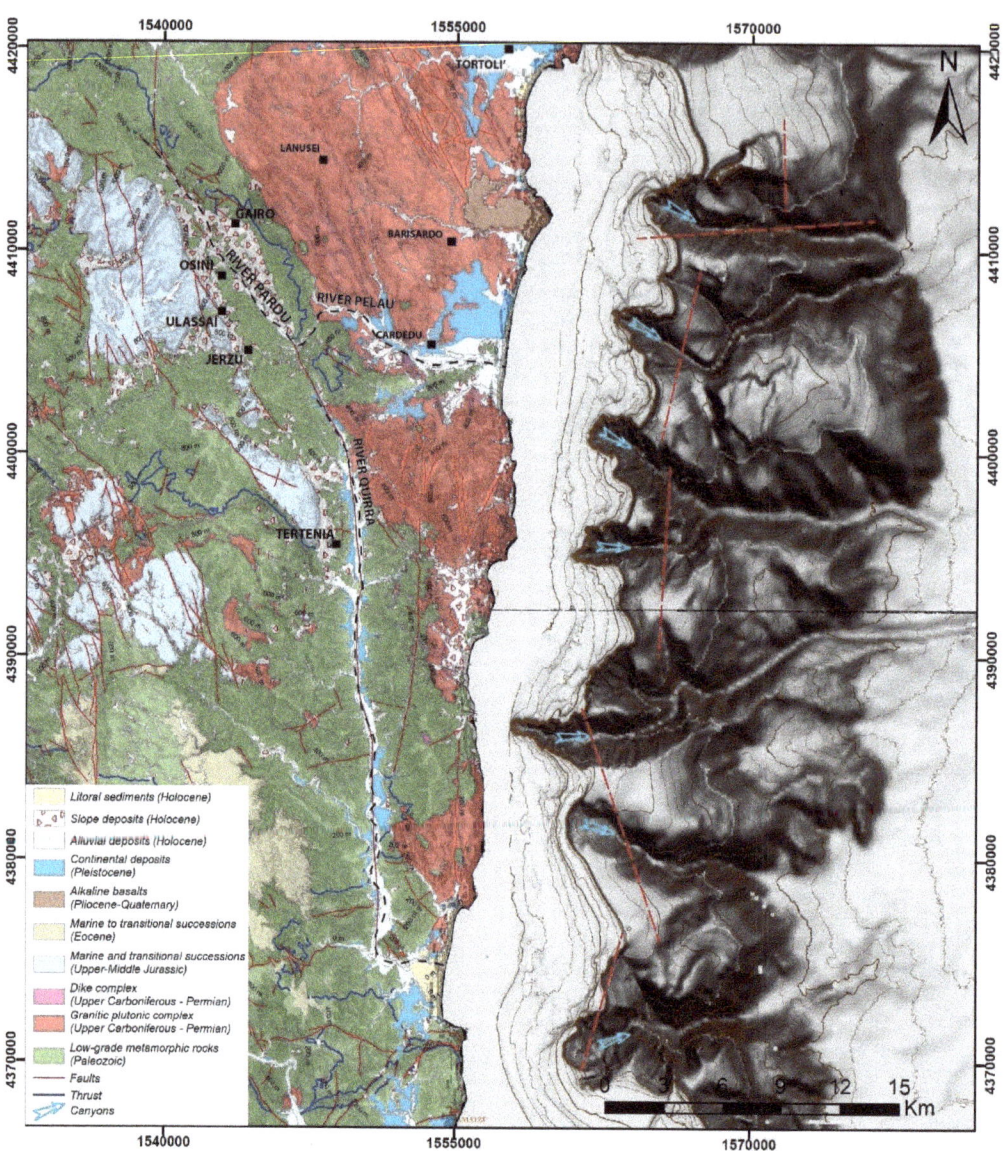

Figure 2. Geolithological sketch map of the study area based on geological data of the Autonomous Region of Sardinia. Continental margin topography by [58]. Black dashed lines show the analyzed rivers.

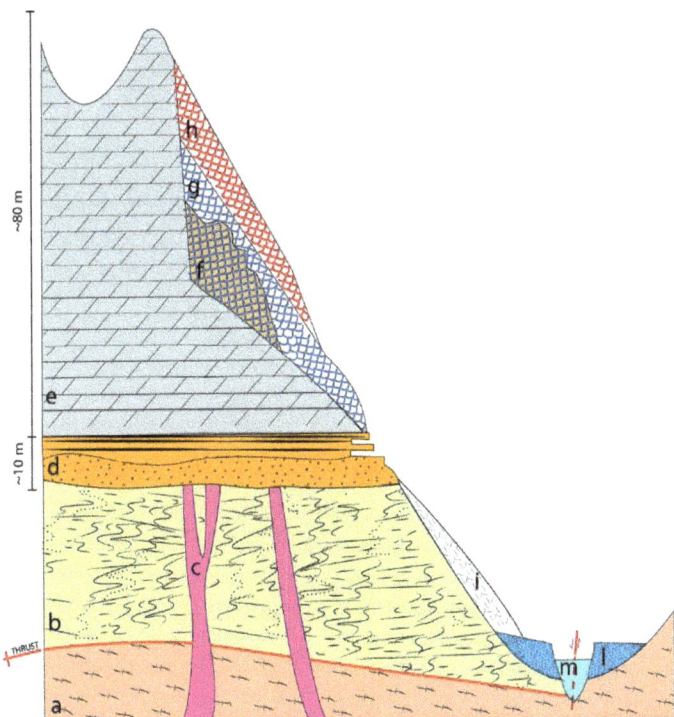

Figure 3. Lithostratigraphic sketch of lithological units: Low-grade metamorphic rocks: (a) Monte Santa Vittoria Fm; (b) Filladi del Gennargentu Formation; (c) granitic plutonic complex and dikes; marine and transitional Mesozoic sedimentary succession; (d) gluvial and deltaic conglomerates, sandstones, and mudstones (Genna SeloleFormation); (e) dolostone (Dorgali Formation); (f) cemented paleo-rockfall deposits; (g) paleo-rockfall deposits; (h) active rockfall; (i) slope deposits; (l) terraced alluvial deposits; (m) alluvial deposits (modified after Demurtas et al. [6]).

The Genna Selole Formation [67,68] represents a mixed succession of siliciclastic to siliciclastic–carbonate deposits. The presence of clay layers is important as a predisposing factor for lateral spread. The Dorgali Formation is represented by dolomitic sequences with thicknesses of up to tens of meters. The lower part, with a thickness of approximately 30 m, is affected by marl intercalations, whereas the upper part is typically massive. The attitude of the strata of the Mesozoic units is sub-horizontal with a dip of approximately N90/0–5°, while at the plateau edges, it can reach a dip of up to 40° and a direction parallel to the slope owing to the DGSDs. Quaternary covers, which are represented by continental deposits, are primarily gravitative and alluvial deposits. The most extensive outcrops are represented by landslide deposits, including rockfalls, toppling, and collapsed DGSDs, and are abundant in the lower part on the right slopes of the Pardu Valley and Quirra Valley (eastern slope of Monte Arbu). Downslope, actual and terraced alluvial deposits have also been identified, and they are well developed and hierarchized in the Rio Quirra [57].

The deposits of the rockfalls and toppling landslides have been characterized by their different sedimentological features based on age (Figure 3f–h). These deposits are associated with rockfalls affecting the plateau edge wall and the collapse of some parts of the DGSDs [6].

3. Geodynamic Setting

The river networks' geometry and gravity processes show a young conformation of the landscape, which is typical of a recent tectonic setting. The geodynamic setting is associated with the collisional dynamics between the African and European plates [2] (Figure 1). The structural setting is associated with the Alpine cycle, which first appeared with a strike-slip fault in the Oligo–Miocene, and then in the Pliocene and Quaternary with an extensional component [1–3,56,69–72].

The major features in the study area are the NW–SE and N–S faults on which, respectively, the Pardu Valley and Quirra River are engraved, and the secondary fault directions include ENE–WSW and NNE–SSW [57].

The Plio-Quaternary tectonic phase is associated with conspicuous N–S faults [73]. These rectilinear and normal faults are also evident in the continental margin and control its morphology (Figure 2). In the continental region, these N–S faults are associated with alkaline basalts with an age of approximately 3.9 Ma—Pleistocene [74]. Especially during the upper Pliocene, a general areal elevation occurred throughout the island, highlighted by the traces of the paleo-surfaces and by the numerous and superimposed paleo-hydrographies; moreover, the Neogenic sediments, which were already affected by Oligo-Miocene Tectonics, are currently also found at altitudes of 700 m, such as on the Tacco di Laconi, and are widely found above 500 m of altitude in various locations on the island. The reasons for these events are related to the more general distensive tectonics that affect the whole Tyrrhenian area [75].

Based on preliminary geodetic data from the Peri-Tyrrhenian Geodetic Array network, Ferranti et al. (2008) [76] revealed the presence of low internal deformation in Sardinia. In Sardinia, seismicity is typically scattered and sporadic, except for the dozen tremors detected following the ML4.7 earthquake of 7 July 2011 in the Corsican Sea, which primarily characterized the edges of the continental lithosphere block. Significant seismic events also occurred in the eastern sector—in particular, three events with a magnitude > 4 (26 April 2000, magnitudes ML 4.2 and 4.7, and 18 December 2004, magnitude ML 4.3)—located in the central Tyrrhenian Sea, approximately 60 km east of Olbia in the Comino depression [77].The most recent low-magnitude earthquake events were ML 1.8 (Escalaplano, 4 April 2019) and ML 1.6 (Perdasdefogu, 14 October 2020) [78].

Along the Ogliastra coast, recent movements have acted by conditioning the trend of the hydrographic network and the morphological evolution. The basaltic plateau of the Teccu in Barisardo can be related to these movements along an N–S line.

The Sardinian continental margin started from around 9 Ma, following the opening of the Tyrrhenian Sea, which caused the thinning of the continental crust and the formation of tectonic depressions, which are now sites of deep intra-slope basins.

The continental margin off the Ogliastra is represented by the continental shelf, the continental slope, and the plain called the Ogliastra basin, which reaches the deepest point of the whole Sardinian margin at 1750 m depth. The continental shelf is very narrow with less than 10 km of width, and it is indented by several submarine canyons [58,70].

4. Geomorphological Setting

The landscape, which is characterized by sub-horizontal carbonate plateaus, represents the result of the paleogeographic evolution of the region. The current dolomitic plateaus represent the extensive carbonate sedimentation due to the Jurassic marine transgression on the peneplanated Paleozoic metamorphites during the Permian and the Triassic. The continental phase following the post-Mesozoic emergence determined the setting of a tectonic control hydrographic network represented by deep rectilinear valleys engraved in the Paleozoic basement for several hundreds of meters [4,5] (Figure 4). Erosion primarily acted on the Oligo-Miocene strike-slip faults with an increase in the erosive rate during the Plio-Pleistocene uplift phases [25]. The presence of major regional faults has influenced the watercourses, which maintain a prevalent N–S direction in the Pardu and Quirra Rivers (set on the main fault).

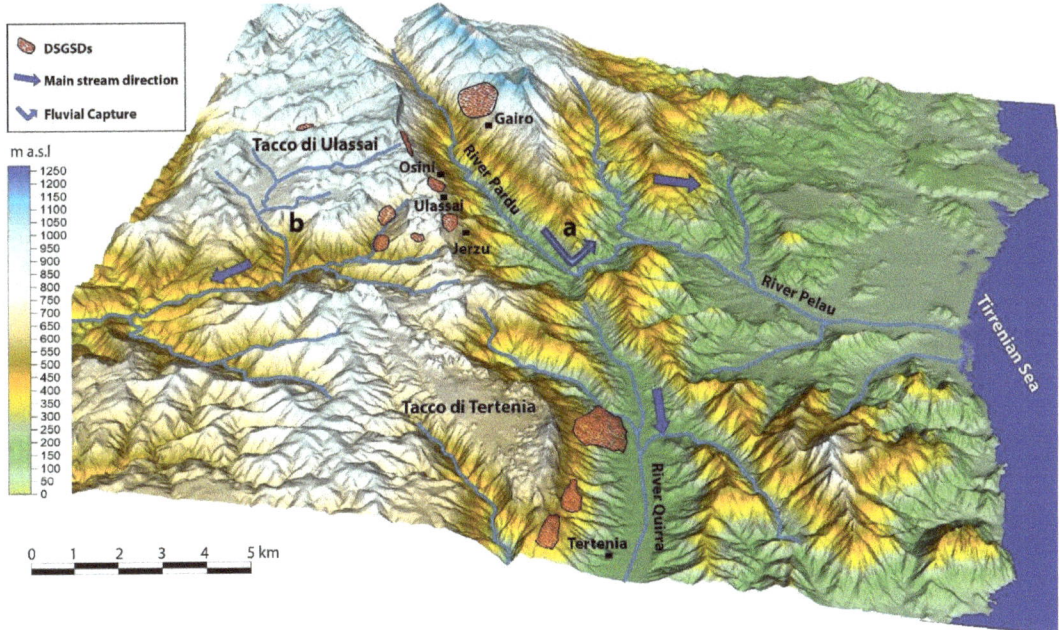

Figure 4. Three-dimensional (3D) model of the Pardu River and Quirra River. Blue lines represent major hydrographic features, and red areas represent the major DGSDs. (a) Fluvial capture elbow; (b) Lequarci waterfall.

The evolution of the Pardu River is closely associated with that of the Quirra River [7,57,79]. The Pardu River flows from the NW toward the SE and then abruptly changes direction toward the NE. At this point, a capture elbow adjacent to the present head of the Quirra River is well developed. The upstream part of the Quirra River flows at an altitude of approximately 200 m higher than the Pardu River. It also presents an over-sized and over-flooded valley with respect to the upstream catchment area. Moreover, there are various orders of river terraces and slope deposits of the Pleistocene. This setting indicates that the Pardu River, previously flowing south along the Quirra River, was captured by the Pelau River [7,79]. Considering the descriptive parameters, longitudinal profile, and the evolutionary conditions, the Pardu Valley is associated with a cycle of underdeveloped fluvial erosion, suggesting a relatively young age of engraving [4,5,25].

DGSDs are present in both river basins and cause collateral landslides. In particular, rockfalls and toppling occur along carbonate cornices, while rotational slide occurs in the metamorphic rocks [6]. We focused on the DGSDs in this study, as they are important in the morphological evolution of the slopes.

A significant karstic process has acted on plateau surfaces, comprising ancient paleoforms and, currently, hypogeal and superficial morphologies [6,7,80]. Karst paleoforms represented by complex cockpit doline types have been characterized, and they belong in a humid and warm paleo-morphoclimatic setting [6,81–83]. These dolines are separated by residual reliefs called Fengcong, which are sorted among the major structural features. The hypogean karst enabled the development of sinkholes, karst springs, cavities, and caves (e.g., Su Marmuri Cave and Is Ianas Cave). The combined action of karst, uplift, river erosion, and gravity has led to the formation and evolution of hanging valleys on the plateau surfaces [5]. The geomorphological analysis of the continental margin off the coast shows that the area occupied by the shelf is rather narrow and is engraved with numerous submarine canyons [58,84,85] (Figure 2). The structural lines coincide with those of the continental part that has emerged—mainly N–S, accompanied by normal tectonic lines in the E–W direction. The shelf break is about −130 m; however, locally, it is at about −60 m due to the erosion of retrogressive canyons. The submerged and emerged morphologies highlight the extremely young landscape conformation, which is associated with the Neogene and Quaternary geodynamic events, implying a series of problems related to the slope process. The control factors of the DGSDs are associated with the geo-structural characteristics and the Neogene and Quaternary geomorphological evolution of the river valley, which is associated with the recent uplift [6].

We can summarize the events that dominated the valleys' evolution [4,6,79]:

- The first stage preceding the capture of the Rio Pardu by the Rio Pelau associated with the uplift and the Plio-Quaternary tectonics. This phase involves a general incision of the valleys and erosion of the slopes, and it led to a new hydrographic setting, causing river capture (Middle-Lower Pleistocene).
- The second phase was associated with major erosive activity in the Pardu Valley following the capture, which led to complete erosion of the valley (Upper Pleistocene).
- The present evolution of the slopes through widespread landslides and DGSDs.

5. Materials and Methods

A morphotectonic analysis of the River Pardu and River Quirra was carried out based on an integrated approach that incorporated a cartographic and morphometric analysis [86–88]. Remote sensing analysis and geological and geomorphological field mapping in slopes and the valley floor of the Rio Quirra and Rio Pardu were performed from the head to the mouth on a scale of 1:10,000. The field surveys were based on the interpretation of data from remote sensing on a large scale. Particular attention was paid to the study of morphologies related to river dynamics (fluvial and orographic terraces) and slope gravitational process (DGSDs and collateral landslides).

Multi-scale field surveys were carried out to analyze the geological and structural setting of the slopes—in particular, the plateaus' edges and the left slope of the Pardu Valley [89–94].

The DGSDs were surveyed in detail by reconstructing the structural setting and analyzing the relationships with the surrounding collateral landslide and alluvial deposits. The study areas were often not accessible due to their steep slopes; therefore, they required remote sensing survey systems to complete the field investigations. Uncrewed aerial vehicle digital photogrammetry (UAV-DP) is a robust methodology for the investigation of DGSDs and large landslides. In particular, it was used for the recognition of large lateral spreads in Malta and Tunisia [95,96]. We used UAV-DF and light detection and ranging (LiDAR) to extract high-resolution topographic 3D DGSD models and perform detailed morphometric analyses.

DGSD displacement and rate were evaluated using space-borne interferometric synthetic aperture radar (InSAR). Over the last 30 years, InSAR techniques have been widely used to investigate geological (e.g., volcano activity, earthquakes' ground effects, etc.) and geomorphological processes—in particular, DGSD. In different geological and climatic contexts, this technique allows one to analyze extremely slow DGSDs and to identify displacements of about 1–2 mm in favorable conditions [95–103].

Based on previous studies on the fluvial deposits of Rio Quirra and Rio Pardu [5,6,79], the geological analysis was implemented by using high-resolution topographies based on UAV-DP and LiDAR. Detail-scale field surveys were carried out in the alluvial quaternary deposits with the aim of the identification and mapping of various terraced orders and the reconstruction of the relative chronology among morphostratigraphy and sedimentological indicators. Stratigraphic profiles relating to the various orders of river terraces and landslide deposits were surveyed in the natural outcrops of the alluvial plains.

5.1. Aerial and Uncrewed Aerial Vehicle Remote Sensing

LiDAR and aerial photogrammetric data produced by the Autonomous Region of Sardinia were used to perform visual and morphometric analysis of DGSDs and fluvial morphologies. A detailed orthophoto dating from 2016 was used together with LiDAR data with a cell size of 1 × 1 m and vertical resolution of 30 cm.

The aerial surveys were performed using UAVs (DJI Phantom 4 and DJI Matrix 200) flying at altitudes of 50–60 m above ground level. The acquired images were analyzed and processed using the photogrammetric Agisoft MetaShape software and constrained by 10–12 ground control points using GEODETIC LEICA GNSS for each area. The resulting orthorectified mosaic and DEM (WGS 84 datum and UTM 32N projection) had a cell size of 5 cm/pixel and were considered sufficiently precise to be used for the geomorphological analysis.

To analyze the DGSDs at the local scale, we used high-resolution digital elevation models (DEMs) acquired via structure from motion from a UAV-DF [8,103–106].

The 3D high-resolution UAV-DF models were used to develop interpretative superficial models by using geomorphological evidence and stratigraphic and structural data of the DGSDs. Geological interpretative cross-sections of geologic features crossing the major DGSDs were also generated to define the movement kinematics, deformative style, and deep geometries of the DGSDs.

The DTMs were used to analyze the morphometric parameters of the hydrographic basins under analysis. The longitudinal and transversal profiles of the valleys were extracted in such a way as to highlight different erosive structures in relation to river capture and to analyze the different altitudes of the various river terraces.

5.2. InSAR Analysis

Space-borne interferometric synthetic aperture radar (InSAR) data were used to analyze the slope deformation [107–110]. Interferometric permanent scatters (PSs) are used to investigate the temporal and spatial superficial slope deformation. To detect ground displacement, we used only high-PS coherence (0.6–1) located on built dolomitic blocks and the metamorphic rock outcrops. Low-coherence PSs, which are not useful, are located on rockfall deposits and in vegetated areas. We used the Sentinel-1 and European Remote Sensing (ERS) satellites (Table 1) and took into account the line-of-sight (LOS) velocities. We used a dataset from 1992 to 2000 from the ERS satellite and a dataset from 2014 to 2020 from Sentinel 1. The processed data from ERS and Sentinel 1 were provided, respectively, by Ministero dell'Ambiente e della Tutela del Territorio e del Mare (Italy) and the Geological Survey of Norway. The total area analyzed covered the entire Pardu Valley and Quirra Valley. Four focus areas (Table 1) that showed interesting results were analyzed by using time series of PSs to understand the landslides' temporal evolution.

Table 1. Parameters of the InSAR data on the sectors in focus.

Area	Satellite	Acquisition Geometry	Acquisition Interval	TrackAngle	Inc Angle
Ulassai	Sentinel 1	Ascending	Oct 2014–Feb 2020	−9.6	42.4
Osini	Sentinel 1	Ascending	Oct 2014–Feb 2020	−9.6	42.4
San Giorgio	Sentinel 1	Ascending	Oct 2014–Feb 2020	−9.6	42.4
Gairo	Sentinel 1	Descending	Oct 2014–Feb 2020	−169.6	36.3
	ERS		May 1992–Dec 2000	- - - - -	- - - - -

6. Results

6.1. InSAR, PS, and Time Series Analysis

The results of the large-scale InSAR analysis showed that most PSs were located in stable areas, while high deformation rates were recorded in the slopes of Pardu Valley, where slope-failure processes—in particular, rockfalls and DGSDs—were widespread. All four focus areas were analyzed in detail with the Sentinel 1 data (from 2014 to 2020). For the left flank of Rio Pardu, the ERS data (from 1992 to 2000) were also used in descending order of acquisition. The data from the periods 1992–2000 and 2014–2020 indicated areas with large slopes that were identified as DGSDs that were active in Pardu Valley. We used only PSs with high coherence (0.6–1) that were located in the rocky outcrops and in the urban structures, while low-coherence points located in rockfall deposits and in vegetated areas were not considered. The PS analysis allowed the recognition of active DGSDs and the measurement of their movement rates, which turned out to be extremely slow, ranging from 6 to 20 mm/year (Figures 5 and 6). We identified a downslope movement of up to 1 cm/y in the right slope of the Pardu Valley and a movement of up to 2 cm/y in the left slope. Continuous movements that did not change over years with both linear and seasonal trends were observed (Figure 6). The InSAR analysis showed no perceptible movements on the slopes of Rio Quirra.

In the Ulassai area, the PS analyses showed a stable surface in the urban area and on the west slope of the main extensional trench of Pranedda Canyon (Figures 5a and 6a). However, in accordance with the geomorphological evidence, downstream from the main trench, the speeds of the PSs showed LOS displacements of up to 1 mm/y. In this sector, the PSs were located in rocky dolomitic outcrops on the top edge of the plateau, in the total absence of vegetation and in excellent exposure conditions. No movements were detected in the DGSD downstream from Bruncu Pranedda due to the low PS coherence due to dense vegetation. Using Sentinel data from 2014 to 2020, we measured a total of 5 cm (orange star in Figure 5). It was possible to observe seasonal deformation trends with an excellent correlation among all of the PSs analyzed. Generally, no movement was observed during the winter and spring, but an acceleration was observed during the summer and autumn.

In Osini, a cluster of PSs were well defined within the inhabited center, particularly in the northwest and southeast sectors, where there was a speed of between 4 and 6 mm/y, with a maximum of 8 mm/y (Figures 5c and 6b) with a seasonal trend. Spotlights were located on the roofs of the buildings. In the surroundings of the inhabited center, the dense vegetation resulted in a low coherence of the PSs; therefore, they were not considered.

In the Gairo sector, the InSAR data showed a large area that was greater than 1 km^2 with a high diffusion of PSs. Based on the high-resolution field surveys, the PSs are located on rocky metamorphic outcrops. The speeds were, on average, greater than 8 mm/y, with a maximum of 2 cm/y. The cluster identified a well-defined area with a circular shape that was delimited by PSs with zero or negligible speed (Figures 5b and 6d1,d2). The higher speeds were located in the central and basal part of the DGSD, while towards the top and lateral flanks, the speeds decreased. In the lateral and top parts, the DGSD was delimited by stable PSs (speeds of 0–2 mm/y), which allowed the deformed area to be circumscribed in detail. in the PSs on the foot slope with a low coherence due to the continuous movement of slope deposits and the vegetation were not considered. The deformation's progression was continuous and linear, and an excellent correlation was found between the Sentinel 1 and ERS data. In the southern part of the DGSD, a high concentration of PSs were located in the abandoned village of Old Gairo with speeds that were sometimes greater than 1 cm/y. The town of New Gairo, which was built after the 1951 catastrophe, showed displacements limited to 2–4 mm/y.

In the San Giorgio sector, scattered PSs were identified with speeds greater than 10 mm/y on the large blocks of the rock avalanche on the slope (Figures 5d and 6c). These blocks, with dimensions of up to 30 m per side, were collateral landslides related to the collapsed DGSD located at the edge of the plateau above. All of the PSs showed a linear trend with a slowdown in the winter and spring between 2016 and 2017. This slowdown, which was observed in all of the PSs, indicates that the causes of the movement are to be found in processes that involve a greater portion of the slope, and not only in the large blocks. The surrounding area did not allow a PS analysis due to the importance of the wooded vegetation, but evidence of deformation was visible in the road infrastructure.

6.2. Deep-Seated Gravitational Slope Deformation

Various areas affected by DGSDs and landslides that were located of the slopes of Pardu Valley and on the slope of Monte Arbu of Tertenia were identified (Table 2).

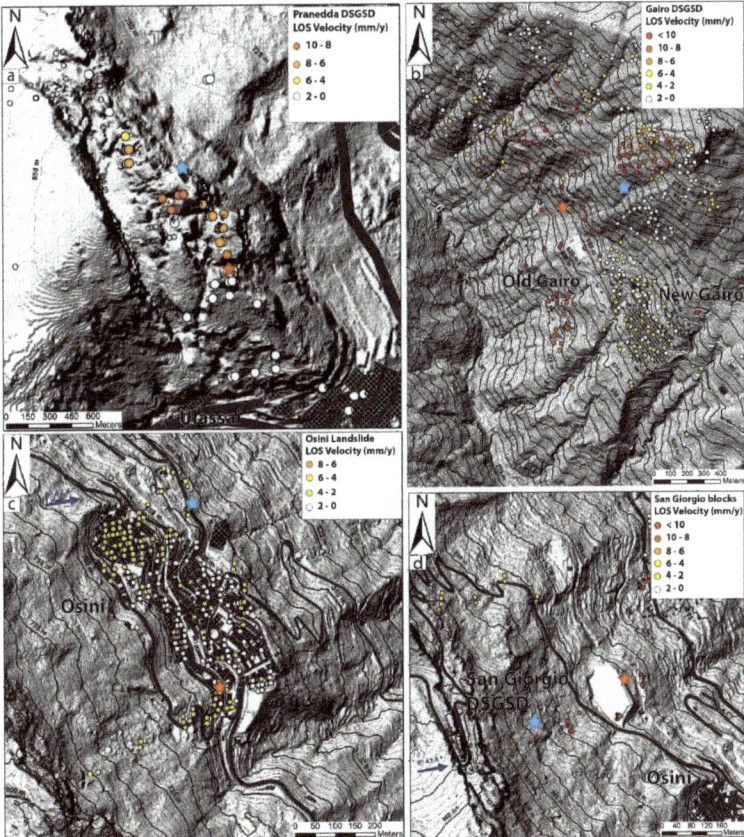

Figure 5. Analysis of the focus areas with InSAR data. The points represent high-coherence permanent scatters located on buildings, rocky outcrops, and blocks of large rock avalanches. The stars represent the PSs used to analyze the time series shown in Figure 6. (**a**) Bruncu Pranedda lateral spread. (**b**) Gairo DGSD. (**c**) Osini landslide. (**d**) San Giorgio paleo-rock avalanche.

On the east side of Tacco di Ulassai and Tisiddu Mountain, three DGSDs were analyzed (Figure 7). The main structures that indicated deep gravitational phenomena were large and deep extensional trenches that were evident in the dolomitic lithotypes. The extensional trenches had lengths of several hundreds of meters and a decametric opening and depth. This slope was characterized by the Mesozoic marine deposits resting on the Paleozoic metamorphites.

The Bruncu Pranedda DGSD (Figure 7b2,c1) is constituted by two regions with different settings located on the top and middle slopes. On the top slope, toward the east of the largest extensional trenches in the area called the Pranedda Canyon, the rock mass fracturing increased, and the attitude of the Dorgali Formation was toward the east, with a dip of up to 40°. In this area, both facies of the Dorgali Formation were visible, with the summit comprising dolomitic banks and the lower part being characterized by an alternation of well-stratified dolomites and marls. This subdivision was not observed in the middle slope, where basal marly levels did not appear on the surface. This indicates that the basal facies (approximately 30 m) were partially covered by slope deposits; however, they also sank a few meters inside the fractured and altered Paleozoic metamorphic basement. This could be correlated with the field observations at the same altitude, as well as with the basement and the massive facies of the Dorgali Formation [6].

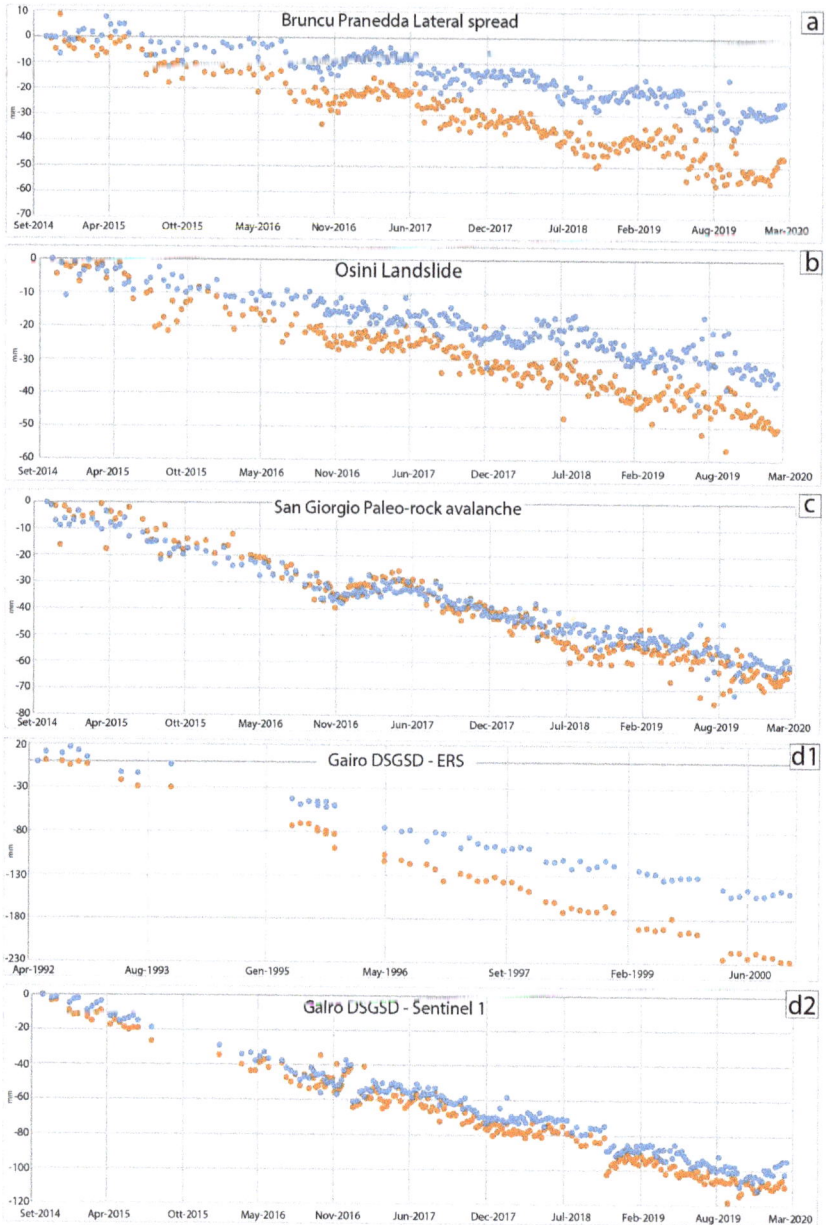

Figure 6. Time series extracted with the representative permanent scatters. The vertical axes represent the cumulative LOS displacement; the horizontal axes represent the time. (**a**) Bruncu Pranedda lateral spread—seasonal displacement trend, maximum displacement of 5 cm from 2014 to 2020; (**b**) Osini landslide—seasonal displacement trend, maximum displacement of 6 cm from 2014 to 2020; (**c**) San Giorgio paleo-rock avalanche—constant movement trend of the large blocks, maximum displacement of 6 cm from 2014 to 2020; (**d**) Gairo DGSD; (**d1**) the ERS data show a constant deformation trend, with a maximum displacement o f23 cm from 1992 to 2000; (**d2**) the Sentinel 1 data show a constant deformation trend that is correlated with the ERS data, with a maximum displacement of 10 cm from 2014 to 2020. The colors of the points agree with the colors of the stars that identifies the location of the PS in Figure 5.

Table 2. Main characteristics of the DGSDs and landslides analyzed.

Location	Landslide	Geology	Landslide Kinematic	Displacement Speed mm/y	Displacement Trend	Area Km²
North Ulassai	Bruncu Pranedda Lateral spread	Dolomitic limestone set on altered and fractured phillites	Lateral spread top slope; sackung middle slope	6–10 mm/y	Seasonal	0.2
South Ulassai	Monte Tisiddu Sackung	Dolomitic limestone set on altered and fractured phillites	Sackung	No movement	——	0.2
North Osini (San Giorgio)	San Giorgio Lateral spread	Dolomitic limestone set on altered and fractured phillites	Lateral spread	No movement	——	0.03
North Osini	San Giorgio paleo-rock avalanche	Megablock rock avalanche deposits set on paleo-rockfalls	Sliding	6–>10	Linear	≈0.1
Osini	Osini Landslide	Cemented paleo-rockfalls set on phillites	Sliding	4–8	Seasonal	≈0.3
Gairo	Gairo DGSD	Phillites on metavolcanites. Slope involved in Hercinical thrust	Sackung	6–20	Linear	1.2
South Tertenia	Tertenia DGSD	Dolomitic limestone set on altered and fractured phillites. Slope involved in Hercinical thrust	Sackung	No movement	——	1.5
North Tertenia	Paleo-DGSD	Dolomitic limestone set on altered and fractured phillites	Sackung	No movement Fossilized by Pleistocenic alluvium	——	1.5

The Scala San Giorgio DGSD (Figure 7b1,c2,d1) is located north of Osini Village and is characterized by two major extensional trenches that are parallel to the slope affecting the Dorgali Formation with a dip amount of up to 20°. All of the sequences of the Dorgali Formation are exposed; however, the Genna Selole Formation is covered by rockfall deposits.

The Tisiddu Mountain DGSD (Figure 7b3,d2) to the south of Ulassai Village is characterized by a highly fractured segment of the Dorgali Formation located tens of meters downstream. Only the tops of the massive banks of dolostones are visible. The basal level partially sank into the metamorphic basement.

In all cases, the shear zones are located in different geological units that represent structural weaknesses (Figure 7d1,d2). (I) The top of the metamorphites was affected by sub-horizontal foliation and advanced weathering, which was highlighted by the reddish or whitish color of the rocks. This type of alteration could be linked to the pre-transgressive Mesozoic period [65]. (II) The Genna Selole Formation was characterized by plastic clay layers; (III) basal levels of the Dorgali Formation were characterized by the alternation of marl and dolomite.

A large landslide that affected the town of Osini and the northernmost slope downstream of the San Giorgio DGSD was identified by using InSAR data. The inhabited center of Osini is built over an extensive cemented paleo-rockfall deposit that rests on the Paleozoic basement. Geomorphological evidence is difficult to observe due to the extensive vegetation around the village.

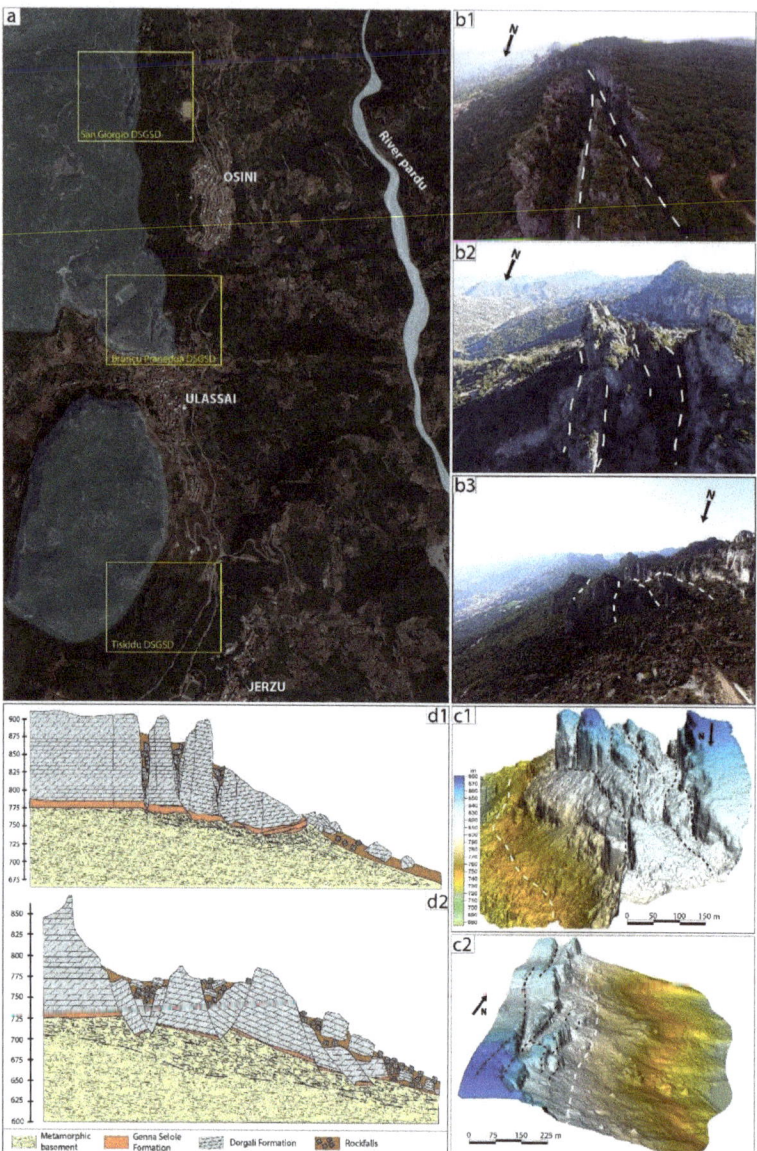

Figure 7. DGSD on the right slope of the Pardu River. (**a**) Orthophoto of the area of Ulassai, Osini, and Jerzu. The Jurassic dolostone plateaus on the metamorphic basement are shown in blue. The yellow square represents the analyzed DGSD. (**b**) UAV images of the DGSD showing the major geomorphological and structural features. The white dashed lines represent the major extensional trenches. (**b1**) San Giorgio lateral spread, (**b2**) Bruncu Pranedda lateral spread, (**b3**) Monte Tisiddu sackung. (**c**) Three-dimensional LiDAR model of the DGSDs with a colored elevation scale. The black dashed lines represent the major extensional trenches. The white dashed lines represent the major stratigraphic discontinuity between the marine Mesozoic sequence and the metamorphic basement. (**c1**) Bruncu Pranedda lateral spread. (**c2**) San Giorgio lateral spread. (**d**) Interpretative geological cross-sections passing through the DGSD in the study area. The hypothetical basal shear zone is highlighted with black dotted lines. (**d1**) San Giorgio lateral spread. (**d2**) Monte Tisiddu sackung.

The left side of the Rio Pardu is characterized by a different geological and structural context compared to the opposite side (Figure 8a). There are metamorphic lithologies belonging to the formation of Monte Santa Vittoria and the Filladi del Gennargentu. The slope is affected by a dip-slope Hercynian thrust that brings the two formations belonging to two different tectonic units into contact. This structure plays a fundamental role in the deep gravitational processes, as it is marked by intense fracturing and alteration of the lithotypes. Based on the geomorphological evidence and the analysis of the InSAR data, a large landslide with a DGSD character was identified in the northwest sector with respect to the town of Gairo (Figure 8b–f). The DGSD extends from the top slope to the middle-lower part of the slope and is about 1 km wide. The crown is circular (Figure 8c,d) and joins laterally rectilinear structural flanks (Figure 8e). Analyzing the profile of the slope along the DGSD, the concave upstream part and the convex downstream part are clearly evident. The foot of the landslide is covered by landslide and slope deposits that reach the valley floor, where lateral erosion by the Rio Pardu is affected (Figure 8).

On the right side of the Rio Quirra, in correspondence with the Tacco di Tertenia, complex gravitational morphologies linked to paleo-DGSDs are evident (Figure 9). The morphology of Mount Arbu is also affected by the complex tectonic structure, which is characterized by a sub-horizontal thrust that brings the Pyllades del Gennargentu Formation into contact with the Metavolcanites of the Monte Santa Vittoria Formation (Figure 9a). The morphological analysis of the slope shows convexity and concavity linked to different DGSDs that are distributed at various altitudes of the slope. The DGSDs consist of portions of the Dorgali Formation, which is tilted up to 30–40° and is translated along the slope. The most complex and evolved movement was identified in the NE sector (Figure 9b,c1,c2). The area extends for a length of about 1800 m from the top of the plateau to the valley floor. The fan-shaped landslide body has a foot with a length of 2 km. The crown is located in the plateau edge, which is affected by faults and distension trenches. The latter delimit mega-blocks of the Dorgali Formation With a prismatic shape and inclination of up to 40°. The foot of the DGSD, which is represented by the Dorgali Formation, is marked by dolomitic outcrops with vertical heights of up to 40 m with sometimes sub-horizontal attitudes of the strata. On these walls, terraced alluvial deposits rest in onlap. Paleo-DGSDs are widespread in the upper part of the slope, with greater diffusion in the southern part of Mount Arbu, but they do not evolve until reaching the valley floor (Figure 9b,d).

6.3. River Capture Analysis

The area has a deep cut made by the Rio Pardu Valley and Rio Quirra Valley, which extend in an NNW–SSE direction, following a major Tertiary fault. For most of the Pardu River's course, the talweg is set on rock, indicating its predominantly erosive nature. Downstream, the river is captured, turning in an eastward direction, and its name changes to Rio Pelau; then, it flows into the Tyrrhenian Sea. South of the capture, the abandoned Rio Pardu Valley continues southward as Rio Quirra. This valley is characterized by a bottom filled with Pleistocene and Holocene terraced alluvial deposits and slope deposits, which are currently undergoing erosion. It is clear that in the past, Rio Pardu was captured by Rio Pelau (Figure 10), causing a rapid incision upstream. Longitudinal profiles were constructed for Rio Pardu, Rio Quirra, and Rio Pelau. Rio Pardu flows up to 750 m below the dolostone near Ulassai, where the main active DGSDs are located. The evolutionary hypotheses are related to the Pliocene and Quaternary uplift, which led to an important erosive phase.

The triggering process can be justified in the following ways:

- An erosive increase caused by a generalized uplift that led to the retreat towards the inland by the head of the Rio Pelau until it connected with the Rio Pardu.
- Another hypothesis foresees the presence of a direct fault with an east–west course along which the Rio Pelau is set. In this case, the differential uplift of the block on which the Rio Quirra is currently set could justify the capture process as tectonogenic.

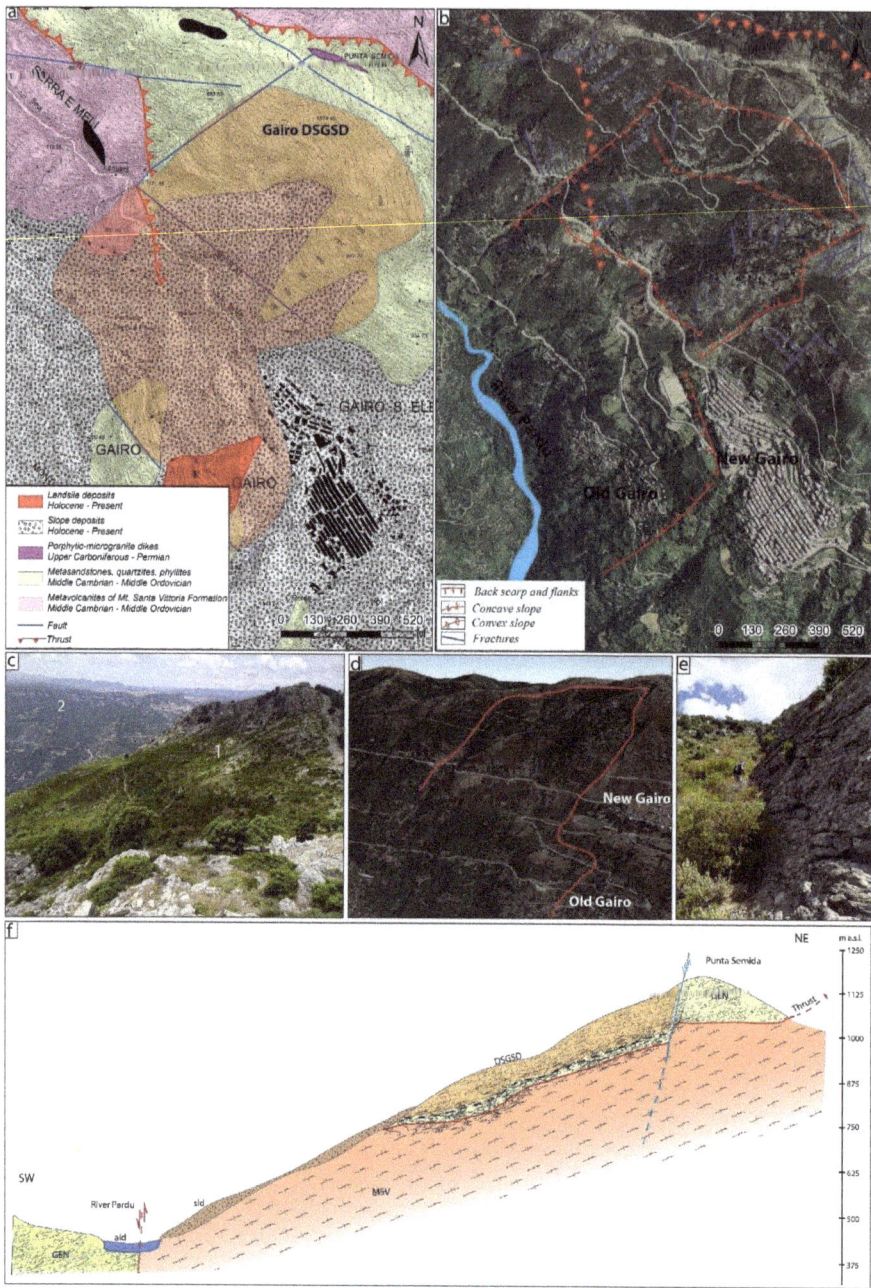

Figure 8. (**a**) Geological map of the Gairo slope with the DGSD localization. (**b**) Orthophoto with the main geomorphological feature of the DGSD. (**c**) Photo of the DGSD head. (1) Crown; (2) right slope of the Pardu River; (**d**) photo showing a 3D view with the DGSD border marked in red; (**e**) linear flank of the DGSD; (**f**) interpretative geological cross-section of the DGSD showing it (in transparent orange) sliding on the highly fractured rock due the underlying dip-slope Paleozoic thrust. Geolithological legend: MSV—Monte Santa Vittoria Formation; GEN—Filladi del Gennargentu Formation; ald—current alluvial deposits; sld—slope deposits.

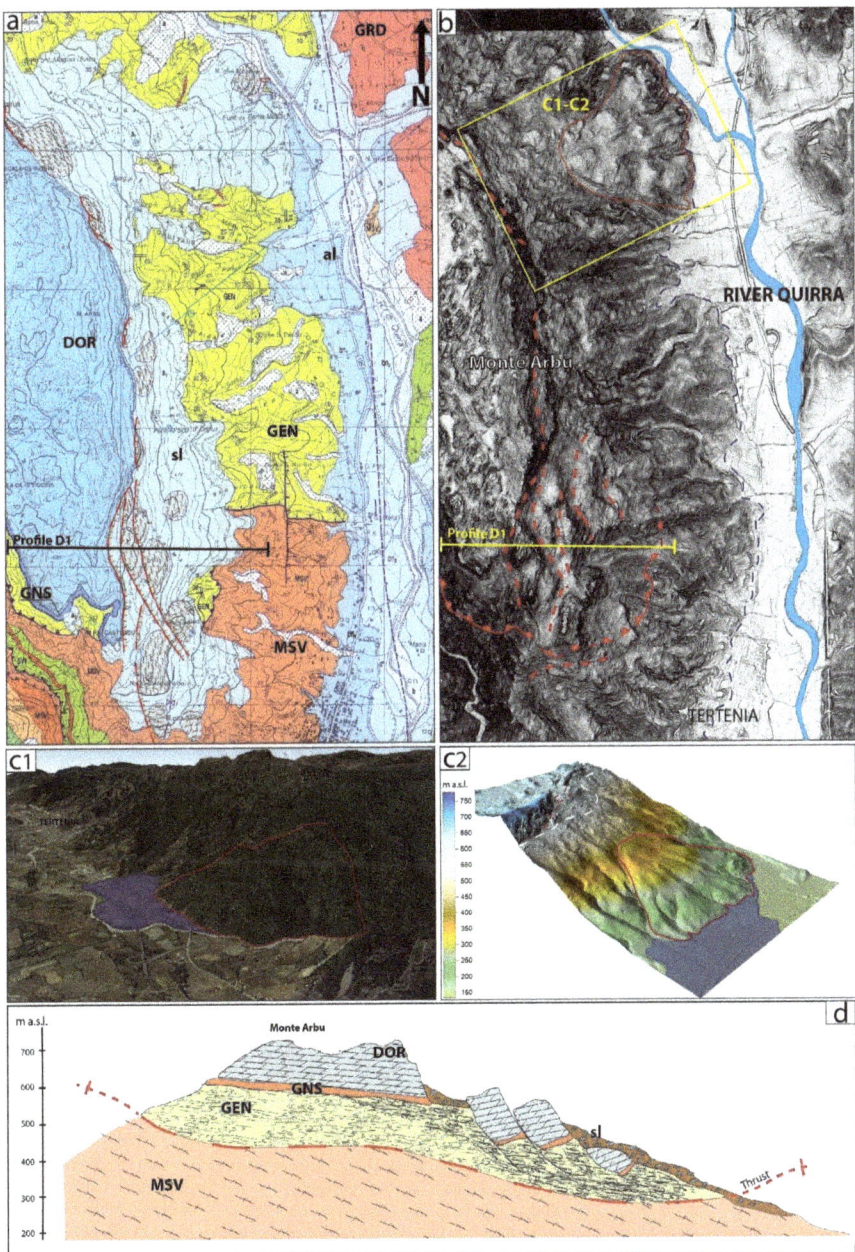

Figure 9. (**a**) Geological map of the eastern slope of Monte Arbu (Tertenia). Geolithological legend: MSV—Monte Santa Vittoria Formation; GEN—Filladi Del Gennargentu Formation; GNS—Genna Selole Formation; DOR—Dorgali Formation; al—terraced and current alluvial deposits; sl—slope deposits. [57]. (**b**) LiDAR hillshade with the main geomorphological feature of the DGSD. (**c1**) Photographic 3D view with the DGSD border marked in red and the terraced alluvial deposit in blue. (**c2**) Three-dimensional LiDAR of the Tertenia paleo-DGSD with the border marked in red and the terraced alluvial deposit in blue. (**d**) Interpretative geological cross-section of the DGSD showing it sliding on highly fractured rock due the underlying Paleozoic thrust.

Figure 10. Three-dimensional LiDAR model of the river capture sector. In the north, the Pardu River flows eastwards, taking the name of Rio Pelau. The blue and light blue show the Holocene alluvial deposit of the Pardu River. South of Genna and Crexia is the head of the Rio Quirra. In red is shown the paleo-slope and paleo-alluvial deposits of the Quirra River.

6.4. Fluvial Morphostratigraphic Analysis

A morphostratigraphic analysis was performed first on Rio Quirra and later on Rio Pardu, which isolated it following the capture (Table 3).

Table 3. Morphostratigraphic synthesis.

Deposits	Characteristic	Elevation from Talweg	Distribution
T0	Pebbles and clastosustained gravels with a scarce sandy matrix	0	Actual embrided riverbed
T1	Heterometric and polygenic pebbles with a scarce dark matrix	0.20/0.30–1.5	Pardu-Quirra
T2	The matrix is decidedly prevalent in the coarse fraction	2–5/6	Quirra
T3	Non-constant matrix–skeleton relationship. Reddish matrix (Fe oxides)	6/7–10	Quirra
Paleo-conoid C1	Clastosustained pebbles up to 40–50 cm in size. Scarce matrix	30	Pardu
Paleo-conoid C2	Reddish pebbles and gravels in sandy, silty, reddish matrix	15	Quirra
Paleo-slope deposits		20–40	Quirra

In the valley of the Rio Quirra, above the current riverbed, the following were identified (Figure 11):

T0—Actual flood surface consisting of pebbles and clastosustained gravels with a scarce sandy matrix (Holocene).

T1—Sub-current Holocene terrace with a maximum height on the riverbed of about 20–30 cm up to 1.5–2 m. The dark brown matrix is subordinate to the coarse fraction, which is represented by heterometric and polygenic pebbles. This terrace often forms alluvial islands in the upstream part of the river; they reach a good stability due to the dense vegetation that has settled there (Upper Pleistocene–Holocene).

T2—In this terrace, the matrix, which is decidedly prevalent in the coarse fraction, has a dark brown color. There is no evidence of prolonged chemical alterations due to climatic conditions other than the current ones. The pebbles are less varied: mainly quartz with, subordinately, granite and schistose. On average, the height of T1 with respect to the riverbed is about 2 m, with a maximum of 5–6 m and a minimum of 50 cm. The deposits that form this terrace show forms of erosion linked to secondary climatic pulsations (Upper Pleistocene).

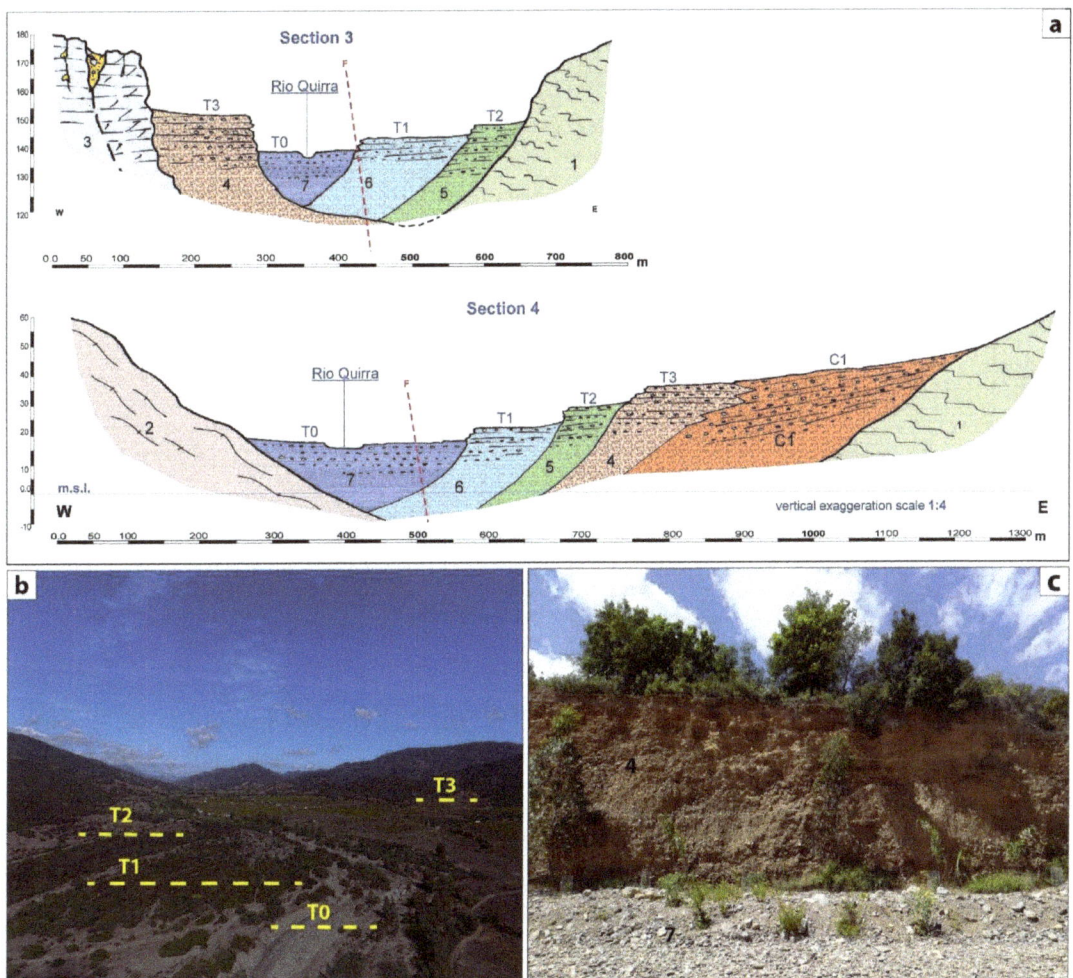

Figure 11. (a) Morphostratigraphic profiles of the Quirra River. (b) UAV photo in the river alluvial plain. (c) Outcrop of Terrace T3. Lithological legend: (1) Filladi Grigie del Gennargentu Formation; (2) Monte Santa Vittoria Formation; (3) paleo-DGSD; (C1) paleo-conoid; (4) T3; (5) T2; (6) T1; (7) T0.

T3—This is the oldest terrace, with an average height of 6–7 m and a maximum of 10 m (Figure 10c). The matrix–skeleton relationship is not constant. The depository is made up of alternations of fine and large sediments that testify to the variations in the river's energy. The matrix is red and sometimes whitish. In the first case, the color derives from Fe oxides, indicating a warm, humid climate typical of tropical and sub-tropical regions; in the second case, the oxides have been leached and for an eluviation horizon. The pebbly fraction does not have a varied lithological composition. It is mainly schistose and, subordinately, quartz. The deposit is well cemented. This terrace rests directly on the slope. The frame of erosion along the riverbed is clear, and the lower terraces rest on it (Middle Pleistocene).

In Rio Pardu, the alluvial deposits cover a valley floor characterized by a well-defined flood bed, which is limited by banks that are intensely affected by landslides. Two orders of alluvial terraces up to 2 m above the current level were detected (Figure 12). The maturity of the flood clasts is very low due to the continuous supply of material from the slopes, while the grain size distribution along the longitudinal profile reflects the trend characterized by the high slope. By analyzing the longitudinal profile of the Rio Pardu, it can be observed that it is divided into two well-defined parts separated by the knickpoint in Ponti Mannu. In the initial part, near the steeply sloping trunk, there is the head of the valley, which continues until an area with a low slope where alluvial deposits appear. Downstream of the Ponte Mannu, after a section of the river in which the waters flow on the rock, the river becomes slightly sloped and establishes an alluvial plain with anastomotic channels and river islands.

Active and quiescent dejection cones are distributed over the Quirra and Pardu Valleys. The active conoids are well highlighted by the morphology, and they have a poorly elaborated clastic component and an uncemented dark brown matrix. In the terminal part of the Quirra, a terraced dejection cone (C1) assumes a certain importance due to its size and evolutionary stage (Inner–Middle Pleistocene). The often large pebbles are very elaborate and have blackish patinas of Mn oxides on their surfaces. Oxides also accumulate inside the matrix, which presents an intense redness. In the upper part of the Rio Pardu, a paleo-conoid (C2) with large pebbles and a brown matrix is currently engraved by the current course of the river (Upper Pleistocene and Holocene).

The paleo-slope deposits are characterized by coarse, elaborate, and sharp-edged components. The matrix is very abundant, strongly cemented, and bright red in color due to the accumulation of pockets of Mn oxides. These deposits are located at the same altitude as that of T3 or sometimes at higher altitudes, and they are connected to the base of the slope (Inner-Middle Pleistocene).

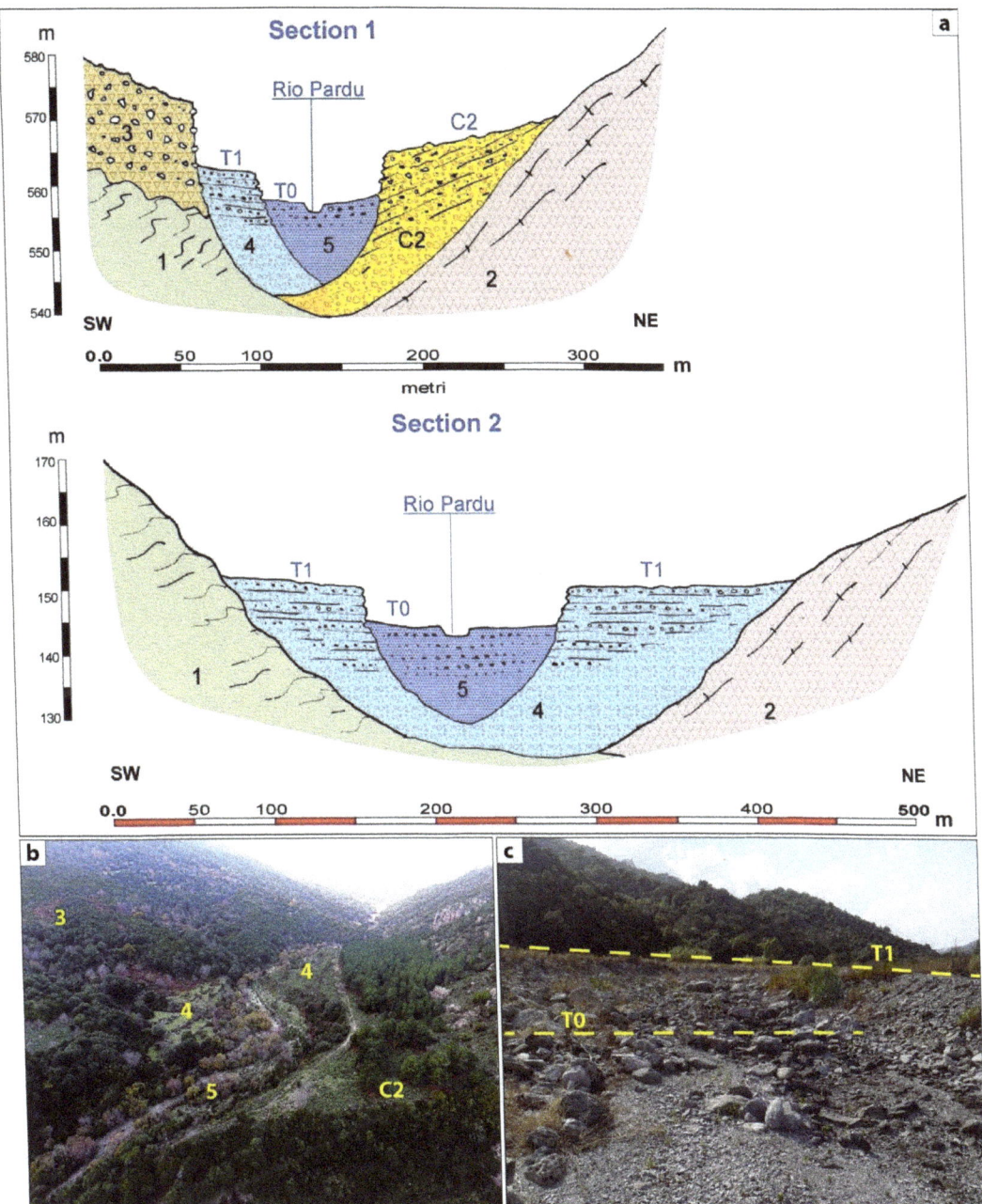

Figure 12. (a) Morphostratigraphic profiles of the Pardu River. (b) UAV photo near the head of the Pardu River. (c) The bottom of the Pardu Valley near the capture elbow. Lithological legend: (1) Filladi Grigie del Gennargentu Formation; (2) Monte Santa Vittoria Formation; (3) paleo-rockfall deposits; (C2) paleo-conoid; (4) Terrace T1; (5) Terrace T0.

7. Discussion

7.1. River Analysis

The hydrographic network engraved by torrential watercourses possesses a tectonic control linked to the Cenozoic structural features. The main incisions that cross the basin of the Rio Pardu give rise to deep valleys with a mainly erosive and only locally depositional character. The Pardu Valley has a transverse "V" profile, which is more or less open depending on the evolutionary stage, the distance from the point of origin, and the competence of the lithotypes in which the river incision takes place. Sometimes, the profiles show marked asymmetries due to the different positions of the layers or the different exposure, which influences the vegetation. A lower steepness can be observed on the left side, which probably due to a lower vegetation cover, which favors erosion. The valley has developed in the formation of the Filladi grigie del Gennargentu. Only in the southeast is the formation of Monte Santa Vittoria affected.

As regards the evolutionary conditions of the Pardu Valley, considering the descriptive parameters, the geometric conditions, and the hypsometric curve, it is noted that was not able to develop to the point of acquiring characteristics that are attributable to a cycle of river erosion in an evolved phase, which suggests a relatively young age for engraving [4,5,25].

The longitudinal profile of the Rio Quirra differs from the normal profile of a river; it has an initial concave part with a strong steepness within the first kilometer from the head and a regular decrease in the slope along the rest of the watercourse. The evolutionary stage of the Quirra appears to have advanced; however, it must be considered that it represents the middle and final parts of the original Rio Pardu–Quirra, which are divided in two by the capture of the Rio Pelau. Currently, the Rio Quirra does not have a catchment basin at the head, and its feeding is mainly given by certain tributaries. The valley is oversized and over-flooded with respect to the current basin (Figure 13).

7.2. DGSD Dynamics

The Rio Pardu and Quirra River represent two of the most susceptible areas to landslides in Sardinia, as well as to rockfalls and rainstorm-induced superficial landslides [4,5,26]. This sector is also interesting due to the fact that extreme rainfall over the last centuries has led to the evacuation and reconstruction of the towns of Osini and Gairo [5,7]. Recent studies have highlighted the presence of deep landslides with sackung-type kinematics and lateral spreads on the right side of the Rio Pardu [6].

In this paper, by analyzing integrated geomorphological, geo-structural, high-resolution topography and InSAR displacement data, we identified diffuse DGSDs on both sides of the valleys of the Pardu River and Quirra River, which are characterized by different kinematics.

DGSDs are commonly found in orogenetic environments with high tectonic and seismic activity and in areas affected by slope decompression due to post-deglaciation. The present work aimed to contribute to the knowledge on the influence of evolution of valleys—in particular, with high incision—on the triggering of large landslides or DGSDs in relation the Quaternary uplift.

Lateral spreads were developed at the edge of the plateau in relation to the favorable stratigraphy (dolostone on clays and altered metamorphites). The slope deformation generates vertical fractures in the carbonate and a zone of ductile basal deformation that affects the Genna Selole Formation and the summit, which thus altered the metamorphites (Bruncu Pranedda and San Giorgio DGSD). DGSDs with a higher vertical shift represent a more advanced stage with sackung features (Tisiddu Mountain and Tertenia DGSDs). The latter evolves in relation to the thrust that affects the median part of the slope. A large part of the deformation affects the Paleozoic basement, which was evidenced by the sinking of the carbonate sequence into the metamorphites.

On the left side of Rio Pardu, Gairo's DGSD shows a different behavior in relation to the different stratigraphic and structural setup. The DGSD has sackung-type kinematics with an important translational component linked to the thrust.

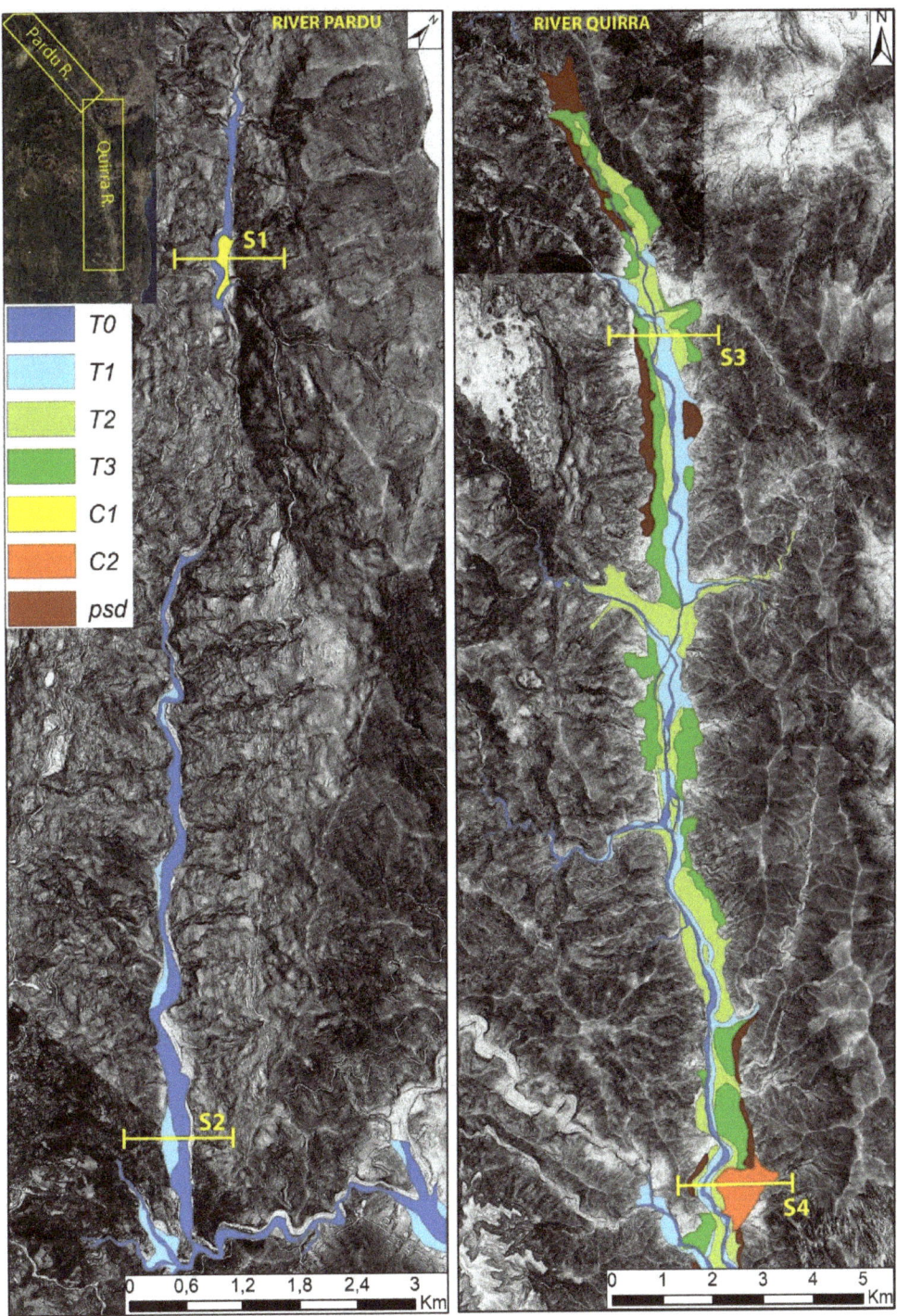

Figure 13. Map of the distribution of alluvial deposits in Rio Pardu and Rio Quirra.

From the structural viewpoint, the major faults in the NW–SE and NE–SW directions were in concordance with the main trenches and back-scarps in all sectors, indicating an important structural control. The secondary trenches and the joints did not exhibit a good correlation with the large-scale structures because they were associated with the features inside the deformation rock mass.

The Rio Pardu shows a straight valley with steep slopes, a valley bottom with a mainly erosive character, and two orders of terraces. This is linked to the intense erosive phase following the capture by the Rio Pelau. The valley of Rio Quirra shows a flat-bottomed valley with an actual riverbed of the braided channel type. The valley is over-sized and over-flooded with four orders of terraces, the result of an evolution prior to the capture of the Rio Pardu. The T3 terrace shows sedimentological characteristics related to a subtropical climate, which is probably linked to the warm climatic phase of MIS 5. The InSAR and morphostratigraphic analyses made it possible to define the state of activity of the DGSDs in the two hydrographic basins. In the valley of Rio Pardu, various areas of the slope that are affected by movements that can be classified as active DGSDs were identified, with movements of up to 2 cm/y on the left slope and up to 1 cm/y on the right slope. However, in the Quirra River, paleo-DGSD bodies are fossilized by the alluvial deposits of the T3 terrace. This indicates that the river capture led to an intense erosive phase in the Rio Pardu, leading to the recent instability of the slopes, thus justifying the active DGSD (Figure 14).

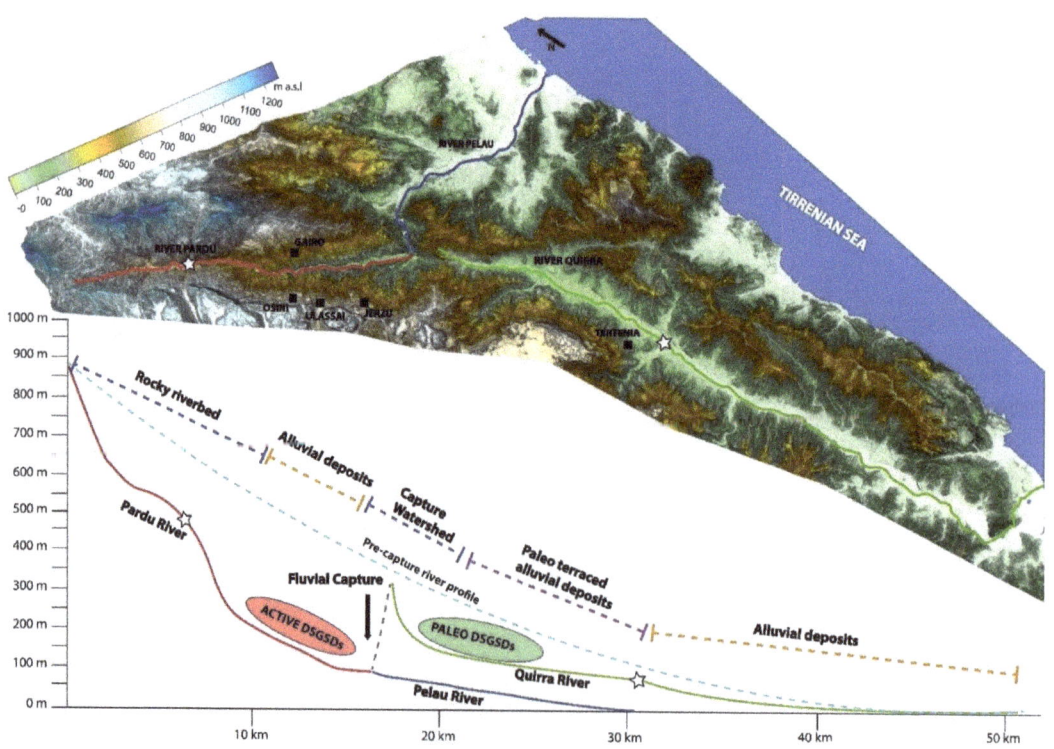

Figure 14. Relation between DGSD activity and river parameters.

These DGSDs were associated with numerous large collateral rockfalls and toppling landslides that affected the slopes. Dolomitic blocks with sizes of up to 30 m on each side were identified; these moved up to 900 m away from the detachment points, which were linked to mega-rockfall events with rock avalanche features. We also identified paleo-

DGSDs on the downslope that were associated with the collapsed slope side. Currently, a reactivation of quiescent DGSDs or an acceleration of movements can be triggered by extreme weather events or earthquakes.

Therefore, an acceleration of slope movements leading to a potential catastrophic failure poses a threat to communities, and the monitoring of these slopes is important for early warning and risk reduction. So, we studied the DGSDs and landslides in the inhabited areas of Pardu Valley in detail by using integrated remote sensing techniques, field mapping, and InSAR in order to understand the temporal evolution. The historical InSAR deformation rate supports our model of rock slope deformation. However, for risk reduction in a populated area, a 24/7 monitoring system could become an essential component of an early-warning system that is aimed at preparing evacuation protocols [55,111–118].

8. Conclusions

The connection between Plio-Pleistocenic tectonic activity and geomorphological evolution in the Pardu Valley and Quirra Valley (Ogliastra, East Sardinia) was studied. The evolutionary conditions of the Pardu Valley are associated with a cycle of undeveloped fluvial erosion, which suggests a relatively young age of the engraving in relation to the capture by the Rio Pelau and the isolation of the Rio Quirra. The intense post-capture erosion has given the Rio Pardu Valley morphometric features that are favorable for the evolution of DGSDs. However, the Rio Quirra Valley presents paleo-DGSDs that have been fossilized by pre-capture terraced alluvial deposits.

The DGSDs' movements are linked to the recent tectonic evolution (areal uplift). The two valleys analyzed are controlled by transcurrent faults that have recently recorded low-magnitude seismic events. Therefore, it is possible that the constant movement of the DGSDs (between about 1 and 2 cm/y) may be susceptible to accelerations due to seismic triggering, causing the partial collapse of the slopes.

In particular, this research highlighted the following:

- The geomorphological and structural setting of Ogliastra is closely linked to the genesis of the east Sardinian continental margin due the opening of the Tyrrhenian basin (Miocene–Pliocene)
- Distensive Pliocene tectonics accompanied by widespread volcanism resulted in a general uplift in Sardinia. The Quaternary uplift rebound manifested itself with an important erosive phase and variations in the hydrographic network. We have evidence of this phase in the Rio Quirra Valley, which is represented by paleo-DGSDs fossilized by pre-Tyrrhenian alluvial deposits (Lower Pleistocene).
- The river capture of Rio Pardu is associated with this important erosive phase and caused an erosive increase that led to a complete emptying of the valley (Upper-Middle Pleistocene).
- The post-capture decompression of the slopes of the Rio Pardu triggered DGSDs in both flanks in the current state of activity.
- Using InSAR data, it was possible to identify and assign displacement rates to the Ulassai, Osini, and Gairo DGSDs.

Author Contributions: Conceptualization, V.D., P.E.O. and G.D.; methodology, V.D. and G.D.; validation, P.E.O.; formal analysis, V.D.; investigation, V.D., P.E.O. and G.D.; resources, P.E.O. and G.D.; data curation, V.D.; writing—original draft preparation, V.D.; writing—review and editing, V.D., P.E.O. and G.D.; visualization, V.D., P.E.O. and G.D.; supervision, G.D. and P.E.O.; project administration, P.E.O.; funding acquisition, P.E.O. All authors have read and agreed to the published version of the manuscript.

Funding: Valentino Demurtas acknowledges the support from the University of Cagliari—Dottorato di Ricerca XXXIV ciclo—Earth and Environmental Sciences and Technologies.

Institutional Review Board Statement: Not applicable.

Informed Consent Statement: Not applicable.

Data Availability Statement: Not applicable.

Acknowledgments: We are thankful to Reginald Hermanns for the precious suggestions. We are also grateful to Marie Bredal for the technical support in the InSAR processing. The precious contributions of the two anonymous reviewers and of the journal editors are acknowledged.

Conflicts of Interest: The authors declare no conflict of interest.

References

1. Carmignani, L.; Oggiano, G.; Barca, S.; Conti, P.; Salvadori, I.; Eltrudis, A.; Funedda, A.; Pasci, S. Geologia della Sardegna: Note Illustrative della Carta Geologica della Sardegna in scala 1:200.000. In *Memorie Descrittive della Carta Geologica d'Italia*; ISPRA—Servizio Geologico d'Italia: Rome, Italy, 2001; Volume 60.
2. Carminati, E.; Doglioni, C. Mediterranean tectonics. In *Encyclopedia of Geology*; Selley, R., Cocks, R., Plimer, I., Eds., Elsevier: Amsterdam, The Netherlands, 2005; pp. 135–146.
3. Cherchi, A.; Montadert, L. Oligo-Miocene rift of Sardinia and the early history of the Western Mediterranean Basin. *Nature* **1982**, *298*, 736–739. [CrossRef]
4. Maxia, C.; Ulzega, A.; Marini, C. Studio idrogeologico dei dissesti nel bacino del Rio Pardu (Sardegna centro-orientale). *Pubblicazione dell'Istituto di Geologia, Paleontologia e Geografia Fisica* **1973**, *12*, 1–9.
5. Ulzega, A.; Marini, A. L'évolution des versants dans la vallée du Rio Pardu (Sardaigne centre-orientale). *Zeitschrift fur Geomorphologie* **1973**, *21*, 466–474.
6. Demurtas., V.; Orrù, P.E.; Deiana, G. Deep-seated gravitational slope deformations in central Sardinia: Insights into the geomorphological evolution. *J. Maps* **2021**, *17*, 594–607.
7. De Waele, J.; Ferrarese, F.; Granger, D.; Sauro, F. Landscape evolution in the Tacchi area (central-east Sardinia, Italy) based on karst and fluvial morphology and age of cave sediments. *Geografia Fisica e Dinamica Quaternaria* **2012**, *35*, 119–127. [CrossRef]
8. Deiana, G.; Lecca, L.; Melis, R.T.; Soldati, M.; Demurtas, V.; Orrù, P.E. Submarine geomorphology of the southwestern Sardinian continental shelf (Mediterranean Sea): Insights into the last glacial maximum sea-level changes and related environments. *Water* **2021**, *13*, 155. [CrossRef]
9. Suhail, H.A.; Yang, R.; Chen, H.; Rao, G. The impact of river capture on the landscape development of the Dadu River drainage basin, eastern Tibetan plateau. *J. Asian Earth Sci.* **2020**, *198*, 1367–9120. [CrossRef]
10. Carabella, C.; Buccolini, M.; Galli, L.; Miccadei, E.; Paglia, G.; Piacentini, T. Geomorphological analysis of drainage changes in the NE Apennines piedmont area: The case of the middle Tavo River bend (Abruzzo, central Italy). *J. Maps* **2020**, *16*, 222–235. [CrossRef]
11. Fan, N.; Chu, Z.; Jiang, L. Abrupt drainage basin reorganization following a Pleistocene river capture. *Nat. Commun.* **2018**, *9*, 3756. [CrossRef]
12. Bishop, P. Drainage rearrangement by river capture, beheading and diversion. *Prog. Phys. Geogr.* **1995**, *19*, 449–473. [CrossRef]
13. Willett, S.D.; McCoy, S.W.; Perron, J.T.; Goren, L.; Chen, C.Y. Dynamic reorganization of river basins. *Science* **2014**, *343*, 1117.
14. Whipple, K.X.; Forte, A.M.; Di Biase, R.A. Timescales of landscape response to divide migration and drainage capture: Implications for the role of divide mobility in landscape evolution. *J. Geophys. Res. Earth Surf.* **2017**, *122*, 248–273.
15. Prince, P.S.; Spotila, J.A.; Henika, W.S. Stream capture as driver of transient landscape evolution in a tectonically quiescent setting. *Geology* **2011**, *39*, 823–826.
16. Aslan, A.; Hood, W.C.; Karlstrom, K.E.; Kirby, E.; Granger, D.E.; Kelley, S.; Asmerom, Y. Abandonment of Unaweep Canyon (1.4–0.8 Ma), western Colorado: Effects of stream capture and anomalously rapid Pleistocene river incision. *Geosphere* **2014**, *10*, 428–446. [CrossRef]
17. Shugar, D.; Clague, J.; Best, J.; Schoof, C.; Willis, M.; Copland, L.; Roe, G. River piracy and drainage basin reorganization led by climate-driven glacier retreat. *Nat. Geosci.* **2017**, *10*, 370–375. [CrossRef]
18. Bracciali, L.; Najman, Y.; Parrish, R.R.; Akhter, S.H.; Millar, I. The Brahmaputra tale of tectonics and erosion: Early Miocene river capture in the Eastern Himalaya. *Earth Planet. Sci. Lett.* **2015**, *415*, 25–37. [CrossRef]
19. Antón, L.; De Vicente, G.; Munoz-Martin, A.; Stokes, M. Using river long profiles and geomorphic indices to evaluate the geomorphological signature of continental scale drainage capture, Duero basin (NW Iberia). *Geomorphology* **2014**, *206*, 250–261. [CrossRef]
20. Pánek, T.; Šilhán, K.; Tábořík, P.; Hradecký, J.; Smolková, V.; Lenart, J.; Pazdur, A. Catastrophic slope failure and its origins: Case of the May 2010 Girová Mountain long-runout rockslide (Czech Republic). *Geomorphology* **2011**, *130*, 352–364. [CrossRef]
21. Quesada-Román, A.; Fallas-López, B.; Hernández-Espinoza, K.; Stoffel, M.; Ballesteros-Cánovas, J.A. Relationships between earthquakes, hurricanes, and landslides in Costa Rica. *Landslides* **2019**, *16*, 1539–1550. [CrossRef]
22. Volpe, E.; Ciabatta, L.; Salciarini, D.; Camici, S.; Cattoni, E.; Brocca, L. The impact of probability density functions assessment on model performance for slope stability analysis. *Geosciences* **2021**, *11*, 322.
23. Shou, K.J.; Chen, J. On the rainfall induced deep-seated and shallow landslide hazard in Taiwan. *Eng. Geol.* **2021**, *288*, 106156. [CrossRef]
24. Quesada-Román, A. Landslide risk index map at the municipal scale for Costa Rica. *Int. J. Disaster Risk Reduct.* **2021**, *56*, 102144. [CrossRef]

25. Marini, A.; Ulzega, A. Osservazioni geomorfologiche sul tacco di ulassai. *Rendiconti Seminario Facoltà Scienze Università di Cagliari* **1977**, *47*, 192–208.
26. Proggetto AVI Aree Vulnerate Italiane. Available online: http://avi.gndci.cnr.it/ (accessed on 1 September 2020).
27. Moretti, A. Sui movimenti franosi degli abitati di Osini e di Gairo (Nuoro). *Bollettino del Servizio Geologico d'Italia* **1953**, *75*, 2.
28. Zischinsky, U. On the deformation of high slopes. In Proceedings of the 1st Conference International Society for RockMechanics, Lisbon, Portugal, 25 September–1 October 1966; Volume 2, pp. 179–185.
29. Zischinsky, U. Über Sackungen. *Rock Mech.* **1969**, *1*, 30–52. [CrossRef]
30. Radbruch-Hall, D.; Varnes, D.J.; Savage, W.Z. Gravitational spreading of steep-sided ridges ("sackung") in western United States. *Bull. Int. Assoc. Eng. Geol.* **1976**, *13*, 23–35. [CrossRef]
31. Radbruch-Hall, D. Gravitational creep of rock masses on slopes. In *Rockslides and Avalanches—Natural Phenomena: Developments in Geotechnical Engineering*; Voight, B., Ed.; Elsevier: Amsterdam, The Netherlands, 1978; Volume 14, pp. 607–658.
32. Bisci, C.; Dramis, F.; Sorriso-Valvo, M. Rock flow (sackung). In *Landslide Recognition: Identification, Movement and Causes*; Dikau, R., Brunsden, D., Schrott, L., Ibsen, M.L., Eds.; John Wiley & Sons, Inc.: Hoboken, NJ, USA, 1996; pp. 150–160.
33. Jahn, A. Slope morphological feature resulting from gravitation. *Z. Für Geomorphol.* **1964**, *5*, 59–72.
34. Cruden, D.M.; Varnes, D.J. Landslide Types and Processes. *Spec. Rep. Natl. Res. Counc. Transp. Res. Board* **1996**, *247*, 36–75.
35. Pasuto, A.; Soldati, M. Lateral spreading. In *Landslide Recognition: Identification, Movement and Causes*; Dikau, R., Brunsden, D., Schrott, L., Ibsen, M.-L., Eds.; John Wiley & Sons, Inc.: Hoboken, NJ, USA, 1996; pp. 122–136.
36. Dramis, F.; Sorriso-Valvo, M. Deep-seated gravitational slope deformations, related landslides and tectonics. *Eng. Geol.* **1994**, *38*, 231–243. [CrossRef]
37. Dramis, F.; Farabollini, P.; Gentili, B.; Pambianchi, G. Neotectonics and large scale gravitational phenomena in the Umbria–Marche Apennines, Italy. In *Seismically Induced Ground Ruptures and Large Scale Mass Movements, Field Excursion and Meeting*; Comerci, V., D'Agostino, N., Fubelli, G., Molin, P., Piacentini, T., Eds.; APAT: Rome, Italy, 2002; Volume 21, pp. 17–30.
38. Agliardi, F.; Scuderi, M.M.; Fusi, N.; Cristiano, C. Slow-to-fast transition of giant creeping rockslides modulated by undrained loading in basal shear zones. *Nat. Commun.* **2020**, *11*, 1352. [CrossRef]
39. Crosta, G.B.; Agliardi, F. Failure forecast for large rock slides by surface displacement measurements. *Can. Geotech. J.* **2003**, *40*, 176–191. [CrossRef]
40. Nemčok, A. Gravitational slope deformation in high mountains. In Proceedings of the 24th International Geology Congress, Montreal, QC, Canada, 1972; Volume 13, pp. 132–141.
41. Ostermann, M.; Sanders, D. The Benner pass rock avalanche cluster suggests a close relation between long-term slope deformation (DSGSDs and translational rock slides) and catastrophic failure. *Geomorphology* **2017**, *289*, 44–59. [CrossRef]
42. Soldati, M. Deep-seated gravitational slope deformation. In *Encyclopedia of Natural Hazards*; Bobrowsky, P.T., Ed.; Encyclopedia of Earth Sciences Series; Springer: Berlin/Heidelberg, Germany, 2013.
43. Agliardi, F.; Crosta, G.; Zanchi, A. Structural constraints on deep seated slope deformation kinematics. *Eng. Geol.* **2001**, *59*, 83–102. [CrossRef]
44. Chigira, M. Long-term gravitational deformation of rocks by mass rock creep. *Eng. Geol.* **1992**, *32*, 157–184. [CrossRef]
45. Crosta, G.B.; Frattini, P.; Agliardi, F. Deep seated gravitational slope deformations in the European Alps. *Tectonophysics* **2013**, *605*, 13–33. [CrossRef]
46. Mariani, G.S.; Zerboni, A. Surface geomorphological features of deep-seated gravitational slope deformations: A look to the role of lithostructure (N Apennines, Italy). *Geosciences* **2020**, *10*, 334. [CrossRef]
47. Pánek, T.; Klimeš, J. Temporal behavior of deep-seated gravitational slope deformations: A review. *Earth-Sci. Rev.* **2016**, *156*, 14–38. [CrossRef]
48. Crosta, G.B.; Zanchi, A. Deep-seated slope deformations. Huge, extraordinary, enigmatic phenomena. In *Landslides in Research, Theory and Practice*; Bromhead, E., Dixon, N., Ibsen, M., Eds.; Thomas Telford Ltd.: London, UK, 2000; pp. 351–358.
49. Gentili, B.; Pambianchi, G. Gravitational morphogenesis of the Apennine chain in Central Italy. In Proceedings of the 7th International Congress International Association of Engineering Geology, Lisboa, Portugal, 5–9 September 1994; Volume 3, pp. 1177–1186.
50. Iovine, G.; Tansi, C. Gravity-accommodated 'structural wedges' along thrust ramps: A kinematic scheme of gravitational evolu-tion. *Nat. Hazards* **1998**, *17*, 195–224. [CrossRef]
51. Frigerio, S.; Schenato, L.; Bossi, G.; Cavalli, M.; Mantovani, M.; Marcato, G.; Pasuto, A. A web-based platform for automatic and continuous landslide monitoring: The Rotolon (Eastern Italian Alps) case study. *Comput. Geosci.* **2014**, *63*, 96–105.
52. Sestras, P.; Bilașco, Ș.; Roșca, S.; Dudic, B.; Hysa, A.; Spalević, V. Geodetic and UAV monitoring in the sustainable management of shallow landslides and erosion of a susceptible urban environment. *Remote Sens.* **2021**, *13*, 385.
53. Zhang, L.; Wang, X.; Xia, T.; Yang, B.; Yu, B. Deformation characteristics of Tianjiaba landslide induced by surcharge. *ISPRS Int. J. Geo-Inf.* **2021**, *10*, 221. [CrossRef]
54. Bianchini, S.; Solari, L.; Bertolo, D.; Thuegaz, P.; Catani, F. Integration of satellite interferometric data in civil protection strategies for landslide studies at a regional scale. *Remote Sens.* **2021**, *13*, 1881. [CrossRef]
55. Demurtas, V.; Orrù, P.; Deiana, G. Multi-source and multi-scale monitoring system of deep-seated gravitational slope deformation in east-central sardinia. *Planet Care Space* **2021**, *2*, 28–32. [CrossRef]

56. Carmignani, L.; Oggiano, G.; Funedda, A.; Conti, P.; Pasci, S. The geological map of Sardinia (Italy) at 1:250,000 scale. *J. Maps* **2016**, *12*, 826–835. [CrossRef]
57. Pertusati, P.C.; Sarria, E.; Cherchi, G.P.; Carmignani, L.; Barca, S.; Benedetti, M.; Chighine, G.; Cincotti, E.; Oggiano, G.; Ulzega, A.; et al. *Geological Map of Italty. Scale 1:50.000*; Scheet 541 "Jerzu"; ISPRA-Servizio Geologico Nazionale: Rome, Italy, 2002.
58. Chiocci, F.L.; Budillon, F.; Ceramicola, S.; Gamberi, F.; Orrù, P. *A Tlante dei Lineamenti di Pericolosità Geologica dei Mari Italiani-Risultati del Progetto MaGIC*; CNR Edizioni: Rome, Italy, 2021.
59. Carmignani, L.; Carosi, R.; Di Pisa, A.; Gattiglio, G.; Musumeci, G.; Oggiano, G.; Pertusati, P.C. The Hercynian chain in Sardinia (Italy). *Geodin. Acta* **1994**, *7*, 31–47. [CrossRef]
60. Elter, F.M.; Corsi, B.; Cricca, P.; Muzio, G. The south-western Alpine foreland: Correlation between two sectors of the Variscan chain belonging to "stable Europe": Sardinia (Italy) Corsica and Maures Massif (south-eastern France). *Geodin. Acta* **2004**, *17*, 31–40. [CrossRef]
61. Elter, F.M.; Padovano, M.; Kraus, R.K. The emplacement of Variscan HT metamorphic rocks linked to the interaction between Gondwana and Laurussia: Structural constraints in NE Sardinia (Italy). *Terra Nova* **2010**, *22*, 369–377. [CrossRef]
62. Meloni, M.A.; Oggiano, G.; Funedda, A.; Pistis, M.; Linnemann, U. Tectonics, ore bodies, and gamma-ray logging of the Variscan basement, southern Gennargentu massif (central Sardinia, Italy). *J. Maps* **2017**, *13*, 196–206. [CrossRef]
63. Vai, G.B.; Cocozza, T. Il "postgotlandiano" sardo, unità sinorogenica ercinica. *Boll. Della Soc. Geol. Ital.* **1974**, *93*, 61–72.
64. Costamagna, L.G.; Barca, S. Stratigraphy, facies analysis, paleogeography and regional framework of the Jurssic succession of the "tacchi" area (Middle-Eastern Sardinia). *Bollettino della Societa Geologica Italiana* **2004**, *123*, 477–495.
65. Marini, C. Le concentrazioni residuali post-erciniche di Fe dell'Ogliastra (Sardegna orientale): Contesto geologico e dati mineralogici. *Rendiconti della Societa Italiana di Mineralogia e Petrologia* **1984**, *39*, 229–238.
66. Costamagna, L.G.; Kustatscher, E.; Scanu, G.G.; Del Rio, M.; Pittau, P.; Van Konijnenburg-van Cittert, J.H.A. A palaeoenvironmental reconstruction of the Middle Jurassic of Sardinia (Italy) based on integrated palaeobotanical, palynological and lithofacies data assessment. *Palaeobiodivers. Palaeoenviron.* **2018**, *98*, 111–138. [CrossRef]
67. Dieni, I.; Fischer, J.C.; Massari, F.; Salard-Cheboldaeff, M.; Vozenin-Serra, C. La succession de Genna Selole (Baunei) dans le cadre de la paléogéographie mésojurassique de la Sardaigne orientale. *Memorie della Società Geologica Italiana* **1983**, *36*, 117–148.
68. Costamagna, L.G. Middle Jurassic continental to marine transition in an extensional tectonics context: The Genna Selole Fm depositional system in the Tacchi area (central sardinia. Italy). *Geol. J.* **2015**, *51*, 722–736. [CrossRef]
69. Gattacceca, J.; Deino, A.; Rizzo, R.; Jones, D.S.; Henry, B.; Beaudoin, B.; Vadeboin, F. Miocene rotation of Sardinia: New paleomagnetic and geochronological constraints and geodynamic implications. *Earth Planet Sci. Lett.* **2007**, *258*, 359–377. [CrossRef]
70. Ulzega, A.; Orrù, P.E.; Pintus, C.; Pertusati, P.C.; Sarria, E.; Cherchi, G.P.; Carmignani, L.; Barca, S.; Benedetti, M.; Chighine, G.; et al. *Geological Map of Italty. Scale 1:50.000*; Scheet 541 "Jerzu"; ISPRA-Servizio Geologico Nazionale: Rome, Italy, 2002.
71. Gueguen, E.; Doglioni, C.; Fernandez, M. Lithospheric boudinage in the Western Mediterranean back-arc basin. *Terra Nova* **1997**, *9*, 184–187. [CrossRef]
72. Oggiano, G.; Funedda, A.; Carmignani, L.; Pasci, S. The Sardinia-Corsica microplate and its role in the northern Apennine geodynamics: New insights from the tertiary intraplate strike-slip tectonics of Sardinia. *Ital. J. Geosci.* **2009**, *128*, 527–541. [CrossRef]
73. Casula, G.; Cherchi, A.; Montadert, L.; Murru, M.; Sarria, E. The cenozoic graben system of Sardinia (Italy): Geodynamic evolution from new seismic and field data. *Mar. Pet. Geol.* **2001**, *18*, 863–888. [CrossRef]
74. Lustrino, M.; Melluso, L.; Morra, V. The geochemical peculiarity of Plio-Quaternary volcanic rocks of Sardinia in the circum-Mediterranean area. *Geol. Soc. Am.* **2007**, *418*, 277–301.
75. Marini, A.; Murru, M. Movimenti tettonici in Sardegna fra il Miocene Superiore ed il Pleistocene. *Geografia Fisica e Dinamica Quaternaria* **1983**, *6*, 39–42.
76. Ferranti, L.; Oldow, J.S.; D'Argenio, B.; Catalano, R.; Lewis, D.; Marsella, E.; Avellone, G.; Maschio, L.; Pappone, G.; Pepe, F.; et al. Active deformation in southern Italy, Sicily and southern Sardinia from GPS velocities of the Peri-Tyrrhenian Geodetic Array (PTGA). *Boll. Soc. Geol. Ital.* **2008**, *127*, 299–316.
77. Cimini, G.B.; Marchetti, A.; Silvestri, M. *L'esperimento Sardinia Passive Array (spa): Acquisizione Dati Sismici Per lo Studio Della Geodinamica e Della Sismotettonica Dell'area Mediterranea*; Istituto Nazionale di Geofisica e Vulcanologia, Centro Nazionale Terremoti (INGV): Rome, Italy, 2016.
78. INGV. INGV Special, the Earthquakes of 2020 in Italy. 2021. Available online: http://terremoti.ingv.it/ (accessed on 30 September 2021).
79. Palomba, M.; Ulzega, A. Geomorfologia dei depositi quaternari del Rio Quirra e della piattaforma continentale antistante (Sardegna Orientale). *Rendiconti Saminario Facoltà Scienze Università Cagliari* **1984**, *54*, 109–121.
80. De Waele, J.; Di Gregorio, F.; Follesa, R.; Piras, G. Geosites and landscape evolution of the "tacchi": An example from central-east Sardinia. *Il Quaternario* **2005**, *18*, 211–220.
81. Fleurant, C.; Tucker, G.E.; Viles, H.A. A model of cockpit karst landscape, Jamaica. In *Géomorphologie: Relief, Processus, Environnement*; Groupe Français de Géomorphologie: Paris, France, 2008; pp. 3–14.
82. Liang, F.; Xu, B. Discrimination of tower-, cockpit-, and non-karst landforms in Guilin, southern China, based on morphometric characteristics. *Geomorphology* **2014**, *204*, 42–48. [CrossRef]

83. Waltham, T. Fengcong, fenglin, cone karst and tower karst. *Cave Karst Sci.* **2008**, *35*, 77–88.
84. Gamberi, F.; Leidi, E.; Dalla Valle, G.; Rovere, M.; Marani, M.; Mercorella, A. Foglio 57 Arbatax. In *A Tlante dei Lineamenti di Pericolosità Geologica dei Mari Italiani-Risultati del Progetto MaGIC*; Chiocci, F.L., Budillon, F., Ceramicola, S., Gamberi, F., Orrù, P., Eds.; CNR Edizioni: Rome, Italy, 2021.
85. Marani, M.; Gamberi, F. Structural framework of the Tyrrhenian Sea unveiled by seafloor morphology. In *From Seafloor to Deep Mantle: Architecture of the Tyrrhenian Backarc Basin*; Memorie Descrittive della Carta Geologica d'Italia; Marani, M., Gamberi, F., Bonatti, E., Eds.; ISPRA: Rome, Italy, 2004; Volume 44, pp. 97–108.
86. ISPRA & AIGEO. *Aggiornamento ed Integrazione Delle linee Guida Della Carta Geomorfologica d'Italia in Scala 1:50,000*; Quaderni Serie III del; Servizio Geologico Nazionale: Rome, Italy, 2018.
87. Miccadei, E.; Carabella, C.; Paglia, G.; Piacentini, T. Paleo-drainage network, morphotectonics, and fluvial terraces: Clues from the verde stream in the Middle Sangro River (central Italy). *Geosciences* **2018**, *8*, 337. [CrossRef]
88. Miccadei, E.; Carabella, C.; Paglia, G. Morphoneotectonics of the Abruzzo Periadriatic area (central Italy): Morphometric analysis and morphological evidence of tectonics features. *Geosciences* **2021**, *11*, 397. [CrossRef]
89. Guzzetti, F.; Carrara, A.; Cardinali, M.; Reichenbach, P. Landslide hazard evaluation: A review of current techniques and their application in a multi-scale study, central Italy. *Geomorphology* **1999**, *31*, 181–216. [CrossRef]
90. Dragićević, S.; Lai, T.; Balram, S. GIS-based multicriteria evaluation with multiscale analysis to characterize urban landslide susceptibility in data-scarce environments. *Habitat Int.* **2015**, *45*, 114–125.
91. Shi, W.; Deng, S.; Xu, W. Extraction of multi-scale landslide morphological features based on local Gi* using airborne LiDAR-derived DEM. *Geomorphology* **2018**, *303*, 229–242.
92. Yi, Y.; Zhang, Z.; Zhang, W.; Jia, H.; Zhang, J. Landslide susceptibility mapping using multiscale sampling strategy and convolutional neural network: A case study in Jiuzhaigou region. *Catena* **2020**, *195*, 104851.
93. Deiana, G.; Melis, M.T.; Funedda, A.; Da Pelo, S.; Meloni, M.; Naitza, L.; Orrù, P.; Salvini, R.; Sulis, A. Integrating remote sensing data for the assessments of coastal cliffs hazard: MAREGOT project. *Earth Obs. Adv. Chang. World* **2019**, *1*, 176–181.
94. Melis, M.T.; Da Pelo, S.; Erbì, I.; Loche, M.; Deiana, G.; Demurtas, V.; Meloni, M.A.; Dessì, F.; Funedda, A.; Scaioni, M.; et al. Thermal remote sensing from UAVs: A review on methods in coastal cliffs prone to landslides. *Remote Sens.* **2020**, *12*, 1971. [CrossRef]
95. Devoto, S.; Macovaz, V.; Mantovani, M.; Soldati, M.; Furlani, S. Advantages of using UAV digital photogrammetry in the study of slow-moving coastal landslides. *Remote Sens.* **2020**, *12*, 3566.
96. Gaidi, S.; Galve, J.P.; Melki, F.; Ruano, P.; Reyes-Carmona, C.; Marzougui, W.; Devoto, S.; Pérez-Peña, J.V.; Azañón, J.M.; Chouaieb, H.; et al. Analysis of the geological controls and kinematics of the chgega landslide (Mateur, Tunisia) exploiting photogrammetry and InSAR technologies. *Remote Sens.* **2021**, *13*, 4048. [CrossRef]
97. Delgado, J.; Vicente, F.; García-Tortosa, F.; Alfaro, P.; Estévez, A.; Lopez-Sanchez, J.M.; Tomás, R.; Mallorquí, J.J. A deep seated compound rotational rock slide and rock spread in SE Spain: Structural control and DInSAR monitoring. *Geomorphology* **2011**, *129*, 252–262. [CrossRef]
98. Oliveira, S.C.; Zêzere, J.L.; Catalão, J.; Nico, G. The contribution of PSInSAR interferometry to landslide hazard in weak rock-dominated areas. *Landslides* **2015**, *12*, 703–719. [CrossRef]
99. Crosetto, M.; Monserrat, O.; Cuevas-González, M.; Devanthéry, N.; Crippa, B. Persistent scatterer interferometry: A review. *ISPRS J. Photogramm. Remote Sens.* **2016**, *115*, 78–89. [CrossRef]
100. Mantovani, M.; Devoto, S.; Piacentini, D.; Prampolini, M.; Soldati, M.; Pasuto, A. Advanced SAR interferometric analysis to support geomorphological interpretation of slow-moving coastal landslides (Malta, Mediterranean Sea). *Remote Sens.* **2016**, *8*, 443. [CrossRef]
101. Frattini, P.; Crosta, G.B.; Rossini, M.; Allievi, J. Activity and kinematic behaviour of deep-seated landslides from PS-InSAR displacement rate measurements. *Landslides* **2018**, *15*, 1053–1070. [CrossRef]
102. Novellino, A.; Cesarano, M.; Cappelletti, P.; Di Martire, D.; Di Napoli, M.; Ramondini, M.; Sowter, A.; Calcaterra, D. Slow-moving landslide risk assessment combining machine learning and InSAR techniques. *Catena* **2021**, *203*, 105317. [CrossRef]
103. Eker, R.; Aydın, A. Long-term retrospective investigation of a large, deep-seated, and slow-moving landslide using InSAR time series, historical aerial photographs, and UAV data: The case of Devrek landslide (NW Turkey). *Catena* **2021**, *196*, 104895. [CrossRef]
104. Peternal, T.; Kumelj, S.; Ostir, K.; Komac, M. Monitoring the Potoška planina landslide (NW Slovenia) using UAV photogrammetry and tachymetric measurements. *Landslides* **2017**, *14*, 395–406. [CrossRef]
105. Valkaniotis, S.; Papathanassiou, G.; Ganas, A. Mapping an earthquake-induced landslide based on UAV imagery; case study of the 2015 Okeanos landslide, Lefkada, Greece. *Eng. Geol.* **2018**, *245*, 141–152. [CrossRef]
106. Clapuyt, F.; Vanacker, V.; Oost, K.V. Reproducibility of UAV-based earth topography reconstructions based on structure-from-motion algorithms. *Geomorphology* **2016**, *260*, 4–15. [CrossRef]
107. Ietto, F.; Perri, F.; Fortunato, G. Lateral spreading phenomena and weathering processes from the Tropea area (Calabria, southern Italy). *Environ. Earth Sci.* **2015**, *73*, 4595–4608. [CrossRef]
108. Mateos, R.M.; Ezquerro, P.; Azañón, J.M.; Gelabert, B.; Herrera, G.; Fernández-Merodo, J.A.; Spizzichino, D.; Sarro, R.; Garcia-Moreno, I.; Bejar-Pizarro, M. Coastal lateral spreading in the world heritage site of the Tramuntana Range (Majorca, Spain). The use of PSInSAR monitoring to identify vulnerability. *Landslides* **2018**, *15*, 797–809. [CrossRef]

109. Moretto, S.; Bozzano, F.; Mazzanti, P. The role of satellite InSAR for landslide forecasting: Limitations and openings. *Remote Sens.* **2021**, *13*, 3735. [CrossRef]
110. Mondini, A.C.; Guzzetti, F.; Chang, K.-T.; Monserrat, O.; Martha, T.R.; Manconi, A. Landslide failures detection and mapping using synthetic aperture radar: Past, present and future. *Earth-Sci. Rev.* **2021**, *216*, 103574.
111. Zhang, Y.; Li, H.; Sheng, Q.; Wu, K.; Chen, G. Real time remote monitoring and pre-warning system for highway landslide in mountain area. *J. Environ. Sci. China* **2011**, *23*, S100–S105.
112. Barla, G.; Antolini, F.; Barla, M.; Mensi, E.; Piovano, G. Monitoring of the Beauregard landslide (Aosta Valley, Italy) using advanced and conventional techniques. *Eng. Geol.* **2010**, *116*, 218–235.
113. Intrieri, E.; Gigli, G.; Mugnai, F.; Fanti, R.; Casagli, N. Design and implementation of a landslide early warning system. *Eng. Geol.* **2012**, *147*, 124–136.
114. Naidu, S.; Sajinkumar, K.S.; Oommen, T.; Anuja, V.J.; Samuel, R.A.; Muraleedharan, C. Early warning system for shallow landslides using rainfall threshold and slope stability analysis. *Geosci. Front.* **2018**, *9*, 1871–1882. [CrossRef]
115. Piciullo, L.; Calvello, M.; Cepeda, J.M. Territorial early warning systems for rainfall-induced landslides. *Earth-Sci. Rev.* **2018**, *179*, 228–247.
116. Guzzetti, F.; Gariano, S.L.; Peruccacci, S.; Brunetti, M.T.; Marchesini, I.; Rossi, M.; Melillo, M. Geographical landslide early warning systems. *Earth-Sci. Rev.* **2020**, *200*, 102973. [CrossRef]
117. Xu, Q.; Peng, D.; Zhang, S.; Zhu, X.; He, C.; Qi, X.; Zhao, K.; Xiu, D.; Ju, N. Successful implementations of a real-time and intelligent early warning system for loess landslides on the Heifangtai terrace, China. *Eng. Geol.* **2020**, *278*, 105817. [CrossRef]
118. Tzouvaras, M. Statistical time-series analysis of interferometric coherence from sentinel-1 sensors for landslide detection and early warning. *Sensors* **2021**, *21*, 6799. [CrossRef]

Article

Geomorphological Hazard in Active Tectonics Area: Study Cases from Sibillini Mountains Thrust System (Central Apennines)

Domenico Aringoli [1,*], Piero Farabollini [1], Gilberto Pambianchi [1], Marco Materazzi [1], Margherita Bufalini [1], Emy Fuffa [1], Matteo Gentilucci [1] and Gianni Scalella [2]

[1] School of Sciences and Technology, Geology Division, University of Camerino, Viale Gentile III da Varano, 7, 62032 Camerino, Italy; piero.farabollini@unicam.it (P.F.); gilberto.pambianchi@unicam.it (G.P.); marco.materazzi@unicam.it (M.M.); margherita.bufalini@unicam.it (M.B.); emy.fuffa@unicam.it (E.F.); matteo.gentilucci@unicam.it (M.G.)

[2] Office of the Extraordinary Commissary for Central Italy Earthquake 2016, via della Ferratella in Laterano, 51, 00184 Rome, Italy; gianni.scalella@regione.marche.it

* Correspondence: domenico.aringoli@unicam.it

Citation: Aringoli, D.; Farabollini, P.; Pambianchi, G.; Materazzi, M.; Bufalini, M.; Fuffa, E.; Gentilucci, M.; Scalella, G. Geomorphological Hazard in Active Tectonics Area: Study Cases from Sibillini Mountains Thrust System (Central Apennines). *Land* **2021**, *10*, 510. https://doi.org/10.3390/land10050510

Academic Editor: Oren Ackermann

Received: 28 March 2021
Accepted: 7 May 2021
Published: 11 May 2021

Publisher's Note: MDPI stays neutral with regard to jurisdictional claims in published maps and institutional affiliations.

Copyright: © 2021 by the authors. Licensee MDPI, Basel, Switzerland. This article is an open access article distributed under the terms and conditions of the Creative Commons Attribution (CC BY) license (https://creativecommons.org/licenses/by/4.0/).

Abstract: In many areas of the Umbria-Marche Apennines, evident traces of huge landslides have been recognized; these probably occurred in the Upper Pleistocene and are conditioned by the tectonic-structural setting of the involved Meso-Cenozoic formations, in a sector of the Sibillini Mountains (central Italy). The present work aimed to focus on a geomorphological hazard in the tectonic-structural setting of a complex area that is the basis of several gravitational occurrences in different types and mechanisms, but nonetheless with very considerable extension and total destabilized volume. An aerophoto-geological analysis and geomorphological survey allowed verification of how the main predisposing factor of these phenomena is connected with the presence in depth of an important tectonic-structural element: the plane of the Sibillini Mountains thrust, which brings the pre-evaporitic member of the Laga Formation in contact with the Cretaceous-Eocene limestone lithotypes (from the Maiolica to the Scaglia Rosata Formations) of the Umbria-Marche sedimentary sequence. Another important element for the mass movements activation is the presence of an important and vast water table and related aquifer, confined prevalently by the different structural elements and in particular by the thrust plane, which has acted and has continued to act, weakening the rocky masses and the overlaying terrains.

Keywords: large-scale landslides; DSGSDs; normal faults and overthrusts; Sibillini Mts.; Central Apennines

1. Introduction

The Central Apennines (Italy) is a young and tectonically active mountain range characterized by a high structural complexity. The single structures related to the different tectonic phases interact with each other, favoring the reactivation of the old ones or leading to the segmentation of the recently formed ones. The result is a strong stress, both on the seismic characteristics of the area and on the geomorphological evolution of the Apennine reliefs, especially those characterized by the presence of important tectonic elements. In general, it is accepted that tectonics can play a dual role in influencing the gravitational evolution of the slopes: (i) a passive role, related to the influence on the structural setting of the slopes, which can be inherited from a tectonic phase no longer active; and (ii) an active role, represented by the changes that it can determine on the slopes, producing increases in the relief energy and tensional stress suffered by the rock volumes [1].

In this perspective, a study was carried out on the southern slopes of the Sibillini Mountains ridge in correspondence with the area most affected by the recent seismic sequence where, through detailed geological and geomorphological surveys, several large landslides and deep-seated gravitational slope deformations (DSGSDs) have been mapped.

Analyses and kinematic verifications allow us to hypothesize an important correlation both with the present tectonic structures and with the hydrogeological characteristics of the rocky masses and the contained aquifers, in analogy with what has already been verified in other areas of the Alps and of the Apennines chain [1–10] or in other areas of the Earth's surface on various mountain ridges, such as Bethic Belt, Carpathian Belt, Tien Shan Mountains, Pamir Mountains, and Williams Range in Colorado [11–18].

The present work therefore aims to highlight how the combination of the above-mentioned factors is fundamental in many gravitational processes ranging from large mass movements to huge landslides and to DSGSD sensu [2], as well as in more superficial phenomena and in "classical" ones [19].

The relationship between active tectonics and DSGSDs and/or large landslides along the tectonic slopes of the Sibillini Mountains also plays an extremely important role in differentiating the risk associated with seismic and/or hydrogeological events occurring in the area.

2. Data and Methods

2.1. Geological and Structural Setting

In the area outcrops the typical Umbria-Marche Succession, a sedimentary sequence consisting of a stratified sedimentary succession of pelagic environment, in which calcareous and marly-calcareous lithotypes alternate with siliceous ones (Figure 1). This succession is set on massive limestones, disjointed by an extensional tectonic phase in the Middle Lias. This situation gave rise both to the deposition of a complete succession, characterized by considerable thicknesses, and to the deposition of reduced/condensed succession, characterized by considerable variations in thickness and with singular lithological features [20–22].

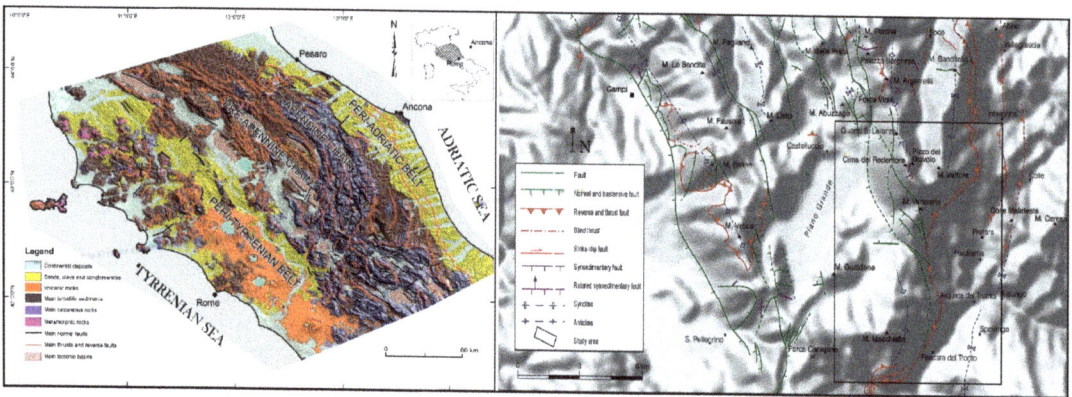

Figure 1. Geological and structural sketch of the Sibillini Mountains area.

The typical structural arrangement is that of thrusts and folds chain with a prevalently eastern vergence, with axes in approximately N–S direction (Figure 1). The tectonic movement that gave rise to the relief has produced a series of overthrusts and inverse faults, very evident on the eastern flanks of the anticlines, which caused the shortening and overlapping of older terms on more recent deposits. The main structural element that delimits to the east the carbonatic ridge is the Sibillini Mountains thrust, which extends southward until it joins the "Ancona-Anzio Line" [23–25].

The structure of the thrust front is characterized by a northeastern arched geometry and Adriatic convexity; in the northern part the trend is "apenninic" (NNW–SSE), while in the southern part it is approximately N–S. This regional tectonic style presents a complex articulation: in some areas it is realized through a main single surface, while in others through two intersecting surfaces delimiting an intermediate body more advanced than the lower one but more backward than the upper one [24,26,27].

The Sibillini Mts. thrust brought the internal Meso-Cenozoic formations on the more external Cretaceous-Paleogenic terms in the northern sector with respect to the Aso River and on essentially Miocene formations in the southern sector (Figure 2). This compressive style, which delimits the carbonatic ridge, is followed by extensional tectonics, whose disjunctive faults have profoundly modified the original stratigraphic structure [28]. This extensional phase then gave rise to the intermontane depressions (Castelluccio di Norcia, Norcia, Cascia, etc.; Figure 1) filled by quaternary continental sediments, characterized by lacustrine and alluvial deposition since the Pliocene [25,29,30]. At the southwestern edge of the study area, a set of compressive tectonic lines, with Apennine direction and northeastern vergence, produced an intense deformation of the different lithotypes, so much so as to totally alter their original hydrogeological characteristics: in fact, locally has been formed a hydraulic barrier, separating the considered area from the adjacent ones.

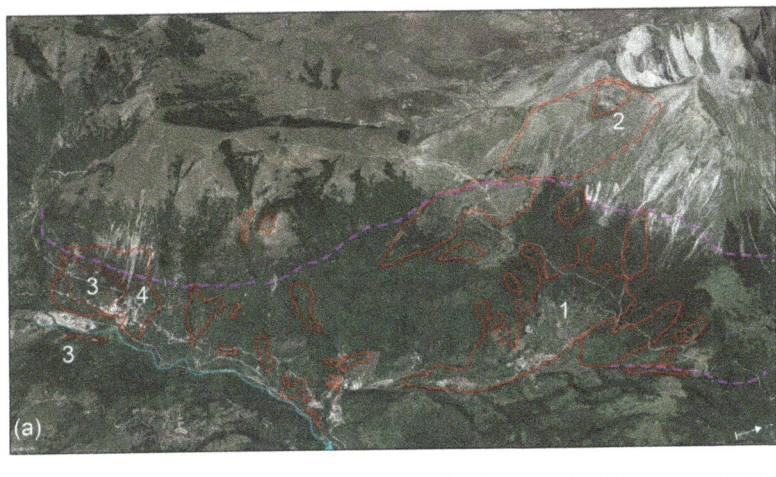

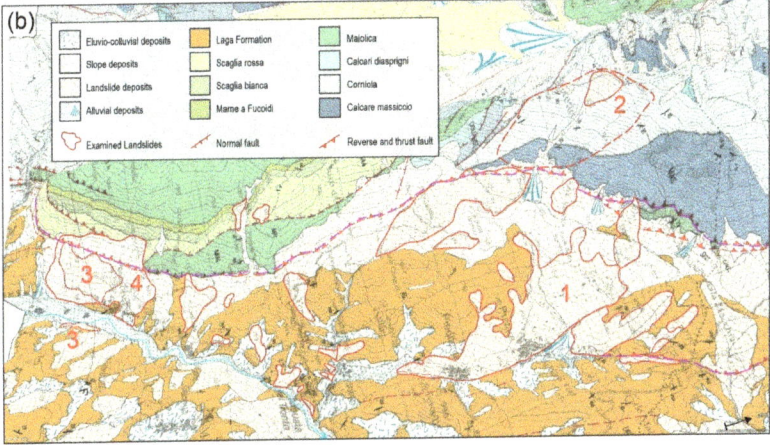

Figure 2. Three-dimensional physiographic schemes of the study area (**a**) (source Google Earth) and corresponding geological sketch (**b**) (after [28], modified). The numbers refer to the gravitational phenomena described in the text: 1, Pretare-Piedilama; 2, Mount Vettoretto; 3 and 4, Pescara del Tronto.

The persistence of the extensional activity in the Apennines is documented in several papers on active tectonics and paleoseismology [31,32] and by the occurrence of earthquakes with M ≥ 6, the last of which occurred in 2016.

2.2. Geomorphological Phenomena and Evolution

The geomorphological structure of the area results from the action of morphogenetic processes that have shaped and still shape the region, strongly conditioned by several factors interacting each other: (i) the litho-structural and geomechanical characteristics of the bedrock that have exercised an important control over the morphological structure of slopes and valley floors, favoring the activation of gravitational phenomena in different types and sizes, directly conditioning the setting of drainage networks; (ii) Plio-Quaternary tectonics and up-lift, which have affected the Apennine belt since the Upper Miocene (in correspondence with the main tectonic phase of the chain) and, subsequently, in the Quaternary (starting from the end of the Lower Pleistocene) and testified by the strong seismicity of the area; (iii) the Quaternary climatic variations there were in the area after its emergence, which occurred from the Upper Miocene, coinciding with the tectogenetic acme of the Apennine chain that has activated specific morphogenetic processes, such as those responsible for the formation of glacial masses (glacial processes) and the frost action in the ground (periglacial processes) and which control the rates of surface alteration, slope degradation, and debris production; and iv) human activity (agriculture, urbanization, water regulation, excavation of gravels from the riverbeds, etc.), which has represented and continues to represent one of the main morphogenesis factors responsible for the activation of erosion and accumulation processes, considerably faster and more intense than those due to natural causes [2,3,33–35].

Studies on the area and new research following the 2016 seismic events [28–31,36–39] have allowed us to establish how the gravity in the Apennines chain area has its main expression in the genesis of large landslides and DSGSD [3,40,41]. In fact, the gravitational morphogenesis is particularly evident on the eastern slopes of the carbonatic ridges, along the folded sides of the structures and/or on the thrust fronts. Along these slopes can be observed steps, trenches, undulations, and fractures, together with large, sometimes rounded, landslide scarps; at the foot of the slopes, vast and thick debris deposits, sometimes coalescent, cover the more recent marly-clayey formations of the Umbria-Marche Succession. It has also been possible to identify important correlations between the morphometric and physiographic characteristics of the slopes and the geological structure of the reliefs [6,42,43]. The analysis of the morphometric and plano-altimetric characteristics of specific morphological elements (old planation surfaces, etc.), their "freshness", and their position with respect to the continental deposits has also allowed their chronological location and sometimes the understanding of their different evolutive stages. As regards the overall activity of the studied processes, considering the morphoclimatic context of the area and the position of the deposits on the slopes, it is possible to verify how almost all the processes are dormant [10,39,44].

The largest landslides present in the study area (Figure 2) were connected to mixed kinematics and generally involved reactivation of ancient debris deposits or removal of thick landslide accumulations—in some cases, the same debris from slopes sutures of ancient landslide bodies that were resting on Miocene marly-clayey bedrock (e.g., palaeolandslide of Pescara del Tronto) [6,10,39] were subsequently involved in new surficial reactivations.

In this stratigraphic framework, important instability elements are represented by the general hydrogeological setting and by the permeability of the lithotypes affected by gravity; this condition favors the activation of shear surfaces in the most clayey lithotypes with consequent passive transport of the overlying masses, sometimes even of several kilometers [6].

3. Case Histories

Along fault scarps and steep erosional slopes, deep-seated gravitational deformations and large-scale landslides, up to several square kilometers wide, may be observed. These phenomena are very common in the axial part of the Apennines (Figure 1), which was affected by strong uplift and extensional tectonics during the Late Quaternary. Therefore, the predisposing factors were identified not only in the recent tectonic uplift of the area, which determine high relief value, but also in the intense fracturing of the bedrock as well as in the presence of rigid rocky bodies overlying more plastic marly-clayey levels. Among the stratigraphic discontinuities, the pre-transgressive erosional surface plays a leading role because the main sliding surfaces coincide with it [2,3,44,45].

Important deep-seated gravitational phenomena are surveyed in the Sibillini Mountains on the eastern side of the calcareous ridge, in particular along the eastern oversteepened sides of anticline fold and overthrust fronts [6,10,46]. Along these slopes, steps, trenches, undulations, and fractures may be observed, together with large-scale landslide scarps and wide debris deposits at the footslopes. These materials are often covered by Upper Pleistocene stratified slope-waste deposits, which sometimes are tilted counterslope.

The above phenomenologies are particularly diffused in the area between the calcareous and the turbiditic lithotypes, along the overthrust front of the Sibillini Mts. that, in the present case studies, emerge just above of the alignment of Pescara del Tronto-Arquata del Tronto towns (Figure 2). Below are the descriptions of some of the most important phenomena recognized and represented in the geomorphological scheme of Figure 2.

3.1. The Pretare–Piedilama Landslides

At the base of the eastern flank of Mount Vettore, on the SW of Mount Pianello della Macchia, there is a large landslide phenomenon, partly obliterated by thick slope debris and debris flow deposits (Figure 3), whose accumulation extends as far as the built-up area of Pretare and Piedilama. The two inhabited centers are in fact located on the accumulation of an ancient landslide with huge calcareous blocks, fed by the hanging wall of the Sibillini Mountains, which occupy the bottom of the Fosso di Morricone valley, engraved on the flysch of the Laga Formation.

From a structural point of view, the flysch sequence on the hydrographic left of the Morricone stream constitutes the western side of an anticline (Figure 3a,b), roughly N 5–10° E oriented, dipping to the East with inclinations generally higher than 60°. On the hydrographic right of the Morricone stream, the reverse side of a further anticline fold is exposed, with an axis approximately oriented like the one previously described with oriental vergence, which is connected to the buried overthrusts but whose continuation is immediately recognizable northward of Pretare, in the direction of Montegallo town. This tectonic-structural setting (cross-section in Figure 3b) makes the Morricone stream an asymmetrical valley, downcutted into the nucleus of the above-mentioned anticline, with a hydrographic side on the left steeper than the one on the right, originating a subsequent ortho-clinal valley [47] and representing an excellent example of geomorphological convergence with the fault slopes [48].

The wide and complex landslide phenomenon (Figure 2) [10] originates in fact at the base of the Mount Vettore slope, on the footwall of the Sibillini Mountains thrust; it extends for several kilometers along the slope up to the Tronto River, with different overlapping mechanisms, essentially related to the stratigraphic-bedding conditioning and to the physical-mechanical characteristics of the bedrock materials.

The oldest continental deposits are represented by chaotic accumulations of mainly carbonate materials, with sharp or poorly rounded edges, in a reddish sandy-silty matrix, with metric levels of locally cemented carbonatic levels, and containing heterometric blocks of carbonate nature. Generally, these deposits are covered by stratified slope waste deposits attributable to the Upper Pleistocene, sometimes tilted mountainward; the two units are locally interspersed, showing how the movement is rather ancient and has undergone successive reactivations [1,3].

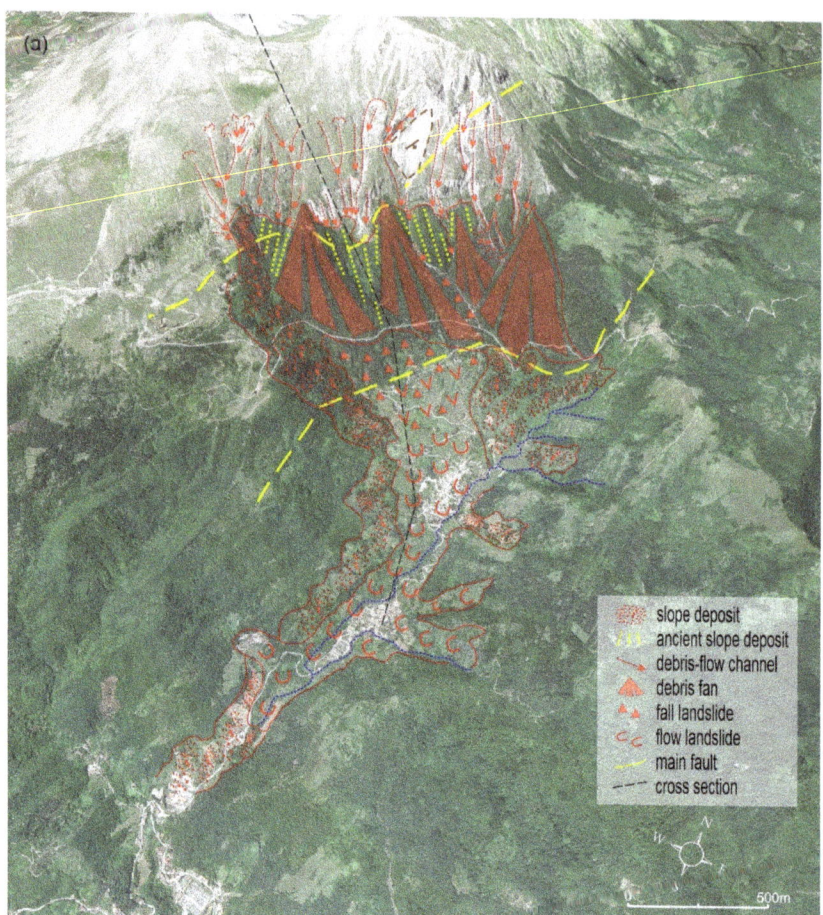
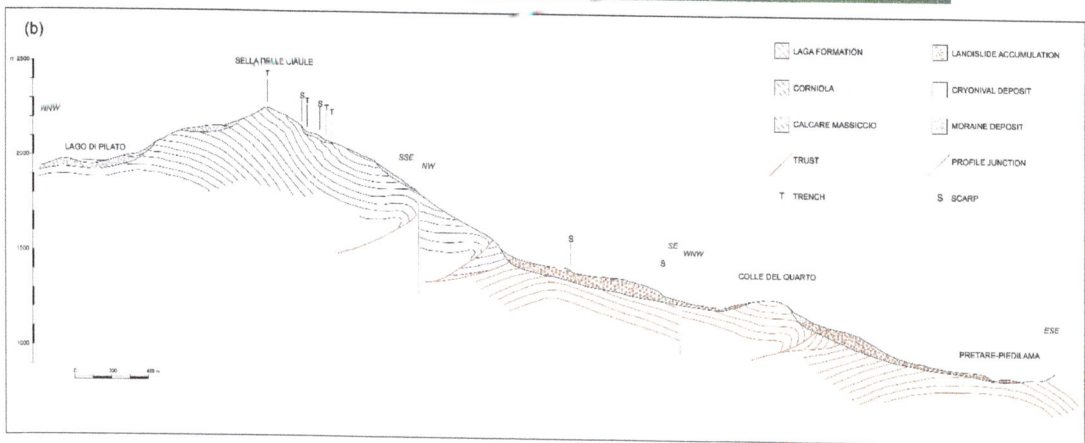

Figure 3. (**a**) Panoramic view of the southeastern slope of Mount Vettore area with the Pretare and Piedilama towns, which are found in the ancient and largest earthflow along the Morricone stream valley. (**b**) Geological and geomorphological schematic cross-section of the phenomena.

The crown is rather irregular because the bedding of the Maiolica formation, which here verticalizes and overturns, has determined and continues to determine frequent collapsing and toppling phenomena (Figure 4); furthermore, the calcareous or marly-calcareous blocks and boulders are transported up to 4 km away from the origin. Along the whole southeastern side of the Mount Vettore–Il Pizzo alignment, in fact, extensive vertical escarpments, still fresh, with an articulated development, are in any case recognizable; this testifies to the repeated and continuous collapses, whose blocks and debris are recognizable at the foot of the slope, constituting an extensive debris layer given by coalescent cones, locally vegetated (Figures 2 and 4).

Figure 4. Southeastern slope of Mount Vettore. The decameter-sized blocks in front are related to the fall phenomena on this flank of the mount. In the top center, it is possible to recognize the scarp of one of the falls that occurred in the past, and at the foot of the slope, the extensive debris accumulation, completely vegetated, partially generated by debris flow processes and by fall landslides.

The large amount of debris accumulated at the foot of the slope, once the load limit has been reached, has triggered flows that were probably facilitated both by the bedding of the Laga Formation and by the sandy-clayey deposits resulting from the alteration of the above formation. The medium-terminal portion of the landslide body, which from the III level seismic microzonation investigations was estimated to be about 30 m thick [49,50], was channeled into the valley of the Moricone stream due to the strong acclivity of the slope and evolved into a real landslide on which large blocks of Maiolica floated, like real rafts. The impressive tongues of boulders and amassed blocks, extended for several kilometers along the slope, are evidence of the passive transport operated by the casting phenomena generally set on the mostly marly-clayey lithotypes (Figure 3).

3.2. The Mt Vettoretto DSGSD

At the top of Mount Vettore, at about 2300 m above sea level, along the alignment of Punta di Prato Pulito–Mount Vettore, near the Zilioli hut, numerous trenches and ridge splits were observed (Figure 5); these are oriented about E–W, with lengths of over 100 m and widths of a few meters. Inside these depressions, more than one meter deep, partially filled with debris and soil, cracks and fractures of about 20 cm deep and 5 cm wide and free-face in rock are visible, denoting compaction following the seismic events of 2016.

Figure 5. The upper part of the DSGSD with the trenches/double ridge near the Zilioli hut.

The southwestern slope of Mount Vettore is strongly deformed (Figure 6) and presents a morphology with a convex profile and small counter slopes, many of which are rounded and filled with soil and debris; the stratification along the slope, although the lithology is the same, presents undulations and micro folds, and the position changes with respect to the summit of the slope.

The above-mentioned morphological elements allow us to hypothesize the presence of a deep mass rock creep [51,52] whose evidences in the upper part of the slope can be masked by debris; the deformation could reach hundreds of meters. The genesis of the phenomenon can be attributed to the high relief energy due to the quaternary tectonic uplift, associated with the particular structural configuration of the area given by the interaction between the fault system of Mount Vettore and the Sibillini Mts. thrust (Figure 1).

These phenomena are generally set in depth [2,3,53,54] along the most ductile levels of the stratigraphic sequence, while on the surface they are the same discontinuities in the rigid rocky masses created by the tectonics that guide them. The interferometric data allowed, even if with some doubt, indications of the possible kinematics of the deformation, allowing delineation of the complex tectono-gravitative scenario that has affected the area of Mount Vettore and has triggered the gravitational phenomenon [55]. In fact, triggering factors can be investigated, as moreover verified in similar situations in the Apennines area, in seismic events or in particular and extreme rainfall events [6]; nevertheless, even in Apennines areas at this altitude, the mentioned gravitational movements can also be associated with glacial decompression stresses [56].

The complex geological setting is characterized by the presence of two thrust planes close together, that overthrust, respectively, the oldest lithotypes of the "Sequence" (Calcare massiccio, Corniola, etc.) onto the more recent ones of (Scaglia formations), and these latter onto the pelitic-arenaceous association of the Laga Formation. In the area being examined, meso-structural analyses have evidenced a considerable reduction of the inclination of the thrust planes toward the west with respect to the characteristic values of the eastern front, which normally are around 30°. The calcareous lithotypes show an attitude that goes from subhorizontal to roughly dipping, the same as the slope. Locally, these lithotypes are covered by strongly cemented and stratified slope deposits.

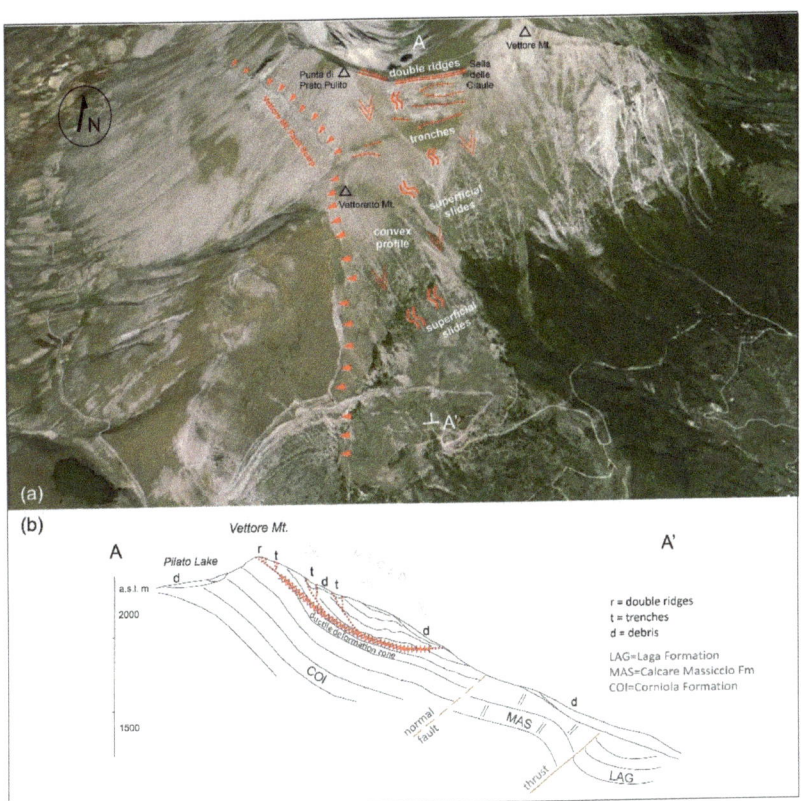

Figure 6. Panoramic view of the southwestern slope of Mount Vettore: (**a**) main geomorphological elements (source Google Earth); (**b**) schematic cross section (after [55], modified).

At the top of the slope (2000–2200 m a.s.l.), in the zone of the narrow watershed that divides it from the Pilato valley, modelled by the Pleistocene glaciation, it is possible to find trenches parallel to the direction of the watershed and that run along its entire length (double ridge). On the steep slope modelled in limestones beneath, certain landslide crowns can be recognized above which fissures, steps, trenches, undulations, and intense fracturing of the strata are present. These elements are ideally "joined" to each other by a linear scarp parallel to the slope and decimetric to metric in height. The more or less straight setting of the landform, though it is connected to polygenetic processes, allows us to hypothesize a tectonic control for its genesis, exerted by an extensive dislocation, without displacement, parallel to the slope.

The most evident landslide accumulations, where the attitude of bedrock is clearly visible, show evident rotation counterslopes in the zone of the crown and prevalent translations in the middle portion. Less extended fall phenomena also recur; their accumulations, at the foot, often evolve into debris-flows, which in turn evolve into alluvial fans. The "freshness" of all the elements permits us to associate the landforms along the calcareous slope with relatively "superficial" and recent gravitational phenomena. They are still in slow evolution, activated by roto-translational slides that take place, respectively, along the shear planes of neo-formation in the thickly stratified and subhorizontal rock masses and along shear planes predisposed by an attitude roughly dipping as the slope and by the presence of interspersed pelitic levels. The elevated frequency and density of these phenomena, anomalous with respect to that found in other sectors of the physiographic unit, allow us to hypothesize a control of the phenomena by a zone (or more than one zone) of ductile deformation. This would be predisposed by way of the probable listric geometry of the cited extensive tectonic dislocation, whose plane loses its own identity in the zone of intense tectonic fracturing between the two thrusts or in the marly levels below, which are even more ductile. On the basis of morphometric and structural considerations, these zones of deformation can be placed at a medium depth. A wider and deeper deformation, in addition, should have affected both the flank of the ridge in question, as is demonstrated by the presence of double crests. As a whole, it could fall within the category of the lateral spreads indicated by [19]. In more detail, a macrocambering phenomenon could affect the limestones of the eastern flank due to the differential dislocation realized in the pelitic-arenaceous lithotypes below the thrust, following strong weathering processes, squeezing out [57], and fluvial erosion. The phenomenon, in part associated with that described by [58] and by [59], is testified to by a considerable reduction of inclination of the thrust plane, connected, probably, to its deformation by gravity. Almost certainly, on the western flank the effects of the glacial decompression are felt.

3.3. The Pescara del Tronto Landslide and the Tronto River Occlusion

Studies carried out over a long time in the area [2–4,10,39] have made it possible to verify how mass movements can create interference with river dynamics. One of the clearest and most impressive phenomena of this type is represented by the ancient movement of Pescara del Tronto, which involved enormous volumes of marly limestones and limited levels of terrigenous sediments. Only a part of the original accumulation, consisting of large masses of intensely fractured reddish limestone and gravels mixed with debris, is preserved on the hydrographic right of the Holocene plain of the Tronto River (Figure 7), while the detachment zone is located on the medium-low section of the slope on the hydrographic left [6,10,39].

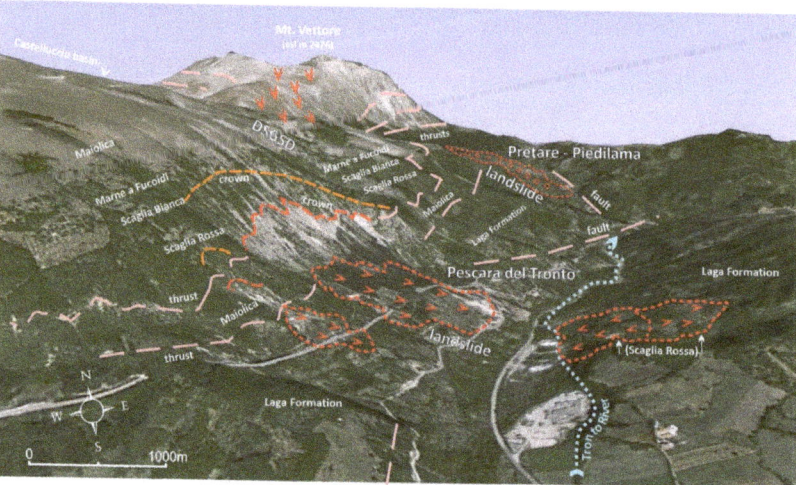

Figure 7. The topographic interpretative 3D scheme of the ancient landslide of Pescara del Tronto and, on the bottom, of the other analyzed phenomena (after [10], modified).

In this sector, the valley is narrow and deeply incised due to the convergence of the two mountain ridges: the calcareous one of the Sibillini Mountains and the turbiditic one of the Laga Mountains, oriented respectively NNE–SSW and NW–SE.

The welding of these two structures is realized through the great Sibillini thrust that overlaps the calcareous lithotypes onto the torbiditic ones. In this zone, a lateral contact must have also occurred between the two lithotypes along a roughly SW–NE alignment, as a result of a limited backthrusting. The backthrust, moreover, drove the Tronto River path in this portion of the valley; its incision, mainly connected to lower-middle Pleistocene tectonic uplifting, occurred into the arenaceous-pelitic lithotypes of the Laga formation. This is testified both by the strong asymmetry of the valley and by the lithological composition of the different generations of alluvial deposits, made up of mainly arenaceous pebbles and sandy-silty matrix.

Along the high slope on the hydrographic left, at a height of about 1000 m. a.s.l., the mentioned thrust plane outcrops. In this zone it shows southeast verging and generally subhorizontal setting, which is interrupted by a modest upward concavity close to the built-up area of Pescara del Tronto. The geometric setting of the thrust plane and the contrasting hydrogeological characteristics of the involved lithotypes probably have caused a particular concentration of the water drained from the huge calcareous aquifer overlapped onto the turbiditic one.

Mountainward with respect to the Pescara del Tronto village, on the calcareous lithotypes, it is possible to observe a large landslide crown between the heights of 900 and 1150 m a.s.l. It seems to have been generated, given its articulated configuration, by the union of several gravitational phenomena [6,10,39], the evidence of which (benches, counterslopes, and high scarps) is still present along its slope. On the hydrographic right, detailed geotematic surveys developed on the valley bottom and at the foot of the slope, have evidenced, starting from the riverbed (660 m a.s.l.) up to a height of about 850 m a.s.l., the presence of a large calcareous "plate" made up mostly of the Scaglia Rossa Formation. It lays at the base and on the southern side on the arenaceous pelitic turbidites of the Laga Formation and in numerous outcroppings still has a well-preserved stratification. The base is an exception to this because it shows an intense fracturing (levels of "breccia"). The summit portion is partially covered by a landslide accumulation made up of turbidic lithotypes of the Laga Formation. The collocation in height of the upper portion of the "plate" is a bit above that of the end of middle Pleistocene (second order) alluvial deposit limbs, which outcrop on both sides of the river.

This presence, unique for the slope on the hydrographic right, is completely anomalous from a stratigraphic and structural point of view. In fact, on the hydrographic left, the same turbidite layers overthrusted due to the thrust plane onto the limestones of the Maiolica formation (1000 m a.s.l.), while the Scaglia Rossa outcrops about 200 m mountainward (at 1100 m a.s.l.). The data shown so far have a possible complete interpretation only through the hypothesis of a large mass movement that involved a much larger volume of material and whose calcareous plate represents a residual limb.

It is possible to schematize by four different steeps the complex kinematics of the phenomena (Figure 8a): the first stage corresponds to a rotational-translational slide that overlapped the calcareous plate onto the colluvial "bed" made of mainly pebbly-sandy-silty material; the second stage is represented by the activation of slide–flow movements with shear planes occurring within the colluvial cover and/or in the top levels of the turbidic bedrock; the third stage is represented by a passive transport of the "plate" by the above-mentioned slide-flow phenomena along a gently dipping slope that was jointing the valley bottom at the end of the middle Pleistocene; there, the deposition of the second order alluvial terraces was being completed. The successive phases of valley incision, during the Late Pleistocene and the Holocene, produced erosion processes at the foot of the "plate" and triggered gravitational relaxing of the plate itself. Temporary fluvial damming could be associated with this phenomenon, but it is only possible to recognize the remnants of the landslide body, constituted by fragments of "Scaglia Rossa" limestone (Figure 8a,b).

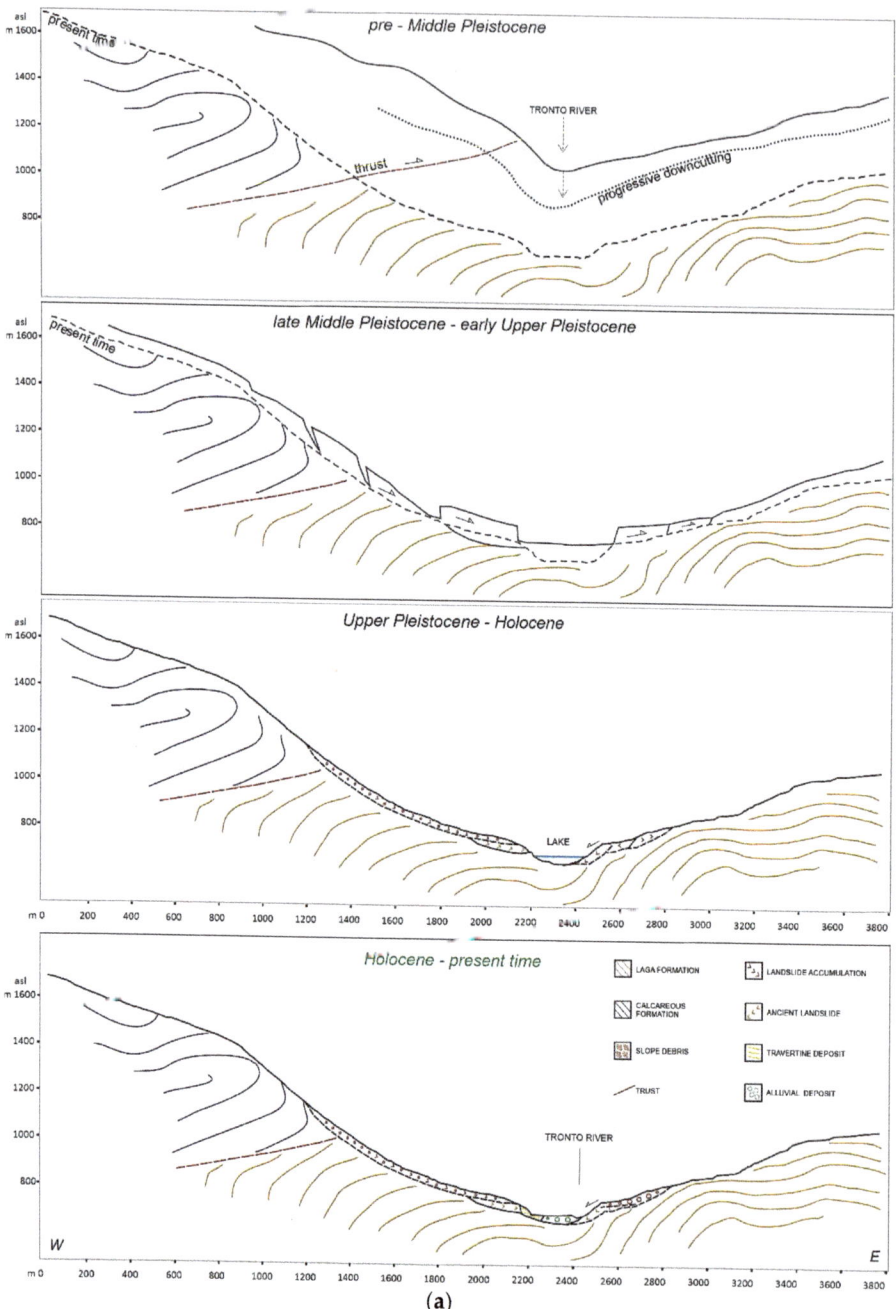

Figure 8. Cont.

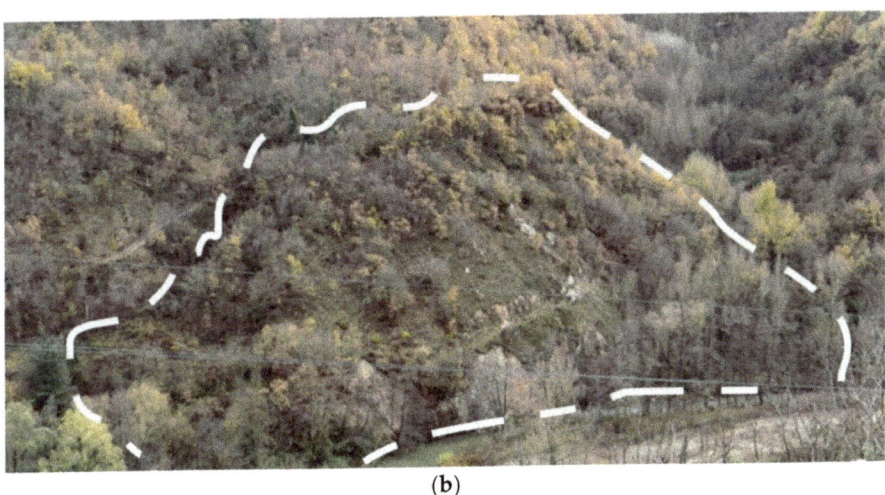

(b)

Figure 8. (a) Evolutionary sketch cross-sections of the Pescara del Tronto landslide (after [10], modified). (b) Remnants (white box) of the ancient body landslide of Pescara del Tronto in the hydrographic right of the Tronto River.

Chronologically it is possible to collocate the phenomena at the end of the mid-Pleistocene; this is demonstrated by the stratified slope deposit of the Late Pleistocene that regularizes the lower portion of the crown and, partially, the third order of alluvial deposits. The main conditioning factors of the mass movement can definitely be recognized: (i) in the intense tectonic fracturing of the rocks, realized both during the building of the compressive structures and during the uplifting and the Pleistocene extensional tectonics; (ii) in the gravitational fracturing, as a result both of the tension release within the calcareous mass following the fluvial erosion of the turbidites below and of the genesis of differential sinking in the same ductile turbidites; (iii) in the high relief; and (iv) in the decay of geomechanical parameters of the calcareous lithotypes related to the genesis of karst levels developed according to the fluvial deepening phases. With regard to activation factors, the following hypothesis is formed. Initial phase was favored by the significant increase of oriented accelerations connected to the intense seismicity of the area and the high-water pressure exercised at the base of the aquifer because it is fed by the endorheic basin of Castelluccio (over 300 m of difference in height). The following phases were favored by the great quantity of water present on the slope and supplied by numerous springs whose capacity must have been much greater than at present. In fact, in that period, the cataglacial phase at the beginning of Early Pleistocene was ongoing, with consequent dismantling of the glacial systems of the Sibillini Mountains and the great water increase in the aquifers.

3.4. The Debris Flows Phenomena of Pescara del Tronto

The geological-stratigraphic setting of the area is rather complex: the bedrock is represented by the lithotypes belonging to the pre-Evaporitic member of the Laga Formation. Slightly above the built-up area of Pescara del Tronto, these lithotypes are in contact, through a very important tectonic-structural element such as the Sibillini Mountains thrust, with the Cretaceous-Eocene calcareous lithotypes (from Maiolica to Scaglia Rossa) of the Umbria-Marche sedimentary sequence [28].

The above-described lithologies are obliterated by continental deposits of different origin. The most ancient sediments of this succession are composed of accumulations of inactive landslides, such as the one located upstream of the inhabited center of Pescara del Tronto, an object, even today, of intense mining activity. Specifically, these deposits are characterized by a chaotic structure and made up of heterogeneous, heterometric elements (also the size of blocks), with sharp edges, both matrix and clast supported, naturally stabilized, and locally covered by stratified slope debris attributed to the Late Pleistocene. The higher zone is covered by recent slope deposits, which in part overlap an older stratified phase of deposition (Figures 2 and 9). The colluvial deposits at the foot of the slopes, which often serve as a connection to the valley floor, are also limited in extent.

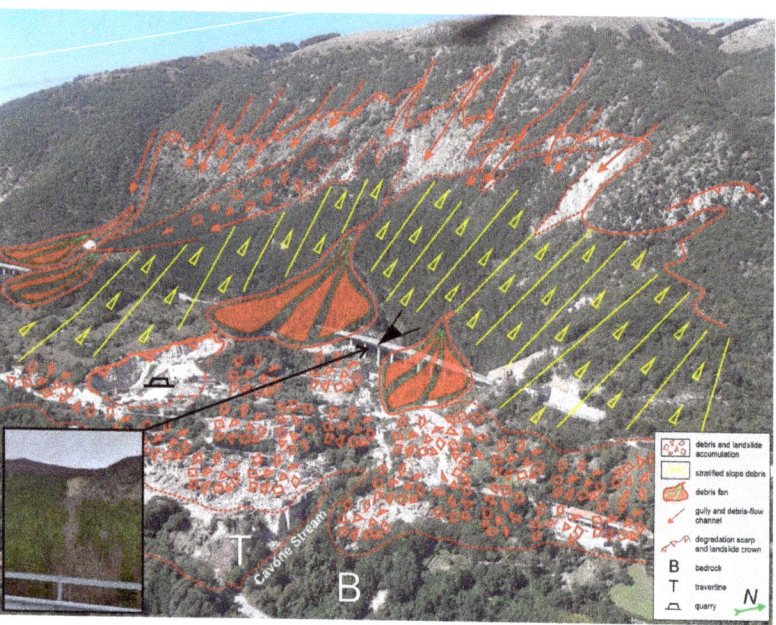

Figure 9. Overview of Pescara del Tronto, the town that was completely destroyed by the seismic events of 24 August and 30 October 2016.

The medium-high sector of the slope upstream of Pescara del Tronto is also crossed by numerous erosion channels filled, in most cases, by debris flow deposits (gravel and subangular polygenic blocks in a silty-sandy matrix). Near the bottom of the valley, gravelly-sandy and sandy-loamy alluvial deposits outcrop, engraved and terraced by the Tronto River (terrace of the first order), which along the current riverbed mainly deposits gravel and sand and subordinately blocks. At the base of the town of Pescara del Tronto and along the valley of the Cavone ditch, travertine deposits emerge in cascade and basin facies and anthropic deposits that are highly heterogeneous and characterized by complex geometrical relationships with the lithotypes present (Figures 2 and 9).

In the southwestern part of the built-up area, there is a quarry used for the extraction of aggregates, with quarry fronts given by steep sub-vertical walls even over twenty meters high. The detritus deposit has a massive structure, except in the central part of the outcrop where it is possible to recognize a pseudo-stratification slightly tilted toward the mountain. The carbonate clasts are heterometric and sharp-edged, from centimeter to meter; the sandy matrix is variable in content, giving the deposit a non-homogeneous texture. The deposit is weakly cemented. At the top of the quarry front there is a paleosoil and above it the debris from the more recent slope. On the upstream side of the built-up area, there is a landslide escarpment set on a deposit made up of intensely cataclasized limestone; the base of the landslide body presents evident consolidation interventions [39].

The debris flows phenomena present dynamics not very different from those of the manure conoids that have shaped many of the wide flat-bottomed valleys within the calcareous Apennines and the area surrounding Mount Vettore, activated on the steep slopes by the conspicuous detritus cover or the regolitic fraction of the rocky substratum. These are extremely rapid processes, generated by the saturation of the debris masses by the water which, unlike the fluvial-denudational processes, passively participates in the movement because it is not responsible for the transport of the materials. In fact, the latter do not show any hint of stratification, and their texture is extremely variable (from clays to pebbles to blocks). For historical documentation and "freshness" of the forms, the activity of these processes is placed in the recent Holocene until the beginning of the 20th century [38,41,60,61]. In particular, the continuous runoff and debris flows that occurred upstream of the town of Pescara del Tronto forced the Consorzio di Bonifica del Tronto to protect the slope through planting and reforestation, the most important of which was carried out between 1960 and 1972.

4. Geomorphological Hazard and Resilience

The aspects related to the geomorphological hazard of large-scale mass movements located along the study area can be summarized in two main points: (1) the possible catastrophic evolution when a stability threshold is overcome, due to natural events and caused by sudden climate variations; (2) the coseismic surface breaks which, similar to the coseismic surface faulting, can affect the territory during important seismic sequences such as that of August 2016 in Central Italy.

The evolution of large-scale landslides characterized by low average slip rates may involve the rapid slide of portions of an unstable slope or the catastrophic movement of entire unstable rock mass [3,15]. The main factor hindering the hazard evaluations for large-scale mass movements is due to the difficulty in the estimation of the recurrence time of the single displacement events. Indeed, the activation of the shear planes seems to be episodic, and the subsequent displacements are probably separated by long time intervals [6,9,62–64].

The study area, both for its tectonic-structural conditions and its morpho-evolutive scenario, is very interesting from the perspective of the gravitational phenomena hazard. The vast area is located across slopes including the lines of the overthrust and of the river deepening; these large slopes present very diversified scenarios but of high geomorphological hazard—proceeding from east to west we can verify the presence of different and complex geomorphological situations with increasing hazard conditions.

At the highest altitudes there is, in fact, the DSGSD of Mount Vettoretto [39,55], which in fact could constitute the lowest hazard situation among those analyzed, with the distinctive characteristics of DSGSDs that also in this case manifest its irregular evolution; these, especially during earthquakes of particular intensity, in fact, lead to an increase in the openings of the trenches of the upper portion of the slope. It is clear that the profile, even more convex than normal, favors an intense movement of the slope debris; this accumulates at lower altitudes with a consequent increase in volume of loose rock in the areas of potential triggering of further and different gravitational movements and/or rapid mass processes (debris flows, debris avalanches, falls, etc.).

Going down in altitude, one reaches the Pretare settlement, an area with a lower slope where numerous rocky blocks ("prete" in dialect) are found scattered among the debris accumulations with a smaller grain size and with rare outcrops of the bedrock in place. This area is affected by a high landslide hazard, since there may be a concomitance of various gravitational phenomena, starting in the upper portions of the slope and expanding and accumulating in this valley bottom. These are the aforementioned debris flows but also the collapse/rollover phenomena of large boulders whose areas of influence can involve significantly lower altitudes up to the Pretare village. A further element of hazard is given by these boulders that, once the initial kinematics are exhausted, can continue to move according to a conveyor belt type mechanism and give rise to phenomena similar to lateral-spread [10,39,41].

The geophysical surveys conducted in the settlements of Piedilama and Pretare for the III level of seismic microzonation studies where carried out following the 2016 seismic events (https://sisma2016data.it/microzonazione/); these highlight some areas susceptible to seismic motion amplification, determined by the impedance contrast between the accumulation of the "paleolandslide" and the bedrock. Moreover, in this general framework, the variability of the litho-technical characteristics of the bedrock under the two settlements and consisting of thick arenaceous banks within a sub-vertical sequence of pelitic-arenaceous formation, constitutes a further element of hazard [10,39,41].

Going to the west, following the same main valley floor, once the settlement of Piedilama (whose toponym means "at the foot of the landslide") is reached, the scenario is similar to the previous one but with significant variations. It is an area of confluence/accumulation of a greater number of gravitational movements both of flow type (not only debris-flows), and also of sliding type, essentially due to the remobilization of the vast accumulations above. Along the valley floor, being more hydrologically structured (presence of a stable hydrographic network), movements of remobilization of the various materials can also occur due to surface run-off water, which is able to transport mainly the finer granulometries. Finally, it is important to say that, as an area of influence of the collapsing phenomena deriving from the Mount Vettore "walls", due to the major distance this is characterized by less hazard.

The highest hazard levels are found in correspondence with the built-up area of Pescara del Tronto (Figure 9). The geological and morphological evolution of this section of the slope has contributed to the creation of various hydro-geomorphological hazard situations concentrated in a restricted area, which all together delineate a high-risk scenario for the entire zone, both slope and valley floor of the Tronto River [10]. In the highest parts of the slope, the debris layers are gradually thicker and more extensive, sometimes ancient and cemented, and in some cases strongly reincised; these, especially in heavy rainfall conditions and given the considerable dip, facilitate the activation of frequent and great mass transports, such as debris-flow and debris-avalanche [39].

Lower down, the thickness of movable materials increases due to both the dynamics of normal growth along the slope and the coalescence of the various landslide accumulations described above. Numerous water springs are also located in this middle section [10] so that both the morphology of the slope and the considerable emergence of water often cause the above-mentioned deposits to become saturated, a mechanism that in turn leads to the triggering of further and much larger gravitational phenomena of different types and evolutions. In fact, not only further debris-flow movements are triggered from this area but also extensive roto-translational slides that succeed in remobilizing the debris accumulations, sometimes even reaching the main valley floor. It should also be noted that these phenomena are often accompanied by local siphoning and sinking of the various thick accumulations, with triggering of subsidence phenomena.

Finally, at the end of the slope, near the built-up area of Pescara del Tronto, in correspondence with the Covone ditch, travertine deposits are recognizable [39], resulting from carbonate precipitation by the drain waters fed by the basal aquifer of Mount Vettore [10]. This deposit, more than 20 m thick and attributable to the Upper Pleistocene-Holocene, is composed of phytoclastic and phytohermal lithofacies and of concretions typical of riffle and pool morphologies [65] with very different degrees of cohesion, sometimes mixed with material deriving from debris flows that have characterized this area [39]. The built-up area of Pescara del Tronto rises on these deposits, and it is the different degree of cohesion of these materials to which most of the damage that the village suffered following the 2016–2017 seismic crisis can be precisely attributed.

In addition, these deposits, although of limited extension [39], are often subject to collapse/tilting phenomena that, while collapsing or tilting only a short distance, can strongly interfere with the valley floor dynamics, with the hydrographic network, and with the existing infrastructure system.

Finally, it should be remembered that the seismicity of the area contributes significantly to increasing the already high level of hazard—not only because of the oriented acceleration that reduces the trigger thresholds but also because of the numerous landslides and instabilities that can also occur simultaneously, determining what in the most serious cases is unfortunately already known.

Therefore, these situations also obligate careful reflection on the modalities that can be implemented to make the population aware of the risk propensity in the inhabited areas and in the territory in general, both to help plan interventions and also to address perceptions of risk [38,61,66] that could consequently increase resilience [61,67].

5. Discussion and Concluding Remarks

The present work aimed to contribute to the knowledge on the conditioning of active tectonic structures on the triggering of large landslides or DSGSD in a mountain front, generated by the Quaternary uplift and characterized by the presence of a very important basal aquifer.

Considering the international scientific literature, the studies that have focused on this topic are rare, as we consider the passive role exerted by tectonic structures that, generating strong disarticulation of the bedrock and in conjunction with high relief energy, favors more or less complex gravitational mechanisms. Therefore, as there are few international studies that have the linkage we discuss, the present work can make an important contribution in a poorly understood field that concerns the relationship between important tectonic structures, critical hydrogeological conditions, and high relief energies versus huge gravitational movements.

The first activation and the following evolution of the observed deep gravitational phenomena can be dated only through relative methods, since they do not have any good element of dateability. Particularly useful for this purpose are the relationships between the elements on the deformed slopes (involved in large landslides) and the stratified slope debris deposited in a periglacial environment, the terraced alluvial deposits, the lacustrine deposits, and the moraines.

These different relationships allow us to place the first activation of many phenomena before the last cold phases of the Pleistocene (or, at least, simultaneously with them), and, in many cases, to recognize their reactivations. Only rarely in historical or recent times have these later phenomena been observed, especially under strong earthquakes or extreme precipitations in the terrigenous lithotypes of the Adriatic belt.

Furthermore, the main causes of these complex gravitational movements are to be found in the intense tectonic deformation of the bedrock, which has been folded and overlapped, and in the existence of residual compressive stresses. The recent uplifting with the associated gradients favorable to erosional processes has also facilitated the activation of lateral spreading phenomena, according to the type described by [68]; these mechanisms have given rise to ridge splitting, lateral expansions, and sackungs, mainly along slopes modelled on stratified limestone formations. Additionally, slides (deep-seated block-slide) along pre-existing planes or shear zones or in marly-clayey layers occurred; these have also evolved into further gravitational phenomena, especially at the margins of the large deformed masses.

The variation of the geomechanical parameters relating to the landslide bodies' rocks is a consequence of the displacements due to the internal deformation of the rock mass [69]; it follows that the progressive reduction of the intrinsic shear strength at the base of the rock mass depends on the lithology and its internal structure, on the type and stage of slope evolution, as well as on the presence of water [2,8,9,70]. The combination of these factors conditions, in fact, the post-collapse stress-strain behavior of the rocky mass at the considered moment of the slope "history"; therefore, in case of large and complex landslide phenomena, although climatic variations [71] may play a primary role in reducing the recurrence interval between displacement events, the rock mass features may have a minor influence.

Due to the long recurrence intervals in the activity of capable faults, such as the Mount Vettore fault, which has conditioned gravitational movements in the past and still does, the occurrence of catastrophic phenomena may take tens of thousands of years or may not occur if other factors (particularly the climate) have a negligible impact on the evolution of the phenomena. If the long recurrence interval for fault activation can be a reassuring factor, the seismic amplification registered on the ground along a fault is a process that can condition the evolution of landslides. Thus, is evident that anomaly amplified seismic waves can accelerate the evolution of instability phenomena in sudden and catastrophic events.

The systematic geomorphological analysis carried out in the study area has made it possible to understand the main genetic factors of surveyed mass movements, which can be summarized as follows:

(a) lithology of the bedrock characterized by high thicknesses of massive limestone rocks or arenaceous formations overlying less resistant marls or thinly stratified levels with pelitic intercalations;
(b) geological structures deriving from compressive tectonics to which is connected the presence of strong residual tensions and shear zones, along which gravitational shifts that may occur on a large scale;
(c) recent strong uplift that has characterized the ridge areas, where the presence of resistant rocks and the river downcutting has given rise to a high relief energy with extended and steep slopes, limited by deep transverse valleys or overthrust fronts;
(d) extensional tectonics that have given rise to high fault scarps that sometimes allowed the outcropping of potential sliding planes along which huge gravitational phenomena have set in;
(e) high seismicity of the area, particularly active along the axes of the chain.

Therefore, the location of many landslides along the seismogenic faults represents a factor that can considerably increase the local hazard related to gravitational phenomena, especially if these are conditioned by an important relief energy.

Author Contributions: Conceptualization, D.A. and P.F.; methodology, D.A. and P.F.; software, D.A. and P.F.; validation, D.A., P.F., M.M. and G.P.; formal analysis, D.A., P.F., M.M., G.P., M.B., E.F. and M.G.; investigation, D.A., P.F. and G.S.; resources, D.A., P.F. and G.P.; data curation, D.A. and P.F.; writing—original draft preparation, D.A. and P.F.; writing—review and editing, D.A. and P.F.; visualization, D.A., P.F., M.M., G.P., M.B., E.F. and M.G.; supervision, D.A., P.F., M.M. and G.P.; project administration, D.A. and P.F.; funding acquisition M.M. and G.P. All authors have read and agreed to the published version of the manuscript.

Funding: This research received no external funding.

Institutional Review Board Statement: Not applicable.

Informed Consent Statement: Not applicable.

Conflicts of Interest: The authors declare no conflict of interest.

Software: All data processing necessary to produce the maps was performed using QGIS (v.3.16). Figures in the document were created using CorelDraw Home&Student 2019.

References

1. Gentili, B.; Pambianchi, G. *Deep-Seated Gravitational Slope Deformations and Large Landslides in the Central Apennines (F.124 Macerata)*; Interlinea Editors: Teramo, Italy, 1993.
2. Dramis, F.; Sorriso Valvo, M. Deep seated gravitational slope deformations, related landslides and tectonics. *Eng. Geol.* **1994**, *38*, 231–243. [CrossRef]
3. Dramis, F.; Farabollini, P.; Gentili, B.; Pambianchi, G. Neotectonics and large-scale gravitational phenomena in the Umbria–Marche Apennines, Italy. In *Steepland Geomorphology*; Slaymaker, O., Ed.; J. Wiley & Sons: Chichester, UK, 1995; pp. 199–217.
4. Farabollini, P.; Folchi Vici D'Arcevia, C.; Gentili, B.; Luzi, L.; Pambianchi, G.; Viglione, F. The gravitational morphogenesis in central Apennine lithoid formations (in Italian: La morfogenesi gravitativa nelle formazioni litoidi dell'Appennino centrale). *Mem. Soc. Geol. It.* **1995**, *50*, 123–136.
5. Coltorti, M.; Farabollini, P.; Gentili, B.; Pambianchi, G. Geomorphological evidence for anti-Apennine faults in the Umbro-Marchean Apennines and in the peri-Adriatic basin, Italy. *Geomorphology* **1996**, *15*, 33–45. [CrossRef]
6. Dramis, F.; Gentili, B.; Pambianchi, G.; Aringoli, D. The gravitational morphogenesis in the Adriatic side of the Marche region (in Italian: La morfogenesi gravitativa nel versante adriatico marchigiano). *Studi Geol. Camerti* **2002**, *1*, 103–125.
7. Di Luzio, E.; Saroli, M.; Esposito, C.; Bianchi-Fasani, G.; Cavinato, G.P.; Scarascia Mugnozza, G. Influence of structural framework on mountain slope deformation in the Maiella anticline (Central Apennines, Italy). *Geomorphology* **2004**, *60*, 417–432. [CrossRef]

8. Martino, S.; Moscatelli, M.; Scarascia Mugnozza, G. Quaternary mass movements controlled by a structurally complex setting in the Central Apennines (Italy). *Eng. Geol.* **2004**, *72*, 33–55. [CrossRef]
9. Galadini, F. Quaternary tectonics and large-scale gravitational deformations with evidence of rock-slide displacements in the Central Apennines (central Italy). *Geomorphology* **2006**, *82*, 201–228. [CrossRef]
10. Aringoli, D.; Gentili, B.; Materazzi, M.; Pambianchi, G. Mass movements in Adriatic Central Italy: Activation and evolutive control factors. In *Landslides: Causes, Types and Effects*, 1st ed.; Werner, E.D., Friedman, H.P., Eds.; Nova Science Publisher, Inc.: New York, NY, USA, 2010; pp. 1–71.
11. Kellogg, K.S. Tectonic controls on a large landslide complex: Williams Fork Mountains near Dillon, Colorado. *Geomorphology* **2001**, *41*, 355–368. [CrossRef]
12. Ambrosi, C.; Crosta, G.B. Large sackung along major tectonic features in the Central Italian Alps. *Eng. Geol.* **2006**, *83*, 183–200. [CrossRef]
13. Hippolyte, J.C.; Brocard, G.; Tardy, M.; Nicoud, G.; Bourlès, D.; Braucher, R.; Ménard, G.; Souffaché, B. The recent fault scarps of the Western Alps (France): Tectonic surface ruptures or gravitational sackung scarps? A combined mapping, geomorphic, levelling, and 10Be dating approach. *Tectonophysics* **2006**, *418*, 255–276. [CrossRef]
14. Hradecky, J.; Panek, T. Deep-seated gravitational slope deformations and their influence on consequent mass movements (case studies from the highest part of the Czech Carpathians). *Nat. Hazards* **2008**, *45*, 235–253. [CrossRef]
15. Zerathe, S.; Lebourg, T. Evolution stages of large deep-seated landslides at the front of a subalpine meridional chain (Maritime-Alps, France). *Geomorphology* **2012**, *138*, 390–403. [CrossRef]
16. Panek, T.; Klimes, J. Temporal behavior of deep-seated gravitational slope deformations: A review. *Earth Sci. Rev.* **2016**, *156*, 14–38. [CrossRef]
17. Alfaro, P.; Delgado, J.; Esposito, C.; Tortosa, F.G.; Marmoni, G.M.; Martino, S. Time-dependent modelling of a mountain front retreat due to a fold-to-fault controlled lateral spreading. *Tectonophysics* **2019**, *773*, 228233. [CrossRef]
18. Teshebaeva, K.; Echtler, H.; Bookhagen, B.; Strecker, M. Deep-seated gravitational slope deformation (DSGSD) and slow-moving landslides in the southern Tien Shan Mountains: New insights from InSAR, tectonic and geomorphic analysis Kanayim Teshebaeva. Earth Surf. Process. *Landforms* **2019**, *44*, 2333–2348. [CrossRef]
19. Varnes, D.J. *Slope Movement Types and Processes*; Landslides: Analysis and Control (U.S.), National Res. Council, Transportation Res. Board Special Report; National Academy of Sciences: Washington, DC, USA, 1978; Volume 76, pp. 11–33.
20. Centamore, E.; Deiana, G. The Geology of Marches (in Italian: La Geologia delle Marche). *Studi Geol. Camerti* **1986**, *13*, 9–27.
21. Centamore, E.; Cantalamessa, G.; Micarelli, A.; Potetti, M.; Berti, D.; Bigi, S.; Morelli, C.; Ridolfi, M. Stratigraphy and facies analysis of Miocene and lower Pliocene deposits of the Marches-Abruzzo foreland and adjacent areas (in Italian: Stratigrafia e analisi di facies dei depositi del Miocene e del Pliocene inferiore dell'avanfossa marchigiano-abruzzese e delle zone limitrofe). *Studi Geol. Camerti* **1991**, *11*, 125–131.
22. Centamore, E.; Micarelli, A. Stratigraphy (in Italian: Stratigrafia). In *The Physical Environment of the Marches (Italian Version: L'ambiente Fisico delle Marche)*; S.E.L.C.A.: Florence, Italy, 1991; pp. 5–58.
23. Castellarin, A.; Colacicchi, A.; Praturlon, A. Distensive phases, transcurrences and overthrusts along the "Ancona-Anzio" line from the Middle Lias to the Pliocene (in Italian: Fasi distensive, trascorrenze e sovrascorrimenti lungo la linea "Ancona-Anzio" dal Lias medio al Pliocene). *Geol. Romana* **1978**, *17*, 161–189.
24. Calamita, F.; Deiana, G. Correlation between neogeneic-quaternary deformation events in the Tuscan-Umbrian-Marches sector (in Italian: Correlazione tra gli eventi deformativi neogenico-quaternari del settore tosco-umbro-marchigiano). *Studi Geol. Camerti* **1995**, *1*, 137–152.
25. Calamita, F.; Coltorti, M.; Farabollini, P.; Pizzi, A. Quaternary normal faults in the Umbria-Marches Apennines Ridge. Proposal of an inversion tectonics model. (in Italian: Le faglie normali quaternarie nella Dorsale appenninica umbro-marchigiana. Proposta di un modello di tettonica d'inversione). *Studi Geol. Camerti* **1994**, 211–225. [CrossRef]
26. Lavecchia, D. The Sibillini Mountains overthrust: Kinematic and structural analysis (in Italian: Il sovrascorrimento dei Monti Sibillini: Analisi cinematica e strutturale). *Boll. Soc. Geol. Ital.* **1985**, *104*, 161–194.
27. Cooper, C.J.; Burbi, L. The geology of the central Sibillini mountains. *Mem. Soc. Geol. It.* **1988**, *35*, 323–347.
28. Pierantoni, P.P.; Deiana, G.; Galdenzi, S. Stratigraphic and structural features of the Sibillini Mountains (Umbria-Marche Apennines, Italy). *Ital. J. Geosci.* **2013**, *132*, 497–520. [CrossRef]
29. Coltorti, M.; Farabollini, P. Quaternary Evolution of the "Castelluccio di Norcia" Basin (Umbro-Marchean Apennines, Central Italy). *Il Quat.* **1995**, *8*, 149–166.
30. Aringoli, D.; Cavitolo, P.; Farabollini, P.; Galindo-Zaldivar, J.; Gentili, B.; Giano, S.I.; Lopez-Garrido, A.C.; Materazzi, M.; Nibbi, L.; Pedrera, A.; et al. Morphotectonic characterization of the quaternary intermontane basins of the Umbria-Marche Apennines (Italy). *Rend. Lincei* **2014**, *25*, 111–128. [CrossRef]
31. Cello, G.; Mazzoli, S.; Tondi, E.; Turco, E. Active tectonics in the Central Apennines and possible implications for seismic hazard analysis in peninsular Italy. *Tectonophysics* **1997**, *272*, 43–68. [CrossRef]
32. Galadini, F.; Galli, P. Active tectonics in the Central Apennines (Italy)—Input data for seismic hazard assessment. *Nat. Hazards* **2000**, *22*, 225–268. [CrossRef]

33. Ambrosetti, P.; Carraro, F.; Deiana, G.; Dramis, F. Il sollevamento dell'Italia centrale tra il Pleistocene inferiore ed il Pleistocene medio. In *Progetto Finalizzato "Geodinamica", Contributi Conclusivi per la Realizzazione della Carta Neotettonica d'Italia*; CNR: Roma, Italy, 1982; Volume 1, pp. 219–223.
34. Dramis, F. The role of wide-range tectonic uplifts in the genesis of the Apennines relief. (in Italian: Il ruolo dei sollevamenti tettonici a largo raggio nella genesi del rilievo appenninico). *Studi Geol. Camerti* **1992**, *1*, 9–15.
35. Coltorti, M.; Dramis, F. *Geology and Geomorphology of the Sibillini Mountains (in Italian: Geologia e Geomorfologia dei Monti Sibillini)*; Tecnoprint: Ancona, Italy, 1990; pp. 132–137.
36. Amato, G.; Devoti, R.; Fubelli, G.; Aringoli, D.; Bignami, C.; Galvani, A.; Moro, M.; Polcari, M.; Saroli, M.; Sepe, V.; et al. Step-like displacements of a deep seated gravitational slope deformation observed during the 2016–2017 seismic events in Central Italy. *Eng. Geol.* **2018**, *246*, 337–348. [CrossRef]
37. Di Naccio, D.; Kastelic, V.; Carafa, M.M.C.; Esposito, C.; Milillo, P.; Di Lorenzo, C. Gravity versus tectonics: The case of 2016 Amatrice and Norcia (central Italy) earthquakes surface coseismic fractures. *J. Geophys. Res. Earth Surf.* **2019**, *124*, 994–1017. [CrossRef]
38. Farabollini, P.; Angelini, S.; Fazzini, M.; Lugeri, F.R.; Scalella, G.; Aringoli, D.; Materazzi, M.; Pambianchi, G. The Central Italy seismic sequence of August 24th and after: Contributions to knowledge and database of surface effects (in italian: La sequenza sismica dell'Italia centrale del 24 agosto e successive: Contributi alla conoscenza e la banca dati degli effetti di superficie). *Rend. Online Soc. Geol. It.* **2018**, *46*, 9–15.
39. Farabollini, P.; De Pari, P.; Discenza, M.E.; Minnillo, M.; Carabella, C.; Paglia, G.; Miccadei, E. Geomorphological evidence of debris flows and landslides in the Pescara del Tronto area (Sibillini Mts., Marche Region, Central Italy). *J. Map* **2020**. [CrossRef]
40. Dramis, F.; Maifredi, P.; Sorriso-Valvo, M. Deep-seated gravitational slope deformations. Geomorphological aspects and diffusion in Italy (in Italian: Deformazioni gravitative profonde di versante. Aspetti geomorfologici e loro diffusione in Italia). *Geol. Appl. Idrogeol.* **1983**, *18*, 255–268.
41. Gentili, B. *Notes on Geomorphology of the Monti Sibillini National Park (in Italian: Note di geomorfologia del Parco Nazionale dei Monti Sibillini)*; Aniballi Grafiche: Ancona, Italy, 2001; 50p.
42. Folchi Vici, C.; Gentili, B.; Luzi, L.; Pambianchi, G.; Viglione, F. Deep-seated gravitational slope deformation in the central-southern Umbro-Marchean Apennines:,morphometric and macrostructural analysis. *Geogr. Fis. Dinam. Quat.* **1996**, *19*, 335–341.
43. Aringoli, D.; Gentili, B.; Pambianchi, G.; Piscitelli, A.M. The contribution of the 'Sibilla Appenninica' legend to karst knowledge in the Sibillini Mountains (Central Apennines, Italy). *Geol. Soc. Lond. Spec. Publ.* **2007**, *273*, 329–340. [CrossRef]
44. Aringoli, D.; Gentili, B.; Materazzi, M.; Pambianchi, G. Deep-seated gravitational slope deformations in active tectonics areas of the umbria-marche apennine (central Italy). *Geogr. Fis. Din. Quat.* **2010**, *33*, 127–140.
45. Gentili, B.; Pambianchi, G.; Aringoli, D.; Cilla, G.; Farabollini, P.; Materazzi, M. Relationships between Plio-Quaternary brittle deformation and gravitational morphogenesis in the high hills of central-southern Marche region (in Italian: Rapporti tra deformazioni fragili plio-quaternarie e morfogenesi gravitativa nella fascia alto-collinare delle Marche centro-meridionali). *Studi Geol. Camerti* **1995**, *1*, 421–435.
46. Aringoli, D.; Bisci, C.; Blumetti, M.; Buccolini, M.; Ciccacci, S.; Cilla, G.; Coltorti, M.; De Rita, D.; D'Orefice, M.; Dramis, F.; et al. Geomorphology and Quaternary evolution of central Italy. Guide for excursion. *Geogr. Fis. Dinam. Quat.* **1997**, *III*, 79–103.
47. Zernitz, E.R. Drainage patterns and their significance. *J. Geol.* **1932**, *40*, 498–521. [CrossRef]
48. Panizza, M.; Piacente, S. Geomorphological convergence of heterogeneous morphosculptures: Fine-tuning for neotectonics research (in Italian: Convergenza geomorfologica di morfosculture eterogeniche: Messa a punto per ricerche di neotettonica). *Gruppo Studio Quat. Padano Torino* **1976**, *3*, 39–44
49. Catalano, S.; Grassi, S.; Imposa, S.; Lombardo, G.; Panzera, F.; Romagnoli, G.; Tortorici, G. Geological and geophysical evidence of the Pretare-Piedilama normal fault (Arquata del Tronto, Central Italy). In Proceedings of the GNGTS Conference, Lecce, Italy, 22–24 November 2016; pp. 30–33.
50. Tortorici, G.; Romagnoli, G.; Grassi, S.; Imposa, S.; Lombardo, G.; Panzera, F.; Catalano, S. Quaternary negative tectonic inversion along the Sibillini Mts. thrust zone: The Arquata del Tronto case history (Central Italy). *Environ. Earth Sci.* **2019**, *78*, 1–12. [CrossRef]
51. Radbruch-Hall, D. Gravitational Creep of Rock Masses on Slopes. *Dev. Geotech. Eng.* **1978**, *14*, 607–657.
52. Chigira, M. Long-term gravitational deformation of rocks by mass rock creep. *Eng. Geol.* **1992**, *32*, 157–184. [CrossRef]
53. Gentili, B.; Pambianchi, G. Gravitational morphogenesis of the Apennine chain in central Italy. In Proceedings of the International Congress International Association of Engineering Geology, Lisboa, Portugal, 5–9 September 1994; pp. 1171–1186.
54. Tolomei, C.; Taramelli, A.; Moro, M.; Saroli, M.; Aringoli, D.; Salvi, S. Analysis of the deep-seated gravitational slope deformations over Mt. Frascare (Central Italy) with geomorphological assessment and DInSAR approaches. *Geomorphology* **2016**, *201*, 281–292. [CrossRef]
55. Aringoli, D.; Farabollini, P.; Giacopetti, M.; Materazzi, M.; Paggi, S.; Pambianchi, G.; Pierantoni, P.P.; Pistolesi, E.; Pitts, A.; Tondi, E. The August 24th 2016 Accumoli earthquake: Surface faulting and deep-seated gravitational slope deformation (DSGSD) in the Monte Vettore area. *Ann. Geophys.* **2016**, *59*. [CrossRef]
56. Aringoli, D.; Gentili, B.; Materazzi, M.; Pambianchi, G.; Sciarra, N. DSGSDs induced by post-glacial decompression in Central Apennine (Italy). *Landslide Sci. Pract. Glob. Environ. Chang.* **2013**, *4*, 417–423.
57. Zaruba, Q.; Mencl, V. Landslides and their control. *Earth-Sci. Rev.* **1969**, *6*, 205.

58. Beck, A.C. Gravity faulting as a mechanism of topographic adjustment. *N. Z. J. Geol. Geophys.* **1968**, *11*, 191–199. [CrossRef]
59. Nemcok, A. Gravitational slope deformation in high mountains. In Proceedings of the 24th International Geological Congress, Montreal, QC, Canada, 21–30 August 1972; Volume 13, pp. 132–141.
60. Farabollini, P.; Spurio, E. GIS applications in the study of debris flows phenomena in the Umbria-Marches Apennines (Marche region, Italy) (in Italian: Applicazione dei GIS nello studio dei fenomeni di debris flows dell'appennino umbro-marchigiano (regione Marche, Italia)). *Rend. Sgi Online* **2009**, *8*, 50–58.
61. Farabollini, P.; Angelini, S.; Fazzini, M.; Lugeri, F.R.; Scalella, G.; Aringoli, D.; Materazzi, M.; Pambianchi, G. Earthquakes and Society: The 2016 Central Italy Reverse Seismic Sequence. In *Earthquake Risk Perception, Communication and Mitigation Strategies across EUROPE, Geographies of the Anthropocene*; Farabollini, P., Lugeri, F.R., Mugnano, S., Eds.; Il Sileno Edizioni: Lago, Italy, 2019; pp. 249–266. ISSN 2611–3171.
62. Sorriso-Valvo, M. Studies on deep-seated gravitational slope deformations in Italy (in Italian: Studi sulle deformazioni gravitative profonde di versante in Italia). *Mem. Soc. Geol. It.* **1988**, *41*, 877–888.
63. Petley, D. *Landslide Hazards. Geomorphological Hazards and Disaster Prevention*; Cambridge University Press: Cambridge, UK, 2010; pp. 63–73.
64. Arai, N.; Chigira, M. Distribution of gravitational slope deformations and deep-seated landslides controlled by thrust faults in the Shimanto accretionary complex. *Eng. Geol.* **2019**, *260*, 105236. [CrossRef]
65. Farabollini, P.; Materazzi, M.; Miccadei, M.; Piacentini, T. The travertines of Central Adriatic Italy: Genesis, chronology, and geomorphological and paleoenvironmental significance (in Italian: I travertini dell'Italia centrale adriatica: Genesi, cronologia e significato geomorfologico e paleoambientale). *Il Quat.* **2004**, *17*, 259–272.
66. Lugeri, F.R.; Farabollini, P. Landscape analysis as a tool for risk reduction. *Aims Geosci.* **2019**, *5*, 617–630. [CrossRef]
67. Farabollini, P. The 2016 Earthquake in Central Italy. The alphabet of reconstruction. In *Earthquake Risk Perception, Communication and Mitigation Strategies across Europe*; Farabollini, P., Lugeri, F.R., Mugnano, S., Eds.; Geographies of the Anthropocene Book Series; Sileno Edizioni: Lago, Italy, 2019; Volume 2, pp. 145–171.
68. Jahn, A. Slope morphological features resulting from gravitation. *Zeit. Geomorph.* **1964**, *5*, 59–72.
69. Hoek, E.; Bray, J. *Rock Slope Engineering*, 3rd ed.; Institution Mining and Metallurgy: London, UK, 1981.
70. Esposito, E.; Martino, S.; Scarascia Mugnozza, G. Mountain slope deformations along thrust fronts in jointed limestone: An equivalent continuum modelling approach. *Geomorphology* **2007**, *90*, 55–72. [CrossRef]
71. Aringoli, D.; Farabollini, P.; Gentili, B.; Materazzi, M.; Pambianchi, G. Climatic influence on slope dynamics and shoreline variations: Examples from Marche region (Central Italy). *Physio-Géo Géogr. Phys. Environ.* **2007**, *1*, 1–20. [CrossRef]

Article

Landslide Hazard Assessment in a Monoclinal Setting (Central Italy): Numerical vs. Geomorphological Approach

Marco Materazzi, Margherita Bufalini *, Matteo Gentilucci, Gilberto Pambianchi, Domenico Aringoli and Piero Farabollini

Geology Division, School of Sciences and Technology, University of Camerino, 62032 Camerino, Italy; marco.materazzi@unicam.it (M.M.); matteo.gentilucci@unicam.it (M.G.); gilberto.pambianchi@unicam.it (G.P.); domenico.aringoli@unicam.it (D.A.); piero.farabollini@unicam.it (P.F.)
* Correspondence: margherita.bufalini@unicam.it; Tel.: +39-0737402603

Abstract: A correct landslide hazard assessment (LHA) is fundamental for any purpose of territorial planning. In Italy, the methods currently in use to achieve this objective alternate between those based on mainly qualitative (geomorphological) and quantitative (statistical–numerical) approaches. The present study contributes to the evaluation of the best procedure to be implemented for LHA, comparing the results obtained using two different approaches (geomorphological and numerical) in a territorial context characterized by conditioning and triggering factors, favorable to the instability of the slopes. The results obtained, although preliminary, evidence the respective limitations of the methods and demonstrate how a combined approach can certainly provide mutual advantages, by addressing the choice of the best numerical model through direct observations and surveys.

Keywords: landslides; factor of safety; numerical models; Hoek–Brown method; monoclinal setting

Citation: Materazzi, M.; Bufalini, M.; Gentilucci, M.; Pambianchi, G.; Aringoli, D.; Farabollini, P. Landslide Hazard Assessment in a Monoclinal Setting (Central Italy): Numerical vs. Geomorphological Approach. *Land* 2021, *10*, 624. https://doi.org/10.3390/land10060624

Academic Editor: Massimiliano Alvioli

Received: 7 March 2021
Accepted: 7 June 2021
Published: 11 June 2021

Publisher's Note: MDPI stays neutral with regard to jurisdictional claims in published maps and institutional affiliations.

Copyright: © 2021 by the authors. Licensee MDPI, Basel, Switzerland. This article is an open access article distributed under the terms and conditions of the Creative Commons Attribution (CC BY) license (https://creativecommons.org/licenses/by/4.0/).

1. Introduction

Landslide hazard assessment (LHA) is a challenging task for the prevention and prediction of a territory for local land management and security services. Proper management of landslide hazard, as well as saving human lives, can minimize socioeconomical impact that in many developing countries may equal a large percentage of the gross national product [1,2]. LHA is usually based on the spatial and temporal probability of landslide occurrences and is performed following three main steps: (i) the creation of a phenomenon inventory, (ii) a landslide susceptibility analysis and (iii) a landslide hazard analysis [3–5].

In Italy, the National Plan for Hydrogeological Risk Assessment (PAI) [6] is the cognitive, regulatory and technical–operational tool through which actions, interventions and rules concerning the defense against hydrogeological risk of the territory are planned and scheduled. Although the term LHA (in its Italian translation) is often mentioned within the PAI, the significance is at times contradictory, and the products of the Plan (maps, inventory sheets, analyses, etc.) rarely come from the steps described above; on the contrary, they are often realized based on an empirical approach and basic available data. In particular, hazard, vulnerability and risk degree (the latter in terms of exposed value) is closely linked to the quality of the expertise, while only in a few cases have specific studies (numerical models, statistical/probabilistic approaches) been applied [7–11]. As a consequence, the representation of the landslide hazard that emerges in some areas of the Italian territory can be over/underestimated, and divergent opinions may arise among technicians and public administrators. This problem, among other aspects, has also been highlighted in other countries of the European Union such as France [3,4,12,13].

In a more general context, the methods currently in use in Italy for LHA include two main types of approach: the field-based qualitative approach and the data-driven quantitative one [14]. The former type includes the so-called "geomorphological" methods [3,5–8,15–17], while the latter includes the statistical analyses (i.e., bivariate and multivariate statistical techniques, [7,18–22]) and the deterministic methods that involve, among others, the analysis of specific sites or slopes based on numerical models [23–30].

The present work compares the results of a landslide analysis carried out in a sample area using both geomorphological and numerical approaches. The area chosen for the analysis is located in a high hilly sector of the Adriatic side of the Central Apennines (Italy), characterized by the presence of monoclinal reliefs and typical cuesta morphologies, formed by differential tectonic movements in a recent uplift area [31,32]. Despite the relative simplicity of the geological model, these contexts can generate complex mass movements, both for characteristics (type of movement) and size (extension and depth of the failure zone) and kinematics (velocity and return time) and consequently represent high hazard conditions in the presence of built-up areas and/or infrastructures. The LHA (following the significance given by Italian PAI) is typically conducted based on a geomorphological approach and an expert judgment as regards the attribution of the degree of vulnerability and the exposed value. For any reclassification of the degree of risk, site-specific analyses (i.e., instrumental monitoring, geognostic bore-holes, geophysical prospecting) or the use of numerical models (slope stability analyses) are usually required by Italian guidelines. Nevertheless, both types of investigations proposed have limitations. In the first case, the non-negligible costs of direct surveys and prospecting limit the representativeness of the surveys themselves; in the case of numerical models, on the other hand, reliability is mainly linked to a correct choice of input parameters, is sometimes not homogeneous in the area considered and is often deduced from bibliographic data.

In this study, using finite-difference software (ITASCA FLAC/Slope v.8.0 [33]) the factor of safety (FoS) was calculated on representative slope sections where detailed geomorphological surveys highlighted the presence of gravitational phenomena or stability conditions. This analysis, in particular, refers to medium-to-low depth landslides, while more complex phenomena (as specified in the following paragraphs) recognized in the area and associated with deep-seated gravitational slope deformations were not included in the study [34,35].

The objective of this study was to clarify the role, usefulness and limits of the different methods to be used in the LHA and also provide useful information for their correct use in any context of territorial planning where specific indications have not been provided. A further aim was to demonstrate a combined and intelligent use of the two methods, pending clearer and universally accepted regulatory indications on the methods to be used for the LHA, which seems at the moment the most suitable choice both in economic and safety terms.

2. Geological and Geomorphological Setting of the Study Area

The study area (around 13 km^2) is located east of the Sibillini Mts. Massif (Central Apennines) (Figure 1a). This sector corresponds to a vast sedimentary basin where, starting from the early Pliocene, thick levels of sandstones and conglomerates alternated with pelitic-arenaceous levels are deposited in transgression over a Miocene (Messinian) turbidite bedrock, mainly consisting of alternating arenaceous-pelitic and pelitic-arenaceous levels (Laga formation) [36,37] (Figure 1b,c). The contact between post- and pre-transgressive sediments is marked by an erosion surface approximately parallel to the Miocene levels [32,38]. The area is particularly characterized by the presence of weak levels and ductile deformation zones, corresponding to the weathered levels of the pre-transgressive clayey basement. Such strong weathering is probably due to the long period of immersion of the Messinian sediments as well as the lithostatic charges and the constant presence of water in the arenaceous-calcarenitic aquifer.

The structuring of the pre-transgressive bedrock was essentially carried out in the early Pliocene when, after intense compressive tectonics, east-verging folds and thrusts (the latter emerging or buried) developed within the Messinian and pre-Messinian formations. Pliocene sediments, on the other hand, show a generalized monoclinal setting, linked to the subsequent tectonic uplift that affected the whole area starting from the early Pleistocene [39–41]; the strata generally dip between 15° and 20° (Figure 1b).

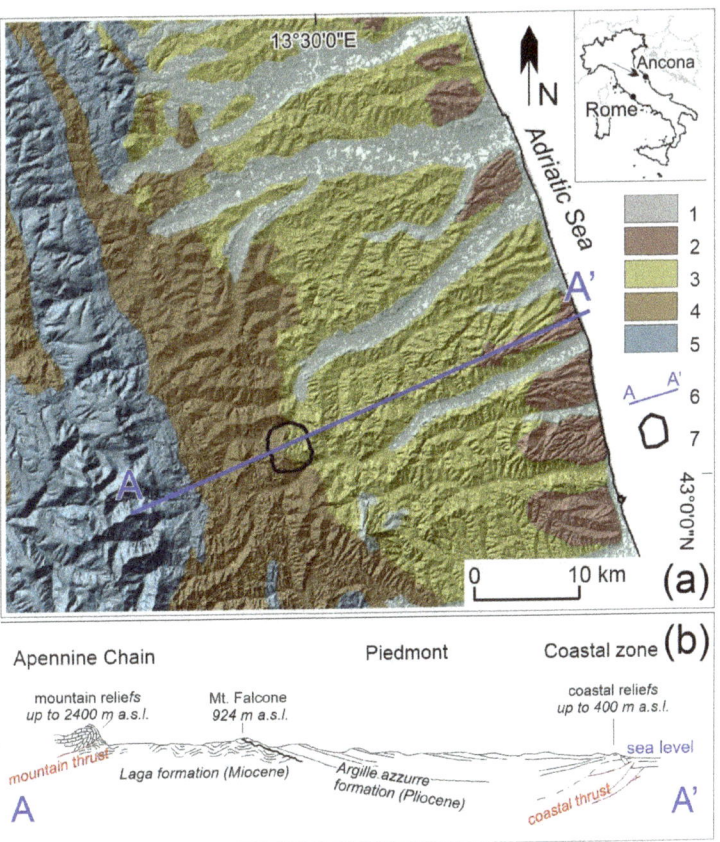

Figure 1. (**a**) Schematic geological map of the study sector: 1—main continental deposits (Pleistocene–Holocene); 2—sands and conglomerates (Pliocene–Pleistocene); 3—clays and sands (Pliocene–Pleistocene); 4—arenaceous-marly-clayey turbidites (late Miocene); 5—limestones, marly limestones and marls (early Jurassic–Oligocene); 6—trace of cross-section shown in Figure 1b; 7—study area (see Figure 4). (**b**) Schematic geological cross-section from the Apennine chain to the Adriatic Sea, modified from [42].

The resulting landscape, characterized by alignments of strongly asymmetrical and NNW–SSE oriented reliefs, is typical of "cuestas", with the main element consisting of the Mount Falcone relief (Figure 2a,b). Selective erosion, due to the presence of tough and massive lithotypes (sandstones and conglomerates) overlying less resistant clayey formations, creates steep escarpments between 50 and 300 m on the southwestern flanks.

The monoclinal structure, as a whole, is displaced by direct faults, mainly oriented NNW–SSE and WSW–ENE, the displacement of which can exceed 10 m [36,42]. Micro- and meso-structural analyses carried out on middle Pliocene and upper Pleistocene formations highlighted intense fracturing according to two main joint systems, N70 ± 15, N150 ± 15, and N20 ± 15, N100 ± 10, both compatible with the abovementioned fault systems. In the arenaceous-conglomeratic body of Mount Falcone, a third system of joints, N–S and E–W oriented, has also been observed. The former, dipping W of 70–80°, completely crosses the rigid plate with a spacing of the order of a few tens of meters; the latter, characterized by less frequency and continuity, is found at the edges of the plate itself. The genesis of this third system is attributed to the expansion processes of the relief resulting from the Pleistocene tectonic uplift [43,44] to the passive action of discontinuities developed with the same direction within the pre-transgressive bedrock and, in general, to the high seismicity of the area.

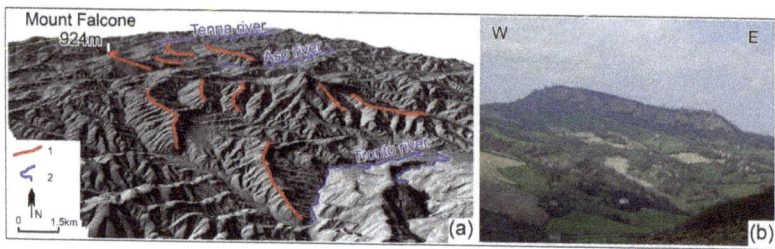

Figure 2. (**a**) 3D digital elevation model of the study sector: 1—edge of cuesta; 2—river. (**b**) The arenaceous-conglomeratic body of Mount Falcone.

From a geomorphological point of view, the joint systems described above, particularly developed within the arenaceous-conglomeratic body of Mount Falcone (924 m a.s.l.), create strong instability, especially corresponding to the W–SW portion, where the high structural scarp (about 150 m) is affected by retreat processes due to past and ongoing falls and topples (Figure 3). The accumulations of these processes, mainly consisting of pebbles and blocks, constitute an extensive and continuous talus at the base of the slope itself but can also be found further downstream, through rolling and/or passive transport processes induced by slow deformations of debris material; isolated blocks of decametric dimensions were found within the Tenna and Aso riverbeds (north and south of the relief, respectively).

Figure 3. Open fractures and toppling phenomena affecting the arenaceous plate of Mount Falcone.

3. Data and Methods

3.1. Geomorphological Approach

The information and the references related to the geological setting of the study area (already included in the text) are included in the caption of Figure 1b.

The detailed study of the morphodynamic processes active in the study area, with particular regard to landslides, was initially carried out following the classical principles of a detailed field geomorphological survey. In the first phase, all the available geological information sources (maps, profiles, stratigraphic sections, stratigraphic logs, etc.) were acquired. The base cartography was the geological map at 1:10,000 scale, which presents an almost complete coverage even at a national level; nevertheless, it is not uniform in symbolism and the legend as it is the product of autonomous regional projects. Furthermore, since this is a relatively old document (about 20 years old), the perimeters and the state of activity of the gravitational processes were updated through a detailed survey, through which soil samples were taken for subsequent geotechnical tests. In this regard, it should be emphasized that specific geotechnical data for outcropping formations are quite rare, and in some cases, it was necessary to use the results of tests conducted on samples taken from the same formations but in different (albeit neighboring) locations.

The synthesis of these surveys, mainly addressing the characterization of the type and evolutionary mechanisms of landslides, is shown in Figure 4a.

From a lithological point of view, the area can be divided into two main sectors, one to the west and one to the east of the relief of Mt. Falcone, consisting of arenaceous and arenaceous-conglomeratic lithotypes associated with a coastal transition environment. In the western sector of the relief, as previously mentioned, the Messinian arenaceous-pelitic and pelitic-arenaceous members of the Laga Formation outcrop with a counter-dip-slope attitude, ranging between 30° and 40° (Figure 4b). Pliocene arenaceous-pelitic and pelitic-arenaceous lithotypes also emerge east of the Mount Falcone relief; the frequency and consistency of the arenaceous levels is, however, less marked as along with the strata inclination (generally dip-slope and between 15–25°). A common feature of some clayey lithotypes is the presence of more or less thick and frequent weathered (weak) levels; these layers, characterized by poor geotechnical properties, are observed mostly in the western sector, within the pelitic-arenaceous member of the Laga formation.

The bedrock is often masked by powerful thicknesses (up to 30 m) of fine colluvial deposits; the greater thicknesses are concentrated corresponding to the numerous minor valleys, which originate from the arenaceous-conglomeratic plate.

The hydrographic network is fairly developed, and because of the low permeability of bedrock and covering soils, the major gullies and streams radially develop from the top of the relief, although there is important tectonic conditioning especially in the lower hierarchical reaches. Concerning water circulation, runoff is limited to a few days/weeks after considerable meteoric events, while the groundwater circulation, due to the predominantly clayey nature of bedrock, is generally limited. However, the presence of widespread perched aquifers within the colluvial deposits, with the water table close to the surface, is a crucial element for the stability of the slopes.

Landslides, as mentioned, are widespread over the area, both for the lithological nature of bedrock and the morphological–structural setting of the slopes, characterized by discrete slope angle and by strata dip often favorable to the activation of gravitational phenomena. The typology of movement is also very highly variable, with falls, topples, flows and slides with different styles and states of activity being present (Figures 4a and 5). Although the stratigraphic setting is favorable to the activation of slides (rotational/planar) particularly on the eastern side, these are less frequent, while flows are dominant; this apparently contradictory aspect can be associated, as described in the following, with the good strength and deformation properties of bedrock and with the presence, on the other hand, of important thicknesses of unconsolidated continental deposits often hosting a perched water table. It is not uncommon, however, to observe complex phenomena characterized by rotational slides in the uppermost portion and flows in the median-terminal one.

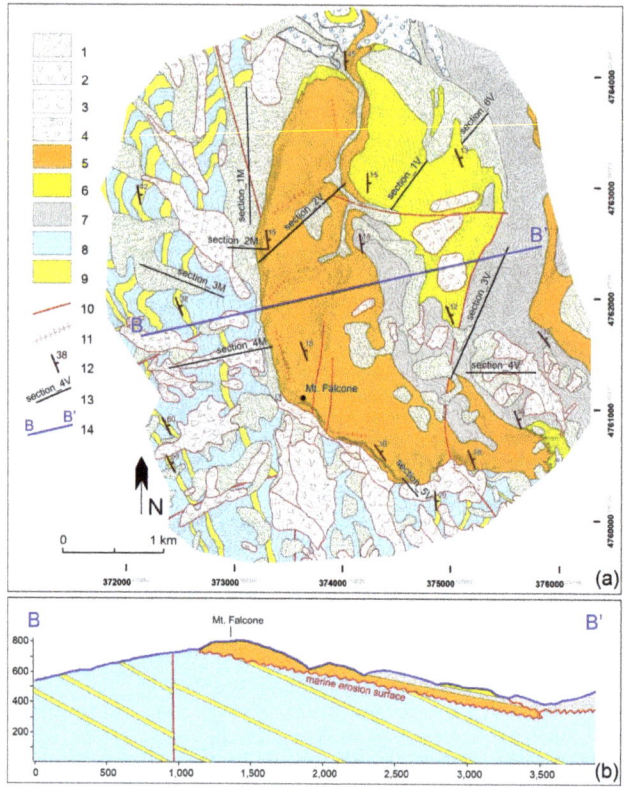

Figure 4. (**a**). Geological and geomorphological map of the study area: 1—slope and colluvial deposits (Holocene); 2—rotational and planar slides (Holocene); 3—flows (Holocene); 4—old and recent fluvial deposits (Pleistocene–Holocene); 5—arenaceous-conglomeratic bedrock (Mount Falcone body—Middle Pliocene); 6—mainly arenaceous-pelitic bedrock (Argille Azzure Formation—Middle Pliocene); 7—mainly clayey bedrock (Argille Azzure Formation—Middle Pliocene); 8—mainly pelitic-arenaceous bedrock (Laga formation—Late Messinian); 9—mainly arenaceous-pelitic bedrock (Laga formation—Late Messinian); 10—main faults; 11—gravitational trench; 12—strata attitude; 13—trace of cross-section used for the numerical simulation; 14—trace of geological cross-section described in (**b**).

The type of movement was mainly attributed on the basis of geomorphological considerations, taking into account size, morphology, typology of material involved, strata dip and damage eventually observed along roads or to infrastructures (Figure 5).

Figure 5. Typical landslides in the study area. (**a**) Rotational slide affecting clayey bedrock and (**b**) consequences on the road conditions; (**c**) example of earthflow involving fine slope deposits and (**d**) secondary road deformed by a landslide.

3.2. Numerical Approach

The numerical code used to perform the FoS calculations is the FLAC/Slope v. 8.0. In contrast to other "limit equilibrium" programs, which test several assumed failure surfaces (method of the slices) thus choosing the one with the lowest FoS, FLAC/Slope uses the procedure known as the "strength reduction technique", a method commonly applied with the Mohr–Coulomb failure criterion in FoS calculations by progressively reducing the shear strength of the material in stages until the slope fails [45–50]. The main advantage of a slope stability analysis performed with FLAC/Slope is the possibility to determine a broad variety of failure mechanisms with no prior assumptions concerning their type, shape or location. Moreover, FLAC/Slope is able to combine slip along joints with failure through intact material, thus offering clear advantages in the modelling of jointed rock masses.

The material failure can be defined by either the "Mohr–Coulomb", the "Modified Hoek–Brown" or the "ubiquitous-joint" plasticity models.

The Mohr–Coulomb model is the conventional model used to represent shear failure in soils and rocks. It assumes that failure is controlled by the maximum shear stress which in turn depends on the normal stress. The solution is obtained by plotting Mohr's circle for states of stress at failure in terms of the maximum and minimum main stresses: the Mohr–Coulomb failure line is the best straight line tangent to the Mohr's circles.

The Mohr–Coulomb criterion can be written as

$$\tau = c + \sigma_n \tan \phi \tag{1}$$

where τ is the shear stress, c is the cohesion of the material, σ_n is the normal stress (negative in compression) and ϕ is the angle of friction.

The ubiquitous joint in FLAC/Slope is an anisotropic plasticity model that includes the presence of an oriented weak plane (such as weathering joints) in a Mohr–Coulomb model, failure may occur in either the intact rock, along the weak plane or both and depends on the stress state, the material properties of the rock and weak plane and the orientation of the latter. The failure of the weak plane (ubiquitous joint) may occur by shear, for which the envelope criterion is:

$$\tau = c_j + \tan \varphi_j \sigma_n \quad (2)$$

or by tension, for which the criterion is:

$$\sigma_3 = -T_j \quad (3)$$

with:

$$T_j \leq \frac{c_j}{\tan \varphi_j} \quad (4)$$

where τ and σ_n are shear and normal stresses respectively, and σ_3 is the minimum principal stress. c_j, φ_j and T_j are the cohesion, friction angle and the tensile strength of the ubiquitous joints, respectively.

Both the Mohr–Coulomb and the ubiquitous-joint models require another parameter, the dilation angle ψ, usually assumed as a fraction of the friction angle and ranging between $\varphi/4$ (very good quality rocks) and 0 (very poor quality rocks) [51–53].

The Hoek–Brown failure model for jointed rock masses is defined by the following:

$$\sigma_1 = \sigma_3 + \sigma_c \left(m_b \frac{\sigma_3}{\sigma_c} + s \right)^a \quad (5)$$

where σ_1 and σ_3 (Pa) are the maximum and minimum stresses at failure respectively. Concerning the other parameters, m_b is the value of the Hoek–Brown constant for the rock mass, s and a are constants that depend upon the characteristics of the rock mass and σ_c (MPa) is the uniaxial compressive strength of the intact rock.

The Modified Hoek–Brown model, sometimes referred to as the "Mhoek model" [54] includes a tensile yield criterion, similar to that used by the Mohr–Coulomb model and can specify a dilation angle ψ. Compared to the original version, the Mhoek model provides a simplified flow rule for both tensile and compressive regions.

The value of σ_c is usually obtained by laboratory analyses even though several field estimates exist in literature (e.g., Table 1 in [51]).

The constants m_b, s and a are usually calculated starting from the evaluation of the geological strength index (GSI) of the rock mass [55–59]. The GSI is a system of rock-mass characterization particularly suitable for use in engineering rock mechanics and input into numerical analysis; through a visual assessment of the geological characters of the rock material, it allows the prediction of the rock-mass strength and deformability.

The GSI estimation is carried out using specific charts (see Tables 4 and 5 in [51]): once the index has been evaluated, the constants s, a and m_b, can be derived with the following equations:

$$s = \exp\left(\frac{GSI - 100}{9}\right) \text{ and } a = 0.5 \quad \text{for } GSI > 25 \quad (6)$$

$$s = 0 \text{ and } a = 0.65 - \frac{GSI}{200} \quad \text{for } GSI < 25 \quad (7)$$

And

$$m_b = m_i \exp\left(\frac{GSI - 100}{28}\right) \quad (8)$$

where m_i is the Hoek–Brown constant for intact rock pieces estimated using GSI, σ_c and the chart of Figure 7 in [51].

The Mhoek model properties can be entered in FLAC/Slope in two different ways; the s, a and m_b constants can be input, or GSI, m_i and ψ can be entered, and the Hoek–Brown strength properties are calculated automatically from the software.

The FoS calculation model for each cross-section was performed choosing a proper resolution and failure criterion (Mohr–Coulomb or Hoek–Brown) for each numerical mesh. In the case of the arenaceous-conglomeratic body of Mt. Falcone, the ubiquitous-joint model, which considers the characteristics (orientation, tensile strength, cohesion and friction angle) of the previously described joint systems, was used.

Taking into account the type of landslides and the fact that no significant erosion phenomena at the expense of the clayey bedrock are known in the area, the initial state stress, in contrast to other models [59], was assumed as lithostatic as a first approximation. Concerning the presence of water, as it was impossible to implement a discontinuous water table in the models (i.e., to simulate a real perched aquifer), a static water table close to the topographic surface was applied.

The input parameters of the geological formations (Table 1) were partly obtained from laboratory analyses, performed by professional geologists and provided privately, and partly from direct observations in the field (through GSI evaluation) (Figure 6); a minor number of data were obtained from literature or neighboring territories [51,57,60–64]. Uniform geotechnical properties were assumed throughout the slope, and all the parameters were then used individually or in association with the simulations; in the case of friction angle and cohesion, linked in every failure envelope, the software takes into account only the pair of values specified in the table.

Figure 6. Arenaceous-conglomeratic bedrock outcropping on the eastern side of the relief of Mount Falcone.

The estimation of the Mohr–Coulomb geotechnical parameters starting from the GSI evaluation was performed by means of the open-source software RocLab v.1.0 (Rockscience Inc., Toronto, ON, Canada) which can plot the Hoek–Brown and the Mohr–Coulomb failure envelopes on the same x–y plane. The results of these calculations are shown in Figure 7.

Table 1. Material properties, related to different geological units and implemented within the numerical simulations.

Failure Model	Material Properties (Mohr–Coulomb Method)						Material Properties (Mohr–Coulomb and Modified Hoek–Brown Methods)								
Parameter	Friction (φ) Deg.			Cohesion (c) kPa			Density (ρ) kg/m³			Tension (T) kPa			Dilation (ψ) Deg.		
Units Range of Values	Min	Mean	Max	Min	Mean	Max	Min	Mean	Max	Min	Mean	Max	Min	Mean	Max
Arenaceous-conglomeratic bedrock (Mount Falcone body)	17	19	20	13,000	15,000	17,000	2050	2335	2620	185	281	377	4.0	6.5	9.0
Mainly arenaceous-pelitic bedrock (Laga formation)	15	17	18	10,500	11,800	13,000	1900	2100	2300	127	147	166	2.0	2.5	3.0
Mainly pelitic-arenaceous bedrock (Laga formation)	14	15	15	5500	6420	7340	1700	1850	2000	56.8	86	11.5	1.0	1.5	2.0
Mainly arenaceous-pelitic bedrock (Argille Azzurre formation)	15	16	17	9500	11,000	12,500	1850	2025	2200	111	132	152	2.0	2.5	3.0
Mainly clayey bedrock (Argille Azzurre formation)	12	20	27	3400	4600	5800	1700	1825	1950	45.9	76.9	108	1.0	1.5	2.0
Weak layer (Laga formation)	n.a	16	n.a	n.a	62.5	n.a	n.a	1650	n.a	n.a	10.5	n.a	0.0	0.0	0.0
Slope and colluvial deposits	17	19	20	8	9	10	1650	1725	1800	0	0	0	n.a	n.a	n.a

Table 1. Cont.

Failure Model							Material Properties (Modified Hoek–Brown Method)										
Parameter	Geological Strength Index			Hoek–Brown Constant			m_b			s			a			σ_c	
Units	(GSI)			(mi)												(MPa)	
Range of Values	Min	Mean	Max	Min	Mean	Max	Min	Mean	Max	Min	Mean	Max	Min	Mean	Max	Mean	Max
Arenaceous-conglomeratic bedrock (Mount Falcone body)	45	55	65	n.a	n.a	n.a	n.a	n.a	n.a	n.a	n.a	n.a	n.a	n.a	35	45	55
Mainly arenaceous-pelitic bedrock (Laga formation)	40	44	48	18.0	20.0	22.0	2.11	2.77	3.43	0.0012	0.0021	0.0030	0.50	0.50	29.8	32.3	34.8
Mainly pelitic-arenaceous bedrock (Laga formation)	31	35.5	40	21.0	22.5	24.0	1.78	2.30	2.81	0.0004	0.0008	0.0012	0.50	0.50	23.5	25.5	27.5
Mainly arenaceous-pelitic bedrock (Argille Azzurre formation)	40	42.5	45	19.0	21.0	23.0	2.22	2.72	3.22	0.0007	0.0015	0.0022	0.50	0.50	27.4	30.2	32.9
Mainly clayey bedrock (Argille Azzurre formation)	27	31	35	22.0	23.5	25.0	1.62	2.04	2.45	0.0003	0.0005	0.0007	0.50	0.50	21.6	23.6	25.6
Weak layer (Laga formation)	n.a	20	n.a	n.a	30.0	n.a	n.a	1.72	n.a	n.a	0.0005	n.a	n.a	0.53	n.a	18	n.a
Slope and colluvial deposits	n.a	n.a	n.a	n.a	n.a	n.a	n.a	n.a	n.a	n.a	n.a	n.a	n.a	n.a	n.a	n.a	n.a

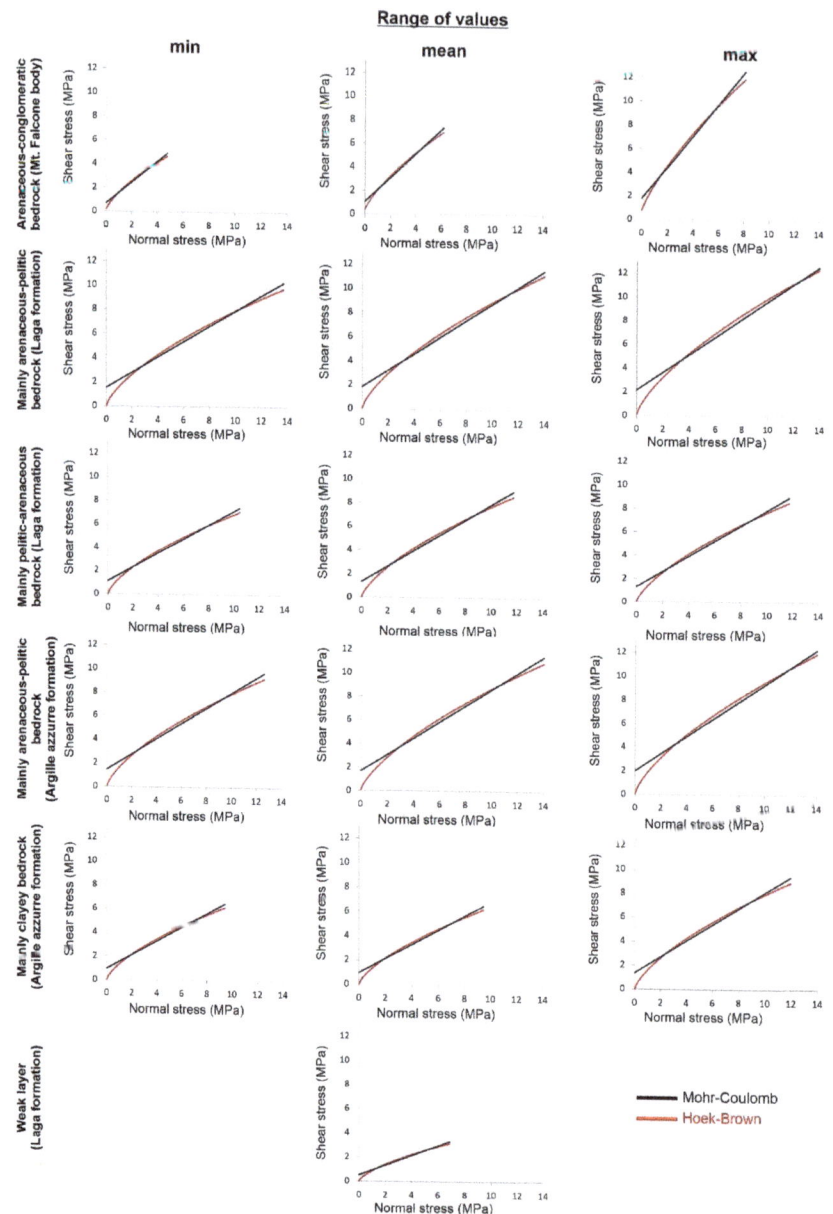

Figure 7. Shear vs. normal stress plots of Hoek–Brown and Mohr–Coulomb failure envelopes, obtained for each lithotechnical unit and different range of values.

A shear vs. normal stress plot was obtained for each bedrock unit and for each class of values (min, mean and max).

The range of values (minimum, mean and maximum) was also used for the development of numerical models for each geotechnical cross-section.

4. Results

4.1. Geomorphological Approach

The following numerical analysis was carried out along 10 geotechnical cross-sections: Six are located on the eastern and southeastern sides of the relief of Mount Falcone and four on the western side (Figure 4a). The choice of the location took into account the different morphological–structural conditions of the relief and the results of the geomorphological surveys carried out in the field. To verify all the possible situations observed in the field, the sections were traced to include both portions of the slope affected by gravitational movements and others considered stable.

The geomorphological processes observed along the individual cross-sections can be summarized as follows.

- Section 1V. The slope is fairly regular, with a generally mild angle (10–20°). The outcropping bedrock, consisting of the mainly arenaceous-pelitic member of the Argille Azzurre formation, shows a general dip-slope trend with an inclination of about 15° towards E–NE. Almost absent are the continental eluvial–colluvial deposits, which never exceed 1–2 m in thickness and are therefore not shown in the map of Figure 4. The observed geomorphological processes are also not very significant, limited to weak and localized soil erosion processes (sheet and/or rill erosion)
- Section 2V. This section is traced in the SSW–NNE direction, along the maximum slope and with an angle ranging between 10° and 20° in the first part to over 40° in the final one. The bedrock consists of the mainly arenaceous-conglomeratic body of Mt. Falcone, dipping about 10–15 ° towards ENE. Eluvial–colluvial deposits with low thickness (of a few meters) are present in the central section, but no significant geomorphological processes were detected.
- Section 3V. The slope, which follows a general SSW–ENE with a moderate angle exceeding 20° only in the lowermost part, is characterized by the presence of the mainly clayey bedrock, dipping 15° towards NE and belonging to the Argille Azzurre formation. In this case, neither appreciable thicknesses of continental deposits nor significant geomorphological processes (mass movements) were observed.
- Section 4V. This section is traced in a W–E direction, just south of Section 3V. The mainly clayey bedrock (Argille Azzurre formation) here is almost totally covered by thick eluvial–colluvial deposits and is characterized by the presence of several mudflows, which coalesce corresponding to the valley floor. The typology of movement was attributed based on geomorphological considerations such as the material size, the elongated shape of the landslide body (deposited within minor U-shaped valleys), the presence of slight undulations on the surface and the possibility of a concentrated runoff due to the morphology of the slope. The estimate of the landslide depth is, however, uncertain, although it should be linked to the thickness of the continental deposits.
- Section 5V. This cross-section is oriented NW–SE and is characterized by the presence of two different litho-technical units: the arenaceous-conglomeratic body of Mt. Falcone in the uppermost part and the pelitic-arenaceous member of the Laga formation in the lowermost one. The morphology of the slope reflects the different nature of the lithotypes with an almost vertical cliff corresponding to the most resistant unit and a gentle slope (15–20°) in the second part of the section. As discussed previously, the Mt. Falcone arenaceous conglomeratic body, especially in this sector, shows a particular joint system, N–S and E–W oriented and associated with tectonic and/or static deformation processes. This unit is affected by rockfalls and toppling phenomena (Figure 3) mainly located along the borders of the plate. The final part of the section, on the other hand, corresponds to the head of an active mudflow roughly WSW–ENE oriented; as with Section 4V this landslide is probably linked to the presence of thick colluvial deposits with a perched water table partially fed by the contact with the permeable arenaceous-conglomeratic plate.

- Section 6V. This section is located in the northeastern sector of the study area and crosses, with a NE direction, the arenaceous-pelitic and the clayey members (in the uppermost and lowermost part of the section respectively) of the Argille Azzurre formation, here dipping 13–15° towards NE. The slope angle is moderate and ranges between 10° and 25°. The upper part of the section is characterized by the presence of thick eluvial–colluvial deposits, although no evident geomorphological processes (neither mass movements nor intense soil erosion processes) are visible.
- Section 1M. This section is traced in the NS direction along a gentle slope (15–20°) almost totally characterized by the presence of thick eluvial–colluvial deposits (10–15 m thick estimated from geomorphological survey and stratigraphic reconstruction); these materials cover a bedrock consisting of alternations of arenaceous-pelitic and pelitic-arenaceous members both belonging to the Laga formation, with apparent horizontal dip. No evident geomorphological processes (landslides or intense soil erosion processes) were observed.
- Section 2M. Located just south of the previous section, it runs roughly E–W and, similarly to the Section 5V, is characterized by a complex morphology: The first part is strongly conditioned by the presence of the arenaceous conglomeratic body of Mt. Falcone, while the second one, where the pelitic-arenaceous member of the Laga formation is present, shows a low-to-moderate slope (between 15° and 30°). The uppermost part is characterized by rockfalls and toppling phenomena analogous to those observed in Section 5V; the final part corresponds to the head of an active complex landslide (rotational-translational) recognized on the basis of typical morphologies (scarps and counterslopes in the upper portion and minor ridges and scarps in the lower one).
- Section 3M. This section is traced in a WNW–ESE direction over a slope characterized by a moderate angle (25–30°). Thick colluvial deposits cover the bedrock, visible only in the uppermost part of the section and consisting of alternations of arenaceous and pelitic lithotypes (Laga formation) that dip upslope of 35–40°. No landslides or intense erosion processes were detected.
- Section 4M. This section is traced in a roughly WSW–ENE direction and, as in Section 3M, runs over a slope characterized by the presence of alternations of arenaceous-pelitic and pelitic-arenaceous lithotypes with the same dip. The geomorphological survey evidenced an active complex landslide (rotational-flow); scarps, counterslopes and minor ridges on a general concave shape are visible, corresponding to the head, while elongated shape and undulations characterize the foot of the landslide. Conditioning and triggering factors can be associated with the presence of colluvial deposits with uncertain thickness and the possible presence of a perched water table fed at the contact with the arenaceous-conglomeratic plate, respectively.

A gravitational movement of greatest scientific and technical interest, given the elevated level of risk connected to the presence of the historical centers mentioned in the previous chapter, is present corresponding to the relief of Mount Falcone (Figure 4). This phenomenon, already studied in the past and verified by numerical modeling both in static and dynamic (seismic) conditions [34,35,58], has been classified as lateral spread and affects, with different intensity, the whole "plate" at the top of the relief. Taking into account the objectives of the present work (mainly addressing the study of medium-to-low depth processes), it was not included in the analysis.

4.2. Numerical Approach

Six simulations were carried out for each section: three using the conditions described above (min, mean and max) and three including the presence of a shallow water table (as often evidenced during field surveys).

The results of the simulations are shown in Figures 8 and 9 and Table 2. More specifically, Figures 8 and 9 show the plots of the simulations that evidenced the lowest FoS values (i.e., worse geotechnical parameters and presence of groundwater); Table 2 includes all the FoS values obtained from the simulations and any correspondences with what was observed in the field (geomorphological model). The same table also reports the failure models used by the software for any simulation.

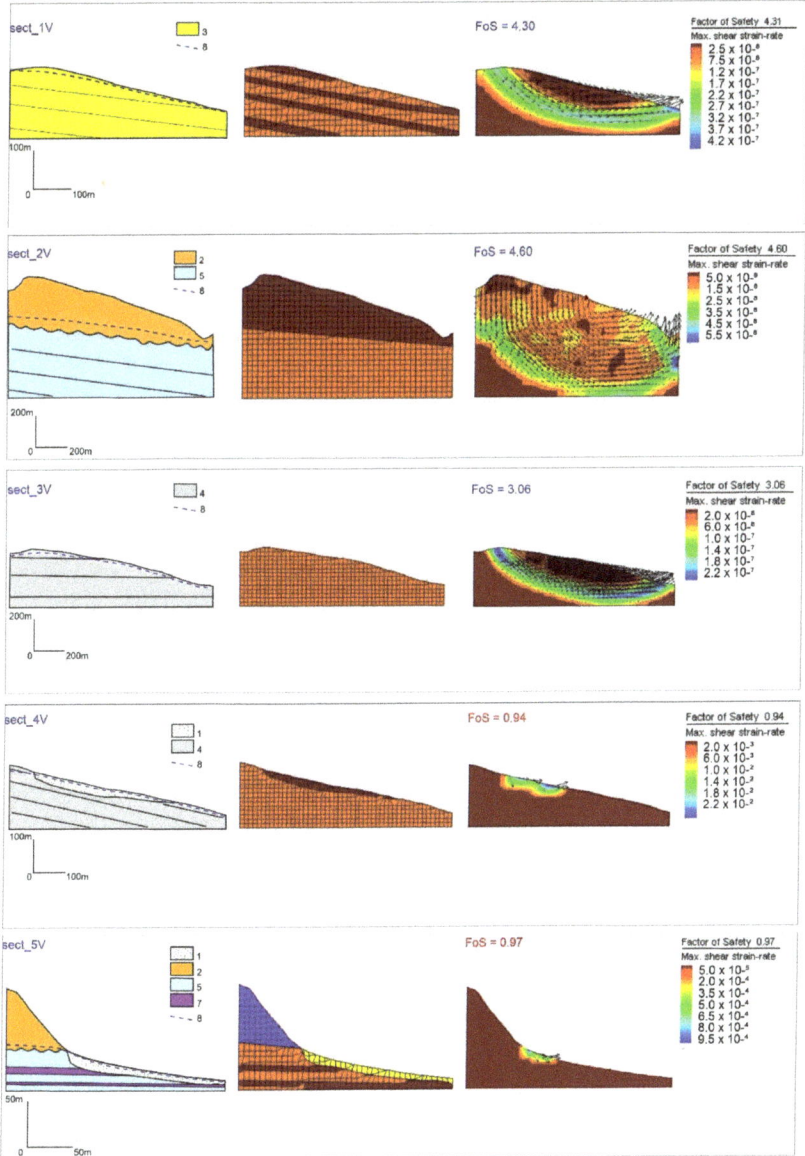

Figure 8. Results of numerical simulation (Sections 1V, 2V, 3V, 4V, 5V). From left to right: geotechnical cross-section, mesh and model result. 1—slope and colluvial deposits; 2—arenaceous-conglomeratic bedrock; 3—arenaceous-pelitic and pelitic-arenaceous bedrock; 4—mainly clayey bedrock; 5—mainly pelitic-arenaceous bedrock; 6—mainly arenaceous-pelitic bedrock; 7—weak layer/ductile deformation zone; 8—water table.

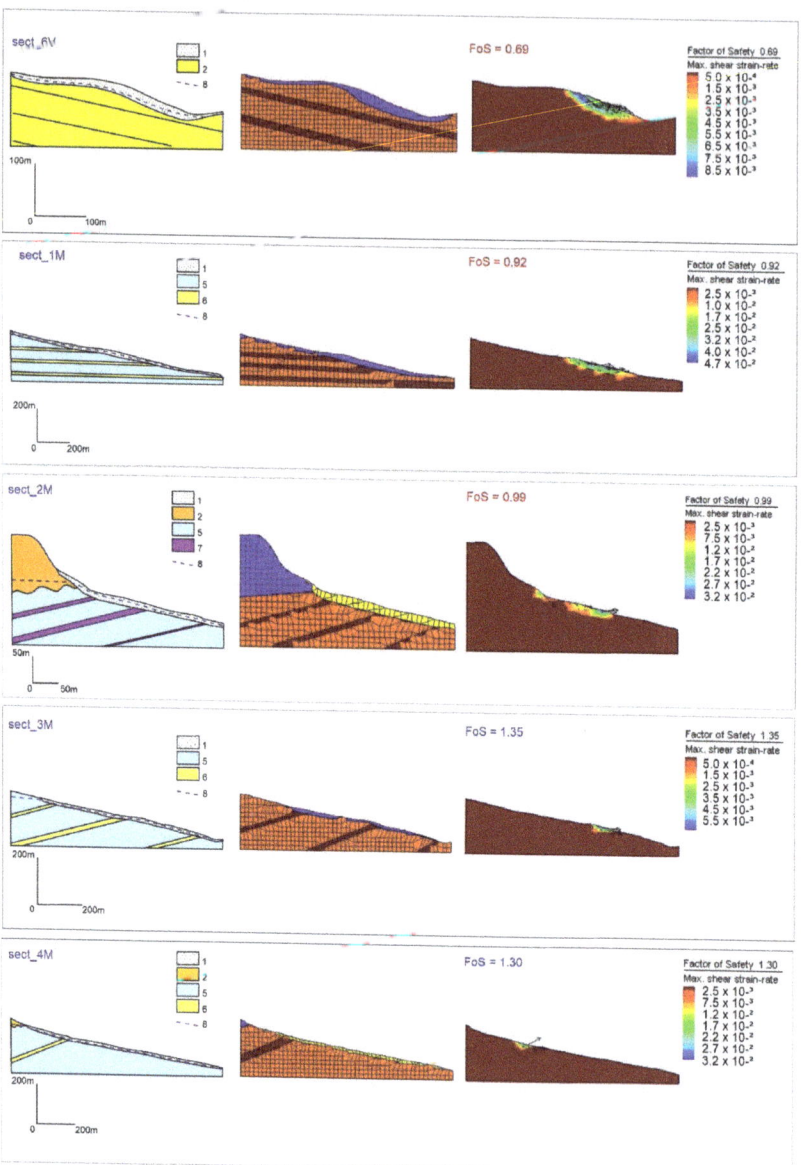

Figure 9. Results of numerical simulation (Sections 6V, 1M, 2M, 3M, 4M). From left to right: geotechnical cross-section, mesh and model result. 1—slope and colluvial deposits; 2—arenaceous-conglomeratic bedrock; 3—arenaceous-pelitic and pelitic-arenaceous bedrock; 4—mainly clayey bedrock; 5—mainly pelitic-arenaceous bedrock; 6—mainly arenaceous-pelitic bedrock; 7—weak layer/ductile deformation zone; 8—water table.

Table 2. Factor of safety resulting from the numerical modeling and correspondence with field evidence.

Cross-Section	Failure Model	Factor of Safety						In Agreement with the Geomorphological Model
		No Water			Water			
		Min	Mean	Max	Min	Mean	Max	
Sect_1V	Hoek–Brown	7.46	8.01	8.47	4.30	4.93	5.48	yes
Sect_2V	Hoek–Brown/Mohr–Coulomb/Ubiquitous	5.57	5.85	6.09	4.60	4.96	5.22	yes
Sect_3V	Hoek–Brown	5.42	5.86	6.22	3.06	3.51	3.92	yes
Sect_4V	Hoek–Brown/Mohr–Coulomb	1.28	1.42	1.50	0.94	1.06	1.15	yes
Sect_5V	Hoek–Brown/Mohr–Coulomb/Ubiquitous	1.47	1.62	1.72	0.97	1.08	1.17	only partially
Sect_6V	Hoek–Brown/Mohr–Coulomb	0.96	1.07	1.13	0.71	0.93	0.99	no
Sect_1M	Hoek–Brown/Mohr–Coulomb	1.22	1.38	1.45	0.89	1.01	1.08	no
Sect_2M	Hoek–Brown/Mohr–Coulomb/Ubiquitous	1.28	1.42	1.51	0.96	1.08	1.16	yes
Sect_3M	Hoek–Brown/Mohr–Coulomb	1.49	1.66	1.76	1.35	1.50	1.60	yes
Sect_4M	Hoek–Brown/Mohr–Coulomb/Ubiquitous	1.56	1.74	1.85	1.30	1.46	1.55	no

Sections 1V, 2V and 3V (Figure 8) show high stability in all lithological conditions, with or without the presence of a water table, with FoS values ranging between 3.06 and 8.47 (Table 2).

Nevertheless, the obtained values of FoS as well as shape and depth of the failure surfaces, as evidenced by the model results, are unrealistic and clearly conditioned by the assumed boundary conditions; therefore, for these specific sections, we can only hypothesize a high stability condition and a perfect congruence with the results of the field surveys, which did not show any appreciable phenomenon.

Section 4V (Figure 8) shows a clear instability only in the presence of groundwater, when the FoS drops below 1; in dry conditions the FoS ranges between 1.28 and 1.50, resulting in a condition of moderate stability. The failure surface is located very close to the surface, within the debris cover, while the bedrock, even when dip-sloping, remains almost stable in all conditions. In this case, the model reflects quite faithfully the reality, as a shallow mudflow was observed within a secondary valley filled by colluvial deposits.

Models relating to the section 5V (Figure 8 and Table 2) evidenced highly unstable conditions (FoS between 0.97 and 1.17) only in the presence of water and in dry conditions. As with the previous model, the failure surface is located in the upper portion of the deposits, near the contact with the arenaceous-conglomeratic body of Mount Falcone; the presence of weak levels and ductile deformation zones, as previously described, does not seem to influence the FoS. Unlike the previous case, however, the geomorphological model only partially reflects the numerical one. The field surveys showed the presence of a complex gravitational movement compatible with a slide (probably rotational) in the upper portion of the slope and with a flow in the medium–low portion. The presence of water within the colluvial deposits, observed mainly in autumn and spring, originates at the contact between the overlying arenaceous-conglomeratic body acting as an aquifer and the underlying low-permeability pelitic-arenaceous formation.

Section 6V (Figure 9) shows high instability in all conditions, with and without the presence of water, with FoS between 0.71 and 1.13. Although the stratigraphic setting of the clayey bedrock is favorable to the occurrence of gravitational movements, the failure surface is localized in the medium–low portion of the slope, at the contact between the bedrock itself and the colluvial deposits above. These simulations, however, do not find correspondence in the geomorphological model: any significant phenomenon was observed in the field.

This could be linked to an incorrect assessment of the real thicknesses of the deposits, the latter having been estimated in this sector without the aid of geognostic surveys.

The simulations carried out with regard to sections 1M and 2M (Figure 9) yielded similar results, with evidence of instability in the presence of water (FoS between 0.89 and 1.16) and moderate stability in dry conditions (FoS between 1.22 and 1.51). The shear-strain belts are located in positions similar to the previous case, in the middle portion of the slope and inside the colluvial deposits. In these two simulations, the setting of bedrock (i.e., the presence of the weak levels within the pelitic-arenaceous formation in section 2M) does not seem to affect the stability of the slope. The correspondence with the field evidence is different: none in section 1M and good correspondence in section 2M. Taking into account previously mentioned factors, the reason, in the first case, could be found in an incorrect evaluation of the overall thickness of continental deposits.

Finally, sections 3M and 4M (Figure 9) also provided similar results, this time in favor of stability, with the FoS always higher than 1.30 (max = 1.85): The presence of a favorable stratigraphic setting (sub-horizontal or slightly counter dip-slope strata), lower slope angle and limited thickness of colluvial deposits certainly affected the result of the simulations. However, a fair correspondence with the field data was found only in section 3M; in the case of section 4M, on the contrary, a fairly evident mudflow was observed inside the valley, E–W oriented, which originates from the arenaceous-conglomeratic body of Mount Falcone.

By analyzing all the simulations, it is possible to form some general considerations:
- The presence of colluvial deposits of fine grain size and discrete extension and thickness, associated with medium–high slopes angles, generally induces instability conditions;
- In the presence of unconsolidated continental deposits and shallow water table, the FoS tends to reach values close to 1 and consequently activate gravitational movements of discrete extent that are generally not very deep;
- Dip-slope strata and/or clayey bedrock are not sufficient requirements to activate gravitational movements, even in the presence of a water table;
- Complex and medium depth landslides, not highlighted by the simulations, can be explained by the presence of particularly weathered levels within the bedrock neither evidenced during field surveys nor "captured" with typical geognostic surveys.

5. Discussion and Conclusions

Numerical models and, in particular, finite difference programs represent a powerful resource for the study and analysis of gravitational phenomena. Specifically, software such as FLAC/Slope, having characteristic geotechnical parameters of soils and rocks, can be used to carry out important assessments on the stability of the slopes and provide an estimate of the FoS. These assessments can then be used successfully both for purposes related to civil engineering (construction of buildings and infrastructures, effectiveness of slope reinforcement works, etc.) and, more generally, for the assessment of landslide susceptibility of variously sized sectors of the slope. Although they provide numerical results that are indispensable for any type of design and planning, the limits of the models are closely linked to the availability and correctness of the input parameters, which are often missing and limited to single laboratory analyses or estimation through direct observations.

The geomorphological model, based on field observations of the processes active in an area, allows a broader and certainly more articulated evaluation of gravitational phenomena; nevertheless, since it consists of an exclusively qualitative analysis, it cannot provide indices and parameters for performing numerical calculations.

A combined approach that integrates the two models can certainly provide mutual advantages: the ability to confirm and quantify the phenomena observed in the field (the geomorphological model) and to verify and modify model parameters and geometry (the numerical model).

The combined approach, however, when compared to the standard methods (i.e., the statistical methods currently used in Italy for LHA), requires a greater effort both in economic terms and in terms of human resources, since it is necessary to proceed with an update and, often, implementation of the field data. This is all the more onerous in the case of a similar methodology as a standard at the national level where there are large disparities (not only economic) between different regional realities. On the other hand, the possibility of having an updated product that is more functional for professional needs or for the planning of particularly critical areas cannot be separated from an approach that provides for a continuous synthesis between real data and numerical models.

The present study, through this type of approach, provides a more objective evaluation of the mechanisms governing landsliding in a typical geological–structural context, characterized by a monoclinal setting and the presence of lithotypes of different nature and consistency.

More specifically, it was possible to verify that:

- Medium to low GSI values, favorable morphological–structural setting (i.e., dip-slope strata and moderate slope angle) and the presence of a water table are sometimes not evaluated by the numerical model as potentially unstable; on the contrary, they give rise to gravitational phenomena of discrete thickness and extension. This suggests the presence of weak layers at a depth not detectable by generic field surveys or highlighted by specific geognostic investigations;
- The presence of medium–fine colluvial deposits of moderate extension and thickness along medium-to-low slope angles (20–35%) constitutes a predisposing factor to the activation of mass movements (flows as prevalent) with or without the presence of water.

The above considerations could provide further confirmation and perhaps be extended to different morphological–structural contexts, through new detailed surveys and a precise characterization of the buried or outcropping lithotypes.

In conclusion, the proposed approach, which can be defined as semi-quantitative, can be proposed as a valid alternative for LHA in all those countries where specific regulatory indications are absent. The obtained results evidence usefulness and limits of the methods currently used in Italy and, in particular, suggest how a combined use of geomorphological surveys and numerical simulations, pending clearer and universally accepted regulatory indications on the methods to be used for the LHA, seems at the moment the most suitable choice both in economic and safety terms.

Author Contributions: Conceptualization, M.M. and M.B.; methodology, M.M.; software, M.M. and M.G.; validation, G.P., P.F. and D.A.; formal analysis, M.M.; investigation, M.M. and M.B.; resources, M.M. and M.B.; data curation, M.M. and M.B.; writing—original draft preparation, M.M. and M.B.; writing—review and editing, M.M. and M.B.; visualization, M.M., M.B. and G.P.; supervision, M.M., M.B. and G.P.; project administration, N/A; funding acquisition, N/A. All authors have read and agreed to the published version of the manuscript.

Funding: This research received no external funding.

Institutional Review Board Statement: Not applicable.

Informed Consent Statement: Not applicable.

Conflicts of Interest: The authors declare no conflict of interest.

References

1. Schuster, R.L.; Highland, L.M. *Socioeconomic Impacts of Landslides in the Western Hemisphere*; United States Geological Survey: Reston, VA, USA, 2001.
2. Winter, M.G.; Shearer, B.; Palmer, D.; Peeling, D.; Harmer, C.; Sharpe, J. The Economic Impact of Landslides and Floods on the Road Network. *Procedia Eng.* **2016**, *143*, 1425–1434. [CrossRef]
3. Corominas, J.; Van Westen, C.; Frattini, P.; Cascini, L.; Malet, J.-P.; Fotopoulou, S.; Catani, F.; Eeckhaut, M.V.D.; Mavrouli, O.; Agliardi, F.; et al. Recommendations for the quantitative analysis of landslide risk. *Bull. Int. Assoc. Eng. Geol.* **2013**, *73*, 209–263. [CrossRef]

4. Thiery, Y.; Terrier, M.; Colas, B.; Fressard, M.; Maquaire, O.; Grandjean, G.; Gourdier, S. Improvement of landslide hazard assessments for regulatory zoning in France: STATE–OF–THE-ART perspectives and considerations. *Int. J. Disaster Risk Reduct.* **2020**, *47*, 101562. [CrossRef]
5. Vandromme, R.; Thiery, Y.; Bernardie, S.; Sedan, O. ALICE (Assessment of Landslides Induced by Climatic Events): A single tool to integrate shallow and deep landslides for susceptibility and hazard assessment. *Geomorphology* **2020**, *367*, 107307. [CrossRef]
6. Trigila, A.; Iadanza, C.; Bussettini, M.; Lastoria, B. *Dissesto Idrogeologico in Italia: Pericolosità e Indicatori Di Rischio*, 2018 ed.; ISPRA: Rome, Italy, 2015.
7. Guzzetti, F.; Mondini, A.C.; Cardinali, M.; Fiorucci, F.; Santangelo, M.; Chang, K.-T. Landslide inventory maps: New tools for an old problem. *Earth Sci. Rev.* **2012**, *112*, 42–66. [CrossRef]
8. Bufalini, M.; Materazzi, M.; De Amicis, M.; Pambianchi, G. From Traditional to Modern " Full Coverage " Geomorphological Mapping: A Study Case in the Chienti River Basin (Marche Region, Central Italy). *J. Maps* **2021**, *17*, 1–12. [CrossRef]
9. Cascini, L. Applicability of landslide susceptibility and hazard zoning at different scales. *Eng. Geol.* **2008**, *102*, 164–177. [CrossRef]
10. Cascini, L. Geotechnics for Urban Planning and Land Use Management. *Riv. Ital. Geotech.* **2015**, *49*, 7–62.
11. Salvati, P.; Bianchi, C.; Rossi, M.; Guzzetti, F. Landslide Risk to the Population and Its Temporal and Geographical Variation in Italy. In *Extreme Events: Observations, Modeling, and Economics*; Chavez, M., Ghil, M., Urrutia-Fucugauchi, J., Eds.; John Wiley & Sons: Hoboken, NJ, USA, 2012; Volume 14, pp. 177–194.
12. Van Westen, C.J.; van Asch, T.W.J.; Soeters, R. Landslide Hazard and Risk Zonation—Why Is It Still so Diffi-cult? *Bull. Eng. Geol. Environ.* **2006**, *65*, 167–184. [CrossRef]
13. Thiery, Y.; Terrier, M. Évaluation de l'aléa Glissements de Terrain: État de l'art et Perspectives Pour La Car-tographie Réglementaire En France. *Rev. Française Géotechnique* **2018**, *156*, 3. [CrossRef]
14. Aleotti, P.; Chowdhury, R. Landslide hazard assessment: Summary review and new perspectives. *Bull. Int. Assoc. Eng. Geol.* **1999**, *58*, 21–44. [CrossRef]
15. Galli, M.; Ardizzone, F.; Cardinali, M.; Guzzetti, F.; Reichenbach, P. Comparing landslide inventory maps. *Geomorphology* **2008**, *94*, 268–289. [CrossRef]
16. Soeters, R.; van Westen, C.J. Slope Instability Recognition, Analysis, and Zonation. In *Landslides, Investigation and Mitigation*; Turner, A.K., Schuster, R.L., Eds.; Special Report; National Research Council, Transportation Research Board: Washington, DC, USA, 1996; Volume 247, pp. 129–147.
17. Cotecchia, F.; Santaloia, F.; Lollino, P.; Vitone, C.; Pedone, G.; Bottiglieri, O. From a phenomenological to a geomechanical approach to landslide hazard analysis. *Eur. J. Environ. Civ. Eng.* **2016**, *20*, 1004–1031. [CrossRef]
18. Guzzetti, F.; Carrara, A.; Cardinali, M.; Reichenbach, P. Landslide Hazard Evaluation: A Review of Current Techniques and Their Application in a Multi-Scale Study, Central Italy. *Geomorphology* **1999**, *31*, 181–216. [CrossRef]
19. Nandi, A.; Shakoor, A. A GIS-Based Landslide Susceptibility Evaluation Using Bivariate and Multivariate Sta-tistical Analyses. *Eng. Geol.* **2010**, *110*, 11–20. [CrossRef]
20. Pourghasemi, H.R.; Kerle, N. Random Forests and Evidential Belief Function-Based Landslide Susceptibility Assessment in Western Mazandaran Province, Iran. *Environ. Earth Sci.* **2016**, *75*, 185. [CrossRef]
21. Chen, X.; Chen, W. GIS-Based Landslide Susceptibility Assessment Using Optimized Hybrid Machine Learn-ing Methods. *Catena* **2021**, *196*, 104833. [CrossRef]
22. Pourghasemi, H.R.; Mohammady, M.; Pradhan, B. Landslide Susceptibility Mapping Using Index of Entropy and Conditional Probability Models in GIS: Safarood Basin, Iran. *Catena* **2012**, *97*, 71–84. [CrossRef]
23. Barla, G. Numerical Modeling of Deep-Seated Landslides Interacting with Man-Made Structures. *J. Rock Mech. Geotech. Eng.* **2018**, *10*, 1020–1036. [CrossRef]
24. Cuomo, S. New Advances and Challenges for Numerical Modeling of Landslides of the Flow Type. *Procedia Earth Planet. Sci.* **2014**, *9*, 91–100. [CrossRef]
25. Fustos, I.; Abarca-del-Río, R.; Mardones, M.; González, L.; Araya, L.R. Rainfall-Induced Landslide Identifica-tion Using Numerical Modelling: A Southern Chile Case. *J. S. Am. Earth Sci.* **2020**, *101*, 102587. [CrossRef]
26. Marcato, G.; Mantovani, M.; Pasuto, A.; Zabuski, L.; Borgatti, L. Monitoring, Numerical Modelling and Hazard Mitigation of the Moscardo Landslide (Eastern Italian Alps). *Eng. Geol.* **2012**, *128*, 95–107. [CrossRef]
27. Viero, A.; Kuraoka, S.; Borgatti, L.; Breda, A.; Marcato, G.; Preto, N.; Galgaro, A. Numerical Models for Plan-ning Landslide Risk Mitigation Strategies in Iconic but Unstable Landscapes: The Case of Cinque Torri (Dolomites, Italy). *Eng. Geol.* **2018**, *240*, 163–174. [CrossRef]
28. Griffiths, D.V.; Lu, N. Unsaturated Slope Stability Analysis with Steady Infiltration or Evaporation Using Elasto-Plastic Finite Elements. *Int. J. Numer. Anal. Methods Geomech.* **2005**, *29*, 249–267. [CrossRef]
29. Sorbino, G.; Sica, C.; Cascini, L. Susceptibility Analysis of Shallow Landslides Source Areas Using Physically Based Models. *Nat. Hazards* **2010**, *53*, 313–332. [CrossRef]
30. Formetta, G.; Rago, V.; Capparelli, G.; Rigon, R.; Muto, F.; Versace, P. Integrated Physically Based System for Modeling Landslide Susceptibility. *Procedia Earth Planet. Sci.* **2014**, *9*, 74–82. [CrossRef]
31. Centamore, E.; Nisio, S. Effects of Uplift and Tilting in the Central-Northern Apennines (Italy). *Quat. Int.* **2003**, *101–102*, 93–101. [CrossRef]

32. Gentili, B.; Pambianchi, G.; Aringoli, D.; Materazzi, M.; Giacopetti, M. Pliocene-Pleistocene Geomorphological Evolution of the Adriatic Side of Central Italy. *Geol. Carpathica* **2017**, *68*, 6–18. [CrossRef]
33. Itasca. *FLAC—Fast Lagrangian Analysis of Continua, Version 8.0.*; Itasca Consulting Group Inc., Ed.; Itasca Consulting Group, Inc.: Minneapolis, MN, USA, 2016.
34. Crescenti, U.; Sciarra, N.; Gentili, B.; Pambianchi, G. Modeling of Complex Deep-Seated Mass Movements in the Central-Southern Marches (Central Italy). In *Landslides*; Rybář, J., Stemberk, J., Wagner, P., Eds.; Balkema: Lisse, The Netherlands, 2002; pp. 149–155. [CrossRef]
35. Aringoli, D.; Gentili, B.; Materazzi, M. Mass movements in Adriatic central Italy: Activation and evolutive control factors. In *Landslides: Causes, Types and Effects*; Nova Science Publishers, Inc.: New York, NY, USA, 2010; pp. 1–71.
36. Cantalamessa, G.; Centamore, E.; Chiocchini, U.; Colalongo, M.L.; Micarelli, A.; Nanni, T.; Pasini, G.; Potetti, M.; Ricci Lucchi, F.; Cristallini, C.; et al. Il Plio-Pleistocene Delle Marche. *Studi Geologici Camerti (Special Volume "La Geologia delle Marche")* **1986**, 61–81. Available online: https://www.semanticscholar.org/paper/Il-Plio-Pleistocene-delle-Marche.-Cantalamessa-Centamore/1f38e45fcff286816bcdedfb2da9ea0146477f9d (accessed on 20 March 2021).
37. Ori, G.G.; Serafini, G.; Visentin, C.; Lucchi, F.R.; Casnedi, R.; Colalongo, M.L.; Mosna, S. The Pliocene-Pleistocene Adriatic Foredeep (Marche and Abruzzo, Italy): An Integrated Approach to Surface and Subsur-face Geology. In *3rd EAPG Conference, Adriatic Foredeep Field Trip Guide Book*; EAPG and AGIP: Florence, Italy, 1991; pp. 26–30.
38. Gentili, B.; Materazzi, M.; Pambianchi, G.; Scalella, G.; Aringoli, D.; Cilla, G.; Farabollini, P. The Slope Deposits of the Ascensione Mount (Southern Marche, Italy). *Geogr. Fis. Din. Quat.* **1998**, *21*, 205–214.
39. Cantalamessa, G.; Di Celma, C. Sequence Response to Syndepositional Regional Uplift: Insights from High-Resolution Sequence Stratigraphy of Late Early Pleistocene Strata, Periadriatic Basin, Central Italy. *Sediment. Geol.* **2004**, *164*, 283–309. [CrossRef]
40. Buccolini, M.; Bufalini, M.; Coco, L.; Materazzi, M.; Piacentini, T. Small Catchments Evolution on Clayey Hilly Landscapes in Central Apennines and Northern Sicily (Italy) since the Late Pleistocene. *Geomorphology* **2020**, *363*, 107206. [CrossRef]
41. Ambrosetti, P.; Carraro, F.; Deiana, G.; Dramis, F. Il Sollevamento Dell'Italia Centrale Tra Il Pleistocene Infe-riore e Il Pleistocene Medio. *P.F. Geodin.* **1982**, *513*, 219–223.
42. Buccolini, M.; Gentili, B.; Materazzi, M.; Piacentini, T. Late Quaternary Geomorphological Evolution and Ero-sion Rates in the Clayey Peri-Adriatic Belt (Central Italy). *Geomorphology* **2010**, *116*, 145–161. [CrossRef]
43. Wise, D.U.; Funiciello, R.; Parotto, M.; Salvini, F. Topographic Lineament Swarms: Clues to Their Origin from Domain Analysis of Italy. *Bull. Geol. Soc. Am.* **1985**, *96*, 952–967. [CrossRef]
44. Calamita, F. Extensional Mesostructures in Thrust Shear Zone: Example from the Umbro-Marchean Appen-nines. *Boll. Soc. Geol. Ital.* **1991**, *110*, 649–660.
45. Zienkiewicz, C.; Humpheson, C.; Lewis, R.W. Associated and Non-Associated Visco-Plasticity and Plasticity in Soil Mechanics. *Geotechnique* **1975**, *25*, 671–689. [CrossRef]
46. Naylor, D.J. Finite Elements and Slope Stability. In *Numerical Methods in Geomechanics*; Springer: Dordrecht, The Netherlands, 1982; Volume 8, pp. 229–244.
47. Matsui, T.; San, K.-C. Finite Element Slope Stability Analysis by Shear Strength Reduction Technique. *Soils Found.* **1992**, *32*, 59–70. [CrossRef]
48. Ugai, K. A Method of Calculation of Total Factor of Safety of Slopes by Elasto-Plastic FEM. *Soils Found.* **1989**, *29*, 190–195. [CrossRef]
49. Ugai, K.; Leshchinsky, D. Three-Dimensional Limit Equilibrium and Finite Element Analyses: A Comparison of Results. *Soils Found.* **1995**, *35*, 1–7. [CrossRef]
50. Dawson, E.M.; Roth, W.H.; Drescher, A. Slope Stability Analysis by Strength Reduction. *Geotechnique* **1999**, *49*, 835–840. [CrossRef]
51. Hoek, E.; Brown, E.T. Practical Estimates of Rock Mass Strength. *Int. J. Rock Mech. Min. Sci.* **1997**, *34*, 1165–1186. [CrossRef]
52. Alejano, L.R.; Alonso, E. Considerations of the Dilatancy Angle in Rocks and Rock Masses. *Int. J. Rock Mech. Min. Sci.* **2005**, *42*, 481–507. [CrossRef]
53. Zhao, X.G.; Cai, M. A Mobilized Dilation Angle Model for Rocks. *Int. J. Rock Mech. Min. Sci.* **2010**, *47*, 368–384. [CrossRef]
54. Cundall, P.A.; Carranza-Torres, C.; Hart, R.D. A New Constitutive Model Based on the Hoek-Brown Criteri-on. FLAG Numer. Model. In *FLAC and Numerical Modeling in Geomechanics, Proceedings of the third International Flac Symposium, Sudbury, ON, Canada, 21–24 October 2003*; Balkema: Lisse, The Netherlands, 2003.
55. Hoek, E. Strength of Rock and Masses. *ISRM News J.* **1994**, *2*, 4–16.
56. Hoek, E.; Kaiser, P.K.; Bawden, W.F. Support of Underground Excavations in Hard Rock. *Environ. Eng. Geosci.* **1996**, *2*, 609–615. [CrossRef]
57. Marinos, P.; Marinos, V.; Hoek, E. Geological Strength Index (GSI): A characterization tool for assessing engineering properties for rock masses. In *Proceedings of International Workshop on Rock Mass Classification for Underground Mining*; Mark, C., Pakalnis, R., Tuchman, R.J., Eds.; Taylor and Francis Group: Madrid, Spain; London, UK, 2007; pp. 87–94. [CrossRef]
58. Aringoli, D.; Buccolini, M.; Coco, L.; Dramis, F.; Farabollini, P.; Gentili, B.; Giacopetti, M.; Materazzi, M.; Pambianchi, G. The Effects of In-Stream Gravel Mining on River Incision: An Example from Central Adriatic Italy. *Zeitschrift Geomorphol.* **2015**, *59*, 95–107. [CrossRef]
59. Lollino, P.; Cotecchia, F.; Elia, G.; Mitaritonna, G.; Santaloia, F. Interpretation of landslide mechanisms based on numerical modelling: Two case-histories. *Eur. J. Environ. Civ. Eng.* **2016**, *20*, 1032–1053. [CrossRef]

60. Hoek, E.; Marinos, P.; Benissi, M. Applicability of the Geological Strength Index (GSI) Classification for Very Weak and Sheared Rock Masses. The Case of the Athens Schist Formation. *Bull. Eng. Geol. Environ.* **1998**, *57*, 151–160. [CrossRef]
61. Marinos, P.; Hoek, E. Estimating the Geotechnical Properties of Heterogeneous Rock Masses Such as Flysch. *Bull. Eng. Geol. Environ.* **2001**, *60*, 85–92. [CrossRef]
62. Budetta, P.; Nappi, M. Heterogeneous Rock Mass Classification by Means of the Geological Strength Index: The San Mauro Formation (Cilento, Italy). *Bull. Eng. Geol. Environ.* **2011**, *70*, 585–593. [CrossRef]
63. Stück, H.; Koch, R.; Siegesmund, S. Petrographical and Petrophysical Properties of Sandstones: Statistical Analysis as an Approach to Predict Material Behaviour and Construction Suitability. *Environ. Earth Sci.* **2013**, *69*, 1299–1332. [CrossRef]
64. Cantalamessa, G.; Centamore, E.; Didaskalou, P.; Micarelli, A.; Napoleone, G.; Potetti, M. Elementi Di Correlazione Nella Successione Marina Plio-Pleistocenica Del Bacino Periadriatico Marchigiano. *Stud. Geol. Camerti* **2002**, *1*, 33–49.

Article

Relationships between Morphostructural/Geological Framework and Landslide Types: Historical Landslides in the Hilly Piedmont Area of Abruzzo Region (Central Italy)

Gianluca Esposito, Cristiano Carabella, Giorgio Paglia and Enrico Miccadei *

Department of Engineering and Geology, Università degli Studi "G. d'Annunzio" Chieti–Pescara, Via dei Vestini 31, 66100 Chieti Scalo (CH), Italy; gianluca.esposito@unich.it (G.E.); cristiano.carabella@unich.it (C.C.); giorgio.paglia@unich.it (G.P.)
* Correspondence: enrico.miccadei@unich.it

Abstract: Landslides are a widespread natural phenomenon that play an important role in landscape evolution and are responsible for several casualties and damages. The Abruzzo Region (Central Italy) is largely affected by different types of landslides from mountainous to coastal areas. In particular, the hilly piedmont area is characterized by active geomorphological processes, mostly represented by slope instabilities related to mechanisms and factors that control their evolution in different physiographic and geological–structural conditions. This paper focuses on the detailed analysis of three selected case studies to highlight the multitemporal geomorphological evolution of landslide phenomena. An analysis of historical landslides was performed through an integrated approach combining literature data and landslide inventory analysis, relationships between landslide types and lithological units, detailed photogeological analysis, and geomorphological field mapping. This analysis highlights the role of morphostructural features on landslide occurrence and distribution and their interplay with the geomorphological evolution. This work gives a contribution to the location, abundance, activity, and frequency of landslides for the understanding of the spatial interrelationship of landslide types, morphostructural setting, and climate regime in the study area. Finally, it represents a scientific tool in geomorphological studies for landslide hazard assessment at different spatial scales, readily available to interested stakeholders to support sustainable territorial planning.

Keywords: historical landslides; multitemporal analysis; geomorphological mapping; GIS analysis; piedmont area; Abruzzo Region

Citation: Esposito, G.; Carabella, C.; Paglia, G.; Miccadei, E. Relationships between Morphostructural/Geological Framework and Landslide Types: Historical Landslides in the Hilly Piedmont Area of Abruzzo Region (Central Italy). *Land* **2021**, *10*, 287. https://doi.org/10.3390/land10030287

Academic Editor: Wojciech Zgłobicki

Received: 16 February 2021
Accepted: 8 March 2021
Published: 11 March 2021

Publisher's Note: MDPI stays neutral with regard to jurisdictional claims in published maps and institutional affiliations.

Copyright: © 2021 by the authors. Licensee MDPI, Basel, Switzerland. This article is an open access article distributed under the terms and conditions of the Creative Commons Attribution (CC BY) license (https://creativecommons.org/licenses/by/4.0/).

1. Introduction

Landslides are considered, worldwide and in Italy, as one of the most important and frequent natural hazards [1–5] as their occurrence can directly impact humans, infrastructures, economic activities, and the social and environmental systems [6–8]. Landslides are a landscape modelling process inducing geomorphological changes on slopes in mountainous, hilly, and coastal areas. Their occurrence is generally controlled by predisposing factors (i.e., morphology, lithological and structural setting, vegetation cover, land use, climate, etc.) and triggering ones (e.g., heavy rainfall and snowfall events, snow melting, earthquakes, wildfires, human activity, etc.) [9–13]. Many of the triggering factors are only sufficient conditions for the occurrence of landslides, which are occasional and spasmodic. Therefore, it is essential to pay attention to predisposing factors in landslide analyses to set an organic correlation between climate regime, morphostructural/geological framework, and slope instability phenomena [14,15].

Many theories and methods have been proposed about the spatial relationship between landslides and causative factors [16–22] to perform landslide hazard assessment studies [23]. However, the type, extent, magnitude, and direction of the geomorphological processes and the location, abundance, activity, and frequency of landslides in a changing environment are still under debate. Establishing a relationship between climate change and its potential effects on the occurrence of landslides remains an open issue [24]. The role played by projected climate changes in modifying the response of single slopes or entire catchments, the frequency and extent of landslides, and the related variations in landslide hazard, remain to be discussed and understood [25–27]. Most of the current landslides in the Central Apennines are the reactivation by pre-existing ones, which have occurred in periods of climatic and geomorphological conditions different from those of the present. Most dormant slides and/or paleolandslides, in which the strength parameters are reduced to values close to the residual ones, can be reactivated and/or modified by natural causes, such as rainfall or snowmelt, as well as man-made disturbance [28,29].

Geomorphological mapping is a common and fundamental tool for the representation and the comprehension of the spatial and temporal development of landslides. Recent and new methods developed in the last decades have improved landslide analysis with multi-disciplinary approaches including (i) morphometric analysis using very-high-resolution Digital Elevation Models (DEMs), (ii) interpretation and analysis of satellite images, including Synthetic Aperture Radar (SAR) images, and (iii) the use of new tools to facilitate field mapping [30–35]. Moreover, the investigation of geomorphological processes and dynamics, in different and complex morphostructural domains, became necessary for the assessment of the areas prone to landslides with reference to the predisposing and/or triggering factors.

According to national and regional inventories [36,37], the Abruzzo Region (Central Italy) is acknowledged as an area highly exposed to landslide hazards and risks. It is located in the central–eastern part of the Italian peninsula, and it is characterized by a landscape that is the result of a complex cyclic evolution that occurred in succeeding stages with the dominance either of morphostructural factors, linked to the conflicting tectonic activity (compressive, strike–slip, and extensional tectonics) and regional uplift, or morphosculptural factors, linked to drainage network linear down-cutting and slope gravity processes [14,38,39].

For developing the present study, the analysis of historical landslides was carried out following an integrated approach that incorporates literature and landslide inventory analysis, relationships between landslide types and lithological units, detailed photogeological analysis, and geomorphological field mapping. The paper focuses on selected slope instabilities to highlight the multitemporal geomorphological evolution and the interplay between morphostructural/geological framework and landslide dynamics in the hilly piedmont area of Abruzzo Region. The work shows an effective integrated approach in geomorphological studies for landslide hazard modelling at different spatial scales, readily available to interested stakeholders. Furthermore, it could provide a scientific basis for the implementation of sustainable territorial planning and loss-reduction measures in a changing environment.

2. Study Area

The study area is located in the central–eastern part of the Italian peninsula along the hilly piedmont area of Abruzzo Region, between the Apennine chain and the coastal area (Figure 1a). It includes the lower part of the main SW–NE to W–E fluvial valleys (i.e., Vomano, Pescara, and Sangro rivers), and the small tributary catchments of the main rivers and those incising the coastal slopes.

The Apennine chain area is characterized by a mountainous landscape (with reliefs up to 2900 m.a.s.l. high) interrupted by longitudinal and transversal valleys and wide intermontane basins (i.e., Fucino Plain, Sulmona Basin). It is made up of carbonate lithological sequences pertaining to different Meso-Cenozoic palaeogeographical domains. Carbonate shelf limestones, slope limestones, basin limestone, and marls represent the carbonate backbone of the main ridges of the Abruzzo Apennines, and allochthonous pelagic deposits are widespread in the southern sectors featuring a chaotic assemblage on clayey–marly–limestone units. The main tectonic features are represented by NW–SE to

N–S-oriented thrusts, which affected the chain from the Late Miocene to the Early Pliocene. Compressional tectonics was followed by strike–slip tectonics along mostly NW–SE to NNW–SSE-oriented faults that were poorly constrained in age and largely masked by later extensional tectonic events since the Early Pleistocene [40,41].

The hilly piedmont area is a low relief area (heights ranging from ~100 to 800 m.a.s.l.) characterized by a cuesta, mesa, and plateau landscape and a gently NE-dipping homocline, locally cut by fault systems (NW–SE, SW–NE) with low displacement [42–44]. Bedrock lithologies pertain to Neogene sandy-pelitic turbidites and Plio-Pleistocene marine clayey-sandy and conglomeratic deposits. The geological and structural setting is related to the Pliocene–Quaternary evolution of the Adriatic foredeep and the related regional uplifting processes. Since the Middle Pleistocene, the geomorphological evolution has primarily comprised the incision of major dip river valleys (WSW–ENE-oriented), characterized by fluvial deposits arranged in flights of at least four orders of terraces (Middle Pleistocene–Holocene) [44,45]. Quaternary continental deposits are widely present in the alluvial valleys, alluvial plains, and coastal slopes. They can be referred to fluvio-lacustrine, travertine, sandy shore, and eluvial–colluvial deposits (Figure 1b).

The geomorphological framework is mainly related to fluvial and slope processes. Fluvial processes affect the main rivers, alternating between channel incisions and flooding. The slope processes due to running water mostly affect the clayey and arenaceous-pelitic areas of piedmont and coastal sectors, generating minor landforms such as rills, gullies, and mudflows [46,47]. The area is extensively affected by different types of landslides (e.g., mostly rotational–translational slides, earth flows, rockfalls, complex slides), mostly characterizing the hilly piedmont and the chain area and, locally, the coastal area [3,48].

The present-day regional tectonic setting is dominated by extensional tectonics still active in the axial part of the chain, which is characterized by intense seismicity and strong historical earthquakes (up to M 7.0; [49]). The piedmont area is characterized by moderate uplifting and moderate seismicity, while the Adriatic Sea is affected by subsidence and by moderate compression and strike–slip related seismicity, as also documented by the recent seismicity [50] (Figure 1b).

Climatically, the study area belongs to temperate sub-littoral regime with scarce annual rainfall, mainly autumnal, and medium temperatures [51]. It is largely affected by the orographic setting, changing from a Mediterranean type with maritime influence along the coasts and the piedmont area to more continental-like in the inner sectors [52]. The hilly piedmont area is characterized by a maritime Mediterranean climate [53]. The average annual precipitation is 600–800 mm/year, with occasional heavy rainfall events (>100 mm/d and 30–40 mm/h). The mean annual temperature ranges between 12 and 16 °C in the coastal part of the region, with mild winters and hot summers, and from 8 to 12 °C in mountain areas, with more severe (low) temperatures, especially in the winter season [54,55].

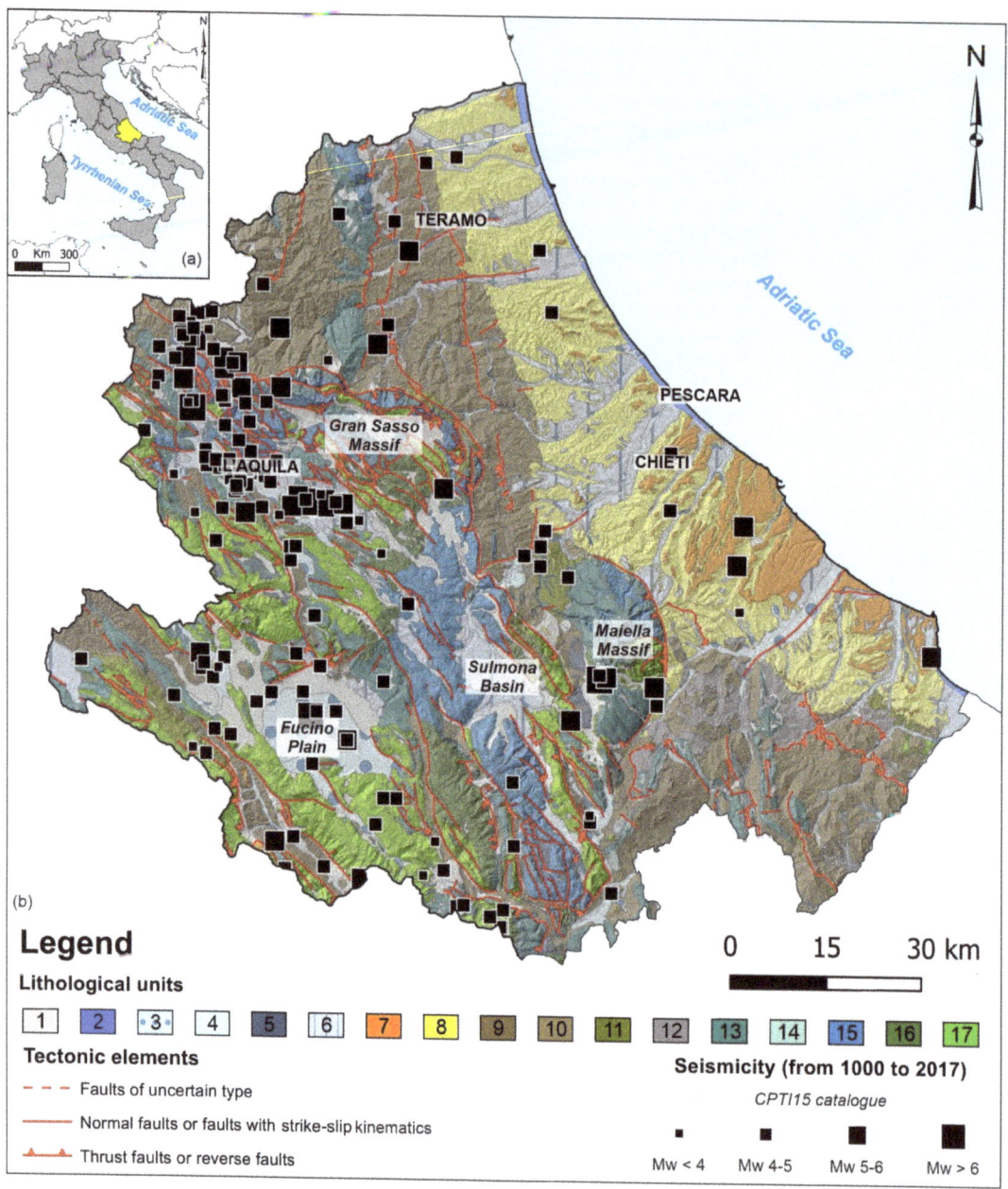

Figure 1. (**a**) Location map of the Abruzzo Region in Central Italy; (**b**) geolithological map of the Abruzzo Region (modified from [56,57]). Legend: (1) eluvial–colluvial deposits; (2) sandy shore deposits; (3) recent fluvio-lacustrine deposits; (4) travertine deposits; (5) morainic deposits; (6) old fluvio-lacustrine deposits; (7) conglomeratic deposits; (8) clayey–sandy deposits; (9) sandy turbidites; (10) pelitic turbidites; (11) carbonate deposits in conglomeratic and calcarenitic facies; (12) allochthonous pelagic deposits; (13) carbonate ramp limestones; (14) basin limestones and marls; (15) slope limestone; (16) open carbonate shelf-edge limestones; (17) carbonate shelf limestones and dolomites. Seismicity derived from [49].

3. Materials and Methods

Landslide analysis was achieved through an integrated approach based on the combination of literature data, landslide inventory analysis, statistical analysis of the relationships between landslide types and lithological units, detailed photogeological analysis, and geomorphological field mapping, supported by multidisciplinary analysis and GIS–based techniques.

3.1. Landslide Inventory Maps and Database Analysis

Landslide inventories and databases represent an important tool to document the extent of landslide phenomena in a region, to investigate the distribution, types, pattern, recurrence, and statistics of slope failures, to determine landslide susceptibility, hazard, and risk, and to study the evolution of landscapes dominated by mass-wasting processes [58].

A preliminary GIS-based analysis was performed to store, organize, and manage available data recorded in four different databases and catalogues, briefly described as follows. The IFFI database (Italian Landslide Inventory—[59,60]) supplies a detailed picture of the distribution of landslide phenomena within Italy. As of today, the IFFI database holds 620,793 landslide phenomena, covering an area of approximately 23,000 km^2, which is equivalent to 7.9% of the Italian territory [37]; for the Abruzzo Region, the database is updated to 2007. The compilation of the catalogue was structured in several phases: (i) collection of bibliographic cartographic data useful to identify areas subject to landslides; (ii) verification by aerial photo interpretation and cartographic transposition; (iii) verification through field-based analysis; (iv) digitization. A total of 6557 events (categorized as rockfalls, lateral spreading, complex landslides, translational and rotational slides, debris flows, earth flows, DSGSDs, and soil creep areas) were included in the inventory used in this study. The CEDIT catalogue (Italian catalogue of earthquake-induced ground failures—[61]) includes more than 150 earthquakes and almost 2000 earthquake-induced effects, which involved almost 1100 localities; the catalogue is updated to the 2016–2017 Central Italy seismic sequence [62,63]. The catalogue implies detailed research of historical documents and reports as well as of already published scientific papers. The analysis of reported seismically induced effects infers that most of them are landslides, which account, alone, for about half of the total (44%). Among all these earthquake-induced landslides, only seven events are located in the hilly piedmont area, and they were selected, recognized, and integrated into the analysis in terms of georeferenced location and detailed information. The EEE catalogue (Earthquake Environmental Effects catalogue—[64]) is aimed to collect in a standard format the wealth of information of environmental/geological effects induced by a seismic event; the catalogue contains tables that include information at site of each EEE, including detailed characteristics on the type of earthquake. The database is updated to the 2016–2017 Central Italy seismic sequence. Among all the documented seismic-induced effects, only landslides (six events falling within the study area) were selected and included in the analysis. The FraneItalia catalogue [65] contains information retrieved from online news sources (especially Google Alerts and Italian Civil Protection press reviews) on landslides that occurred in Italy. It contains all the landslide events reported since 2010 (January 2010–December 2017), not only the ones that caused direct consequences to people or major damage; it is structured as a geo-referenced open-access database containing information on a variety of landslide features and consequences. For this study, all the landslides (162 events falling in the study area) for which it was possible to univocally define the location and the type of movement were selected and included in the inventory.

Available data (i.e., georeferenced location and detailed information) from the above-mentioned catalogues were merged to completely define the landslides' spatial distribution over the Abruzzo Region (Figure 2).

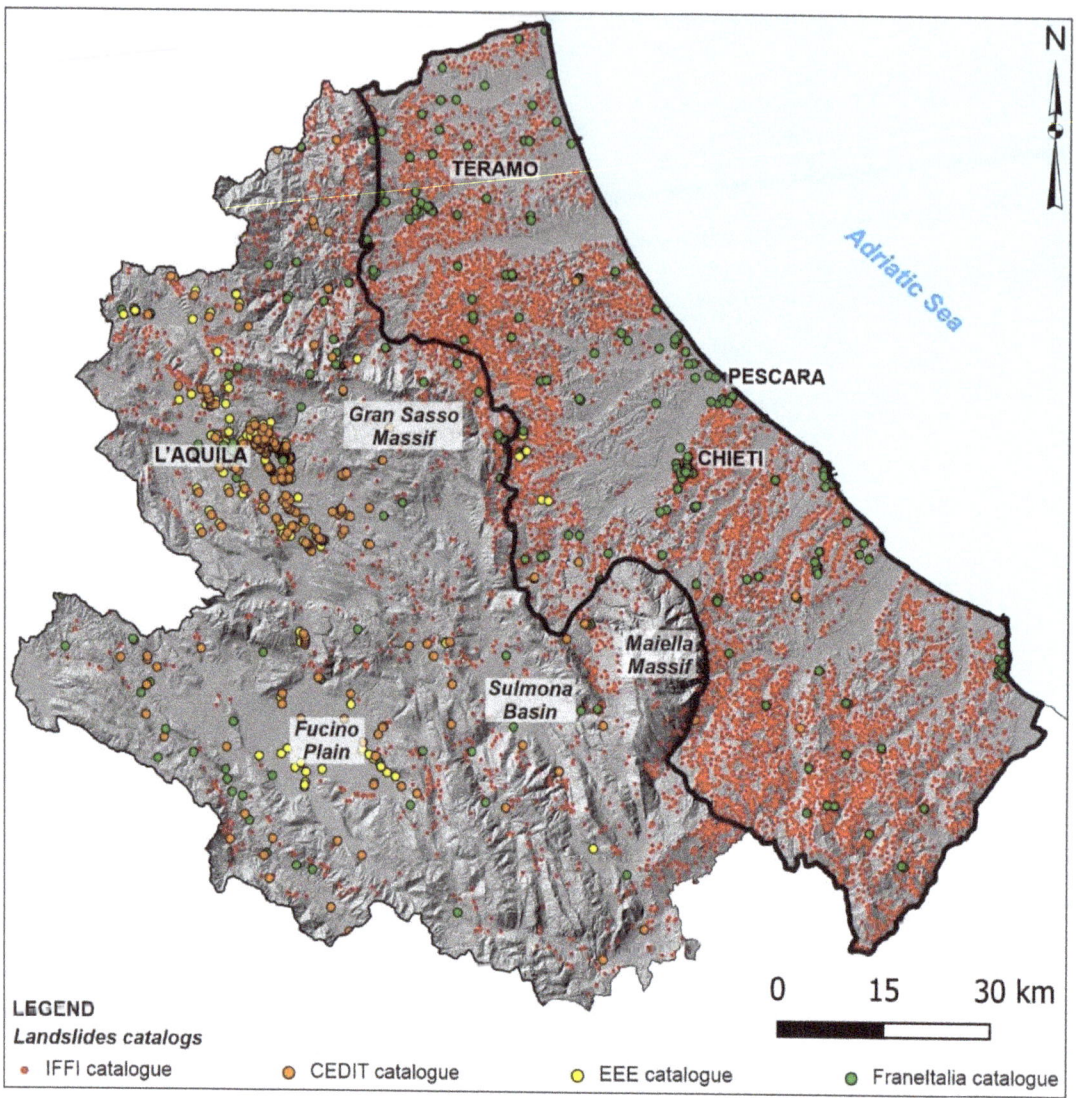

Figure 2. Landslide spatial distribution over the Abruzzo Region. This graphical representation includes the georeferenced location of rockfalls, landslides (lateral spreading, complex landslides, translational and rotational slides), debris flows, earth flows, DSGSDs, and soil creep areas. This general labelling derives from all historical documents, technical reports, and detailed information included in available inventories and databases, such as the Italian Landslide Inventory (IFFI) catalogue [60]; the Italian earthquake-induced ground failures (CEDIT) catalogue [61]; the Earthquake Environmental Effects (EEE) catalogue [64]; the FraneItalia catalogue [65]. The black line represents the study area.

Even if the landslide spatial distribution over the Abruzzo Region is related to rockfalls, landslides (lateral spreading, complex landslides, translational and rotational slides), debris flows, earth flows, DSGSDs, and soil creep area, landslides located in the study area and used for this analysis were categorized and selected according to the type of movement into four categories: rotational and translational slides, complex landslides, earth flows, and rockfalls. This specific labelling was followed to highlight the most characterizing and frequent mass movement types, according to geological–structural setting, location and abundance of landslides. Then, the spatial distribution of each category was evaluated through the creation of density maps, generated using the QGIS (version 3.10, 2019, "A Coruña") HeatMaps (Kernel Density) tool, which calculates a magnitude-per-unit (1 km^2) area from a point or polyline features using a kernel function to fit a smoothly tapered surface to each point. Landslide density maps generally show a synoptic view of landslide distribution for large regions or entire nations in order to portray the first-order overview of landslide abundance. Density is a clearly definable and easily comprehended quantitative measure of the spatial distribution of slope failures. These maps derive from the georeferenced location of each initiation point of landslides (defined as the center of the main headscarp) and assume that landslide density is continuous in space, which may not be the case everywhere.

3.2. Statistical Analysis of the Relationships between Landslides and Lithological Units

Lithology shows a great influence on landslide development since different lithological units may be affected by different landslide types. Moreover, soil cover deposits, mostly exposed to weathering, may influence land permeability and the landslide type, as known from thematic literature [66,67].

In order to stress the role played by lithological units on the development of landslides and build up a statistical relationship with the spatial distribution of landslide type, a vector lithological map (previously categorized into 17 lithological units according to the sedimentation environment and the lithological features of the outcrops) was spatially overlapped with the landslide distribution layer, derived from the selected inventories and databases.

A GIS-based overlay between the georeferenced location of the initiation points of landslides (defined as the center of the main headscarp) and lithological units was performed to understand the influence of lithologies on landslides. This correlation was carried out for different types of landslides (rotational and translational slides, complex landslides, earth flows, and rockfalls) recorded in the hilly piedmont area.

3.3. Detailed Multitemporal and Multidisciplinary Analysis

Multitemporal and multidisciplinary analyses were performed to outline the mass movement types and evolution mechanisms that characterize the different morphostructural domains of the study area. Selected case studies (one for landslide type; about rockfalls, according to a moderate to low spatial distribution, no landslide events have been identified as clearly representative of this mass movement type in the study area, so no case study was reported) have undergone several main movements from the 18th century onwards. These are intended to be representative of the most characterizing and frequent mass movement type, showing significant features useful for understanding the relationships between landslide types, lithologies, and morphostructural setting.

Multitemporal geomorphological analysis was based on detailed analysis of historical maps and literature data, stereoscopic air-photo interpretation, and field mapping. Air-photo interpretation was performed using 1:33,000, 1:20,000, 1:13,000, and 1:5000 scale stereoscopic air-photos (Flight GAI 1954, Flight CASMEZ 1974, Flight Abruzzo Region 1981–1987, and Flight Abruzzo Region 2018–2019), 1:5000-scale orthophoto color images (Flight Abruzzo Region 2010), and Google Earth imagery; this analysis was also supported using high-resolution Digital Elevation Models (DEMs). Field mapping was carried out at an appropriate scale (1:5000–1:10,000), according to international guidelines [68], Italian geomorphological guidelines [69] and the thematic literature concerning geomorphological mapping, fieldbased and numerical analysis [70–73]. It was focused on the definition of lithological and morphostructural features, superficial deposit cover, and the type and distribution of geomorphological landforms with reference to the main landslides affecting the study area.

Rainfall data analysis was carried out to outline the distribution of the climatic parameters and conditions in the hilly piedmont area. The analysis was based on a rainfall dataset obtained from a network of 51 gauges (data provided by the Functional Center and Hydrographic Office of the Abruzzo Region, Pescara, Italy). Using the ArcGIS Kernel Interpolation function, the variation of the distribution of rainfall in the study area was derived for a 65-year time record (1950–2015).

To support the geomorphological dynamic of the area and improve the knowledge of spatial and temporal evolution of landslides, an interferometric analysis (InSAR) was implemented. The approach used is the so-called Persistent Scatterers Interferometry (PSInSAR), which is based on the information achieved by pixels of the SAR images characterized by high coherence over long time intervals [74]. Generally, constructed structures, such as buildings, bridges, dams, railways, pylons, or natural elements, such as outcropping rocks or homogeneous terrain areas, can represent good Persistent Scatterers (PSs). However, these techniques are also affected by some limitations. First, because only objects which are good "radar reflectors" can be analyzed, they cannot attain information over highly vegetated areas. This aspect is not secondary, as landslides often involve non-urban areas [75]. For the present study, we performed analyses of past displacements using data-stacks from the ESA archive ranging in the period 1992–2010. Specifically, Envisat data were selected from the 2003–2010 period, providing quantitative data (i.e., the detection of targets affected by displacements) about displacement information present in both the ascending and descending geometries.

4. Results

4.1. Density Maps (Heatmaps) of Landslide Types over Abruzzo Hilly Piedmont Area

Heatmaps of various slope instability processes over the Abruzzo hilly piedmont area (Figure 3) were produced using GIS technology. These maps allowed us to outline the spatial distribution of landslide phenomena. For this kind of analysis, landslides data were labelled according to the type of movement (rotational and translational slides, complex landslides, earth flows, and rockfalls). Colored areas represent the sites with a higher density of slope instability processes in each category. In the current study, a heterogeneous spatial distribution of landslide types was identified, reflecting the physiographic, geological–structural, and geomorphologic setting of the hilly piedmont area.

The analysis allowed us to identify that (i) rotational and translational slides are most widespread in central and southern sectors (Figure 3a) with high density in correspondence of the mesa-plateau landscape on clayey–sandy and conglomeratic deposits and the incision of the main rivers; (ii) complex landslides are heterogeneously widespread in the study area, with the highest density in the southern sectors following the complex rough topography developed on allochthonous pelagic deposits (Figure 3b); (iii) earth flows mainly characterize the northernmost sectors of the study area reflecting the physical landscape on sandy-pelitic turbidites (Figure 3c). Rockfall density map (Figure 3d) shows a moderate to low spatial distribution as the result of episodic and localized slope instability processes related to the morphostructural setting in the inner sectors [76] and cliff recession processes combined with wavecut and gravity-induced slope processes in coastal areas [77]. Regarding this latter case, no landslide events have been identified as clearly representative of this mass movement type in the study area. In detail, we selected the following case studies intended to be representative of the most characterizing and frequent slope instability processes:

(A). San Martino sulla Marruccina landslide;
(B). Roccamontepiano landslide;
(C). Montebello sul Sangro landslide.

The georeferenced location of selected case studies is graphically shown in Figure 3 with capital letters in white circles.

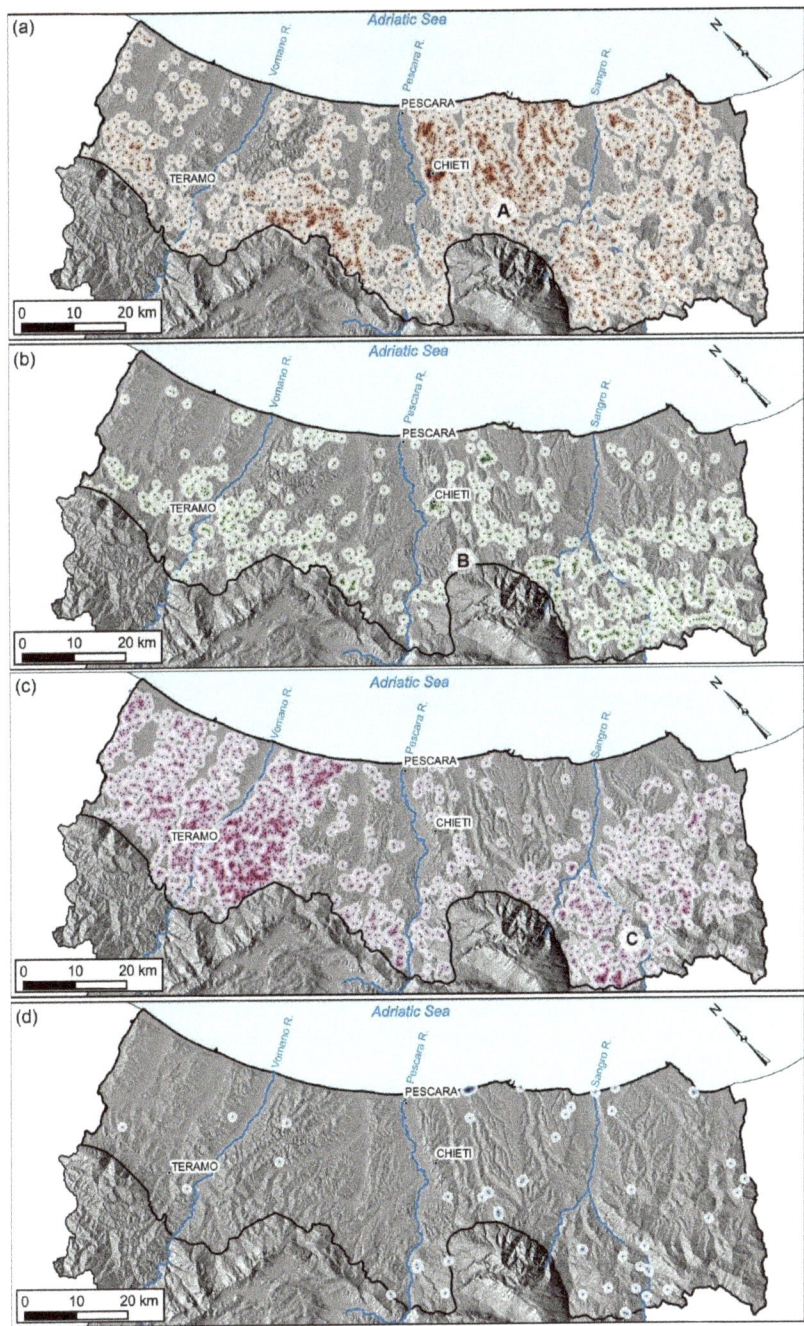

Figure 3. Density maps (heatmaps) of various slope instability processes over the Abruzzo hilly piedmont area: (**a**) rotational and translational slides; (**b**) complex landslides; (**c**) earth flows; (**d**) rockfalls. Colored areas represent the sites with a higher density of slope instability processes in each category (black dots). Capital letters in white circles locate the selected case studies. The black line represents the study area.

4.2. Relationship between Lithology and Spatial Distribution of Landslide Types

A detailed landslide analysis allowed us to differentiate landslide types in order to define the role played by lithological units on landscape development and build up a statistical relationship with the spatial distribution of landslide type.

Preliminary GIS-based analysis of the data derived from available databases (i.e., georeferenced location and detailed landslide information) allowed us to recognize the presence of a large number of landslide phenomena in the study area, reaching 5605 recorded events. In order to promote a relationship between mass movements and lithological units outcropping in the area, recorded landslides were classified according to their typology of movement (e.g., rotational and translational slides, complex landslides, earth flows, and rockfalls). Then, a spatial overlapping between the landslide distribution layer and the vector lithology layer was performed, and a new table of attributes was built (Figure 4).

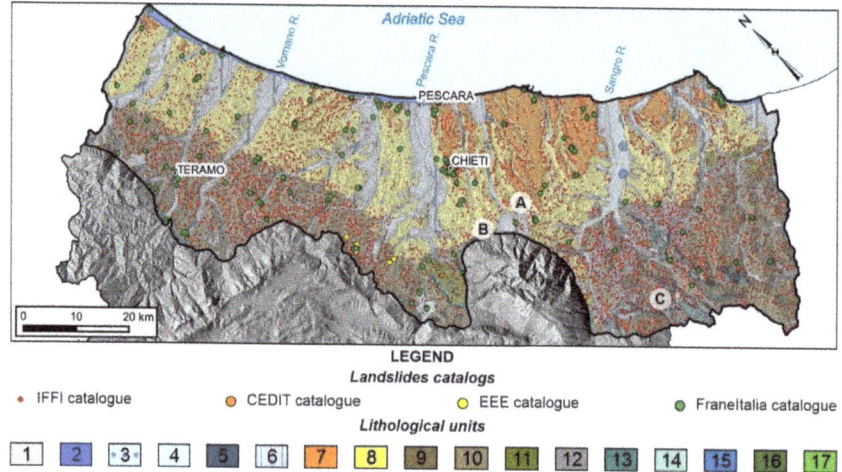

Figure 4. Lithological sketch map of the Abruzzo hilly piedmont area (modified from [56,57]) and spatial landslides distribution [60,61,64,65]. Legend: (1) eluvial–colluvial deposits; (2) sandy shore deposits; (3) recent fluvio-lacustrine deposits; (4) travertine deposits; (5) morainic deposits; (6) old fluvio-lacustrine deposits; (7) conglomeratic deposits; (8) clayey–sandy deposits; (9) sandy turbidites; (10) pelitic turbidites; (11) carbonate deposits in conglomeratic and calcarenitic facies; (12) allochthonous pelagic deposits; (13) carbonate ramp limestones; (14) basin limestones and marls; (15) slope limestone; (16) open carbonate shelf-edge limestones; (17) carbonate shelf limestones and dolomites. Capital letters in white circles locate the selected case studies. The black line represents the study area.

The area of each landslide was obtained from this estimation so that the area ratio of the distribution of landslides in each lithology was derived.

The spatial overlapping allowed us to quantitatively estimate the extension of each lithological unit in the study area in terms of area (km^2) and percentage (Table 1). This GIS-based technique was useful to define the major lithological abundance (both in percentage and area) of clayey–sandy deposits and pelitic turbidites over the study area. Then, the analysis of spatial distribution compared to the outcropping lithologies was carried by comparing the percentage and number of landslides (rotational and translational slides, complex landslides, earth flows, and rockfalls) on each lithological unit as graphically shown by the pie charts and tables in Figure 5.

This overlapping process shows a heterogeneous relationship between lithological units and the distribution of different types of landslides in the Abruzzo hilly piedmont area. Landslides on Quaternary continental deposits were mostly small flows and slides located along the scarp edge of fluvial terraces. Landslides affecting the cuesta and mesa reliefs on the sands and conglomerates on high gradient slopes or else along structural scarps are represented by rapid earth flows affecting surface colluvial cover; falls and topples affecting the edge of structural scarps on sandstones and conglomerates; rotational and translational sliding, which was less frequent but developed for a long time after the event due to deep water infiltration in the permeable conglomerates and sandstones laying on impermeable clays. Landslides on the hilly slopes and cuesta and mesa slopes affecting clayey–sandy deposits were mostly earth flows, from the small to the very wide. Landslides on the arenaceous-pelitic and marly rocks of the turbiditic succession consisted of mostly rapid surface flows and sliding, affecting the eluvial and colluvial cover, particularly where it is more clay-rich. Landslides on the slopes and isolated reliefs on allochthonous pelagic deposits outcropping in the southernmost sectors were mostly flows and complex landslides occurring on all the slopes with a low gradient due to its complex geological–structural setting.

Table 1. Extension of each lithological unit in the study area.

Lithological Unit	Unit Description	Area (km^2)	% of Area
1	Eluvial–colluvial deposits	57.998	1.310
2	Sandy shore deposits	49.622	1.121
3	Recent fluvio-lacustrine deposits	347.175	7.842
4	Travertine deposits	11.571	0.261
5	Morainic deposits	2.333	0.053
6	Old fluvio-lacustrine deposits	464.892	10.501
7	Conglomeratic deposits	380.776	8.601
8	Clayey–sandy deposits	1450.528	32.763
9	Sandy turbidites	38.679	0.874
10	Pelitic turbidites	1228.908	27.757
11	Carbonate deposits in conglomeratic and calcarenitic facies	53.939	1.218
12	Allochthonous pelagic deposits	189.460	4.279
13	Carbonate ramp limestones	125.675	2.839
14	Basin limestones and marls	13.672	0.309
15	Slope limestone	2.127	0.048
16	Open carbonate shelf-edge limestones	0.000	0.000
17	Carbonate shelf limestones and dolomites	9.976	0.225

The study area is characterized by 2694 rotational and translational slides, 851 complex landslides, 2003 earth flows, and 57 rockfalls. In detail, rotational and translational slides mostly develop on pelitic turbidites (31.1%), clayey–sandy deposits (29.8%), and conglomeratic deposits (23.2%), with a higher number of events recorded (839) on pelitic turbidites. Complex landslides mostly develop on pelitic turbidites (47.0%), clayey–sandy deposits (16.5%), carbonate ramp limestones (10.8%), conglomeratic deposits (10.2%), with the higher number of events recorded (400) on pelitic turbidites. Earth flows develop on pelitic turbidites (47.9%) and clayey–sandy deposits (35.9%), with a higher number of events recorded (959) on pelitic turbidites. Rockfalls develop on conglomeratic deposits (31.6%), pelitic turbidites (17.5%), carbonate ramp limestones (15.8%), and clayey–sandy deposits (12.3%) with 18 recorded events on conglomeratic deposits. This latter relationship shows a moderate to low distribution as the result of episodic and localized processes related to morphostructural setting in the inner sectors and cliff recession processes combined with wavecut and gravity-induced slope processes in coastal areas.

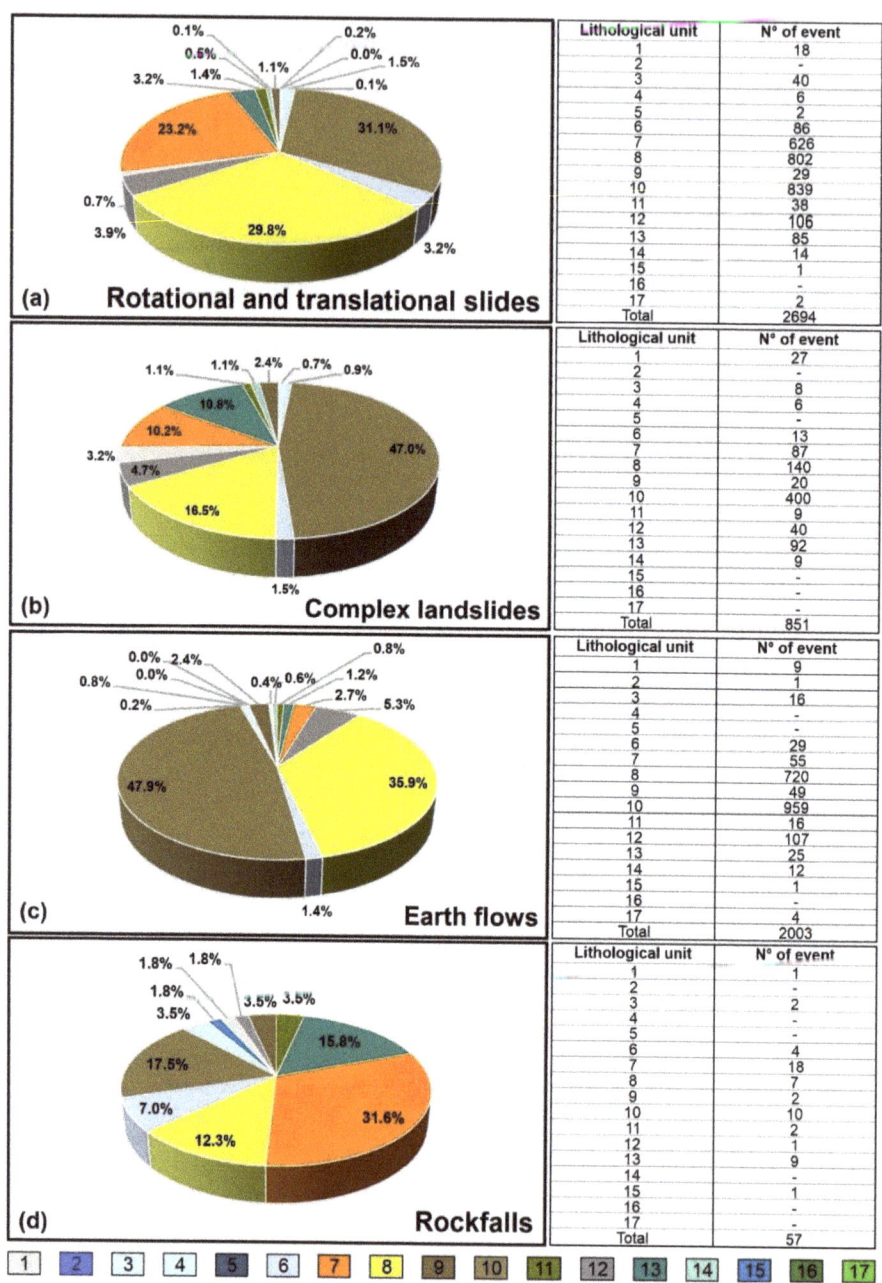

Figure 5. Relationships between lithological units and the distribution of different types of landslides in the Abruzzo hilly piedmont area. (**a**) Pie chart and table showing the percentage and number of rotational and translational slides on each lithological unit; (**b**) pie chart and table showing the percentage and number of complex landslides on each lithological unit; (**c**) pie chart and table showing the percentage and number of earth flows on each lithological unit; (**d**) pie chart and table showing the percentage and number of rockfalls on each lithological unit. Numbers and colors refer to legend in Table 1 and Figure 4.

4.3. Selected Landslide Case Studies

4.3.1. San Martino Sulla Marruccina Landslide

The case study area is located in the central-eastern hilly area of the Abruzzo Region with heights ranging from 200 to 450 m.a.s.l.; this landscape is interrupted by the S–N-oriented Dendalo River valley, where lower altitudes (up to 200 m.a.s.l.) are reached. The study area shows a homogeneous slope distribution (about 5°–15°), with some peaks (>20°) especially in correspondence with the main steep scarps and along the secondary slopes.

From a lithological standpoint, bedrock lithology is composed of a thick marine succession, composed of arenaceous-pelitic and pelitic-arenaceous deposits, known in the literature as the *Mutignano Formation* [78,79]. This succession is composed of clays and silty clays alternated with gray to yellow sands in the lower part, and by gray to yellow sands in medium layers with frequent intercalations of fine-grained sandstone, in the upper part. Quaternary continental deposits include landslide, alluvial, and eluvial–colluvial deposits mainly observed along fluvial incisions and slopes. Strength features of the outcropping rocks are considerably complex, being linked not only to the lithological and structural setting (sub-vertical fracture-sets NNW–SSE to E–W-oriented) but also to the alteration, rearrangement, and loosening processes during complex gravitational phenomena [80]. The landslide phenomenon covers an area of about 2.5 km^2 extending between 400 and 300 m.a.s.l.; it presents a medium length of about 750 m and a significant width of surface rupture area. It is characterized by the main crown of about 2.5 km long, which is locally more than 20 m high. Multitemporal analysis of air-photos, technical cartography, and dendrochronological analysis reveals the first signs of activity in the second half of the 1960s, causing the definition of the first slopes and causing huge damage to roads, buildings, and crops [79,81]. These geomorphological effects, definable in the timespan 1968–1981, are represented by complex landslide bodies with related scarps in the northernmost areas and rotational–translational landslide bodies in the central sector. Nowadays, the movements recorded by the monitoring network are due to a residual activity, but the central sectors are currently affected by a significant local instability due to retrogressive evolution (Figure 6a). Currently, landslides mainly show a rotational and translational sliding surface, as highlighted by counterslopes, counterscarps, and formation of ponds and peatbogs recognized in landslide bodies; smaller instability phenomena are represented by complex landslides and earth flows. Landslide scarps (Figure 7) have different morphological and geomorphological characteristics: where the pelitic deposits outcrop, they are highly degraded, while where sandy deposits are present, they are fresh and evident. The geometrical development of the main and the subordinate crowns are influenced by the spatial disposition of the structural landforms. The planimetric development of the scarps, corresponding in part to the disposition of the families of faults, shows how the geomorphologic processes have been conditioned by the structural setting. The area that surrounds the currently active landslide also presents an old and generalized familiarity with the slope instability processes. Relict shapes and quiescent minor instability phenomena have been observed owing to detailed field surveys and stereoscopic observations [80].

The geomorphological cross-section (Figure 6b) shows how the landslides are in close connection with each other, often presenting several coalescent bodies, also involving landslides activated in the previous time frame. These landslides are characterized by deep failure surfaces, often in the range of several tens of meters. The geometry of the sliding surfaces shows a strong structural control, mainly connected to fault zones and bedding planes; in fact, most of the main landslide scarps and flanks coincide with inferred faults, while the geometry of the sliding surfaces, especially in the middle and lower part of the landslide body, is conditioned by the bedding of the pelitic sequences.

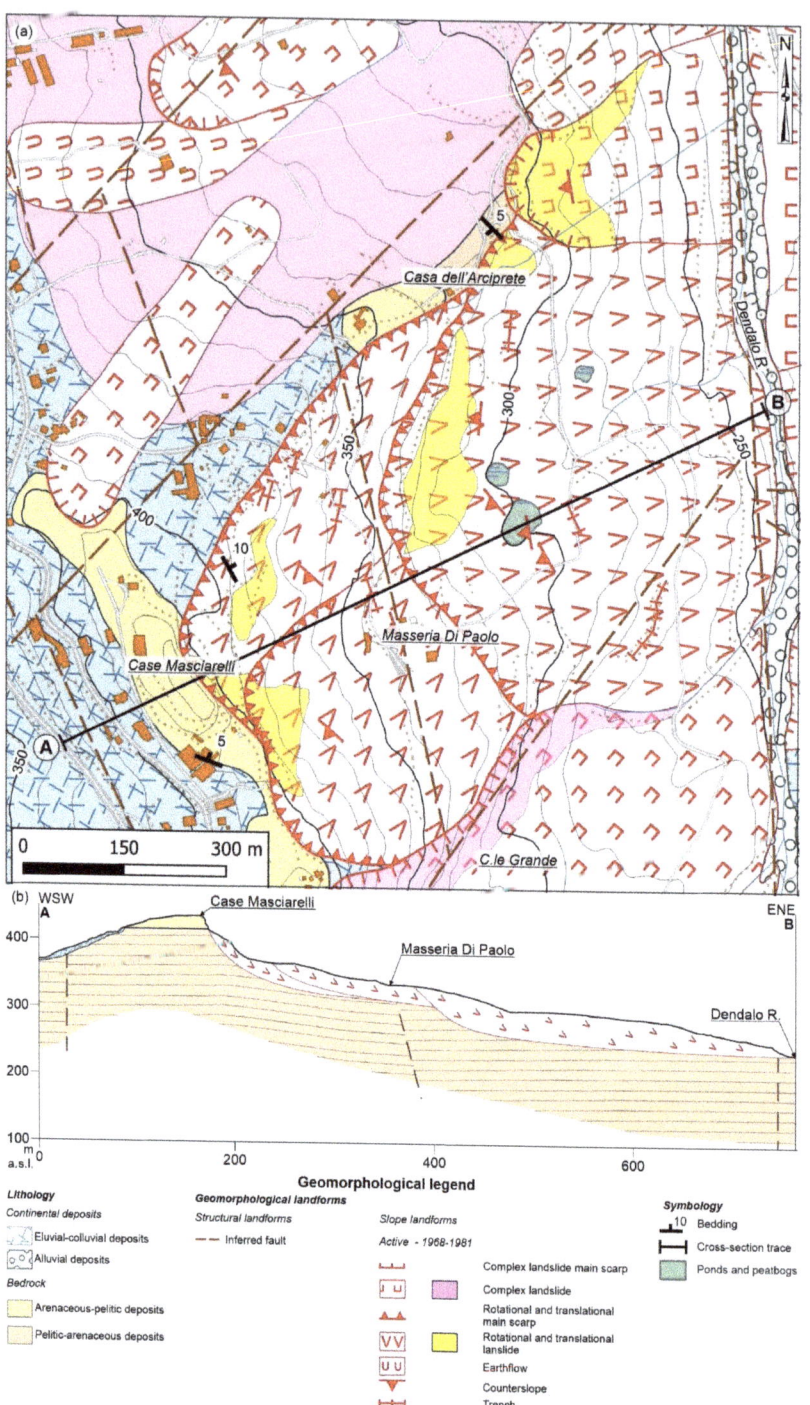

Figure 6. San Martino sulla Marruccina: (**a**) multitemporal geomorphological map (derived from unpublished data and modified and updated from [79–81]); (**b**) geomorphological cross-section.

Figure 7. Photo documentation of geomorphological features of San Martino sulla Marruccina landslide. (**a**) Aerial view of the landslide area; (**b**) panoramic view of the landslide scarp of Casa dell'Arciprete. Red lines show the planimetric development of main landslide scarps.

The complex landslide system could be divided into fairly regular "blocks", dislocated from each other and generally prismatic in form, originally created by the intersection of tectonic fracturing and faulting systems. The main direction of the landslide mass movement is SW–NE, that is, obliquely to the slope.

4.3.2. Roccamontepiano Landslide

The case study area is located next to the northeastern front of the Maiella Massif. It is characterized by the presence of a pseudo-rectangular-shaped travertine plateau (Montepiano) which dominates both topographically and morphologically the landscape of the area. Montepiano is a flat tabular relief 610–650 m.a.s.l. high, with a maximum length of about 2.3 km (along NW–SE direction). It generally dips gently north-east with an average gradient of about 5%, and it is bounded by vertical cliffs and scarps up to 30 m high [82,83]. Furthermore, the landform is cut by a series of small SW–NE-oriented fluvial incisions that raise the relief values along the slopes.

From a lithological standpoint, the area is characterized by an approximately 40 m thick travertine layer that overlies arenaceous-pelitic and pelitic-arenaceous deposits, with thin conglomerate layers in between, pertaining to the *Mutignano Formation* [78]; also, these bedrock layers are gently dipping towards the NE. Physical–mechanical parameters show a significant variability in terms of rock resistance and behavior according to lithological nature (travertine layer and arenaceouspelitic lithotypes) and subsequent loosening and weathering phenomena [84]. Quaternary continental deposits include eluvial–colluvial deposits mainly observed along the southwestern flank of the plateau. The landslide phenomenon covers a wide area (~4 km^2) with high slope gradients (seldom less than 30%) and high variability in width and thickness due to repeated historical landslide events (Figure 8a). The maximum width of more than 2 km can be found downstream of the Ripa Rossa, whilst the maximum thickness of more than 20 m is located immediately above Roccamontepiano village. From historical sources, the first landslide events that occurred in the area took place on 24 June 1765, causing severe damage to the village and 2000 casualties [82,85].

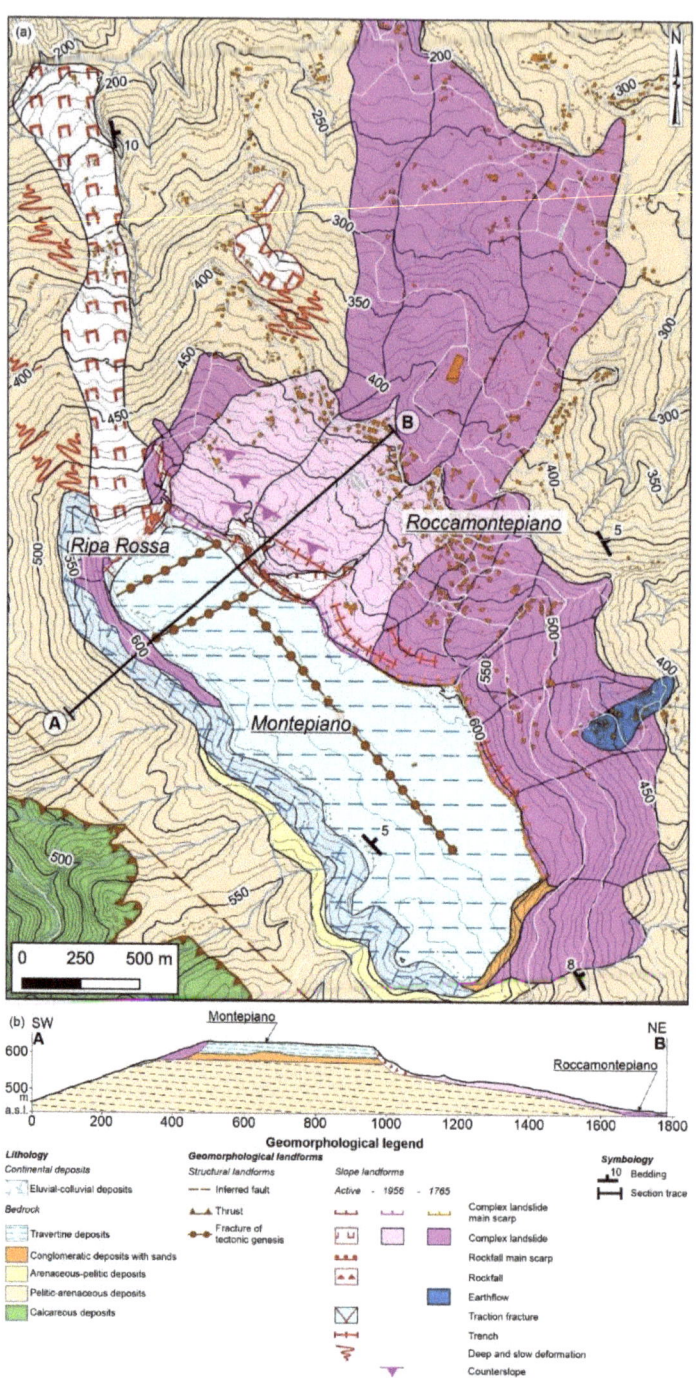

Figure 8. Roccamontepiano landslide: (**a**) multitemporal geomorphological map (derived from unpublished data and modified and updated from [82,83]); (**b**) geomorphological cross-section.

Therefore, the relief is almost surrounded by wide complex landslide bodies and related scarp, which characterize most of the area south of Roccamontepiano village. Other historical movements occurred in the second half of the 1950s, reactivating pre-existing ones and causing extensive damage to the village of Roccamontepiano but this time without the report of victims [86]. Evidence of this second historical event is represented by wide counterslopes located at ~500 m.a.s.l. Actually, the landslide body is formed by a thick heap (up to 17 m) of travertine blocks and fragments with secondarily reworked sandstone-conglomerate deposits, active especially in the northeastern and northern part of the Montepiano plateau (Figure 9a). The overall mechanism could be referred to a complex landslide system, including lateral spreading with rockfalls, rotational, and translational movements.

Figure 9. Photo documentation of geomorphological features of Roccamontepiano landslide. (**a**) Panoramic view of the landslide area. Red lines show the planimetric development of main landslide scarps, red circles show travertinous blocks in the landslide body; (**b**) detail of NE–SW trending traction fractures (red lines) in the travertine cliff scarp near Ripa Rossa.

The geomorphological cross-section (Figure 8b) shows how the landslides are strictly connected with the structural framework of the study area; the mechanism implies the involvement of the plastic clays that underlie the travertines in the mass movement.

The presence of a thick layer of massive rocks over plastic lithologies leads to tension stresses along the edge of the travertine layer and the progressive opening of preexisting fractures. The travertine layer exhibits NW–SE and NE–SW trending fracture systems, probably caused by tectonic activity (Figure 9b). Fracture of tectonic genesis up to 10 m wide and in different stages of evolution are sub-parallel to the plate edge and the major fracture systems all along the cliff scarps. When these fractures reach the clays, large blocks of travertine are isolated over the plastic materials, and lateral spreading accelerates, defining sliding surfaces; the movement evolves as a complex landslide.

4.3.3. Montebello sul Sangro Landslide

The case study area is located in the transition zone between the central Apennines chain front and the piedmont area on the left side of the middle Sangro River valley. It is on a narrow-faulted anticline ridge, more than 900 m.a.s.l. high, trending N–S. The landscape outlines a strongly asymmetric calcareous hogback ridge, with a gentler eastern slope and a steeper western one, resulting from the erosion of the anticline flank; northwards the ridge is deeply incised and separated by a second hogback ridge on which the Pennadomo village is located.

From a lithological standpoint, bedrock lithology is made of rocks pertaining to allochthonous pelagic deposits. Clayey deposits with embedded terrigenous siliciclastic deposits (*Argille varicolori formation*) outcrops in the western side of the ridge; alternating calcareous-marly and calcirudite rocks (*Tufillo formation*) represent the backbone of the ridge; pelitic-arenaceous deposits (*Flysch of Agnone formation*) mostly outcrop in the eastern side of the ridge [39,87]. Physical–mechanical properties of chaotic marly–clayey deposits reflect the great amount of lithological variability within them, and consequently the rock behavior is not constant. Moreover, detailed analysis showed that outer area of the scree slope deposits appears plasticized, and the most superficial zones are at yield in tension [87]. Quaternary continental deposits include eluvial–colluvial and scree deposits mainly observed along fluvial incisions and slopes.

The landslide phenomenon covers an area of ~1.1 km^2, and it is affected by strong variations in the state of activity. Large landslides (mostly dormant and/or abandoned) and small landslides (generally more recent and active) constitute the wide and complex landslide system. Historical documents and chronicles show multiple activations of the main event, involving the western side of the Montebello hogback and spreading out on the eastern side (Figure 10a). These worrying geomorphological dynamics are testified by the involvement of the Montebello village. The first evolution of events occurred in the second half of 1800 (1864, 1891, and 1899); after that, the new village of Montebello sul Sangro was reconstructed in a more western site [86–88]. It was characterized by a complex dynamic including earth flows, complex landslides, rotational and translational landslides, and localized rockfalls. Another significant landslide event occurred in 1971 [89], and it was mainly characterized by earth flows due to the activation of several small mass movements composing the large one. Nowadays, a principal earth flow is present, and the activity of this movement is demonstrated by a range of surface expressions such as irregular mounds, landslip troughs, and several tension fractures that opened both longitudinally and transversely to the main landslide. The main landslide is characterized by a mass that flows down along a narrow channel and then spreads out in a wide accumulation lobe, with depressions and undulations. Thrust features in the accumulation area point out at least three overlapped flows, suggesting an intermittent movement (Figure 11). Moreover, the geomorphological complexity of the area is evidenced by the presence of several families of rotational and translational landslides, complex landslides, and rockfalls, present especially along the steep western side of the hogback.

The geomorphological cross-section (Figure 10b) shows that the landslide movements are strictly controlled by the geological and morphostructural setting of the carbonate hogback (east overturned faulted anticline trending from N–S to NNW–SSE) and chaotic clay rocks; the main earth flow is influenced by the progressive involvement of the clay units in the landslide movement, and the rockfalls in the upper part of the ridge are linked to fractures and jointing in the calcareous strata. The scarp area involves the steep western calcareous slope of the ridge down to the gentle lower slope on clay units; the regressive enlargement of the landslide scarp, close to the Montebello village, involves the western side of the calcareous ridge, with systems of tension fractures and reverse slope areas, affecting the Montebello village (Figure 11b).

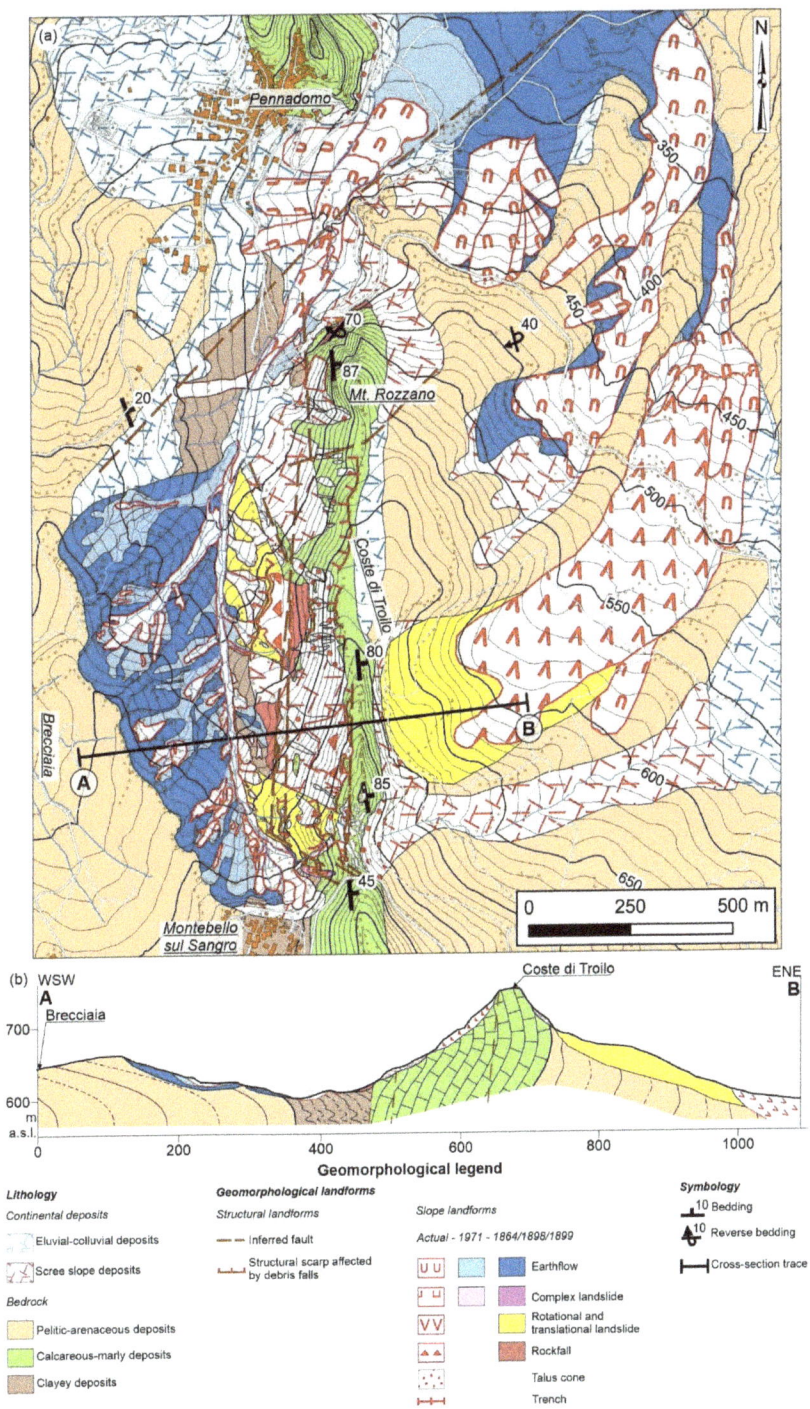

Figure 10. Montebello sul Sangro landslide: (**a**) multitemporal geomorphological map (derived from unpublished data and modified and updated from [87]); (**b**) geomorphological cross-section.

Figure 11. Photo documentation of geomorphological features of Montebello del Sangro landslide. (**a**) Panoramic view of the main earthflow; (**b**) detail of allochthonous pelagic deposits involved in the landslide phenomenon, with regressive enlargement of the landslide scarp near the Montebello sul Sangro village.

5. Discussion

Landslides have been widely considered as principal mass-wasting agents in areas experiencing varied influence of several causative factors (i.e., lithology, geological setting, climate regime, etc.). However, patterns of landslides are rarely addressed as a surface manifestation of interrelationships between morphostructural setting, lithology, and climate. Here, we have attempted to understand such interrelationships in the context of landslide distribution patterns in the hilly piedmont area of Abruzzo Region. Historical landslides analysis allowed us to understand that the distribution, mechanisms, and types of mass movements in the study area strictly correlate with the different physiographic, lithological, and geological–structural settings. The work mostly focuses on three landslide case studies analyzed with the aim of highlighting the multitemporal evolution of the landslide phenomena, emphasizing the role of lithological and morphostructural features on landslide types and the interplay between such processes and the geomorphological evolution. Landslide density maps, directly combined with the inventories and databases from which they were obtained, allowed us to define and graphically show different sectors in the study area. In each sector, we have outlined the landslide types and the mechanisms that mostly determine the slope instability reflecting the geological–structural and geomorphological setting. Selected case studies are representative of the most characterizing and frequent slope instability processes over the hilly piedmont area, showing different influences on geomorphological dynamics according to the physiographic and litho-structural setting. In the northern and central sectors, landslide phenomena affect a gently hilly area made of clayey–sandy deposits (with sandstone-conglomerate sequence on top), gently dipping towards the northeast or horizontal. In the southernmost sector, landslide phenomena affect a landscape derived from exogenous processes (fluvial and slope processes) on mostly chaotic marly–clayey deposits or chaotic succession of calcareous-marly deposits.

Several previous studies in the Abruzzo Region [48,79,90,91] analyzed and described the widespread slope geomorphic processes, showing an organic correlation between the morphostructural/geological setting and landslide types as the result of the dynamic interaction between morphostructural factors, linked to the conflicting tectonic activity and regional uplift, and morphosculptural factors, linked to drainage network linear down-cutting and slope gravity processes. The slope evolution is mainly related to the interplay of different landslide types referable to lateral spreading, rockfall, earth flow, rotational and translational sliding, evolving into complex movements and systems.

In this framework, local features such as lithology and morphostructural framework should be noted to control the occurrence and distribution of landslides. Nonetheless, interrelationships of these factors have been rarely associated with spatiotemporally varying landslide distribution patterns. However, there are many limitations to infer temporally varying landslide distribution, such as delineation of individual failure events on the reactivated landslide, loss of landslide scarp caused by the successive mass movement, etc. Different predisposing and triggering factors can influence the stability of slopes and can cause landslides, among which heavy rainfall events are intended to be a significant one. It is well known that extreme and localized heavy rainfalls constitute the main triggering causal factor of landslides. Rainfall pattern is strongly controlled and influenced by climate regime and its variations. Therefore, it is to be expected that climate changes could influence slope stability at different temporal and geographical scales. The frequency and the intensity of heavy rainfall events are also increasing, although both at local and regional scale the average annual rainfall is not showing significant changes. The assessment of the effects of climate change on the natural environment is an open issue for the scientific community trying to establish a relation between climate change and its potential effects on the occurrence, or lack of occurrence, of landslides. However, the effects of changes in climate regimes on landslides (as on other geo-hydrological hazards) remain difficult to quantify and predict.

This work represents a useful source for investigating landslide behaviors in terms of spatial and temporal distribution, as well as for analyzing and attempting correlation between climate regime, historical landslides, and present-day geomorphological activity.

In order to understand and quantify how climate regime and its variability could affect landslides, a climatic analysis was performed using a 65-year period rainfall gauges data. Figure 12a shows the spatial distribution of annual average rainfall in the study area, with minimum values (~700 mm/year) recorded along with the coastal areas and the southeastern sector of the Maiella Massif; these rainfall values are gradually increasing, moving towards the innermost areas, where the maximum values (about 1150 mm/year) are reached. Similarly, the analysis of the annual average rainfall diagram from 1950 to 2015 shows values ranging from ~530 to ~1130 mm/year, with a clear decreasing trend over the examined period (Figure 12b).

Taking into account the spatial distribution (landslide heatmaps), the location and abundance of landslides, and the geomorphological features of selected case studies, the landscape dynamics and activity of the hilly piedmont area have also been confirmed by the interferometric analysis. Considering that movements recorded by interferometric data can be due to different causes acting at different scales (i.e., uplift, subsidence, landslide, etc.), the PSInSAR technique was here used as a tool for systematic monitoring of ground deformation related to slope instability. The presence and temporal persistence of clusters of anomalies within the main landslide body act as the most important parameters that show present-day landscape changes linked to temporal landslide dynamic. Figure 12c shows the total number of persistent anomalies detected over the period 2002–2010, clipped by landslide bodies (green polygons) mapped by the IFFI project [59,60] over the hilly piedmont area of Abruzzo Region. Analyzed data show a spatial distribution of negative movements (lowering) and positive movements (raising), which reflect the extension of the investigated landslide phenomena with the highest values located in the central-southern sectors and locally in the northernmost coastal slopes.

Moreover, in order to attempt a general correlation between long-term rainfall trends and trends in landslide occurrence, a statistical analysis of the annual distribution of landslides was carried. This kind of analysis was completed collecting data from historical sources, technical reports, and updated catalogues [60,65,86] containing a variety of historical, geographical, geomorphological, and bibliographical information on landslides.

Reported diagram (Figure 12d) stores information regarding dates of occurrences of several landslides, starting from the year 1950 until the present, with non-homogeneous rates of recorded landslides per year. A detailed analysis shows that the frequency remains under the value of 10 landslides per year starting from 1950 to 1990, with unique exception years (i.e., 1954, 1956, 1986). Subsequently, growth rates, from 1991 onwards, clearly increase. Even if the variance of the number of reported landslides over time is also due to the different availability of sources of information and not necessarily linked to the real frequency of landslide occurrences [92], it is possible to consider this analysis a reasonably true reflection of reality for the period 1950–2018. Despite the presence of a timespan with a lack of suitable and univocal data (i.e., year, day, hour, etc.) on landslides' activation–reactivation in the period 2002–2009, it is possible to note that the annual landslide distribution ranges from ~5 to ~75 individual events. Considering the complete distribution of the number of landslides during the years covered, the annual landslide distribution during this period shows different periods of landslide activity and abundance. It is possible to note a nearly stable trend in the first 20-year time record (1950–1970), followed by a general increasing trend in the 1970–2000-time record, also supported and corroborated by a weak increasing trend in the last decade (2010–2018). The identified trend should be considered in relation to both the incremental data availability and the rise in mass-wasting processes, as directly shown by historical information on past and current landslides. Moreover, regarding the study area, it is not correct to conclude that a lack of reported landslides in a given time interval would be due to a minor activity of gravitational mass wasting or to a gap in the documental source, as marked by the present-day geomorphological activity testified by the temporal persistence anomalies of movement related to slope instability (Figure 12c).

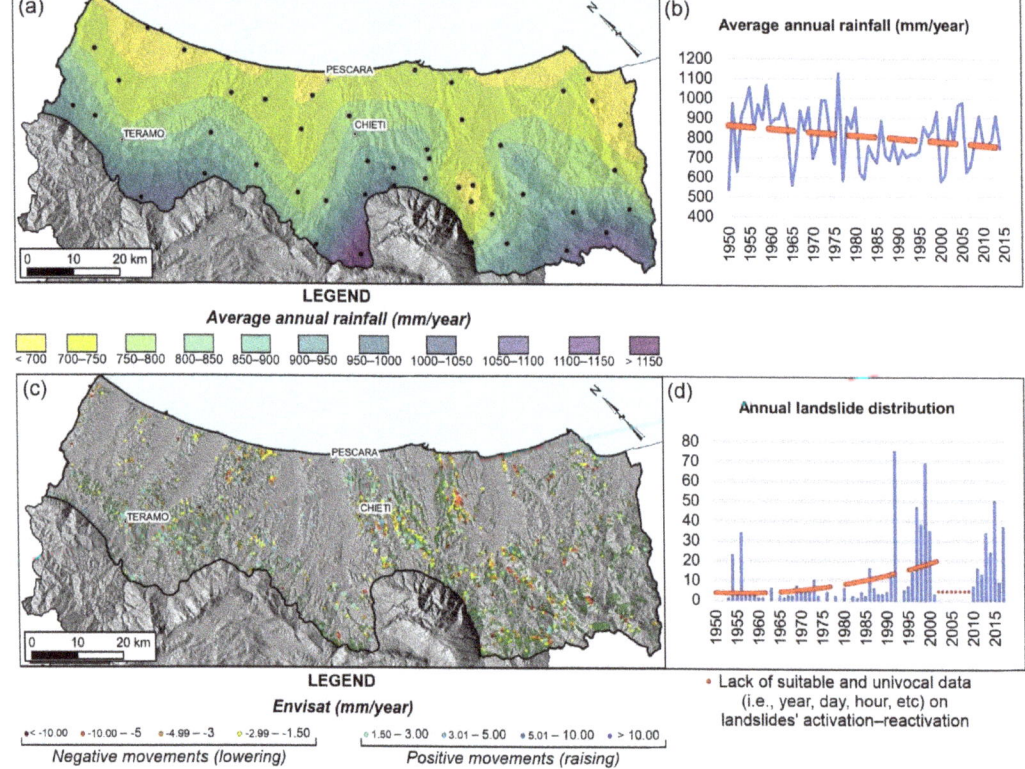

Figure 12. (a) Average annual rainfall map. Black dots represent rainfall gauges. (b) Average annual rainfall diagram from a 65-year time record (1950–2015). (c) InSAR observations for the selected area over the hilly–piedmont area. Mean line-of-sight (LOS) velocity for the period 2002–2010 from Envisat descending track. Only the Persistent Scatterers (PSs) that fall within the landslide areas (dark green polygons) have been selected and are represented as colored dots. Positive values represent the motion of the ground toward the satellite (raising), and negative values represent the motion away from the satellite (lowering). Green polygons represent landslide bodies detected by the IFFI Project [60]. (d) Distribution of annual landslide occurrences over the 1950–2015 period (derived from [60,65,86]).

The final combination and overlapping between the spatial and temporal landslide distribution pattern, the mismatch between landslide areas and sectors characterized by high rainfall density, the lack of correspondence between decreasing annual average rainfall trend and the increasing annual landslide distribution allowed us to highlight the interplay between the morphostructural/geological framework and landslide dynamics in the hilly piedmont area of Abruzzo Region. The present study allowed us to better characterize the present-day landscape setting of the study area, confirming that it is characterized by active geomorphological processes, mostly represented by slope instabilities (i.e., rotational and translational slides, complex landslides, earth flows, and rockfalls). This was obtained from historical information on past and current landslides. Currently, geomorphological activity and landslide dynamics are testified and supported by interferometric data (clusters of persistent anomalies, detected over the period 2002–2010, and clipped by landslide' polygons) with negative movements (values between -10 and -2) and positive ones (values between 2 and 10) heterogeneously distributed over the hilly piedmont area. Detailed multitemporal geomorphological analysis on selected case studies (San Martino sulla Marruccina, Roccamontepiano, and Montebello sul Sangro) show multiple activations of the main event since the 18th century onwards with large landslides (mostly dormant and/or abandoned) and small landslides (generally more recent and active) constituting the wide and complex landslide systems and reflecting the physiographic, geological–structural, and geomorphologic setting.

In conclusion, by summarizing data obtained from multitemporal and multidisciplinary, it is possible to suggest that landslide occurrence and the dynamics of the hilly piedmont area of Abruzzo Region are not directly linked to climate regime variations, but the most influential factors are represented by the lithological and morphostructural setting. These predisposing factors are strictly related to a cuesta, mesa, and plateau landscape in which it is possible to outline the landslide types and the mechanisms that mostly determine the slope morphogenesis and are characterizing of the specific geological–structural setting. To these characterizing landslide types are obviously associated and sometimes super-imposed a set of landslides secondary or however controlled by local conditions, single factors (i.e., extreme heavy rainfall events), and not by the whole morphostructural setting. Moreover, considering the historical landslide events and the geomorphological activity of the area, most of the recorded landslides could be considered as reactivations of pre-existing ones (dormant slides and/or paleolandslides), which have occurred in periods of climatic and geomorphological conditions different from those of the present, evolving in complex movements and systems because of the absence of sustainable land planning and appropriate landslide hazard mitigation measures.

6. Conclusions

This paper presents detailed analyses of the occurrence and distribution of landslides over the hilly piedmont area of Abruzzo Region (Central Italy) in relation to mechanisms and factors that control their evolution in different orographic, lithological, and geological–structural conditions. Historical landslides analysis, supported by GIS-based techniques, was performed through an integrated approach combining literature data and landslide inventories analysis, relationships between landslide types and lithological units, detailed photogeological analysis, and geomorphological field mapping. In detail, the work focuses on three landslide case studies that have undergone several main movements since the 18th century onwards, intending to highlight the multitemporal geomorphological evolution of phenomena and the interplay between morphostructural/geological framework and landslide dynamics. The main landslide cases analyzed and discussed in this paper consist of rotational and translational slide in a complex landslide system on clayey–sandy deposits, characterized by a very rough topography documenting the activity of long-term landslide processes (San Martino sulla Marruccina landslide case); complex landslide system including lateral spreading with rockfalls, rotational and translational movements, characterized by a travertine layer that overlies arenaceous-pelitic and pelitic-arenaceous deposits (Roccamontepiano landslide case); main earth flow on chaotic allochthonous pelagic deposits with several families of rotational and translational landslides, complex landslides, and rockfalls (Montebello sul Sangro landslide case).

A multidisciplinary and multitemporal analysis allowed us to better characterize the present-day landscape setting of the study area, deriving data from historical information on past and current landslides. Furthermore, this work represents an attempt for the understanding of spatial interrelationship of landslide types, morphostructural setting, and climate regime in the study area. It gives a contribution about the location, abundance, activity, and frequency of landslides in a changing environment, by means of the analysis of historical events and a comparison between the long-term rainfall trends and the distribution of annual landslide occurrences, which shows that landslide dynamics are not directly linked to climate regime variations, but that the most influential factors are represented by the lithological and morphostructural setting. Finally, the work could represent a scientific tool for any study in the future concerning susceptibility, hazard, and risk assessment at different spatial scales, readily available to interested stakeholders for sustainable territorial planning and loss-reduction measures.

Author Contributions: Conceptualization, G.E., and E.M.; methodology, C.C., G.P., and E.M.; software, C.C.; validation, G.E., and E.M.; investigation, G.E., C.C., G.P., and E.M.; writing—original draft preparation, C.C., and G.P.; writing—review and editing, E.M.; supervision, G.E., and E.M.; project administration, G.E., and E.M.; funding acquisition, E.M. All authors have read and agreed to the published version of the manuscript.

Funding: This research and the APC was funded by Enrico Miccadei, grant provided by Università degli Studi "G. d'Annunzio" Chieti-Pescara.

Institutional Review Board Statement: Not applicable.

Informed Consent Statement: Not applicable.

Data Availability Statement: The data presented in this study are available on request from the author. The data are not publicly available due to privacy. Images employed for the study will be available online for readers.

Acknowledgments: The authors wish to thank the Cartographic Office of the Abruzzo Region by means of the Open Geodata Portal (http://opendata.regione.abruzzo.it/, accessed on 24 October 2020), the Ministero dell'Ambiente e della Tutela del Territorio e del Mare (http://www.minambiente.it/, accessed on 24 October 2020), and the IGMI (Istituto Geografico Militare Italiano) for providing topographic and interferometric data, aerial photos, and orthophotos used for this work. The rainfall data were provided by the Functional Center and Hydrographic Office of the Abruzzo Region (Centro Funzionale e Ufficio Idrografico Regione Abruzzo, Pescara, Italy). The authors wish to thank the anonymous reviewers for their critical review of the paper and their precious suggestions, which significantly improved this manuscript.

Conflicts of Interest: The authors declare no conflict of interest.

References

1. Haque, U.; Blum, P.; da Silva, P.F.; Andersen, P.; Pilz, J.; Chalov, S.R.; Malet, J.P.; Auflič, M.J.; Andres, N.; Poyiadji, E.; et al. Fatal landslides in Europe. *Landslides* **2016**, *13*, 1545–1554. [CrossRef]
2. Marsala, V.; Galli, A.; Paglia, G.; Miccadei, E. Landslide susceptibility assessment of Mauritius Island (Indian ocean). *Geosciences* **2019**, *9*, 493. [CrossRef]
3. Miccadei, E.; Mascioli, F.; Ricci, F.; Piacentini, T. Geomorphology of soft clastic rock coasts in the mid-western Adriatic Sea (Abruzzo, Italy). *Geomorphology* **2019**, *324*, 72–94. [CrossRef]
4. Calista, M.; Menna, V.; Mancinelli, V.; Sciarra, N.; Miccadei, E. Rockfall and Debris Flow Hazard Assessment in the SW Escarpment of Montagna del Morrone Ridge (Abruzzo, Central Italy). *Water* **2020**, *12*, 1206. [CrossRef]
5. Aleotti, P.; Chowdhury, R. Landslide hazard assessment: Summary review and new perspectives. *Bull. Eng. Geol. Environ.* **1999**, *58*, 21–44. [CrossRef]
6. Kjekstad, O.; Highland, L. Economic and social impacts of landslides. In *Landslides—Disaster Risk Reduction*; Sassa, K., Canuti, P., Eds.; Springer: Berlin/Heidelberg, Germany; Tokyo, Japan, 2009; pp. 573–587. [CrossRef]
7. Petley, D. Global patterns of loss of life from landslides. *Geology* **2012**, *40*, 927–930. [CrossRef]

8. Rossi, M.; Guzzetti, F.; Salvati, P.; Donnini, M.; Napolitano, E.; Bianchi, C. A predictive model of societal landslide risk in Italy. *Earth-Sci. Rev.* **2019**, *196*, 102849. [CrossRef]
9. Segoni, S.; Piciullo, L.; Gariano, S.L. A review of the recent literature on rainfall thresholds for landslide occurrence. *Landslides* **2018**, *15*, 1483–1501. [CrossRef]
10. Carabella, C.; Miccadei, E.; Paglia, G.; Sciarra, N. Post-Wildfire Landslide Hazard Assessment: The Case of The 2017 Montagna Del Morrone Fire (Central Apennines, Italy). *Geosciences* **2019**, *9*, 175. [CrossRef]
11. Farabollini, P.; De Pari, P.; Discenza, M.E.; Minnillo, M.; Carabella, C.; Paglia, G.; Miccadei, E. Geomorphological evidence of debris flows and landslides in the Pescara del Tronto area (Sibillini Mts, Marche Region, Central Italy). *J. Maps* **2020**. [CrossRef]
12. Piacentini, T.; Calista, M.; Crescenti, U.; Miccadei, E.; Sciarra, N. Seismically induced snow avalanches: The central Italy case. *Front. Earth Sci.* **2020**, *8*, 1–27. [CrossRef]
13. Piacentini, T.; Miccadei, E.; Di Michele, R.; Sciarra, N.; Mataloni, G. Geomorphological analysis applied to rock falls in Italy: The case of the san venanzio gorges (Aterno river, Abruzzo, Italy). *Ital. J. Eng. Geol. Environ.* **2013**, *6*, 467–479.
14. D'Alessandro, L.; Miccadei, E.; Piacentini, T. Morphostructural elements of central-eastern Abruzzi: Contributions to the study of the role of tectonics on the morphogenesis of the Apennine chain. *Quat. Int.* **2003**, *101–102*, 115–124. [CrossRef]
15. Kumar, V.; Gupta, V.; Sundriyal, Y.P. Spatial interrelationship of landslides, litho-tectonics, and climate regime, Satluj valley, Northwest Himalaya. *Geol. J.* **2019**, *54*, 537–551. [CrossRef]
16. Zhou, C.H.; Lee, C.F.; Li, J.; Xu, Z.W. On the spatial relationship between landslides and causative factors on Lantau Island, Hong Kong. *Geomorphology* **2002**, *43*, 197–207. [CrossRef]
17. Safaei, M.; Omar, H.; Huat, B.K.; Yousof, Z.B.M. Relationship between lithology factor and landslide occurrence based on information value (IV) and frequency ratio (FR) approaches—Case study in north of Iran. *Electron. J. Geotech. Eng.* **2012**, *17*, 79–90.
18. Guzzetti, F.; Cardinali, M.; Reichenbach, P. The influence of structural setting and lithology on landslide type and pattern. *Environ. Eng. Geosci.* **1996**, *2*, 531–555. [CrossRef]
19. Marchesini, I.; Santangelo, M.; Guzzetti, F.; Cardinali, M.; Bucci, F. Assessing the influence of morpho-structural setting on landslide abundance. *Georisk* **2015**, *9*, 261–271. [CrossRef]
20. Benzougagh, B.; Meshram, S.G.; Baamar, B.; Dridri, A.; Boudad, L.; Sadkaoui, D.; Mimich, K. Relationship between landslide and morpho-structural analysis: A case study in Northeast of Morocco. *Appl. Water Sci.* **2020**, *10*, 175. [CrossRef]
21. D'Amato Avanzi, G.; Giannecchini, R.; Puccinelli, A. The influence of the geological and geomorphological settings on shallow landslides. An example in a temperate climate environment: The 19 June 1996 event in northwestern Tuscany (Italy). *Eng. Geol.* **2004**, *73*, 215–228. [CrossRef]
22. Wang, G.; Chen, X.; Chen, W. Spatial prediction of landslide susceptibility based on GIS and discriminant functions. *ISPRS Int. J. Geo-Inf.* **2020**, *9*, 144. [CrossRef]
23. Fell, R.; Whitt, G.; Miner, T.; Flentje, P. Guidelines for landslide susceptibility, hazard and risk zoning for land use planning. *Eng. Geol.* **2008**, *102*, 83–84. [CrossRef]
24. Gariano, S.L.; Guzzetti, F. Landslides in a changing climate. *Earth-Sci. Rev.* **2016**, *162*, 227–252. [CrossRef]
25. Alvioli, M.; Melillo, M.; Guzzetti, F.; Rossi, M.; Palazzi, E.; von Hardenberg, J.; Brunetti, M.T.; Peruccacci, S. Implications of climate change on landslide hazard in Central Italy. *Sci. Total Environ.* **2018**, *630*, 1528–1543. [CrossRef]
26. Coe, J.A.; Godt, J.W. Review of approaches for assessing the impact of climate change on landslide hazards. In *Landslides and Engineered Slopes: Protecting Society through Improved Understanding, Proceedings of the 11th International and 2nd North American Symposium on Landslides and Engineered Slopes, Banff, AB, Canada, 3–8 June 2012*; Eberhardt, E., Froese, C., Turner, A.K., Leroueil, S., Eds.; Taylor & Francis Group: London, UK, 2012; pp. 371–377. ISBN 9780415621236.
27. Ciervo, F.; Rianna, G.; Mercogliano, P.; Papa, M.N. Effects of climate change on shallow landslides in a small coastal catchment in southern Italy. *Landslides* **2017**, *14*, 1043–1055. [CrossRef]
28. Canuti, P.; Casagli, N.; Ermini, L.; Fanti, R.; Farina, P. Landslide activity as a geoindicator in Italy: Significance and new perspectives from remote sensing. *Environ. Geol.* **2004**, *45*, 907–919. [CrossRef]
29. Guzzetti, F. Landslide fatalities and the evaluation of landslide risk in Italy. *Eng. Geol.* **2000**, *58*, 89–107. [CrossRef]
30. Melis, M.T.; Pelo, S.D.; Erbì, I.; Loche, M.; Deiana, G.; Demurtas, V.; Meloni, M.A.; Dessì, F.; Funedda, A.; Scaioni, M.; et al. Thermal remote sensing from UAVs: A review on methods in coastal cliffs prone to landslides. *Remote Sens.* **2020**, *12*, 1971. [CrossRef]
31. Lundmark, A.M.; Augland, L.E.; Jørgensen, S.V. Digital fieldwork with Fieldmove—How do digital tools influence geoscience students' learning experience in the field? *J. Geogr. High. Educ.* **2020**, *44*, 427–440. [CrossRef]
32. Reddy, G.P.O. Remote Sensing and GIS for Geomorphological Mapping. In *Geospatial Technologies in Land Resources Mapping, Monitoring and Management*; Reddy, G., Singh, S., Eds.; Springer: Cham, Switzerland, 2018; pp. 223–252. [CrossRef]
33. Melelli, L.; Gregori, L.; Mancinelli, L. The Use of Remote Sensed Data and GIS to Produce a Digital Geomorphological Map of a Test Area in Central Italy. In *Remote Sensing of Planet Earth*; Chemin, Y., Ed.; InTech: London, UK, 2012; pp. 97–116. [CrossRef]
34. Jaboyedoff, M.; Oppikofer, T.; Abellán, A.; Derron, M.H.; Loye, A.; Metzger, R.; Pedrazzini, A. Use of LIDAR in landslide investigations: A review. *Nat. Hazards* **2012**, *61*, 5–28. [CrossRef]
35. Antonielli, B.; Mazzanti, P.; Rocca, A.; Bozzano, F.; Cas, L.D. A-DInSAR performance for updating landslide inventory in mountain areas: An example from lombardy region (Italy). *Geosciences* **2019**, *9*, 364. [CrossRef]

36. Herrera, G.; Mateos, R.M.; García-Davalillo, J.C.; Grandjean, G.; Poyiadji, E.; Maftei, R.; Filipciuc, T.C.; Jemec Auflič, M.; Jež, J.; Podolszki, L.; et al. Landslide databases in the Geological Surveys of Europe. *Landslides* **2018**, *15*, 359–379. [CrossRef]
37. Trigila, A.; Iadanza, C.; Bussettini, M.; Lastoria, B. *Dissesto Idrogeologico in Italia: Pericolosità e Indicatori di Rischio—Edizione 2018*; ISPRA: Rome, Italy, 2018; ISBN 9788844809010.
38. Parotto, M.; Cavinato, G.P.; Miccadei, E.; Tozzi, M. Line CROP 11: Central Apennines. In *CROP Atlas: Seismic Reflection Profiles of the Italian Crust*; Scrocca, D., Doglioni, C., Innocenti, F., Manetti, P., Mazzotti, A., Bertelli, L., Burbi, L., D'Offizi, S., Eds.; Memorie Descrittive della Carta Geologica d'Italia: Rome, Italy, 2004; pp. 145–153.
39. Miccadei, E.; Carabella, C.; Paglia, G.; Piacentini, T. Paleo-drainage network, morphotectonics, and fluvial terraces: Clues from the verde stream in the middle Sangro river (central italy). *Geosciences* **2018**, *8*, 337. [CrossRef]
40. Miccadei, E.; Piacentini, T.; Buccolini, M. Long-term geomorphological evolution in the Abruzzo area, Central Italy: Twenty years of research. *Geol. Carpathica* **2017**, *68*, 19–28. [CrossRef]
41. Ascione, A.; Cinque, A.; Miccadei, E.; Villani, F.; Berti, C. The Plio-Quaternary uplift of the Apennine chain: New data from the analysis of topography and river valleys in Central Italy. *Geomorphology* **2008**, *102*, 105–118. [CrossRef]
42. Ori, G.; Serafini, G.; Visentin, C.; Ricci Lucchi, F.; Casnedi, R.; Colalongo, M.L.; Mosna, S. The Pliocene-Pleistocene Adriatic foredeep (Marche and Abruzzo, Italy): An integrated approach to surface and subsurface geology. In Proceedings of the 3rd EAPG Conference, Adriatic Foredeep Field Trip Guide Book, Firenze, Italy, 26–30 May 1991; p. 85.
43. Carabella, C.; Buccolini, M.; Galli, L.; Miccadei, E.; Paglia, G.; Piacentini, T. Geomorphological analysis of drainage changes in the NE Apennines piedmont area: The case of the middle Tavo River bend (Abruzzo, Central Italy). *J. Maps* **2020**, *16*, 222–235. [CrossRef]
44. Bigi, S.; Cantalamessa, G.; Centamore, E.; Didaskalou, P.; Dramis, F.; Farabollini, P.; Gentili, B.; Invernizzi, C.; Micarelli, A.; Nisio, S.; et al. La fascia periadriatica marchigiano-abruzzese dal pliocene medio ai tempi attuali: Evoluzione tettonico-sedimentaria e geomorfologica. *Stud. Geol. Camerti* **1995**, 37–49. [CrossRef]
45. Parlagreco, L.; Mascioli, F.; Miccadei, E.; Antonioli, F.; Gianolla, D.; Devoti, S.; Leoni, G.; Silenzi, S. New data on Holocene relative sea level along the Abruzzo coast (central Adriatic, Italy). *Quat. Int.* **2011**, *232*, 179–186. [CrossRef]
46. Carabella, C.; Boccabella, F.; Buccolini, M.; Ferrante, S.; Pacione, A.; Gregori, C.; Pagliani, T.; Piacentini, T.; Miccadei, E. Geomorphology of landslide–flood-critical areas in hilly catchments and urban areas for EWS (Feltrino Stream and Lanciano town, Abruzzo, Central Italy). *J. Maps* **2020**. [CrossRef]
47. Piacentini, T.; Galli, A.; Marsala, V.; Miccadei, E. Analysis of soil erosion induced by heavy rainfall: A case study from the NE Abruzzo Hills Area in Central Italy. *Water* **2018**, *10*, 1314. [CrossRef]
48. Miccadei, E.; Piacentini, T.; Daverio, F.; Di Michele, R. Geomorphological Instability Triggered by Heavy Rainfall: Examples in the Abruzzi Region (Central Italy). In *Studies on Environmental and Applied Geomorphology*; Piacentini, T., Miccadei, E., Eds.; InTech: London, UK, 2012; pp. 45–62. [CrossRef]
49. Rovida, A.; Locati, M.; Camassi, R.; Lolli, B.; Gasperini, P.; Antonucci, A. *The Italian Earthquake Catalogue CPTI15—Version 3.0*; Istituto Nazionale di Geofisica e Vulcanologia (INGV): Rome, Italy, 2021. [CrossRef]
50. ISIDe Working Group. *Italian Seismological Instrumental and Parametric Database (ISIDe)*; Istituto Nazionale di Geofisica e Vulcanologia (INGV): Rome, Italy, 2007. [CrossRef]
51. Fazzini, M.; Giuffrida, A. Une nouvelle proposition quantitative des régimes pluviométriques dans le territoire de Italie: Premiers résultats. In Proceedings of the Climat Urbain, Ville et Architecture—Actes XVIII Colloque Internationale de Climatologie, Genova, Italy, 7–11 September 2005; pp. 361–364.
52. Peel, M.C.; Finlayson, B.L.; McMahon, T.A. Updated world map of the Köppen-Geiger climate classification. Spatial Data Access Tool (SDAT) OGC Standards-based Geospatial Data Visualization/Download. *Hydrol. Earth Syst.* **2007**, *11*, 1633–1644. [CrossRef]
53. Di Lena, B.; Antenucci, F.; Mariani, L. Space and time evolution of the Abruzzo precipitation. *Ital. J. Agrometeorol.* **2012**, *1*, 5–20.
54. Mariani, L.; Parisi, S.G. Extreme rainfalls in the mediterranean area. In *Storminess and Environmental Change*; Diodato, N., Bellocchi, G., Eds.; Springer: London, UK, 2014; pp. 17–37. [CrossRef]
55. Scorzini, A.R.; Leopardi, M. Precipitation and temperature trends over central Italy (Abruzzo Region): 1951–2012. *Theor. Appl. Climatol.* **2019**, *135*, 959–977. [CrossRef]
56. Accordi, B.; Carbone, F. Carta delle litofacies del Lazio-Abruzzi ed aree limitrofe. CNR—Progetto Finalizzato "Geodinamica": Rome, Italy. *Quad. Ric. Sci.* **1988**, *114*, 223.
57. Vezzani, L.; Ghisetti, F. *Carta Geologica Dell'Abruzzo—Scala 1:100,000*; SELCA: Firenze, Italy, 1998.
58. Guzzetti, F.; Mondini, A.C.; Cardinali, M.; Fiorucci, F.; Santangelo, M.; Chang, K.T. Landslide inventory maps: New tools for an old problem. *Earth-Sci. Rev.* **2012**, *112*, 42–66. [CrossRef]
59. Trigila, A.; Iadanza, C.; Spizzichino, D. Quality assessment of the Italian Landslide Inventory using GIS processing. *Landslides* **2010**, *7*, 455–470. [CrossRef]
60. ISPRA. Progetto IFFI (Inventario dei Fenomeni Franosi in Italia) Dipartimento Difesa del Suolo-Servizio Geologico d'Italia–Regione Abruzzo. 2007. Available online: https://idrogeo.isprambiente.it/app/iffi?@=41.55172525894153,12.57350148381829,1 (accessed on 24 October 2020).
61. Fortunato, C.; Martino, S.; Prestininzi, A.; Romeo, R.W. New release of the italia catalogue of earthquake-induced ground failures (CEDIT). *Ital. J. Eng. Geol. Environ.* **2012**, *2*, 63–74. [CrossRef]

62. Martino, S.; Bozzano, F.; Caporossi, P.; D'Angiò, D.; Della Seta, M.; Esposito, C.; Fantini, A.; Fiorucci, M.; Giannini, L.M.; Iannucci, R.; et al. Impact of landslides on transportation routes during the 2016–2017 Central Italy seismic sequence. *Landslides* **2019**, *16*, 1221–1241. [CrossRef]
63. Martino, S.; Bozzano, F.; Caporossi, P.; D'Angiò, D.; Della Seta, M.; Esposito, C.; Fantini, A.; Fiorucci, M.; Giannini, L.M.; Iannucci, R.; et al. Ground Effects triggered by the 24 August 2016, Mw 6.0 Amatrice (Italy) earthquake: Surveys and inventorying to update the cebit catalogue. *Geogr. Fis. Din. Quat.* **2017**, *40*, 77–95. [CrossRef]
64. Guerrieri, L. Earthquake Environmental Effect for seismic hazard assessment: The ESI intensity scale and the EEE Catalogue. *Mem. Descr. Della Cart. Geol. Italia* **2015**, *97*, 181.
65. Calvello, M.; Pecoraro, G. FraneItalia: A catalog of recent Italian landslides. *Geoenviron. Disasters* **2018**, *5*, 16. [CrossRef]
66. Peruccacci, S.; Brunetti, M.T.; Luciani, S.; Vennari, C.; Guzzetti, F. Lithological and seasonal control on rainfall thresholds for the possible initiation of landslides in central Italy. *Geomorphology* **2012**, *139–140*, 79–90. [CrossRef]
67. Henriques, C.; Zêzere, J.L.; Marques, F. The role of the lithological setting on the landslide pattern and distribution. *Eng. Geol.* **2015**, *189*, 17–31. [CrossRef]
68. Smith, M.J.; Paron, P.; Griffiths, J. *Geomorphological Mapping, Methods and Applications*; Elsevier Science: Oxford, UK, 2011; ISBN 9780444534460.
69. ISPRA. AIGEO Aggiornamento ed Integrazione delle Linee Guida della Carta Geomorfologica D'italia in Scala 1:50,000. In *Quaderni Serie III*; Servizio Geologico d'Italia: Rome, Italy, 2018.
70. Gustavsson, M.; Kolstrup, E.; Seijmonsbergen, A.C. A new symbol-and-GIS based detailed geomorphological mapping system: Renewal of a scientific discipline for understanding landscape development. *Geomorphology* **2006**, *77*, 90–111. [CrossRef]
71. Seijmonsbergen, A.C. The Modern Geomorphological Map. In *Treatise on Geomorphology*; Elsevier: Amsterdam, The Netherlands, 2013; pp. 35–52. [CrossRef]
72. D'Alessandro, L.; De Pippo, T.; Donadio, C.; Mazzarella, A.; Miccadei, E. Fractal dimension in Italy: A geomorphological key to interpretation. *Z. Geomorphol.* **2006**, *50*, 479–499. [CrossRef]
73. Pasculli, A.; Palermi, S.; Sarra, A.; Piacentini, T.; Miccadei, E. A modelling methodology for the analysis of radon potential based on environmental geology and geographically weighted regression. *Environ. Model. Softw.* **2014**, *54*, 165–181. [CrossRef]
74. Crosetto, M.; Monserrat, O.; Cuevas-González, M.; Devanthéry, N.; Crippa, B. Persistent Scatterer Interferometry: A review. *ISPRS J. Photogramm. Remote Sens.* **2016**, *115*, 78–89. [CrossRef]
75. Bozzano, F.; Mazzanti, P.; Perissin, D.; Rocca, A.; De Pari, P. Basin scale assessment of landslides geomorphological setting by advanced InSAR analysis. *Remote Sens.* **2017**, *9*, 267. [CrossRef]
76. Miccadei, E.; Paron, P.; Piacentini, T. The SW escarpment of Montagna del Morrone (Abruzzi, Central Italy): Geomorphology of a fault-generated mountain front. *Geogr. Fis. Din. Quat.* **2004**, *27*, 55–87.
77. Calista, M.; Mascioli, F.; Menna, V.; Miccadei, E.; Piacentini, T. Recent geomorphological evolution and 3d numerical modelling of soft clastic rock cliffs in the mid-western Adriatic Sea (Abruzzo, Italy). *Geosciences* **2019**, *9*, 309. [CrossRef]
78. ISPRA. Geological Map of Italy, Scale 1:50,000, Sheet 361 "Chieti". Available online: http://www.isprambiente.gov.it/Media/carg/361_CHIETI/Foglio.html (accessed on 22 September 2020).
79. Bozzano, F.; Carabella, C.; De Pari, P.; Discenza, M.E.; Fantucci, R.; Mazzanti, P.; Miccadei, E.; Rocca, A.; Romano, S.; Sciarra, N. Geological and geomorphological analysis of a complex landslides system: The case of San Martino sulla Marruccina (Abruzzo, Central Italy). *J. Maps* **2020**, *16*, 126–136. [CrossRef]
80. Buccolini, M.; D'Alessandro, L.; Miccadei, E.; Sciarra, N. Susceptibility assessment of an area subject to a large landslide: The case of San Martino sulla Marrucina (Chieti province—Central Italy). In Proceedings of the IV international conference on computer simulation in risk analysis and hazard mitigation, Rome, Italy, 28–30 September 2021; WIT Press: Southampton, UK, 2004; pp. 245–255.
81. Damiano, E.; Giordan, D.; Allasia, P.; Baldo, M.; Sciarra, N.; Lollino, G. Multitemporal study of the San Martino sulla Marrucina landslide (Central Italy). In *Landslide Science and Practice: Early Warning, Instrumentation and Monitoring*; Margottini, C., Canuti, P., Sassa, K., Eds.; Springer: Berlin, Germany, 2013; pp. 257–263. [CrossRef]
82. Crescenti, U.; D'Alessandro, L.; Genevois, R. La Ripa di Montepiano (Abruzzo): Un primo esame delle caratteristiche geomorfologiche in rapporto alla stabilità. *Mem. Della Soc. Geol. Ital.* **1987**, *37*, 775–787.
83. D'Alessandro, L.; Genevois, R.; Berti, M.; Urbani, A.; Tecca, P.R. Geomorphology, stability analises and stabilization works on the Montepiano travertinous cliff (Central Italy). In *Applied Geomorphology: Theory and Practice*; Allison, R.J., Ed.; John Wiley & Sons Ltd.: Chichester, UK, 2002; pp. 21–38.
84. Pasculli, A.; Sciarra, N. A 3D landslide analyses with constant mechanical parameters compared with the results of a probabilistic approach assuming selected heterogeneities at different spatial scales. *G. Geol. Appl.* **2006**, *3*, 269–280. [CrossRef]
85. Almagià, R. La grande frana di Roccamontepiano (prov. Di Chieti). 24 giugno 1765. *Riv. Abruzz.* **1910**, *25*, 337–346.
86. Guzzetti, F.; Cardinali, M.; Reichenbach, P. The AVI project: A bibliographical and archive inventory of landslides and floods in Italy. *Environ. Manage.* **1994**, *18*, 623–633. [CrossRef]
87. Calista, M.; Miccadei, E.; Pasculli, A.; Piacentini, T.; Sciarra, M.; Sciarra, N. Geomorphological features of the Montebello sul Sangro large landslide (Abruzzo, Central Italy). *J. Maps* **2016**, *12*, 882–891. [CrossRef]
88. Almagià, R. *Studi Geografici Sulle Frane in Italia*; Società Geografica Italiana: Rome, Italy, 1910.

89. D'Alessandro, L.; Genevois, R.; Prestinizi, A. Preliminary report on an earthflow in the Sangro valley (Central Italy). In *Polish-Italian Seminar, Superficial Mass Movements in Mountain Region, Szymbark, Theme 3*: Analysis of the stability of the rocks slopes; Instytut Meteorologii i Gospodarki Wodnej: Warszawa, Poland, 1979; pp. 174–189.
90. D'Alessandro, L.; Pantaleone, A. Caratteristiche geomorfologiche e dissesti nell'Abruzzo sud-orientale. *Mem. Della Soc. Geol. Ital.* **1991**, *37*, 805–821.
91. Buccolini, M.; Crescenti, U.; Sciarra, N. Interazione fra dinamica dei versanti ed ambienti costruiti: Alcuni esempi in Abruzzo. *Alp. Mediterr. Quat.* **1994**, *7*, 179–196.
92. Piacentini, D.; Troiani, F.; Daniele, G.; Pizziolo, M. Historical geospatial database for landslide analysis: The Catalogue of Landslide OCcurrences in the Emilia-Romagna Region (CLOCkER). *Landslides* **2018**, *15*, 811–822. [CrossRef]

MDPI
St. Alban-Anlage 66
4052 Basel
Switzerland
Tel. +41 61 683 77 34
Fax +41 61 302 89 18
www.mdpi.com

Land Editorial Office
E-mail: land@mdpi.com
www.mdpi.com/journal/land